Fixed-Point Algorithms for Inverse Problems in Science and Engineering

For further volumes:
http://www.springer.com/series/7393

Springer Optimization and Its Applications

VOLUME 49

Aims and Scope
Optimization has been expanding in all directions at an astonishing rate during the last few decades. New algorithmic and theoretical techniques have been developed, the diffusion into other disciplines has proceeded at a rapid pace, and our knowledge of all aspects of the field has grown even more profound. At the same time, one of the most striking trends in optimization is the constantly increasing emphasis on the interdisciplinary nature of the field. Optimization has been a basic tool in all areas of applied mathematics, engineering, medicine, economics and other sciences.

The series *Springer Optimization and Its Applications* publishes undergraduate and graduate textbooks, monographs and state-of-the-art expository works that focus on algorithms for solving optimization problems and also study applications involving such problems. Some of the topics covered include nonlinear optimization (convex and nonconvex), network flow problems, stochastic optimization, optimal control, discrete optimization, multi-objective programming, description of software packages, approximation techniques and heuristic approaches.

Heinz H. Bauschke • Regina S. Burachik
Patrick L. Combettes • Veit Elser
D. Russell Luke • Henry Wolkowicz
Editors

Fixed-Point Algorithms for Inverse Problems in Science and Engineering

Springer

Editors
Heinz H. Bauschke
Department of Mathematics and Statistics
University of British Columbia
Okanagan Campus
Kelowna, British Columbia
Canada
heinz.bauschke@ubc.ca

Regina S. Burachik
School of Mathematics & Statistics
Division of Information Technology
Engineering & the Environment
University of South Australia
Mawson Lakes Campus
Mawson Lakes Blvd.
5095 Mawson Lakes
South Australia
regina.burachik@unisa.edu.au

Patrick L. Combettes
Université Pierre et Marie Curie
Laboratoire Jacques-Louis Lions
4, Place Jussieu
75005 Paris
France
plc@math.jussieu.fr

Veit Elser
Laboratory of Atomic and Solid
State Physics
Cornell University
Clark Hall
14853–2501 Ithaca, New York
USA
ve10@cornell.edu

D. Russell Luke
Institut für Numerische und Angewandte
Mathematik
Universität Göttingen
Lotzestr. 16-18, 37073 Göttingen
Germany
r.luke@math.uni-goettingen.de

Henry Wolkowicz
Department of Combinatorics
& Optimization
Faculty of Mathematics
University of Waterloo
Waterloo, Ontario
Canada
hwolkowicz@uwaterloo.ca

ISSN 1931-6828
ISBN 978-1-4419-9568-1 e-ISBN 978-1-4419-9569-8
DOI 10.1007/978-1-4419-9569-8
Springer New York Dordrecht Heidelberg London

Library of Congress Control Number: 2011928237

Printed on acid-free paper

Springer is part of Springer Science+Business Media (www.springer.com)

Preface

This book brings together 18 carefully refereed research and review papers in the broad areas of optimization and functional analysis, with a particular emphasis on topics related to fixed-point algorithms. The volume is a compendium of topics presented at the *Interdisciplinary Workshop on Fixed-Point Algorithms for Inverse Problems in Science and Engineering*, held at the Banff International Research Station for Mathematical Innovation and Discovery (BIRS), on November 1–6, 2009. Forty experts from around the world were invited. Participants came from Australia, Austria, Brazil, Bulgaria, Canada, France, Germany, Israel, Japan, New Zealand, Poland, Spain, and the United States.

Most papers in this volume grew out of talks delivered at this workshop, although some contributions are from experts who were unable to attend. We believe that the reader will find this to be a valuable state-of-the-art account on emerging directions related to first-order fixed-point algorithms.

The editors thank BIRS and their sponsors – Natural Sciences and Engineering Research Council of Canada (NSERC), US National Science Foundation (NSF), Alberta Science Research Station (ASRA), and Mexico's National Council for Science and Technology (CONACYT) – for their financial support in hosting the workshop, and Wynne Fong, Brent Kearney, and Brenda Williams for their help in the preparation and realization of the workshop. We are grateful to Dr. Mason Macklem for his valuable help in the preparation of this volume. Finally, we thank the dedicated referees who contributed significantly to the quality of this volume through their instructive and insightful reviews.

Kelowna (Canada)	Heinz H. Bauschke
Adelaide (Australia)	Regina S. Burachik
Paris (France)	Patrick L. Combettes
Ithaca (U.S.A.)	Veit Elser
Göttingen (Germany)	D. Russell Luke
Waterloo (Canada)	Henry Wolkowicz
December 2010	

Contents

Contributors

Heinz H. Bauschke Department of Mathematics, Irving K. Barber School, University of British Columbia, Kelowna, B.C. V1V 1V7, Canada, heinz.bauschke@ubc.ca

Amir Beck Department of Industrial Engineering, Technion, Israel Institute of Technology, Haifa 32000, Israel, becka@ie.technion.ac.il

Adi Ben-Israel RUTCOR – Rutgers Center for Operations Research, Rutgers University, 640 Bartholomew Road, Piscataway, NJ 08854-8003, USA, adi.benisrael@gmail.com

Jonathan M. Borwein CARMA, School of Mathematical and Physical Sciences, University of Newcastle, NSW 2308, Australia, jonathan.borwein@newcastle.edu.au

Radu Ioan Boţ Faculty of Mathematics, Chemnitz University of Technology, 09107 Chemnitz, Germany, radu.bot@mathematik.tu-chemnitz.de

Jérôme Boulanger Johann Radon Institute for Computational and Applied Mathematics, Austrian Academy of Sciences, Altenbergerstraße 69, 4040 Linz, Austria, jerome.boulanger@ricam.oeaw.ac.at

Andrzej Cegielski Faculty of Mathematics, Computer Science and Econometrics, University of Zielona Góra, ul. Szafrana 4a, 65-514 Zielona Góra, Poland, a.cegielski@wmie.uz.zgora.pl

Yair Censor Department of Mathematics, University of Haifa, Mt. Carmel, Haifa 31905, Israel, yair@math.haifa.ac.il

Patrick L. Combettes UPMC Université Paris 06, Laboratoire Jacques-Louis Lions – UMR CNRS 7598, 75005 Paris, France, plc@math.jussieu.fr

Ernö Robert Csetnek Faculty of Mathematics, Chemnitz University of Technology, 09107 Chemnitz, Germany, robert.csetnek@mathematik.tu-chemnitz.de

Frank Deutsch Department of Mathematics, Pennsylvania State University, University Park, PA 16802, USA, deutsch@math.psu.edu

Peter Elbau Johann Radon Institute for Computational and Applied Mathematics, Austrian Academy of Sciences, Altenbergerstraße 69, 4040 Linz, Austria, peter.elbau@ricam.oeaw.ac.at

Bryan Gardiner Computer Science, I. K. Barber School, University of British Columbia Okanagan, Kelowna, B.C. V1V 1V7, Canada, khumba@interchange.ubc.ca

W. L. Hare Department of Mathematics and Statistics, UBC Okanagan Campus, Kelowna, B.C. V1V 1V7, Canada, warren.hare@ubc.ca

Hein Hundal 146 Cedar Ridge Drive, Port Matilda, PA 16870, USA, hundalhh@yahoo.com

Yuri Levin School of Business, Queen's University, 143 Union Street, Kingston, ON K7L 3N6, Canada, ylevin@business.queensu.ca

Genaro López Department of Mathematical Analysis, University of Seville, 41012 Seville, Spain, glopez@us.es

Yves Lucet Computer Science, I. K. Barber School, University of British Columbia Okanagan, Kelowna, B.C. V1V 1V7, Canada, yves.lucet@ubc.ca

D. Russell Luke Institut für Numerische und Angewandte Mathematik Universität Göttingen, Lotzestr. 16-18, 37073 Göttingen, Germany r.luke@math.uni-goettingen.de

Mason S. Macklem Department of Mathematics, Irving K. Barber School, University of British Columbia, Kelowna, B.C. V1V 1V7, Canada, mason.macklem@ubc.ca

Victoria Martín-Márquez Department of Mathematical Analysis, University of Seville, 41012 Seville, Spain, victoriam@us.es

Sarah M. Moffat Department of Mathematics, Irving K. Barber School, University of British Columbia, Kelowna, B.C. V1V 1V7, Canada, sarah.moffat@ubc.ca

J.-C. Pesquet Laboratoire d'Informatique Gaspard Monge, UMR CNRS 8049, Université Paris-Est, 77454 Marne la Vallée Cedex 2, France, jean-christophe.pesquet@univ-paris-est.fr

Carsten Pontow Department of Mathematics, University Innsbruck, Technikerstr. 21a, 6020 Innsbruck, Austria, Carsten.Pontow@uibk.ac.at

Simeon Reich Department of Mathematics, The Technion – Israel Institute of Technology, 32000 Haifa, Israel, sreich@tx.technion.ac.il

J.P. Revalski Institute of Mathematics and Informatics, Bulgarian Academy of Sciences, Acad. G. Bonchev Street, block 8, 1113 Sofia, Bulgaria, revalski@math.bas.bg

Shoham Sabach Department of Mathematics, The Technion – Israel Institute of Technology, 32000 Haifa, Israel, ssabach@tx.technion.ac.il

Otmar Scherzer Computational Science Center, University Vienna, Nordbergstr. 15, 1090 Vienna, Austria, and Johann Radon Institute for Computational and Applied Mathematics, Austrian Academy of Sciences, Altenbergerstraße 69, 4040 Linz, Austria, otmar.scherzer@univie.ac.at

Brailey Sims CARMA, School of Mathematical and Physical Sciences, University of Newcastle, NSW 2308, Australia, brailey.sims@newcastle.edu.au

Marc Teboulle School of Mathematical Sciences, Tel Aviv University, Tel Aviv 69978, Israel, teboulle@post.tau.ac.il

Xianfu Wang Department of Mathematics, Irving K. Barber School, University of British Columbia, Kelowna, B.C. V1V 1V7, Canada, shawn.wang@ubc.ca

Isao Yamada Department of Communications and Integrated Systems, Tokyo Institute of Technology, S3-60, Tokyo, 152-8550 Japan, isao@sp.ss.titech.ac.jp

Masao Yamagishi Department of Communications and Integrated Systems, Tokyo Institute of Technology, S3-60, Tokyo, 152-8550 Japan, myamagi@sp.ss.titech.ac.jp

Liangjin Yao Department of Mathematics, Irving K. Barber School, University of British Columbia, Kelowna, B.C. V1V 1V7, Canada, ljinyao@interchange.ubc.ca

Masahiro Yukawa Department of Electrical and Electronic Engineering, Niigata University, 8050 Ikarashi Nino-cho, Nishi-ku, Niigata, 950-2181 Japan, yukawa@eng.niigata-u.ac.jp

Chapter 1
Chebyshev Sets, Klee Sets, and Chebyshev Centers with Respect to Bregman Distances: Recent Results and Open Problems

Heinz H. Bauschke, Mason S. Macklem, and Xianfu Wang

Abstract In Euclidean spaces, the geometric notions of nearest-points map, farthest-points map, Chebyshev set, Klee set, and Chebyshev center are well known and well understood. Since early works going back to the 1930s, tremendous theoretical progress has been made, mostly by extending classical results from Euclidean space to Banach space settings. In all these results, the distance between points is induced by some underlying norm. Recently, these notions have been revisited from a different viewpoint in which the discrepancy between points is measured by Bregman distances induced by Legendre functions. The associated framework covers the well known Kullback–Leibler divergence and the Itakura–Saito distance. In this survey, we review known results and we present new results on Klee sets and Chebyshev centers with respect to Bregman distances. Examples are provided and connections to recent work on Chebyshev functions are made. We also identify several intriguing open problems.

Keywords Bregman distance · Chebyshev center · Chebyshev function · Chebyshev point of a function · Chebyshev set · Convex function · Farthest point · Fenchel conjugate · Itakura–Saito distance · Klee set · Klee function · Kullback–Leibler divergence · Legendre function · Nearest point · Projection

AMS 2010 Subject Classification: Primary 41A65; Secondary 28D05, 41A50, 46N10, 47N10, 49J53, 54E52, 58C06, 90C25

H.H. Bauschke (✉)
Department of Mathematics, Irving K. Barber School, University of British Columbia, Kelowna, B.C. V1V 1V7, Canada
e-mail: heinz.bauschke@ubc.ca

H.H. Bauschke et al. (eds.), *Fixed-Point Algorithms for Inverse Problems in Science and Engineering*, Springer Optimization and Its Applications 49,
DOI 10.1007/978-1-4419-9569-8_1,

1.1 Introduction

1.1.1 Legendre Functions and Bregman Distances

Throughout, we assume that

$$X = \mathbb{R}^n \text{ is the standard Euclidean space with inner product } \langle\cdot,\cdot\rangle, \tag{1.1}$$

with induced norm $\|\cdot\|: x \mapsto \sqrt{\langle x,x\rangle}$, and with metric $(x,y) \mapsto \|x-y\|$. In addition, it is assumed that

$$f : X \to \left]-\infty,+\infty\right] \text{ is a convex function of Legendre type,} \tag{1.2}$$

also referred to as a Legendre function. We assume the reader is familiar with basic results and standard notation from Convex Analysis; see, e.g., [33, 34, 40]. In particular, f^* denotes the Fenchel conjugate of f, and $\operatorname{int}\operatorname{dom} f$ is the interior of the domain of f. For a subset C of X, $\overline{C}$ stands for the closure of C, $\operatorname{conv} C$ for the convex hull of C, and ι_C for the indicator function of C, i.e., $\iota_C(x) = 0$, if $x \in C$ and $\iota_C(x) = +\infty$, if $x \in X \smallsetminus C$. Now set

$$U = \operatorname{int}\operatorname{dom} f. \tag{1.3}$$

Example 1.1 (Legendre functions). The following are Legendre functions,[1] each evaluated at a point $x \in X$.

(i) *Halved energy:* $f(x) = \frac{1}{2}\|x\|^2 = \frac{1}{2}\sum_j x_j^2$.

(ii) *Negative entropy:* $f(x) = \begin{cases} \sum_j \big(x_j \ln(x_j) - x_j\big), & \text{if } x \geq 0; \\ +\infty, & \text{otherwise.} \end{cases}$

(iii) *Negative logarithm:* $f(x) = \begin{cases} -\sum_j \ln(x_j), & \text{if } x > 0; \\ +\infty, & \text{otherwise.} \end{cases}$

Note that $U = \mathbb{R}^n$ inss (i), whereas $U = \mathbb{R}^n_{++}$ in (ii) and (iii).

Further examples of Legendre functions can be found in, e.g., [2, 5, 12, 33].

[1] Here and elsewhere, inequalities between vectors in $\mathbb{R}^n$ are interpreted coordinate-wise.

Fact 1.2 (Rockafellar). (See [33, Theorem 26.5].) The gradient map ∇f is a continuous bijection between $\operatorname{int}\operatorname{dom} f$ and $\operatorname{int}\operatorname{dom} f^*$, with continuous inverse map $(\nabla f)^{-1} = \nabla f^*$. Furthermore, f^* is also a convex function of Legendre type.

Given $x \in U$ and $C \subseteq U$, it will be convenient to write

$$x^* = \nabla f(x), \tag{1.4}$$
$$C^* = \nabla f(C), \tag{1.5}$$
$$U^* = \operatorname{int}\operatorname{dom} f^*, \tag{1.6}$$

and similarly for other vectors and sets in U. Note that we used Fact 1.2 for (1.6).

While the Bregman distance defined next is not a distance in the sense of metric topology, it does possess some good properties that allow it to measure the discrepancy between points in U.

Definition 1.3 (Bregman distance). (See [13, 15, 16].) The *Bregman distance* with respect to f, written D_f or simply D, is the function

$$D\colon X \times X \to [0,+\infty] : (x,y) \mapsto \begin{cases} f(x) - f(y) - \langle \nabla f(y), x-y \rangle, & \text{if } y \in U; \\ +\infty, & \text{otherwise.} \end{cases} \tag{1.7}$$

Fact 1.4. (See [2, Proposition 3.2(i) and Theorem 3.7(iv) and (v)].) Let x and y be in U. Then the following hold:

(i) $D_f(x,y) = f(x) + f^*(y^*) - \langle y^*, x \rangle = D_{f^*}(y^*, x^*)$.
(ii) $D_f(x,y) = 0 \Leftrightarrow x = y \Leftrightarrow x^* = y^* \Leftrightarrow D_{f^*}(x^*, y^*) = 0$.

Example 1.5. The Bregman distances corresponding to the Legendre functions of Example 1.1 between two points x and y in X are as follows:

(i) $D(x,y) = \frac{1}{2}\|x-y\|^2$.

(ii) $D(x,y) = \begin{cases} \sum_j \big(x_j \ln(x_j/y_j) - x_j + y_j\big), & \text{if } x \geq 0 \text{ and } y > 0; \\ +\infty, & \text{otherwise.} \end{cases}$

(iii) $D(x,y) = \begin{cases} \sum_j \big(\ln(y_j/x_j) + x_j/y_j - 1\big), & \text{if } x > 0 \text{ and } y > 0; \\ +\infty, & \text{otherwise.} \end{cases}$

These Bregman distances are also known as (i) the *halved Euclidean distance squared*, (ii) the *Kullback–Leibler divergence*, and (iii) the *Itakura–Saito distance*, respectively.

From now on, we assume that C is a subset of X such that

$$C \text{ is closed} \quad \text{and} \quad \varnothing \neq C \subseteq U. \tag{1.8}$$

The *power set* (the set of all subsets) of C is denoted by 2^C.

We are now in a position to introduce the various geometric notions.

1.1.2 Nearest Distance, Nearest Points, and Chebyshev Sets

Definition 1.6 (Bregman nearest-distance function and nearest-points map). The *left Bregman nearest-distance function* with respect to C is

$$\overleftarrow{D}_C : X \to [0,+\infty] : y \mapsto \inf_{x\in C} D(x,y), \tag{1.9}$$

and the *left Bregman nearest-points map*[2] with respect to C is

$$\overleftarrow{P}_C : X \to 2^C : y \mapsto \{x \in C \mid D(x,y) = \overleftarrow{D}_C(y) < +\infty\}. \tag{1.10}$$

The *right Bregman nearest-distance* and the *right Bregman nearest-point map* with respect to C are

$$\overrightarrow{D}_C : X \to [0,+\infty] : x \mapsto \inf_{y\in C} D(x,y) \tag{1.11}$$

and

$$\overrightarrow{P}_C : X \to 2^C : x \mapsto \{y \in C \mid D(x,y) = \overrightarrow{D}_C(x) < +\infty\}, \tag{1.12}$$

respectively. If we need to emphasize the underlying Legendre function f, then we write $\overleftarrow{D}_{f,C}$, $\overleftarrow{P}_{f,C}$, $\overrightarrow{D}_{f,C}$, and $\overrightarrow{P}_{f,C}$.

Definition 1.7 (Chebyshev sets). The set C is a *left Chebyshev set* with respect to the Bregman distance, or simply $\overleftarrow{D}$*-Chebyshev*, if for every $y \in U$, $\overleftarrow{P}_C(y)$ is a singleton. Similarly, the set C is a *right Chebyshev set* with respect to the Bregman distance, or simply $\overrightarrow{D}$*-Chebyshev*, if for every $x \in U$, $\overrightarrow{P}_C(x)$ is a singleton.

Remark 1.8 (Classical Bunt-Motzkin result). Assume that f is the halved energy as in Example 1.1(i). Since the halved Euclidean distance squared (see Example 1.5(i)) is symmetric, the left and right (Bregman) nearest distances coincide, as do the corresponding nearest-point maps. Furthermore, the set C is Chebyshev if and only

---

[2] This operator, which has turned out to be quite useful in Optimization and which has found many applications (for a recent one, see [32]), is often referred to as the Bregman projection.

if for every $z \in X$, the metric[3] projection $P_C(z)$ is a singleton. It is well known that if C is convex, then C is Chebyshev. In the mid-1930s, Bunt [14] and Motzkin [28] showed independently that the following converse holds:

$$C \text{ is Chebyshev} \implies C \text{ is convex.} \tag{1.13}$$

For other works in this direction, see, e.g., [1, 9–11, 17, 22, 24, 25, 35–37]. It is still unknown whether or not (1.13) holds in general Hilbert spaces. We review corresponding results for the present Bregman setting in Sect. 1.3.

1.1.3 Farthest Distance, Farthest Points, and Klee Sets

Definition 1.9 (Bregman farthest-distance function and farthest-points map). The *left Bregman farthest-distance function* with respect to C is

$$\overleftarrow{F}_C \colon X \to [0,+\infty] : y \mapsto \sup_{x\in C} D(x,y), \tag{1.14}$$

and the *left Bregman farthest-points map* with respect to C is

$$\overleftarrow{Q}_C \colon X \to 2^C : y \mapsto \big\{x \in C \mid D(x,y) = \overleftarrow{F}_C(y) < +\infty\big\}. \tag{1.15}$$

Similarly, the *right Bregman farthest-distance function* with respect to C is

$$\overrightarrow{F}_C \colon X \to [0,+\infty] : x \mapsto \sup_{y\in C} D(x,y), \tag{1.16}$$

and the *right Bregman farthest-points map* with respect to C is

$$\overrightarrow{Q}_C \colon X \to 2^C : x \mapsto \big\{y \in C \mid D(x,y) = \overrightarrow{F}_C(x) < +\infty\big\}. \tag{1.17}$$

If we need to emphasize the underlying Legendre function f, then we write $\overleftarrow{F}_{f,C}$, $\overleftarrow{Q}_{f,C}$, $\overrightarrow{F}_{f,C}$, and $\overrightarrow{Q}_{f,C}$.

Definition 1.10 (Klee sets). The set C is a *left Klee set* with respect to the Bregman distance, or simply $\overleftarrow{D}$*-Klee*, if for every $y \in U$, $\overleftarrow{Q}_C(y)$ is a singleton. Similarly, the set C is a *right Klee set* with respect to the right Bregman distance, or simply $\overrightarrow{D}$*-Klee*, if for every $x \in U$, $\overrightarrow{Q}_C(x)$ is a singleton.

Remark 1.11 (Classical Klee result). Assume again that f is the halved energy as in Example 1.1(i). Then the left and right (Bregman) farthest-distance functions

[3] The metric projection is the nearest-points map with respect to the Euclidean distance.

coincide, as do the corresponding farthest-points maps. Furthermore, the set C is Klee if and only if for every $z \in X$, the metric farthest-points map $Q_C(z)$ is a singleton. It is obvious that if C is a singleton, then C is Klee. In 1961, Klee [27] showed the following converse:

$$C \text{ is Klee} \implies C \text{ is a singleton.} \tag{1.18}$$

See, e.g., also [1,11,17,23–25,29,39]. Once again, it is still unknown whether or not (1.18) remains true in general Hilbert spaces. The present Bregman-distance setting is reviewed in Sect. 1.4.

1.1.4 Chebyshev Radius and Chebyshev Center

Definition 1.12 (Chebyshev radius and Chebyshev center). The left $\overleftarrow{D}$-*Chebyshev radius* of C is

$$\overleftarrow{r}_C = \inf_{y \in U} \overleftarrow{F}_C(y) \tag{1.19}$$

and the left $\overleftarrow{D}$-*Chebyshev center* of C is

$$\overleftarrow{Z}_C = \{y \in U \mid \overleftarrow{F}_C(y) = \overleftarrow{r}_C < +\infty\}. \tag{1.20}$$

Similarly, the right $\overrightarrow{D}$-*Chebyshev radius* of C is

$$\overrightarrow{r}_C = \inf_{x \in U} \overrightarrow{F}_C(x) \tag{1.21}$$

and the right $\overrightarrow{D}$-*Chebyshev center* of C is

$$\overrightarrow{Z}_C = \{x \in U \mid \overrightarrow{F}_C(x) = \overrightarrow{r}_C < +\infty\}. \tag{1.22}$$

If we need to emphasize the underlying Legendre function f, then we write $\overleftarrow{r}_{f,C}$, $\overleftarrow{Z}_{f,C}$, $\overrightarrow{r}_{f,C}$, and $\overrightarrow{Z}_{f,C}$.

Remark 1.13 (Classical Garkavi-Klee result). Again, assume that f is the halved energy as in Example 1.1(i) so that the left and right (Bregman) farthest-distance functions coincide, as do the corresponding farthest-points maps. Furthermore, assume that C is bounded. In the 1960s, Garkavi [19] and Klee [26] proved that the Chebyshev center is a singleton, say $\{z\}$, which is characterized by

$$z \in \operatorname{conv} Q_C(z). \tag{1.23}$$

See also [30, 31] and Sect. 1.5. In passing, we note that Chebyshev centers are also utilized in Fixed Point Theory; see, e.g., [20, Chap. 4].

1.1.5 Goal of the Paper

The aim of this survey is threefold. First, we review recent results concerning Chebyshev sets, Klee sets, and Chebyshev centers with respect to Bregman distances. Second, we provide some new results and examples on Klee sets and Chebyshev centers. Third, we formulate various tantalizing open problems on these notions as well as on the related concepts of Chebyshev functions.

1.1.6 Organization of the Paper

The remainder of the paper is organized as follows. In Sect. 1.2, we record auxiliary results which will make the derivation of the main results more structured. Chebyshev sets and corresponding open problems are discussed in Sect. 1.3. In Sect. 1.4, we review results and open problems for Klee sets, and we also present a new result (Theorem 1.27) concerning left Klee sets. Chebyshev centers are considered in Sect. 1.5, where we also provide a characterization of left Chebyshev centers (Theorem 1.31). Chebyshev centers are illustrated by two examples in Sect. 1.6. Recent related results on variations of Chebyshev sets and Klee sets are considered in Sect. 1.7. Along our journey, we pose several questions that we list collectively in the final Sect. 1.8.

1.2 Auxiliary Results

For the reader's convenience, we present the following two results which are implicitly contained in [6] and [7].

Lemma 1.14. *Let x and y be in C. Then the following hold:*

(i) $\overleftarrow{D}_{f,C}(y) = \overrightarrow{D}_{f^*,C^*}(y^*)$ *and* $\overrightarrow{D}_{f,C}(x) = \overleftarrow{D}_{f^*,C^*}(x^*)$.
(ii) $\overleftarrow{P}_{f,C}\big|_U = \nabla f^* \circ \overrightarrow{P}_{f^*,C^*} \circ \nabla f$ *and* $\overrightarrow{P}_{f,C}\big|_U = \nabla f^* \circ \overleftarrow{P}_{f^*,C^*} \circ \nabla f$.
(iii) $\overleftarrow{P}_{f^*,C^*}\big|_{U^*} = \nabla f \circ \overrightarrow{P}_{f,C} \circ \nabla f^*$ *and* $\overrightarrow{P}_{f^*,C^*}\big|_{U^*} = \nabla f \circ \overleftarrow{P}_{f,C} \circ \nabla f^*$.

Proof. This follows from Fact 1.2, Fact 1.4(i), and Definition 1.6. (See also [6, Proposition 7.1].) ■

Lemma 1.15. *Let x and y be in C. Then the following hold:*

(i) $\overleftarrow{F}_{f,C}(y) = \overrightarrow{F}_{f^*,C^*}(y^*)$ *and* $\overrightarrow{F}_{f,C}(x) = \overleftarrow{F}_{f^*,C^*}(x^*)$.
(ii) $\overleftarrow{Q}_{f,C}\big|_U = \nabla f^* \circ \overrightarrow{Q}_{f^*,C^*} \circ \nabla f$ *and* $\overrightarrow{Q}_{f,C}\big|_U = \nabla f^* \circ \overleftarrow{Q}_{f^*,C^*} \circ \nabla f$.
(iii) $\overleftarrow{Q}_{f^*,C^*}\big|_{U^*} = \nabla f \circ \overrightarrow{Q}_{f,C} \circ \nabla f^*$ *and* $\overrightarrow{Q}_{f^*,C^*}\big|_{U^*} = \nabla f \circ \overleftarrow{Q}_{f,C} \circ \nabla f^*$.

Proof. This follows from Fact 1.2, Fact 1.4(i), and Definition 1.9. (See also [7, Proposition 7.1].) ■

The next observation on the duality of Chebyshev radii and Chebyshev centers is new.

Lemma 1.16. *The following hold:*

(i) $\overleftarrow{r}_{f,C} = \overrightarrow{r}_{f^*,C^*}$ *and* $\overrightarrow{r}_{f,C} = \overleftarrow{r}_{f^*,C^*}$.
(ii) $\overleftarrow{Z}_{f,C} = \nabla f^*\big(\overrightarrow{Z}_{f^*,C^*}\big)$ *and* $\overrightarrow{Z}_{f,C} = \nabla f^*\big(\overleftarrow{Z}_{f^*,C^*}\big)$.
(iii) $\overleftarrow{Z}_{f^*,C^*} = \nabla f\big(\overrightarrow{Z}_{f,C}\big)$ *and* $\overrightarrow{Z}_{f^*,C^*} = \nabla f\big(\overleftarrow{Z}_{f,C}\big)$.
(iv) $\overleftarrow{Z}_{f,C}$ *is a singleton* $\Leftrightarrow$ $\overrightarrow{Z}_{f^*,C^*}$ *is a singleton.*
(v) $\overrightarrow{Z}_{f,C}$ *is a singleton* $\Leftrightarrow$ $\overleftarrow{Z}_{f^*,C^*}$ *is a singleton.*

Proof. (i): Using Definition 1.12 and Lemma 1.15(i), we see that

$$\overleftarrow{r}_{f,C} = \inf_{y\in U} \overleftarrow{F}_C(y) = \inf_{y^*\in U^*} \overrightarrow{F}_{C^*}(y^*) = \overrightarrow{r}_{f^*,C^*} \tag{1.24}$$

and that

$$\overrightarrow{r}_{f,C} = \inf_{y\in U} \overrightarrow{F}_C(y) = \inf_{y^*\in U^*} \overleftarrow{F}_{C^*}(y^*) = \overleftarrow{r}_{f^*,C^*}. \tag{1.25}$$

(ii) and (iii): Let $z \in U$. Using (i) and Lemma 1.15(i), we see that

$$z \in \overleftarrow{Z}_{f,C} \Leftrightarrow \overleftarrow{F}_{f,C}(z) = \overleftarrow{r}_{f,C} \Leftrightarrow \overrightarrow{F}_{f^*,C^*}(z^*) = \overrightarrow{r}_{f^*,C^*} \Leftrightarrow z^* \in \overrightarrow{Z}_{f^*,C^*}. \tag{1.26}$$

This verifies $\overleftarrow{Z}_{f,C} = \nabla f^*\big(\overrightarrow{Z}_{f^*,C^*}\big)$ and $\overrightarrow{Z}_{f^*,C^*} = \nabla f\big(\overleftarrow{Z}_{f,C}\big)$. The remaining identities follow similarly.

(iv) and (v): Clear from (ii) and (iii) and Fact 1.2. ■

The following two results play a key role for studying the single-valuedness of $\overrightarrow{P}_{f,C}$ via $\overleftarrow{P}_{f^*,C^*}$ and $\overrightarrow{Q}_{f,C}$ via $\overleftarrow{Q}_{f^*,C^*}$ by duality.

Lemma 1.17. *Let V and W be nonempty open subsets of X, and let $T : V \to W$ be a homeomorphism, i.e., T is a bijection and both T and T^{-1} are continuous. Furthermore, let G be a residual[4] subset of V. Then $T(G)$ is a residual subset of W.*

Proof. As G is residual, there exist sequence of dense open subsets $(O_k)_{k\in\mathbb{N}}$ of V such that $G \supseteq \bigcap_{k\in\mathbb{N}} O_k$. Then $T(G) \supseteq T(\bigcap_{k\in\mathbb{N}} O_k) = \bigcap_{k\in\mathbb{N}} T(O_k)$. Since $T : V \to W$ is a homeomorphism and each O_k is dense in V, we see that each $T(O_k)$ is open and dense in W. Therefore, $\bigcap_{k\in\mathbb{N}} T(O_k)$ is a dense G_δ subset in W. ■

Lemma 1.18. *Let V be a nonempty open subset of X, and let $T : V \to \mathbb{R}^n$ be locally Lipschitz. Furthermore, let S be a subset of V that has Lebesgue measure zero. Then, $T(S)$ has Lebesgue measure zero as well.*

---

[4] Also known as "second category".

Proof. Denote the closed unit ball in X by $\mathbb{B}$. For every $y \in V$, let $r(y) > 0$ be such that T is Lipschitz continuous with constant $c(y)$ on the open ball $O(y)$ centered at y of radius $r(y)$. In this proof, we denote the Lebesgue measure by λ. Let K be a compact subset of X. To show that $T(S)$ has Lebesgue measure zero, it suffices to show that $\lambda(T(K \cap S)) = 0$ because

$$\lambda(T(S)) = \lambda\left(T\left(\bigcup_{k\in\mathbb{N}} S \cap k\mathbb{B}\right)\right) \le \sum_{k\in\mathbb{N}} \lambda(T(S \cap k\mathbb{B})). \tag{1.27}$$

The Heine–Borel theorem provides a finite subset $\{y_1, \dots, y_m\}$ of V such that

$$K \subseteq \bigcup_{j=1}^{m} O(y_j). \tag{1.28}$$

We now proceed using a technique implicit in the proof of [21, Corollary 1]. Set $c = \max\{c_1, c_2, \dots, c_m\}$. Given $\varepsilon > 0$, there exists an open subset G of X such that $G \supseteq K \cap S$ and $\lambda(G) < \varepsilon$. For each $y \in K \cap S$, let $Q(y)$ be an open cubic interval centered at y of semi-edge length $s(y) > 0$ such that

$$(\exists\, j \in \{1, \dots, m\}) \quad Q(y) \subseteq G \cap O(y_j). \tag{1.29}$$

Then for each $x \in Q(y)$, we have

$$\|Tx - Ty\| \le c\|x - y\| \le c\sqrt{n}s(y). \tag{1.30}$$

Hence, the image of $Q(y)$ by T, $T(Q(y))$, is contained in a cubic interval – which we denote by $Q^*(Ty)$ – of center Ty and with semi-edge length $c\sqrt{n}s(y)$. Applying the Besicovitch Covering Theorem, we see that there exists a sequence $(Q_k)_{k\in\mathbb{N}}$ chosen among the open covering $(Q(y))_{y\in K\cap S}$ such that

$$K \cap S \subseteq \bigcup_{k\in\mathbb{N}} Q_k \quad \text{and} \quad \sum_{k\in\mathbb{N}} \chi_{Q_k} \le \theta, \tag{1.31}$$

where χ_{Q_k} stands for the characteristic function of Q_k and where the constant θ only depends on the dimension of X. Thus,

$$T(K \cap S) \subseteq T\left(\bigcup_{k\in\mathbb{N}} Q_k\right) = \bigcup_{k\in\mathbb{N}} T(Q_k) \subseteq \bigcup_{k\in\mathbb{N}} Q_k^*. \tag{1.32}$$

Now set $d = (c\sqrt{n})^n$ so that $\lambda(Q_k^*) \le d\lambda(Q_k)$. Then, using (1.29) and (1.31), we see that

$$\begin{aligned} \lambda\left(\cup_{k\in\mathbb{N}} Q_k^*\right) &\le \sum_{k\in\mathbb{N}} \lambda(Q_k^*) \le d \sum_{k\in\mathbb{N}} \lambda(Q_k) = d \sum_{k\in\mathbb{N}} \int \chi_{Q_k} = d \int \sum_{k\in\mathbb{N}} \chi_{Q_k} \\ &\le d\theta\lambda(G) \\ &\le d\theta\varepsilon. \end{aligned} \tag{1.33}$$

Since ε was chosen arbitrarily, we conclude that $\lambda(T(K \cap S)) = 0$.

Alternatively, one may argue as follows starting from (1.28). We have $K \cap S \subseteq \left(\bigcup_{j=1}^m O(y_j)\right) \cap S = \bigcup_{j=1}^m O(y_j) \cap S$ so that

$$T(K \cap S) \subseteq \bigcup_{j=1}^{m} T(O(y_j) \cap S). \tag{1.34}$$

Since T is Lipschitz on each $O(y_j)$ with constant $c(y_j)$ and since $\lambda(O(y_j) \cap S) = 0$, we apply [18, Proposition 262D, page 286] and conclude that $\lambda(T(O(y_j) \cap S)) = 0$. Therefore, $\lambda(T(K \cap S)) = 0$ by (1.34). ■

1.3 Chebyshev Sets

We start by reviewing the strongest known results concerning left and right Chebyshev sets with respect to Bregman distances.

Fact 1.19 ($\overleftarrow{D}$-Chebyshev sets). (See [6, Theorem 4.7].) Suppose that f is supercoercive[5] and that C is $\overleftarrow{D}$-Chebyshev. Then C is convex.

Fact 1.20 ($\overrightarrow{D}$-Chebyshev sets). (See [6, Theorem 7.3].) Suppose that $\operatorname{dom} f = X$, that $\overline{C^*} \subseteq U^*$, and that C is $\overrightarrow{D}$-Chebyshev. Then C^* is convex.

It is not known whether or not Fact 1.19 and 1.20 are the best possible results. For instance, is the assumption on supercoercivity in Fact 1.19 really necessarily? Similarly, do we really require full domain of f in Fact 1.20?

Example 1.21. (See [6, Example 7.5].) Suppose that $X = \mathbb{R}^2$, that f is the negative entropy (see Example 1.1(ii)), and that

$$C = \left\{ (e^{\lambda}, e^{2\lambda}) \mid \lambda \in [0,1] \right\}. \tag{1.35}$$

Then f is supercoercive and C is a *nonconvex* $\overrightarrow{D}$-Chebyshev set.

Example 1.21 is somewhat curious – not only does it illustrate that the right-Chebyshev-set counterpart of Fact 1.19 fails but it also shows that the conclusion of Fact 1.20 may hold even though f is not assumed to have full domain.

Fact 1.22. (See [4, Lemma 3.5].) Suppose that f is the negative entropy (see Example 1.1(ii)) and that C is convex. Then C is $\overrightarrow{D}$-Chebyshev.

---

[5] By [2, Proposition 2.16] and [33, Corollary 14.2.2], f is supercoercive $\Leftrightarrow \lim_{\|x\| \to +\infty} \frac{f(x)}{\|x\|} = +\infty \Leftrightarrow \operatorname{dom} f^* = X \Rightarrow 0 \in \operatorname{int} \operatorname{dom} f^* \Leftrightarrow \lim_{\|x\| \to +\infty} f(x) = +\infty \Leftrightarrow f$ is coercive.

Fact 1.22 raises two intriguing questions. Apart from the case of quadratic functions, are there instances of f, where f has full domain and where every closed convex subset of U is $\overrightarrow{D}$-Chebyshev? Because of Fact 1.20, an affirmative answer to this question would imply that ∇f is a (quite surprising) *nonaffine yet convexity-preserving* transformation. Combining Example 1.21 and Fact 1.22, we deduce that – when working with the negative entropy – if C is convex, then C is $\overrightarrow{D}$-Chebyshev but *not* vice versa. Is it possible to describe the $\overrightarrow{D}$-Chebyshev sets in this setting?

We also note that C is "nearly $\overleftarrow{D}$-Chebyshev" in the following sense.

Fact 1.23. (See [6, Corollary 5.6].) Suppose that f is supercoercive, that f is twice continuously differentiable, and that for every $y \in U$, $\nabla^2 f(y)$ is positive definite. Then, $\overleftarrow{P}_C$ is almost everywhere and generically[6] single-valued on U.

It would be interesting to see whether or not supercoercivity is essential in Fact 1.23. By duality, we obtain the following result on the single-valuedness of $\overrightarrow{P}_{f,C}$.

Corollary 1.24. *Suppose that f has full domain, that f^* is twice continuously differentiable, and that $\nabla^2 f^*(y)$ is positive definite for every $y \in U^*$. Then, $\overrightarrow{P}_{f,C}$ is almost everywhere and generically single-valued on U.*

Proof. By Lemma 1.14(ii), $\overrightarrow{P}_{f,C}\big|_U = \nabla f^* \circ \overleftarrow{P}_{f^*,C^*} \circ \nabla f$. Fact 1.23 states that $\overleftarrow{P}_{f^*,C^*}$ is almost everywhere and generically single-valued on U^*. Since f^* is twice continuously differentiable, it follows from the Mean Value Theorem that ∇f^* is locally Lipschitz. Since $(\nabla f)^{-1} = \nabla f^*$ is a locally Lipschitz homeomorphism from U^* to U, the conclusion follows from Lemmas 1.17 and 1.18. ∎

1.4 Klee Sets

Previously known were the following two results:

Fact 1.25 ($\overleftarrow{D}$-Klee sets). (See [7, Theorem 4.4].) Suppose that f is supercoercive, that C is bounded, and that C is $\overleftarrow{D}$-Klee. Then C is a singleton.

Fact 1.26 ($\overrightarrow{D}$-Klee sets). (See [8, Theorem 3.2].) Suppose that C is bounded and that C is $\overrightarrow{D}$-Klee. Then C is a singleton.

Fact 1.25 immediately raises the question whether or not supercoercivity is really an essential hypothesis. Fortunately, thanks to Fact 1.26, which was recently proved for general Legendre functions without any further assumptions, we are now able to present a new result which removes the supercoercivity assumption in Fact 1.25.

[6] That is, the set S of points $y \in U$ where $\overleftarrow{P}_C(y)$ is *not* a singleton is very small both in measure theory (S has measure 0) and in category theory (S is meager/first category).

Theorem 1.27 **($\overleftarrow{D}$-Klee sets revisited).** *Suppose that C is bounded and that C is $\overleftarrow{D}$-Klee. Then C is a singleton.*

Proof. On the one hand, since C is compact, Fact 1.2 implies that C^* is compact. On the other hand, by Lemma 1.15(iii), the set C^* is $\overrightarrow{D}_{f^*}$-Klee. Altogether, we deduce from Fact 1.26 (applied to f^* and C^*) that C^* is a singleton. Therefore, C is a singleton by Fact 1.2. ■

Similarly to the setting of Chebyshev sets, the set C is "nearly $\overleftarrow{D}$-Klee" in the following sense.

Fact 1.28. (See [6, Corollary 5.2(ii)].) Suppose that f is supercoercive, f is twice continuously differentiable, for every $y \in U$, $\nabla^2 f(y)$ is positive definite, and that C is bounded. Then, $\overleftarrow{Q}_C$ is almost everywhere and generically single-valued on U.

Again, it would be interesting to see whether or not supercoercivity is essential in Fact 1.28. Similarly to the proof of Corollary 1.24, we obtain the following result on the single-valuedness of $\overrightarrow{Q}_{f,C}$.

Corollary 1.29. *Suppose that f has full domain, f^* is twice continuously differentiable, $\nabla^2 f^*(y)$ is positive definite for every $y \in U^*$, and C is bounded. Then, $\overrightarrow{Q}_{f,C}$ is almost everywhere and generically single-valued on U.*

1.5 Chebyshev Centers: Uniqueness and Characterization

Fact 1.30 **($\overrightarrow{D}$-Chebyshev centers).** (See [8, Theorem 4.4].) Suppose that C is bounded. Then the right Chebyshev center with respect to C is a singleton, say $\overrightarrow{Z}_C = \{x\}$, and x is characterized by

$$x \in \nabla f^*\big(\operatorname{conv} \nabla f(\overrightarrow{Q}_C(x))\big). \tag{1.36}$$

We now present a corresponding new result on the left Chebyshev center.

Theorem 1.31 **($\overleftarrow{D}$-Chebyshev centers).** *Suppose that C is bounded. Then the left Chebyshev center with respect to C is a singleton, say $\overleftarrow{Z}_C = \{y\}$, and y is characterized by*

$$y \in \operatorname{conv} \overleftarrow{Q}_C(y). \tag{1.37}$$

Proof. By Lemma 1.16(ii),

$$\overleftarrow{Z}_{f,C} = \nabla f^*\big(\overrightarrow{Z}_{f^*,C^*}\big). \tag{1.38}$$

Now, C^* is a bounded subset of U^* because of the compactness of C and Fact 1.2. Applying Fact 1.30 to f^* and C^*, we obtain that $\overrightarrow{Z}_{f^*,C^*} = \{y^*\}$ for some $y^* \in U^*$ and that y^* is characterized by

$$y^* \in \nabla f\big(\operatorname{conv} \nabla f^*(\overrightarrow{Q}_{f^*,C^*}(y^*))\big). \tag{1.39}$$

By (1.38), $\overleftarrow{Z}_{f,C} = \nabla f^*\big(\overrightarrow{Z}_{f^*,C^*}\big) = \{\nabla f^*(y^*)\} = \{y\}$ is a singleton. Moreover, using Lemma 1.15(ii), we see that the characterization (1.39) becomes

$$\begin{aligned}\overleftarrow{Z}_{f,C} = \{y\} &\Leftrightarrow y^* \in \nabla f\big(\operatorname{conv} \nabla f^*(\overrightarrow{Q}_{f^*,C^*}(y^*))\big)\\ &\Leftrightarrow \nabla f^*(y^*) \in \operatorname{conv} \nabla f^*(\overrightarrow{Q}_{f^*,C^*}(y^*))\\ &\Leftrightarrow y \in \operatorname{conv} \nabla f^*(\overrightarrow{Q}_{f^*,C^*}(\nabla f(y)))\\ &\Leftrightarrow y \in \operatorname{conv} \overleftarrow{Q}_{f,C}(y),\end{aligned} \tag{1.40}$$

as claimed. ■

Remark 1.32. The proof of Fact 1.30 does not carry over directly to the setting of Theorem 1.31. Indeed, one key element in that proof was to realize that the right farthest distance function

$$\overrightarrow{F}_C = \sup_{y \in C} D(\cdot, y) \tag{1.41}$$

is *convex* (as the supremum of convex functions) and then to apply the Ioffe-Tihomirov theorem (see, e.g., [40, Theorem 2.4.18]) for the subdifferential of the supremum of convex function. In contrast, $\overleftarrow{F}_C = \sup_{x \in C} D(x, \cdot)$ is generally *not convex*. (For more on separate and joint convexity of D, see [3].)

1.6 Chebyshev Centers: Two Examples

1.6.1 Diagonal-Symmetric Line Segments in the Strictly Positive Orthant

In addition to our standing assumptions from Sect. 1.1, we assume in this section that the following hold:

$$X = \mathbb{R}^2; \tag{1.42}$$

$$\mathbf{c}_0 = (1, a) \text{ and } \mathbf{c}_1 = (a, 1), \quad \text{where } 1 < a < +\infty; \tag{1.43}$$

$$\mathbf{c}_\lambda = (1-\lambda)\mathbf{c}_0 + \lambda \mathbf{c}_1, \quad \text{where } 0 < \lambda < 1; \tag{1.44}$$

$$C = \operatorname{conv}\{\mathbf{c}_0, \mathbf{c}_1\} = \{\mathbf{c}_\lambda \mid 0 \le \lambda \le 1\}. \tag{1.45}$$

Theorem 1.33. *Suppose that f is any of the functions considered in Example 1.1. Then the left Chebyshev center is the midpoint of C, i.e., $\overleftarrow{Z}_C = \{\mathbf{c}_{1/2}\}$.*

Proof. By Theorem 1.31, we write $\overleftarrow{Z}_C = \{\mathbf{y}\}$, where $\mathbf{y} = (y_1, y_2) \in U$. In view of (1.37) and Fact 1.4(ii), we obtain that $\overleftarrow{Q}_C(\mathbf{y})$ contains at least two elements. On the other hand, since $\overleftarrow{Q}_C(\mathbf{y})$ consists of the maximizers of the *convex* function $D(\cdot, \mathbf{y})$ over the compact set C, [33, Corollary 32.3.2] implies that $\overleftarrow{Q}_C(\mathbf{y}) \subseteq \{\mathbf{c}_0, \mathbf{c}_1\}$. Altogether,

$$\overleftarrow{Q}_C(\mathbf{y}) = \{\mathbf{c}_0, \mathbf{c}_1\}. \tag{1.46}$$

In view of (1.37),

$$\mathbf{y} \in C. \tag{1.47}$$

On the other hand, a symmetry argument identical to the proof of [8, Proposition 5.1] and the uniqueness of its Chebyshev center show that $\mathbf{y}$ must lie on the diagonal, i.e., that

$$y_1 = y_2. \tag{1.48}$$

The result now follows because the only point satisfying both (1.47) and (1.48) is $\mathbf{c}_{1/2}$, the midpoint of C. ■

Remark 1.34. Theorem 1.33 is in stark contrast with [8, Sect. 5], where we investigated the *right* Chebyshev center in this setting. Indeed, there we found that the right Chebyshev center does depend on the underlying Legendre function used (see [8, Examples 5.2, 5.3, and 5.5]). Furthermore, for each Legendre function f considered in Example 1.1, we obtain the following formula.

$$(\forall \mathbf{y} = (y_1, y_2) \in U) \quad \overleftarrow{Q}_{f,C}(\mathbf{y}) = \begin{cases} \{\mathbf{c}_0\}, & \text{if } y_2 < y_1; \\ \{\mathbf{c}_1\}, & \text{if } y_2 > y_1; \\ \{\mathbf{c}_0, \mathbf{c}_1\}, & \text{if } y_1 = y_2. \end{cases} \tag{1.49}$$

Indeed, since for every $\mathbf{y} \in U$, the function $D(\cdot, \mathbf{y})$ is convex; the points where the supremum is achieved is a subset of the extreme points of C, i.e., of $\{\mathbf{c}_0, \mathbf{c}_1\}$. Therefore, it suffices to compare $D(\mathbf{c}_0, \mathbf{y})$ and $D(\mathbf{c}_1, \mathbf{y})$.

1.6.2 Intervals of Real Numbers

Theorem 1.35. *Suppose that $X = \mathbb{R}$ and that $C = [a, b] \subset U$, where $a \neq b$. Denote the right and left Chebyshev centers by x and y, respectively. Then*[7]

$$x = \frac{f^*(b^*) - f^*(a^*)}{b^* - a^*} \quad \textit{and} \quad y^* = \frac{f(b) - f(a)}{b - a}. \tag{1.50}$$

[7] Recall the convenient notation introduced on page 3!

Proof. Analogously to the derivation of (1.46), it must hold that

$$\overleftarrow{Q}_C(y) = \{a,b\}. \tag{1.51}$$

This implies that y satisfies $D(a,y) = D(b,y)$. In turn, using Fact 1.4(i), this last equation is equivalent to $D_{f^*}(y^*,a^*) = D_{f^*}(y^*,b^*) \Leftrightarrow f^*(y^*) + f(a) - y^*a = f^*(y^*) + f(b) - y^*b \Leftrightarrow f(b) - f(a) = y^*(b-a) \Leftrightarrow y^* = (f(b) - f(a))/(b-a)$, as claimed. Hence,

$$y = \nabla f^*\left(\frac{f(b) - f(a)}{b-a}\right). \tag{1.52}$$

Combining this formula (applied to f^* and $C^* = [a^*, b^*]$) with Lemma 1.16(ii), we obtain that the right Chebyshev center is given by

$$x = \nabla f^*\left(\nabla f^{**}\left(\frac{f^*(b^*) - f^*(a^*)}{b^* - a^*}\right)\right) = \frac{f^*(b^*) - f^*(a^*)}{b^* - a^*}, \tag{1.53}$$

as required. ■

Example 1.36. Suppose that $X = \mathbb{R}$ and $C = [a,b]$, where $0 < a < b < +\infty$. In each of the following items, suppose that f is as in the corresponding item of Example 1.1. Denote the corresponding right and left Chebyshev centers by x and y, respectively. Then the following hold:

(i) $x = y = \dfrac{a+b}{2}$.

(ii) $x = \dfrac{b-a}{\ln(b) - \ln(a)}$ and $y = \exp\left(\dfrac{b\ln(b) - b - a\ln(a) + a}{b-a}\right)$.

(iii) $x = \dfrac{ab\big(\ln(b) - \ln(a)\big)}{b-a}$ and $y = \dfrac{b-a}{\ln(b) - \ln(a)}$.

Proof. This follows from Theorem 1.35. ■

1.7 Generalizations and Variants

Chebyshev set and Klee set problems can be generalized to problems involving functions.

Throughout this section,

$$g\colon X \to \left]-\infty, +\infty\right] \text{ is lower semicontinuous and proper.} \tag{1.54}$$

For convenience, we also set

$$q = \frac{1}{2}\|\cdot\|^2. \tag{1.55}$$

Recall that the *Moreau envelope* $e_\lambda g\colon X \to [-\infty, +\infty]$ and the set-valued *proximal mapping* $P_\lambda g\colon X \rightrightarrows X$ are given by

$$x \mapsto e_\lambda g(x) = \inf_w \left(g(w) + \frac{1}{2\lambda}\|x-w\|^2 \right) \tag{1.56}$$

and

$$x \mapsto P_\lambda g(x) = \operatorname*{argmin}_w \left(g(w) + \frac{1}{2\lambda}\|x-w\|^2 \right). \tag{1.57}$$

It is natural to ask: If $P_\lambda g$ is single-valued everywhere on $\mathbb{R}^n$, what can we say about the function g?

Similarly, define $\phi_\mu g\colon X \to]-\infty, +\infty]$ and $Q_\mu g\colon X \rightrightarrows X$ by

$$y \mapsto \phi_\mu g(y) = \sup_x \left(\frac{1}{2\mu}\|y-x\|^2 - g(x) \right), \tag{1.58}$$

and

$$y \mapsto Q_\mu g(y) = \operatorname*{argmax}_x \left(\frac{1}{2\mu}\|y-x\|^2 - g(x) \right). \tag{1.59}$$

Again, it is natural to ask: If $Q_\mu g$ is single-valued everywhere on X, what can we say about the function g? When $g = \iota_C$, then $P_\lambda g = P_C, Q_\mu g = Q_C$, and we recover the classical Chebyshev and Klee set problems.

Definition 1.37. (i) The function g is *prox-bounded* if there exists $\lambda > 0$ such that $e_\lambda g \not\equiv -\infty$. The supremum of the set of all such λ is the threshold λ_g of the prox-boundedness for g.
(ii) The constant μ_g is defined to be the infimum of all $\mu > 0$ such that $g - \mu^{-1}q$ is bounded below on X; equivalently, $\phi_\mu g(0) < +\infty$.

Fact 1.38. (See [34, Examples 5.23 and 10.32].) Suppose that g is prox-bounded with threshold λ_g, and let $\lambda \in]0, \lambda_g[$. Then, $P_\lambda g$ is everywhere upper semicontinuous and locally bounded on X, and $e_\lambda g$ is locally Lipschitz on X.

Fact 1.39. (See [38, Proposition 4.3].) Suppose that $\mu > \mu_g$. Then, $Q_\mu g$ is upper semicontinuous and locally bounded on X, and $\phi_\mu g$ is locally Lipschitz on X.

Definition 1.40. (i) We say that g is *λ-Chebyshev* if $P_\lambda g$ is single-valued on X.
(ii) We say that g is *μ-Klee* if $Q_\mu g$ is single-valued on X.

Facts 1.41 and 1.43 below concern Chebyshev functions and Klee functions; see [38] for proofs.

Fact 1.41 (Single-valued proximal mappings). Suppose that g is prox-bounded with threshold λ_g, and let $\lambda \in]0, \lambda_g[$. Then the following are equivalent.

(i) $e_\lambda g$ is continuously differentiable on X.
(ii) g is λ-Chebyshev, i.e., $P_\lambda g$ is single-valued everywhere.
(iii) $g + \lambda^{-1}q$ is essentially strictly convex.

If any of these conditions holds, then

$$\nabla\big((g+\lambda^{-1}q)^*\big) = P_\lambda g \circ (\lambda \operatorname{Id}). \tag{1.60}$$

Corollary 1.42. *The function g is convex if and only if $\lambda_g = +\infty$ and $P_\lambda g$ is single-valued on X for every $\lambda > 0$.*

Fact 1.43 (Single-valued farthest mappings). Suppose that $\mu > \mu_g$. Then the following are equivalent.

(i) $\phi_\mu g$ is (continuously) differentiable on X.
(ii) g is μ-Klee, i.e., $Q_\mu g$ is single-valued everywhere.
(iii) $g - \mu^{-1}q$ is essentially strictly convex.

If any of these conditions holds, then

$$\nabla\big((g-\mu^{-1}q)^*\big) = Q_\mu g(-\mu \operatorname{Id}). \tag{1.61}$$

Corollary 1.44. *Suppose that g has bounded domain. Then* $\operatorname{dom} g$ *is a singleton if and only if for all $\mu > 0$, the farthest operator $Q_\mu g$ is single-valued on X.*

Definition 1.45 (Chebyshev points). The set of μ-Chebyshev points of g is

$$\operatorname{argmin} \phi_\mu g.$$

If $\operatorname{argmin} \phi_\mu g$ is a singleton, then we denote its unique element by p_μ and we refer to p_μ as the *μ-Chebyshev point of g*.

The following result is new.

Theorem 1.46 (Chebyshev point of a function). *Suppose that $\mu > \mu_g$. Then, the set of μ-Chebyshev points is a singleton, and the μ-Chebyshev point is characterized by*

$$p_\mu \in \operatorname{conv} Q_\mu g(p_\mu). \tag{1.62}$$

Proof. As $\mu > \mu_g$, Fact 1.39 implies that

$$y \mapsto \phi_\mu g(y) = \frac{1}{2\mu}\|y\|^2 + \left(-\frac{1}{\mu}q + g\right)^*(-y/\mu), \tag{1.63}$$

is finite. Hence, $\phi_\mu g$ is strictly convex and supercoercive; thus, $\phi_\mu g$ has a unique minimizer. Furthermore, we have

$$\partial \phi_\mu g(y) = \frac{1}{\mu}\big(y - \operatorname{conv} Q_\mu g(y)\big) \tag{1.64}$$

by the Ioffe–Tikhomirov Theorem [40, Theorem 2.4.18]. Therefore,

$$0 \in \partial \phi_\mu g(y) \quad \Leftrightarrow \quad y \in \operatorname{conv} Q_\mu g(y), \tag{1.65}$$

which yields the result. ■

We now provide three examples to illustrate the Chebyshev point of functions.

Example 1.47. Suppose that $g = q$. Then $\mu_g = 1$ and for $\mu > 1$, we have

$$\phi_\mu g \colon y \mapsto \sup_x \left(\frac{1}{2\mu}(y-x)^2 - \frac{x^2}{2} \right) = \frac{y^2}{2(\mu - 1)}. \tag{1.66}$$

Hence, the μ-Chebyshev point of g is $p_\mu = 0$.

Example 1.48. Suppose that $g = \iota_{[a,b]}$, where $a < b$. Then $\mu_g = 0$ and for $\mu > 0$, we have

$$\phi_\mu g \colon y \mapsto \sup_x \left(\frac{1}{2\mu}(y-x)^2 - \iota_{[a,b]}(x) \right) = \begin{cases} \dfrac{(y-b)^2}{2\mu} & \text{if } y \le \dfrac{a+b}{2}, \\ \dfrac{(y-a)^2}{2\mu} & \text{if } y > \dfrac{a+b}{2}. \end{cases} \tag{1.67}$$

Hence, $p_\mu = \frac{a+b}{2}$.

Example 1.49. Let $a < b$ and suppose that g is given by

$$x \mapsto \begin{cases} 0 & \text{if } a \le x \le \dfrac{a+b}{2}, \\ 1 & \text{if } \dfrac{a+b}{2} < x \le b, \\ +\infty & \text{otherwise.} \end{cases} \tag{1.68}$$

Then $\mu_g = 0$, and when $\mu > 0$ we have

$$\begin{aligned} \phi_\mu g(y) &= \sup_x \left(\frac{1}{2\mu}(y-x)^2 - g(x) \right) \\ &= \sup_x \begin{cases} \dfrac{1}{2\mu}(y-x)^2 & \text{if } a \le x \le \dfrac{a+b}{2} \\ \dfrac{1}{2\mu}(y-x)^2 - 1 & \text{if } \dfrac{a+b}{2} < x \le b \\ -\infty & \text{otherwise} \end{cases} \\ &= \max \left\{ \frac{(y-a)^2}{2\mu}, \frac{(y-(a+b)/2)^2}{2\mu}, \frac{(y-b)^2}{2\mu} - 1 \right\}, \end{aligned}$$

by using the fact that a strictly convex function only achieves its maximum at the extreme points of its domain. Elementary yet tedious calculations yield the following. When $\mu > (a-b)^2/4$, we have

$$\phi_\mu g(y) = \begin{cases} \dfrac{(y-b)^2}{2\mu} - 1 & \text{if } y < \dfrac{2\mu}{a-b} + \dfrac{a+3b}{4} \\ \dfrac{(y-(a+b)/2)^2}{2\mu} & \text{if } \dfrac{2\mu}{a-b} + \dfrac{a+3b}{4} \leq y < \dfrac{3a+b}{4} \\ \dfrac{(y-a)^2}{2\mu} & \text{if } y > \dfrac{3a+b}{4}; \end{cases}$$

while when $0 < \mu \leq (a-b)^2/4$, one obtains

$$\phi_\mu g(y) = \begin{cases} \dfrac{(y-b)^2}{2\mu} - 1 & \text{if } y < \dfrac{\mu}{a-b} + \dfrac{a+b}{2} \\ \dfrac{(y-a)^2}{2\mu} & \text{if } y \geq \dfrac{\mu}{a-b} + \dfrac{a+b}{2}. \end{cases}$$

Hence, the Chebyshev point of g is

$$p_\mu = \begin{cases} \dfrac{3a+b}{4}, & \text{if } \mu > (a-b)^2/4; \\ \dfrac{\mu}{a-b} + \dfrac{a+b}{2}, & \text{if } 0 < \mu \leq (a-b)^2/4. \end{cases}$$

1.8 List of Open Problems

Problem 1. Is the assumption that f be supercoercive in Fact 1.19 really essential?

Problem 2. Are the assumptions that f have full domain and that $\overline{C^*} \subseteq U^*$ in Fact 1.20 really essential?

Problem 3. Does there exist a Legendre function f with full domain such that f is not quadratic yet every nonempty closed convex subset of X is $\overrightarrow{D}$-Chebyshev? In view of Fact 1.19, the gradient operator ∇f of such a function would be nonaffine and it would preserve convexity.

Problem 4. Is it possible to characterize the class of $\overrightarrow{D}$-Chebyshev subsets of the strictly positive orthant when f is the negative entropy? Fact 1.22 and Example 1.21 imply that this class contains not only all closed convex but also some nonconvex subsets.

Problem 5. Is the assumption that f be supercoercive in Fact 1.23 really essential?

Problem 6. Is the assumption that f be supercoercive in Fact 1.28 really essential?

Problem 7. For the Chebyshev functions and Klee functions, we have used the halved Euclidean distance. What are characterizations of f and Chebyshev point of f when one uses the Bregman distances?

Problem 8. How do the results on Chebyshev functions and Klee functions extend to Hilbert spaces or even general Banach spaces?

1.9 Conclusion

Chebyshev sets, Klee sets, and Chebyshev centers are well known notions in classical Euclidean geometry. These notions have been studied traditionally also in an infinite-dimensional setting or with respect to metric distances induced by different norms. Recently, a new framework was provided by measuring the discrepancy between points differently, namely by Bregman distances, and new results have been obtained that generalize the classical results formulated in Euclidean spaces. These results are fairly well understood for Klee sets and Chebyshev centers with respect to Bregman distances; however, the situation is much less clear for Chebyshev sets.

The current state-of-the-art is reviewed in this paper and several new results have been presented. The authors hope that the list of open problems (in Sect. 1.8) will entice the reader to make further progress on this fascinating topic.

Acknowledgements The authors thank two referees for their careful reading and pertinent comments. Heinz Bauschke was partially supported by the Natural Sciences and Engineering Research Council of Canada and by the Canada Research Chair Program. Xianfu Wang was partially supported by the Natural Sciences and Engineering Research Council of Canada.

References

1. Asplund, E.: Sets with unique farthest points. Israel J. Math. **5**, 201–209 (1967)
2. Bauschke, H.H., Borwein, J.M.: Legendre functions and the method of random Bregman projections. J. Convex Anal. **4**, 27–67 (1997)
3. Bauschke, H.H., Borwein, J.M.: Joint and separate convexity of the Bregman distance. In: D. Butnariu, Y. Censor, S. Reich (ed.) Inherently Parallel Algorithms in Feasibility and Optimization and their Applications (Haifa 2000), pp. 23–36. Elsevier (2001)
4. Bauschke, H.H., Noll, D.: The method of forward projections. J. Nonlin. Convex Anal. **3**, 191–205 (2002)
5. Bauschke, H.H., Borwein, J.M, Combettes, P.L.: Essential smoothness, essential strict convexity, and Legendre functions in Banach spaces. Commun. Contemp. Math. **3**, 615–647 (2001)
6. Bauschke, H.H., Wang, X., Ye, J., Yuan, X.: Bregman distances and Chebyshev sets. J. Approx. Theory **159**, 3–25 (2009)
7. Bauschke, H.H., Wang, X., Ye, J., Yuan, X.: Bregman distances and Klee sets. J. Approx. Theory **158**, 170–183 (2009)
8. Bauschke, H.H., Macklem, M.S., Sewell, J.B., Wang, X.: Klee sets and Chebyshev centers for the right Bregman distance. J. Approx. Theory **162**, 1225–1244 (2010)
9. Berens, H., Westphal, U.: Kodissipative metrische Projektionen in normierten linearen Räumen. In: P. L. Butzer and B. Sz.-Nagy (eds.) Linear Spaces and Approximation, vol. 40, pp. 119–130, Birkhäuser (1980)
10. Borwein, J.M.: Proximity and Chebyshev sets. Optim. Lett. **1**, 21–32 (2007)
11. Borwein, J.M, Lewis, A.S.: Convex Analysis and Nonlinear Optimization, 2nd edn. Springer (2006)

12. Borwein, J.M., Vanderwerff, J.: Convex Functions: Constructions, Characterizations and Counterexamples. Cambridge University Press (2010)
13. Bregman, L.M.: The relaxation method of finding the common point of convex sets and its application to the solution of problems in convex programming. U.S.S.R. Comp. Math. Math **7**, 200–217 (1967)
14. Bunt, L.N.H.: Bijdrage tot de theorie de convexe puntverzamelingen. Thesis, Univ. of Groningen, Amsterdam, 1934
15. Butnariu, D., Iusem, A.N.: Totally Convex Functions for Fixed Points Computation in Infinite Dimensional Optimization. Kluwer, Dordrecht (2000)
16. Censor, Y., Zenios, S.A.: Parallel Optimization. Oxford University Press (1997)
17. Deutsch, F.: Best Approximation in Inner Product Spaces. Springer (2001)
18. Fremlin, D.H.: Measure Theory, vol. 2. Broad Foundations, 2nd edn. Torres Fremlin, Colchester (2010)
19. Garkavi, A.L.: On the Čebyšev center and convex hull of a set. Usp. Mat. Nauk **19**, 139–145 (1964)
20. Goebel, K., Kirk, W.A.: Topics in Metric Fixed Point Theory. Cambridge University Press (1990)
21. De Guzmán, M.: A change-of-variables formula without continuity. Am. Math. Mon. **87**, 736–739 (1980)
22. Hiriart-Urruty, J.-B.: Ensembles de Tchebychev vs. ensembles convexes: l'etat de la situation vu via l'analyse convexe non lisse. Ann. Sci. Math. Québec **22**, 47–62 (1998)
23. Hiriart-Urruty, J.-B.: La conjecture des points les plus éloignés revisitée. Ann. Sci. Math. Québec **29**, 197–214 (2005)
24. Hiriart-Urruty, J.-B.: Potpourri of conjectures and open questions in nonlinear analysis and optimization. SIAM Rev. **49**, 255–273 (2007)
25. Hiriart-Urruty, J.-B., Lemaréchal, C.: Convex Analysis and Minimization Algorithms II. Springer (1996)
26. Klee, V.: Circumspheres and inner products. Math. Scand. **8**, 363–370 (1960)
27. Klee, V.: Convexity of Chebyshev sets. Math. Ann. **142**, 292–304 (1960/1961)
28. Motzkin, T.: Sur quelques propriétés caractéristiques des ensembles convexes. Atti. Accad. Naz. Lincei, Rend., VI. Ser. **21**, 562–567 (1935)
29. Motzkin, T.S., Straus, E.G., Valentine, F.A.: The number of farthest points. Pac. J. Math. **3**, 221–232 (1953)
30. Nielsen, F., Nock, R.: On the smallest enclosing information disk. Inform. Process. Lett. **105**, 93–97 (2008)
31. Nock, R., Nielsen, F.: Fitting the smallest enclosing Bregman ball. In: J. Gama, R. Camacho, P. Brazdil, A. Jorge and L. Torgo (eds.) Machine Learning: 16th European Conference on Machine Learning (Porto 2005), pp. 649–656, Springer Lecture Notes in Computer Science vol. 3720 (2005)
32. Reich, S., Sabach, S.: Two strong convergence theorems for Bregman strongly nonexpansive operators in reflexive Banach spaces. Nonlinear Anal. **73**, 122–135 (2010)
33. Rockafellar, R.T.: Convex Analysis. Princeton University Press, Princeton (1970)
34. Rockafellar, R.T., Wets, R. J.-B.: Variational Analysis. Springer, New York (1998)
35. Singer, I.: Best Approximation in Normed Linear Spaces by Elements of Linear Subspaces. Springer (1970)
36. Singer, I.: The Theory of Best Approximation and Functional Analysis. Conference Board of the Mathematical Sciences Regional Conference Series in Applied Mathematics, No. 13. Society for Industrial and Applied Mathematics (1974)
37. Vlasov, L.P.: Approximate properties of sets in normed linear spaces. Russian Math. Surv. **28**, 1–66 (1973)
38. Wang, X.: On Chebyshev functions and Klee functions. J. Math. Anal. Appl. **368**, 293–310 (2010)
39. Westphal, U., Schwartz, T.: Farthest points and monotone operators. B. Aust. Math. Soc. **58**, 75–92 (1998)
40. Zălinescu, C.: Convex Analysis in General Vector Spaces. World Scientific Publishing (2002)

Chapter 2
Self-Dual Smooth Approximations of Convex Functions via the Proximal Average

Heinz H. Bauschke, Sarah M. Moffat, and Xianfu Wang

Abstract The proximal average of two convex functions has proven to be a useful tool in convex analysis. In this note, we express the Goebel self-dual smoothing operator in terms of the proximal average, which allows us to give a different proof of self duality. We also provide a novel self-dual smoothing operator. Both operators are illustrated by smoothing the norm.

Keywords Approximation · Convex function · Fenchel conjugate · Goebel smoothing operator · Moreau envelope · Proximal average

AMS 2010 Subject Classification: Primary 26B25; Secondary 26B05, 65D10, 90C25

2.1 Introduction

Let X be the standard Euclidean space $\mathbb{R}^n$, with inner product $\langle\cdot,\cdot\rangle$ and induced norm $\|\cdot\|$. It will be convenient to set

$$\mathfrak{q} = \frac{1}{2}\|\cdot\|^2. \tag{2.1}$$

Now let $f\colon X \to \left]-\infty,+\infty\right]$ be convex, lower semicontinuous, and proper. Since many convex functions are nonsmooth, it is natural to ask: How can one approximate f with a smooth function?

The most famous and very useful answer to this question is provided by the *Moreau envelope* [15, 17], which, for $\lambda > 0$, is defined by[1]

$$e_\lambda f = f \,\square\, \lambda^{-1}\mathfrak{q}. \tag{2.2}$$

[1] The symbol "□" denotes *infimal convolution*: $(f_1 \,\square\, f_2)(x) = \inf_y \big(f_1(y) + f_2(x-y)\big)$.

H.H. Bauschke (✉)
Department of Mathematics, Irving K. Barber School, University of British Columbia, Kelowna, B.C. V1V 1V7, Canada
e-mail: heinz.bauschke@ubc.ca

H.H. Bauschke et al. (eds.), *Fixed-Point Algorithms for Inverse Problems in Science and Engineering*, Springer Optimization and Its Applications 49,
DOI 10.1007/978-1-4419-9569-8_2,

It is well known that $e_\lambda f$ is smooth (i.e., continuously differentiable) and that $\lim_{\lambda\to 0^+} e_\lambda f = f$ point-wise; see, e.g., [17, Theorems 1.25 and 2.26]. Parenthetically, other approaches to smoothing are Ghomi's integral convolution method [9], Seeger's ball rolling technique [18], and Teboulle's entropic proximal mappings [19].

Let us now consider the norm, which is nonsmooth at the origin.

Example 2.1 (Moreau envelope of the norm). Let $\lambda \in]0,1[$, set $f = \|\cdot\|$, and denote the closed unit ball by C. Then, for x and x^* in X, we have[2]

$$e_\lambda f(x) = \begin{cases} \dfrac{\|x\|^2}{2\lambda}, & \text{if } \|x\| \le \lambda; \\[2ex] \|x\| - \dfrac{\lambda}{2}, & \text{if } \|x\| > \lambda, \end{cases} \tag{2.3}$$

$(e_\lambda f)^* = \iota_C + \lambda \mathrm{q}$, and $e_\lambda (f^*)(x^*) = (2\lambda)^{-1} \cdot \big(\max\{0, \|x^*\| - 1\}\big)^2$. Consequently, $(e_\lambda f)^* \neq e_\lambda (f^*)$.

Proof. Either a straight-forward computation or [17, Example 11.26(a)] yields

$$f^* = \iota_C. \tag{2.4}$$

Next, if $y \in X$, then

$$e_{1/\lambda}\, \iota_C(y) = \inf_{c\in C} \lambda \mathrm{q}(y-c) \tag{2.5}$$

$$= \frac{\lambda}{2} d_C^2(y) \tag{2.6}$$

$$= \frac{\lambda}{2} \cdot \begin{cases} \big(\|y\| - 1\big)^2, & \text{if } \|y\| > 1; \\[1ex] 0, & \text{if } \|y\| \le 1, \end{cases} \tag{2.7}$$

and thus

$$e_{1/\lambda}\, \iota_C\big(x/\lambda\big) = \frac{\lambda}{2} \cdot \begin{cases} \big(\|x/\lambda\| - 1\big)^2, & \text{if } \|x\| > \lambda; \\[1ex] 0, & \text{if } \|x\| \le \lambda. \end{cases} \tag{2.8}$$

---

[2] Here, ι_C is the *indicator function* defined by $\iota_C(x) = 0$, if $x \in C$; $\iota_C(x) = +\infty$, if $x \notin C$, $f^*(x^*) = \sup_{x\in X} \big(\langle x, x^*\rangle - f(x)\big)$ is the *Fenchel conjugate* of f, and $d_C(x) = \inf_{c\in C} \|x - c\| = \big(\|\cdot\| \Box \iota_C\big)(x)$ is the *distance function*.

By [17, Example 11.26(b) on page 495], we obtain

$$e_\lambda f(x) = \frac{1}{\lambda}\mathfrak{q}(x) - e_{1/\lambda} f^*\left(x/\lambda\right) \tag{2.9}$$

$$= \frac{1}{2\lambda}\|x\|^2 - \frac{\lambda}{2}\cdot\begin{cases} \dfrac{\|x\|^2}{\lambda^2} - \dfrac{2\|x\|}{\lambda} + 1, & \text{if } \|x\| > \lambda; \\ 0, & \text{if } \|x\| \le \lambda \end{cases} \tag{2.10}$$

$$= \begin{cases} \|x\| - \dfrac{\lambda}{2}, & \text{if } \|x\| > \lambda; \\ \dfrac{\|x\|^2}{2\lambda}, & \text{if } \|x\| \le \lambda \end{cases} \tag{2.11}$$

and $\left(e_\lambda f\right)^* = f^* + \lambda\mathfrak{q} = \iota_C + \lambda\mathfrak{q}$. Alternatively, one may use [7, Example 2.16], which provides the proximal mapping of f, and then use the proximal mapping calculus to obtain these results. Finally, a referee pointed out that (2.11) can also be derived by reducing the computation of the Moreau envelope to

$$e_\lambda f(x) = (f \Box \lambda^{-1}\mathfrak{q})(x) \tag{2.12}$$

$$= \inf_y \left(f(y) + \lambda^{-1}\mathfrak{q}(x-y)\right) \tag{2.13}$$

$$= \inf_y \left(\|y\| + \frac{1}{2\lambda}\left(\|x\|^2 + \|y\|^2 - 2\langle x,y\rangle\right)\right) \tag{2.14}$$

$$= \inf_{\eta\ge 0}\inf_{\|y\|=\eta} \left(\|y\| + \frac{1}{2\lambda}\left(\|x\|^2 + \|y\|^2 - 2\langle x,y\rangle\right)\right) \tag{2.15}$$

$$= \inf_{\eta\ge 0}\inf_{\|y\|=\eta} \left(\eta + \frac{1}{2\lambda}\left(\|x\|^2 + \eta^2 - 2\eta\|x\|\right)\right) \tag{2.16}$$

$$= \frac{\|x\|^2}{2\lambda} + \frac{1}{2\lambda}\inf_{\eta\ge 0}\left(\eta^2 + 2(\lambda - \|x\|)\eta\right), \tag{2.17}$$

which can now be treated by one-dimensional calculus. ■

While the Moreau envelope has many desirable properties, we see from Example 2.1 that the smooth approximation $e_\lambda f$ is not *self-dual* in the sense that

$$\left(e_\lambda f\right)^* \ne e_\lambda\left(f^*\right). \tag{2.18}$$

It is perhaps surprising that self-dual smoothing operators even exist. The first example appears in [11]. Specifically, Goebel defined

$$G_\lambda f = (1-\lambda^2)e_\lambda f + \lambda\mathfrak{q} \tag{2.19}$$

and proved that

$$(G_\lambda f)^* = G_\lambda (f^*), \tag{2.20}$$

that is, *Fenchel conjugation and Goebel smoothing commute!* For applications of the Goebel smoothing operator, see [11].

The purpose of this note is twofold. First, we present a different representation of the Goebel smoothing operator which allows us to prove self-duality using the Fenchel conjugation formula for the proximal average. Second, the proximal average is also utilized to obtain a novel smoothing operator. Both smoothing operators are computed explicitly for the norm. The formulas derived show that the new smoothing operator is distinct from the one provided by Goebel.

For f_1 and f_2, two functions from X to $]-\infty,+\infty]$ that are convex, lower semicontinuous and proper, and for two strictly positive convex coefficients ($\lambda_1 + \lambda_2 = 1$), the *proximal average* is defined by

$$\operatorname{pav}(f_1,f_2;\lambda_1,\lambda_2) = \big(\lambda_1(f_1+\mathsf{q})^* + \lambda_2(f_2+\mathsf{q})^*\big)^* - \mathsf{q}. \tag{2.21}$$

The proximal average, which is actually a convex function, has been a useful tool for constructing primal-dual symmetric antiderivatives [4] and for extending monotone operators [2]; see also [3, 5, 6, 11, 12] for further information and applications. One of the key properties is the *Fenchel conjugation formula*

$$\operatorname{pav}(f_1,f_2;\lambda_1,\lambda_2)^* = \operatorname{pav}(f_1^*,f_2^*;\lambda_1,\lambda_2); \tag{2.22}$$

see [3, Theorem 6.1], [5, Theorem 4.3], or [6, Theorem 5.1].

We use standard convex analysis calculus and notation as, e.g., in [16, 17, 21]. In Sect. 2.2, we consider the Goebel smoothing operator from the proximal-average view point. The new smoothing operator is presented in Sect. 2.3.

2.2 The Goebel Smoothing Operator

Definition 2.2 (Goebel smoothing operator). Let $f\colon X \to]-\infty,+\infty]$ be convex, lower semicontinuous and proper, and let $\lambda \in]0,1[$. Then the *Goebel smoothing operator* [11] is defined by

$$G_\lambda f = (1-\lambda^2)e_\lambda f + \lambda \mathsf{q}. \tag{2.23}$$

Note that (2.23) and standard properties of the Moreau envelope imply that point-wise

$$\lim_{\lambda\to 0^+} G_\lambda f = f \tag{2.24}$$

and that each $G_\lambda f$ is smooth.

Our first main result provides two alternative descriptions of the Goebel smoothing operator. The first description, item (i) in Theorem 2.3, shows a pleasing reformulation in terms of the proximal average. The second description, item (ii) in Theorem 2.3, is less appealing but has the advantage of providing a different proof of the *self-duality*, item (iii), observed by Goebel.

Theorem 2.3. *Let $f\colon X \to \left]-\infty,+\infty\right]$ be convex, lower semicontinuous and proper, and let $\lambda \in \left]0,1\right[$. Then the following hold.*[3]

(i) $G_\lambda f = (1+\lambda)\,\mathrm{pav}(f,0;1-\lambda,\lambda) + \lambda\mathfrak{q}$.
(ii) $G_\lambda f = (1+\lambda)^2\,\mathrm{pav}\left(f,\mathfrak{q};\frac{1-\lambda}{1+\lambda},\frac{2\lambda}{1+\lambda}\right)\circ(1+\lambda)^{-1}\mathrm{Id}$.
(iii) ***(Goebel)*** $\left(G_\lambda f\right)^* = G_\lambda(f^*)$.

Proof. Let $x \in X$. Then, using (2.21) and standard convex calculus, we obtain

$$\left((1+\lambda)^2\mathrm{pav}\left(f,\mathfrak{q};\frac{1-\lambda}{1+\lambda},\frac{2\lambda}{1+\lambda}\right)\circ(1+\lambda)^{-1}\mathrm{Id}\right)(x) \tag{2.25}$$

$$= (1+\lambda)^2\left(\left(\frac{1-\lambda}{1+\lambda}(f+\mathfrak{q})^* + \frac{2\lambda}{1+\lambda}(\mathfrak{q}+\mathfrak{q})^*\right)^* - \mathfrak{q}\right)\left(\frac{x}{1+\lambda}\right) \tag{2.26}$$

$$= (1+\lambda)^2\left(\frac{1-\lambda}{1+\lambda}(f+\mathfrak{q})^* + \frac{\lambda}{1+\lambda}\mathfrak{q}\right)^*\left(\frac{x}{1+\lambda}\right) - \mathfrak{q}(x) \tag{2.27}$$

$$= (1+\lambda)\left((1-\lambda)(f+\mathfrak{q})^* + \lambda\mathfrak{q}\right)^*(x) - \mathfrak{q}(x) \tag{2.28}$$

$$= (1+\lambda)\left(\left((1-\lambda)(f+\mathfrak{q})^* + \lambda(0+\mathfrak{q})^*\right)^* - \mathfrak{q}\right)(x) + \lambda\mathfrak{q}(x) \tag{2.29}$$

$$= \left((1+\lambda)\,\mathrm{pav}(f,0;1-\lambda,\lambda) + \lambda\mathfrak{q}\right)(x). \tag{2.30}$$

We have verified that (2.28) as well as the right sides of (i) and (ii) coincide. Starting from (2.28) and again applying standard convex calculus, we see that

$$(1+\lambda)\left((1-\lambda)(f+\mathfrak{q})^* + \lambda\mathfrak{q}\right)^*(x) - \mathfrak{q}(x) \tag{2.31}$$

$$= (1+\lambda)\left(\left((1-\lambda)(f+\mathfrak{q})^*\right)^* \,\square\, (\lambda\mathfrak{q})^*\right)(x) - \mathfrak{q}(x) \tag{2.32}$$

$$= (1+\lambda)\left((1-\lambda)(f+\mathfrak{q})\left(\frac{\cdot}{1-\lambda}\right) \square\, \frac{1}{\lambda}\mathfrak{q}\right)(x) - \mathfrak{q}(x) \tag{2.33}$$

---
[3] Here $\mathrm{Id}\colon X \to X\colon x \mapsto x$ is the *identity operator*.

$$= (1+\lambda)\inf_y\Big((1-\lambda)(f+\mathsf{q})\Big(\frac{y}{1-\lambda}\Big)+\frac{1}{\lambda}\mathsf{q}(x-y)\Big)-\mathsf{q}(x) \tag{2.34}$$

$$= (1+\lambda)\inf_y\Big((1-\lambda)f\Big(\frac{y}{1-\lambda}\Big)+(1-\lambda)\mathsf{q}\Big(\frac{y}{1-\lambda}\Big) + \frac{1}{\lambda}\mathsf{q}(x-y)-\frac{1}{1+\lambda}\mathsf{q}(x)\Big) \tag{2.35}$$

$$= (1-\lambda^2)\inf_y\Big(f\Big(\frac{y}{1-\lambda}\Big)+\mathsf{q}\Big(\frac{y}{1-\lambda}\Big)+\frac{1}{\lambda(1-\lambda)}\mathsf{q}(x-y)-\frac{1}{1-\lambda^2}\mathsf{q}(x)\Big). \tag{2.36}$$

Simple algebra shows that for every $y\in X$,

$$\mathsf{q}\Big(\frac{y}{1-\lambda}\Big)+\frac{1}{\lambda(1-\lambda)}\mathsf{q}(x-y)-\frac{1}{1-\lambda^2}\mathsf{q}(x)=\frac{1}{\lambda}\mathsf{q}\Big(x-\frac{y}{1-\lambda}\Big)+\frac{\lambda}{1-\lambda^2}\mathsf{q}(x). \tag{2.37}$$

Therefore,

$$(1+\lambda)\Big((1-\lambda)(f+\mathsf{q})^*+\lambda\mathsf{q}\Big)^*(x)-\mathsf{q}(x) \tag{2.38}$$

$$= (1-\lambda^2)\inf_y\Big(f\Big(\frac{y}{1-\lambda}\Big)+\mathsf{q}\Big(\frac{y}{1-\lambda}\Big)+\frac{1}{\lambda(1-\lambda)}\mathsf{q}(x-y)-\frac{1}{1-\lambda^2}\mathsf{q}(x)\Big) \tag{2.39}$$

$$= (1-\lambda^2)\inf_y\Big(f\Big(\frac{y}{1-\lambda}\Big)+\frac{1}{\lambda}\mathsf{q}\Big(x-\frac{y}{1-\lambda}\Big)+\frac{\lambda}{1-\lambda^2}\mathsf{q}(x)\Big) \tag{2.40}$$

$$= (1-\lambda^2)\inf_z\Big(f(z)+\frac{1}{\lambda}\mathsf{q}(x-z)+\frac{\lambda}{1-\lambda^2}\mathsf{q}(x)\Big) \tag{2.41}$$

$$= \big((1-\lambda^2)e_\lambda f+\lambda\mathsf{q}\big)(x) \tag{2.42}$$

$$= G_\lambda f(x), \tag{2.43}$$

which completes the proof of (i) and (ii).

(iii): In view of the conjugate formula $(\beta^2 h\circ(\beta^{-1}\mathrm{Id}))^*=\beta^2 h^*\circ(\beta^{-1}\mathrm{Id})$, (ii), and (2.22), we obtain

$$(G_\lambda f)^* = \left((1+\lambda)^2 \operatorname{pav}\left(f, \mathfrak{q}; \frac{1-\lambda}{1+\lambda}, \frac{2\lambda}{1+\lambda}\right) \circ (1+\lambda)^{-1}\mathrm{Id}\right)^* \tag{2.44}$$

$$= (1+\lambda)^2 \left(\operatorname{pav}\left(f, \mathfrak{q}; \frac{1-\lambda}{1+\lambda}, \frac{2\lambda}{1+\lambda}\right)\right)^* \circ (1+\lambda)^{-1}\mathrm{Id} \tag{2.45}$$

$$= (1+\lambda)^2 \operatorname{pav}\left(f^*, \mathfrak{q}^*; \frac{1-\lambda}{1+\lambda}, \frac{2\lambda}{1+\lambda}\right) \circ (1+\lambda)^{-1}\mathrm{Id} \tag{2.46}$$

$$= (1+\lambda)^2 \operatorname{pav}\left(f^*, \mathfrak{q}; \frac{1-\lambda}{1+\lambda}, \frac{2\lambda}{1+\lambda}\right) \circ (1+\lambda)^{-1}\mathrm{Id} \tag{2.47}$$

$$= G_\lambda (f^*). \tag{2.48}$$

The proof is complete. ■

Remark 2.4. Theorem 2.3(i) and (ii) gives two representations of the Goebel smoothing operator in terms of the proximal average. Goebel [10] discovered a converse formula, which we state next without proof:

$$\operatorname{pav}(f, \mathfrak{q}; \lambda, 1-\lambda) = \frac{(2-\lambda)^2}{4} G_{\lambda/(2-\lambda)} f \circ \left(\frac{2}{2-\lambda}\mathrm{Id}\right). \tag{2.49}$$

Example 2.5. Let $\lambda \in \,]0,1[$ and set $f = \|\cdot\|$. Then, for every $x \in X$,

$$G_\lambda f(x) = \begin{cases} \dfrac{\|x\|^2}{2\lambda}, & \text{if } \|x\| \leq \lambda; \\[2ex] \dfrac{\lambda\|x\|^2}{2} + (1-\lambda^2)\|x\| - \dfrac{\lambda(1-\lambda^2)}{2}, & \text{if } \|x\| > \lambda. \end{cases} \tag{2.50}$$

Proof. Combine (2.23) and (2.3). ■

2.3 A New Smoothing Operator

We now provide a novel smoothing operator that has a very simple expression in terms of the proximal average.

Definition 2.6 (New smoothing operator). Let $f \colon X \to \,]-\infty, +\infty]$ be convex, lower semicontinuous and proper, and let $\lambda \in \,]0,1[$. Then the $S_\lambda f$ is defined by

$$S_\lambda f = \operatorname{pav}(f, \mathfrak{q}; 1-\lambda, \lambda). \tag{2.51}$$

Theorem 2.7. *Let $f\colon X \to \left]-\infty,+\infty\right]$ be convex, lower semicontinuous and proper, and let $\lambda \in \left]0,1\right[$. Set $\mu = \lambda/(2-\lambda)$. Then the following hold.*

(i) $S_\lambda f = (1-\lambda)e_\mu f \circ \left(\frac{2}{2-\lambda}\operatorname{Id}\right) + \mu \mathrm{q}$.

(ii) $\left(S_\lambda f\right)^* = S_\lambda (f^*)$.

Proof. (i): Let $x \in X$. Then, using (2.51), (2.21) and standard convex calculus, we obtain

$$
\begin{aligned}
&\left(S_\lambda f\right)(x)\\
&\quad = \left((1-\lambda)(f+\mathrm{q})^* + \lambda(\mathrm{q}+\mathrm{q})^*\right)^*(x) - \mathrm{q}(x) && (2.52)\\
&\quad = \left((1-\lambda)(f+\mathrm{q})^* + \tfrac{\lambda}{2}\mathrm{q}\right)^*(x) - \mathrm{q}(x) && (2.53)\\
&\quad = \left((1-\lambda)(f+\mathrm{q})\left(\frac{\cdot}{1-\lambda}\right) \Box \frac{2}{\lambda}\mathrm{q}\right)(x) - \mathrm{q}(x) && (2.54)\\
&\quad = \inf_y \left((1-\lambda)f\left(\frac{y}{1-\lambda}\right) + (1-\lambda)\mathrm{q}\left(\frac{y}{1-\lambda}\right) + \frac{2}{\lambda}\mathrm{q}(x-y) - \mathrm{q}(x)\right) && (2.55)\\
&\quad = (1-\lambda)\inf_y \left(f\left(\frac{y}{1-\lambda}\right) + \mathrm{q}\left(\frac{y}{1-\lambda}\right) + \frac{2}{\lambda(1-\lambda)}\mathrm{q}(x-y) - \frac{1}{1-\lambda}\mathrm{q}(x)\right). && (2.56)
\end{aligned}
$$

Simple algebra shows that for every $y \in X$,

$$
\begin{aligned}
&\mathrm{q}\left(\frac{y}{1-\lambda}\right) + \frac{2}{\lambda(1-\lambda)}\mathrm{q}(x-y) - \frac{1}{1-\lambda}\mathrm{q}(x)\\
&\quad = \frac{2-\lambda}{\lambda}\mathrm{q}\left(\frac{2x}{2-\lambda} - \frac{y}{1-\lambda}\right) + \frac{\lambda}{(1-\lambda)(2-\lambda)}\mathrm{q}(x). && (2.57)
\end{aligned}
$$

Therefore,

$$
\begin{aligned}
&\left(S_\lambda f\right)(x) && (2.58)\\
&\quad = (1-\lambda)\inf_y \left(f\left(\frac{y}{1-\lambda}\right) + \frac{2-\lambda}{\lambda}\mathrm{q}\left(\frac{2x}{2-\lambda} - \frac{y}{1-\lambda}\right) + \frac{\lambda}{(1-\lambda)(2-\lambda)}\mathrm{q}(x)\right) && (2.59)\\
&\quad = (1-\lambda)\inf_z \left(f(z) + \frac{2-\lambda}{\lambda}\mathrm{q}\left(\frac{2x}{2-\lambda} - z\right)\right) + \frac{\lambda}{2-\lambda}\mathrm{q}(x) && (2.60)\\
&\quad = (1-\lambda)\left(f \Box \frac{1}{\mu}\mathrm{q}\right)\left(\frac{2x}{2-\lambda}\right) + \mu\mathrm{q}(x), && (2.61)
\end{aligned}
$$

as claimed.

(ii): Using (2.51) and (2.22), we get

$$
\begin{aligned}
(S_\lambda f)^* &= \big(\operatorname{pav}(f, \mathsf{q}; 1-\lambda, \lambda)\big)^* && (2.62)\\
&= \operatorname{pav}(f^*, \mathsf{q}^*; 1-\lambda, \lambda) && (2.63)\\
&= \operatorname{pav}(f^*, \mathsf{q}; 1-\lambda, \lambda) && (2.64)\\
&= S_\lambda(f^*). && (2.65)
\end{aligned}
$$

The proof is complete. ■

Note that Theorem 2.7(i) and standard properties of the Moreau envelope imply that point-wise

$$\lim_{\lambda \to 0^+} S_\lambda f = f \tag{2.66}$$

and that each $S_\lambda f$ is smooth.

Example 2.8. Let $\lambda \in\,]0,1[$ and set $f = \|\cdot\|$. Then, for every $x \in X$,

$$
S_\lambda f(x) = \begin{cases} \dfrac{(2-\lambda)\|x\|^2}{2\lambda}, & \text{if } \|x\| \le \dfrac{\lambda}{2};\\[2ex] \dfrac{\lambda\|x\|^2}{2(2-\lambda)} + \dfrac{2(1-\lambda)}{2-\lambda}\|x\| - \dfrac{\lambda(1-\lambda)}{2(2-\lambda)}, & \text{if } \|x\| > \dfrac{\lambda}{2}. \end{cases} \tag{2.67}
$$

Proof. Combine (2.3) and Theorem 2.7(i). ■

Remark 2.9. Let $f = \|\cdot\|$. The explicit formulas provided in Examples 2.5 and 2.8 imply that $G_\alpha f \neq S_\beta f$, for *all* α and β in $]0,1[$. Thus, the smoothing operator defined by (2.51) is indeed new and different from the Goebel smoothing operator.

Remark 2.10. It would be desirable to obtain further explicit formulas beyond the example of the norm. Given a more complicated function f, the explicit computation of the smoothing operators $G_\lambda f$ and $S_\lambda f$ may not be easy. However, computational convex analysis provides tools [8, 13, 14] to compute the Moreau envelope numerically which – due to the Moreau envelope formulations (2.23) and Theorem 2.7(i) – make it possible to compute the smoothing operators $G_\lambda f$ and $S_\lambda f$ numerically. It would also be interesting to extend the present results to infinite-dimensional settings. Promising starting points for this endeavor are [1, 21]. Finally, self-dual regularizations of maximal monotone operators are studied in [20].

Acknowledgements The authors thank the two referees for their constructive comments. Heinz Bauschke was partially supported by the Natural Sciences and Engineering Research Council of Canada and by the Canada Research Chair Program. Sarah Moffat was partially supported by the Natural Sciences and Engineering Research Council of Canada. Xianfu Wang was partially supported by the Natural Sciences and Engineering Research Council of Canada.

References

1. Attouch, H.: Variational Convergence for Functions and Operators. Pitman, Boston (1984)
2. Bauschke, H.H. and Wang, X.: The kernel average for two convex functions and its applications to the extension and representation of monotone operators. Trans. Amer. Math. Soc. **361**, 5947–5965 (2009)
3. Bauschke, H.H., Matoušková, E., and Reich, S.: Projection and proximal point methods: convergence results and counterexamples. Nonlinear Anal. **56**, 715–738 (2004)
4. Bauschke, H.H., Lucet, Y., and Wang, X.: Primal-dual symmetric intrinsic methods for finding antiderivatives for cyclically monotone operators. SIAM J. Control Optim. **46**, 2031–2051 (2007)
5. Bauschke, H.H., Lucet, Y., and Trienis, M.: How to transform one convex function continuously into another. SIAM Rev. **50**, 115–132 (2008)
6. Bauschke, H.H., Goebel, R., Lucet, Y., and Wang, X.: The proximal average: basic theory. SIAM J. Optim. **19**, 766–785 (2008)
7. Combettes, P.L., and Wajs, V.R.: Signal recovery by proximal forward-backward splitting. Multiscale Model. Simul. **4**, 1168–1200 (2005)
8. Gardiner, B. and Lucet, Y.: Graph-matrix calculus for computational convex analysis. Springer Optimization and Its Applications **49**, 243–259 (2011)
9. Ghomi, M.: The problem of optimal smoothing for convex functions. Proc. Amer. Math. Soc. **130**, 2255–2259 (2002)
10. Goebel, R.: Personal communication (2006)
11. Goebel, R.: Self-dual smoothing of convex and saddle functions. J. Convex Anal. **15**, 179–190 (2008)
12. Goebel, R.: The proximal average for saddle functions and its symmetry properties with respect to partial and saddle conjugacy. J. Nonlinear Convex Anal. **11**, 1–11 (2010)
13. Lucet, Y.: Faster than the fast Legendre transform, the linear-time Legendre transform. Numer. Algorithms **16**, 171–185 (1997)
14. Lucet, Y., Bauschke, H.H., and Trienis, M.: The piecewise linear-quadratic model for computational convex analysis. Comput. Optim. Appl. **43**, 95–118 (2009)
15. Moreau, J.J.: Proximité et dualité dans un espace hilbertien. Bull. Soc. Math. France **93**, 273–299 (1965)
16. Rockafellar, R.T.: Convex Analysis. Princeton University Press, Princeton (1970)
17. Rockafellar, R.T. and Wets, R.J-B.: Variational Analysis. Corrected 3rd printing, Springer, Berlin (2009)
18. Seeger, A.: Smoothing a nondifferentiable convex function: the technique of the rolling ball. Rev. Mat. Apl. **18**, 259–268 (1997)
19. Teboulle, M.: Entropic proximal mappings with applications to nonlinear programming. Math. Oper. Res. **17**, 670–690 (1992)
20. Wang, X.: Self-dual regularization of monotone operators via the resolvent average. SIAM J. Optim. (to appear)
21. Zălinescu, C.: Convex Analysis in General Vector Spaces. World Scientific Publishing, River Edge, NJ (2002)

Chapter 3
A Linearly Convergent Algorithm for Solving a Class of Nonconvex/Affine Feasibility Problems

Amir Beck and Marc Teboulle

Abstract We introduce a class of nonconvex/affine feasibility (NCF) problems that consists of finding a point in the intersection of affine constraints with a nonconvex closed set. This class captures some interesting fundamental and NP hard problems arising in various application areas such as sparse recovery of signals and affine rank minimization that we briefly review. Exploiting the special structure of NCF, we present a simple gradient projection scheme which is proven to converge to a unique solution of NCF at a linear rate under a natural assumption explicitly given defined in terms of the problem's data.

Keywords Nonconvex affine feasibility · Inverse problems · Gradient projection algorithm · Linear rate of convergence · Scalable restricted isometry · Mutual coherence of a matrix · Sparse signal recovery · Compressive sensing · Affine rank minimization

AMS 2010 Subject Classification: 90C30, 90C26

3.1 Introduction

Let $\mathbb{E}$ and $\mathbb{V}$ be finite dimensional Euclidean spaces, $\mathscr{A} : \mathbb{E} \to \mathbb{V}$ a given linear mapping, and $\mathbf{b} \in \mathbb{V}$ a vector of observations. Consider the feasibility problem defined by

$$\text{(NCF)} \quad \text{Find } \mathbf{x} \in \mathscr{C} \text{ such that } \mathscr{A}(\mathbf{x}) = \mathbf{b},$$

A. Beck (✉)
Department of Industrial Engineering, Technion Israel Institute of Technology, Haifa 32000, Israel
e-mail: becka@ie.technion.ac.il

H.H. Bauschke et al. (eds.), *Fixed-Point Algorithms for Inverse Problems in Science and Engineering*, Springer Optimization and Its Applications 49,
DOI 10.1007/978-1-4419-9569-8_3,

where $\mathscr{C} \subseteq \mathbb{E}$ is a set which describes some a priori information on the unknown element $\mathbf{x}$. One natural approach for tackling NCF is via the associated minimization problem

$$\text{(NC)} \quad \min\left\{\frac{1}{2}\|\mathscr{A}(\mathbf{x})-\mathbf{b}\|^2 : \mathbf{x} \in \mathscr{C}\right\} \tag{3.1}$$

for some given norm in $\mathbb{V}$.[1]

The above problem formulations are very well known and have been extensively studied over the last several decades, in particular when $\mathscr{C}$ is a closed convex subset of $\mathbb{E}$, giving rise to the so-called convex feasibility problems, see the comprehensive review paper [1] and references therein. Problems of this kind naturally arise in the area of linear inverse problems which covers a wide range of data processing problems, such as imaging sciences, optics, and astrophysics, see e.g., the classical monograph [17] and references therein. In such situations, one has to derive an estimate of some physical quantity of interest (e.g., a signal or an image) from given measurements and some a priori information described through the set $\mathscr{C}$, see for instance the in-depth review paper [12]. Furthermore, it should be noted that non-convex feasibility problems have also been studied in the literature, see for instance, [7, 13]. In particular, the method of successive projections for closed convex sets was extended in [13] to a class of nonconvex compact sets satisfying some hypothesis.

A current trend of research in the data processing areas (e.g., signal processing, machine learning etc.), which has recently attracted a lot of attention focuses on solving problems that can recover *sparse* objects. Finding the sparsest solution of a linear system or the more general problem that consists of finding a low rank matrix satisfying linear matrix equations are at the heart of these current activities. These problems being generally NP hard are often solved by their convex relaxations. The current algorithmic, theoretical and applications literature is vast, and we refer the reader to the excellent very recent survey papers [6] and [21] and references therein.

In this paper, we depart from the convex relaxation approach. We focus on the class of problems (NCF) where the constraint set $\mathscr{C}$ is a closed and *nonconvex* subset of $\mathbb{E}$, which will be defined to naturally captures sparsity features, and we propose to solve NCF via a very simple gradient projection algorithm which under a natural assumption on the problem's data is proven to converge linearly to a global optimal solution of (NC).

The paper is organized as follows. In Sect. 3.2, we define the problem and give some examples arising in fundamental applications that naturally fit our formalism. Section 3.3 first gives some background on the so-called restricted isometry property (RIP), which has been central in the analysis of sparse recovery problems via their *convex* relaxations. This leads us to introduce a natural extension of RIP, called Scalable Restricted Isometry Property (SRIP) for the class of problems under study, and that will play a key role in the analysis of the proposed gradient projection scheme and which here solved directly the nonconvex problem. The analysis is developed in

[1] Throughout the paper, $\|\cdot\|$ will denote the endowed norm of the relevant Euclidean space (either $\mathbb{E}$ or $\mathbb{V}$).

Sect. 3.4, where we prove that despite the nonconvex nature of the problem, if SRIP is satisfied, the gradient projection method converges at a linear rate to a global optimal solution of (NC) also shown to be unique. The convergence is established both for a constant and backtracking stepsize rules, the later being particularly useful in applications as it does not require the knowledge of any unknown parameter. The algorithm is useful and efficient whenever the projection map onto the nonconvex set is easy to compute, this is shown to be the case in the context of sparse recovery problems, for which we also derive a further interesting consequence from our main convergence result.

3.2 Problem Statement, Motivation and Examples

3.2.1 General Problem Statement

In most practical applications, prior knowledge on some desired features of the unknown $\mathbf{x} \in \mathbb{E}$ is available, and can be quantified by some given function, e.g., a norm like function. The motivation for the proposed definition will be described below.

Definition 3.1. $\mathscr{S}$ is the set of all functions $\varphi : \mathbb{E} \to \mathbb{R}_+$ which are lower semi-continuous (lsc) function satisfying the following properties:

$$(i) \quad \varphi(\mathbf{0}) = 0, \tag{3.2}$$
$$(ii) \quad \varphi(\mathbf{x}) = \varphi(-\mathbf{x}) \text{ (symmetry)}, \tag{3.3}$$
$$(iii) \quad \varphi(\mathbf{x}+\mathbf{y}) \leq \varphi(\mathbf{x}) + \varphi(\mathbf{y}) \text{ (subadditivity)}. \tag{3.4}$$

We are interested in the situation where $\varphi \in \mathscr{S}$ is *nonconvex* and we want to solve the nonconvex feasibility problem:

$$\text{(NCF)} \quad \text{Find } \mathbf{x} \in \mathscr{C}_s \text{ such that } \mathscr{A}(\mathbf{x}) = \mathbf{b},$$

where the admissible constraint is defined by the closed *nonconvex* set

$$\mathscr{C}_s := \{\mathbf{x} \in \mathbb{E} : \ \varphi(\mathbf{x}) \leq s\} \tag{3.5}$$

for some fixed given $s > 0$.

To solve NCF, we consider the related nonconvex minimization problem

$$\text{(NC)} \quad \min\left\{ f(\mathbf{x}) \equiv \frac{1}{2}\|\mathscr{A}(\mathbf{x}) - \mathbf{b}\|^2 : \mathbf{x} \in \mathscr{C}_s \right\}, \tag{3.6}$$

where $\|\cdot\|$ is the underlying norm of the Euclidean space $\mathbb{V}$. For example $\|\cdot\|_2$ when $\mathbb{E} = \mathbb{R}^n$ and $\|\cdot\|_F$ (the Frobenius norm) when $\mathbb{E} = \mathbb{R}^{m\times n}$. Given that NCF has a solution, the optimal value of NC is zero and $\bar{\mathbf{x}}$ is an optimal solution of NC if and only if $\bar{\mathbf{x}}$ is a solution to NCF.

This formalism encompasses a wide class of problems, which has attracted considerable interest in the recent literature and which has triggered the motivation of this work, this will be briefly discussed below. We end by noting that well known alternative ways to tackle NCF include the following three closely related problems:

$$\begin{aligned} &\min \{\varphi(\mathbf{x}) : \|\mathscr{A}(\mathbf{x}) - \mathbf{b}\| \le \eta,\ \mathbf{x} \in \mathbb{E}\}\ (\eta > 0,\ \text{perturbed case}),\\ &\min \{\varphi(\mathbf{x}) : \mathscr{A}(\mathbf{x}) = \mathbf{b},\ \in \mathbb{E}\},\\ &\min \{\|\mathscr{A}(\mathbf{x}) - \mathbf{b}\|^2 + \tau\varphi(\mathbf{x}) : \mathbf{x} \in \mathbb{E}\}, \end{aligned} \tag{3.7}$$

where the last formulation corresponds to a penalty approach, with a penalty parameter $\tau > 0$ which measures the tradeoff between the error in the approximation measured by $\|\mathscr{A}(\mathbf{x}) - \mathbf{b}\|^2$ and the desired property of the unknown $\mathbf{x}$ quantified by the function $\varphi(\mathbf{x})$. Note that all these formulations remain essentially NP hard for the choices of the nonconvex function $\varphi \in \mathscr{S}$ which are described in the following sections.

3.2.2 *Motivation and Examples*

We briefly describe three models of interest in applications that naturally fit as special cases of the proposed formalism of this paper.

Example 3.2 (Compressive sensing). Roughly speaking, in the new emerging compressed sensing technology we are interested in recording as much information as possible in a signal or image $\mathbf{x}$ in the "cheapest" way. In other words, under suitable conditions on the problem's data, few measurements are enough to correctly recover a signal, see [14] for more details.

Let $\mathbb{E} = \mathbb{R}^n, \mathbb{V} = \mathbb{R}^m$. Here the mapping $\mathscr{A} : \mathbb{R}^n \to \mathbb{R}^m$ can be represented by an $m \times n$ matrix $\mathbf{A}$ satisfying $\mathscr{A}(\mathbf{x}) = \mathbf{A}\mathbf{x}$ for every $\mathbf{x} \in \mathbb{R}^n$ (for the sake of notation consistency with the other examples, we will often not use the "matrix" notation). A typical approach is to select a sparse vector, namely with many zero components, that solves a linear system of equations $\mathscr{A}(\mathbf{x}) = \mathbf{b}$ for $\mathbf{x} \in \mathbb{R}^n$. Let $\|\mathbf{x}\|_0$ be the l_0-norm[2] of $\mathbf{x}$ which counts the number of nonzero components of $\mathbf{x}$. Given that the observed vector $\mathbf{b} \in \mathbb{R}^m$ and that the number of measurements is smaller than the size of the vector $\mathbf{x}$, i.e., $m < n$, the sparse reconstruction problem amounts to finding an s-sparse solution (with $s \ll n$) of a nonempty linear system, i.e.,

$$\text{find } \mathbf{x} \in \mathbb{R}^n \text{ with } \|\mathbf{x}\|_0 \le s \text{ such that } \mathscr{A}(\mathbf{x}) = \mathbf{b}.$$

Clearly, this problem is a special case of our model (NCF) with $\mathscr{S} \ni \varphi(\mathbf{x}) := \|\mathbf{x}\|_0$, since the l_0-norm satisfies all the premises of Definition 3.1.

Example 3.3 (Affine rank minimization). Let $\mathbb{E} = \mathbb{R}^{m\times n}, \mathbb{V} = \mathbb{R}^p$ and $\mathscr{A} : \mathbb{R}^{m\times n} \to \mathbb{R}^p$ a linear map. The problem consists of finding a matrix $\mathbf{x} \in \mathbb{R}^{m\times n}$ of minimal

[2] This is by some abuse of terminology, since $\|\mathbf{x}\|_0$ is not a norm, as it clearly does not satisfy the the homogeneity property of a norm.

rank that satisfies a given system of linear matrix equations $\mathscr{A}(\mathbf{x}) = \mathbf{b}$. This is a fundamental problem in many diverse areas, see [21]. Recall that the rank of a matrix is the number of its positive singular values. Thus, when $\mathbf{x}$ is a square diagonal matrix with diagonal elements x_j, the rank function coincides with the l_0-norm of $\mathbf{x}$, and the affine rank minimization problem can be viewed as a natural extension of the previous compressive sensing example.

Now, with $\varphi(\mathbf{x}) := \text{rank}(\mathbf{x})$, one has $\varphi \in \mathscr{S}$ since $\text{rank}(\mathbf{0}) = 0$, $\text{rank}(-\mathbf{x}) = \text{rank}(\mathbf{x})$ and $\text{rank}(\mathbf{x}+\mathbf{y}) \leq \text{rank}(\mathbf{x}) + \text{rank}(\mathbf{y})$ and the conditions in Definition 3.1 are thus satisfied, so that the problem of finding a matrix of rank at most s satisfying $\mathscr{A}(\mathbf{x}) = \mathbf{b}$ fits our model NCF.

Note that both problems described in Examples 3.2 and 3.3 are also often tackled through either one of the corresponding three related optimization problems described via (3.7).

Example 3.4 (l_p-pseudo norm minimization, $0 < p < 1$). Let $\mathbb{E} = \mathbb{R}^n, \mathbb{V} = \mathbb{R}^m$ and $\varphi_p(\mathbf{x}) := \|\mathbf{x}\|_p^p = \sum_{j=1}^n |x_j|^p$ $(0 < p < 1)$. The l_p pseudo-norms are connected to the l_0-norm via the relation $\|\mathbf{x}\|_0 = \lim_{p\to 0^+} \|\mathbf{x}\|_p^p$ (with the convention $0^0 = 0$). Thus, for instance, one could try to solve an approximation of the sparse recovery problem by solving the resulting nonconvex minimization models with $\varphi_p(\cdot)$ for small p. This approach is well known, and it has been recently considered by several authors, see e.g., [10] and references therein.

We now verify that $\varphi_p \in \mathscr{S}$. Clearly, we have $\varphi_p(\mathbf{0}) = 0, \varphi_p(-\mathbf{x}) = \varphi_p(\mathbf{x})$. Moreover, it is easy to see that for any $p \in (0,1)$ one has $(u+v)^p \leq u^p + v^p$ for all $u, v \geq 0$, from which it follows that $\|\mathbf{x}+\mathbf{y}\|_p^p \leq \|\mathbf{x}\|_p^p + \|\mathbf{y}\|_p^p$ so that, as in the previous two examples, the conditions of Definition 3.1 are satisfied and thus this problem fits our formalism.

The last example, but now with $p = 1$, that results in the l_1-norm $\varphi_1(\mathbf{x}) = \|\mathbf{x}\|_1 := \sum_{j=1}^n |x_j|$ of $\mathbf{x} \in \mathbb{R}^n$, and which is a *convex* relaxation of the l_0-norm,[3] is of particular interest. It leads us in the next section to first review some of the recent interesting results in sparse recovery problems, which rely on the so-called RIP and also provide the motivation for introducing a natural extension of this notion within our formalism, and that will play an essential role in our analysis.

3.3 A Scalable Restricted Isometry Property (SRIP)

3.3.1 Convex Relaxation and Restricted Isometry

In sparse solutions of linear systems and affine rank minimization, one faces to solve two computationally intractable combinatorial problems [20, 21]:

[3] The l_1-norm of $\mathbf{x} \in \mathbb{R}^n$ is the lowest convex envelope of $\|\mathbf{x}\|_0$ over the l_∞ unit ball.

$$\text{(CS)} \qquad \min\{\|\mathbf{x}\|_0 : \mathscr{A}(\mathbf{x}) = \mathbf{b}, \mathbf{x} \in \mathbb{R}^n\},$$
$$\text{(AR)} \qquad \min\{\text{rank}(\mathbf{x}) : \mathscr{A}(\mathbf{x}) = \mathbf{b}, \mathbf{x} \in \mathbb{R}^{m \times n}\}.$$

Recent and extensive studies (see [6, 21] and their references) have shown that under appropriate assumptions on the data, that will be discussed shortly, it is possible to solve these problems via their *convex* relaxations. More precisely, we replace the l_0-norm and the rank function by their tractable convex counterparts, namely the l_1-norm in (CS) and the Ky-Fan (nuclear) norm in (AR). The nuclear norm of a matrix $\mathbf{x} \in \mathbb{R}^{m \times n}$ is denoted by $\|\mathbf{x}\|_*$ and is defined as the sum of the nonzero singular values of $\mathbf{x}$. It is the convex envelope of the rank function over the set $\{\mathbf{x} \in \mathbb{R}^{m \times n} : \|\mathbf{x}\|_F \leq 1\}$, see [18]. It should be noted that the idea of using the l_1-norm in the context of sparsity is not a new idea, and goes back to some works in geophysics, see [22, 23].

The convex relaxed problems for (CS) and (AR) which provide lower bounds to the original problems then read as two well-known problems:

$$(\mathit{ConvCS}) \qquad \min\{\|\mathbf{x}\|_1 : \mathscr{A}(\mathbf{x}) = \mathbf{b}, \mathbf{x} \in \mathbb{R}^n\} \text{ (Basis Pursuit [11]),}$$
$$(\mathit{ConvAR}) \qquad \min\{\|\mathbf{x}\|_* : \mathscr{A}(\mathbf{x}) = \mathbf{b}, \mathbf{x} \in \mathbb{R}^{m \times n}\} \text{ (Trace minimization [18]).}$$

Both problems above are tractable convex optimization problems that can be efficiently solved by many convex minimization schemes, see for instance the fast and simple optimal gradient based scheme recently developed in [2], and also the recent review [3] and references therein.

The main question that has been extensively investigated in the literature is then

Main question: For which $\mathscr{A}$, a sparse solution (a low rank matrix) can be recovered? That is to say, under which conditions an optimal solution of the original nonconvex problems (CS) and (AR) can be obtained by solving their convex counterparts (ConvCS) and (ConvAR) respectively?

One of the first results to answer that question was for the compressed sensing l_0-minimization problem (CS) and was obtained via the concept of *mutual coherence of a matrix*, which is also related the forthcoming property. For the interested reader, we have briefly summarized some of these pertinent results in the appendix.

Another concept which plays a fundamental role in answering the main stated question is the so-called RIP. Below, we state the definition of RIP for the general matrix rank minimization recently introduced in [21] as a natural generalization of the vector case which is recalled after the definition (and which can be recovered by setting $\mathbf{x}$ to be a diagonal matrix).

Definition 3.5. The linear map $\mathscr{A} : \mathbb{R}^{m \times n} \to \mathbb{R}^p$ with $m < n$ and $1 \leq d \leq m$ is said to satisfy the RIP with the isometry constant δ_d associated to $\mathscr{A}$, if δ_d is the smallest number such that the following holds:

$$(1 - \delta_d)\|\mathbf{x}\|^2 \leq \|\mathscr{A}(\mathbf{x})\|^2 \leq (1 + \delta_d)\|\mathbf{x}\|^2 \text{ for all } \mathbf{x} \in \mathbb{R}^{m \times n} \text{ s.t. } \text{rank}(\mathbf{x}) \leq d,$$

where $\|\cdot\|$ stands here for the Frobenius norm.

In the vector case (originally developed for compressed sensing problems, see e.g., [9]), the linear mapping $\mathscr{A} : \mathbb{R}^n \to \mathbb{R}^m$ reads $\mathscr{A}(\mathbf{x}) = \mathbf{A}\mathbf{x}$, and the RIP condition reduces to:

$$(1-\delta_d)\|\mathbf{x}\|^2 \leq \|\mathscr{A}(\mathbf{x})\|^2 \leq (1+\delta_d)\|\mathbf{x}\|^2 \text{ for all } \mathbf{x} \in \mathbb{R}^n \text{ s.t. } \|\mathbf{x}\|_0 \leq d.$$

The following two results answer the main question stated above. In the sequel we use the terminology "$\mathbf{x}$ is s-sparse" for all vectors such that $\|\mathbf{x}\|_0 \leq s$.

In [9], the following result has recently been proven for problem (CS).

Theorem 3.6 ([9]). *Consider problem (CS). Let* $\mathbf{b} = \mathscr{A}(\bar{\mathbf{x}})$ *for some s-sparse vector* $\bar{\mathbf{x}} \in \mathbb{R}^n$ *with* $s \geq 1$*. Then,*

(i) if $\delta_{2s} < 1$*, the* l_0 *problem (CS) has a unique s-sparse solution;*
(ii) if $\delta_{2s} < \sqrt{2}-1$*, the optimal solution of the* l_1*-problem (ConvCS) is the same as of the* l_0 *problem.*

In a similar vein, in [21], the previous result has been extended for the rank minimization problem.

Theorem 3.7 ([21]). *Consider problem (AR). Let* $\mathbf{b} = \mathscr{A}(\bar{\mathbf{x}})$ *for some matrix* $\bar{\mathbf{x}} \in \mathbb{R}^{m\times n}$ *of rank* $s \geq 1$*. Then,*

(i) if $\delta_{2s} < 1$*, then* $\bar{\mathbf{x}}$ *is the unique matrix of rank at most s.*
(ii) if $\delta_{5s} < 1/10$*, then the optimal solution of the convex problem (ConvAR) coincides with the minimum rank solution of problem (AR).*

If either of the above RIP assumptions are satisfied for $\mathscr{A}$, for some given d, and with the requested upper bound on δ_d, we will simply write that RIP(d,δ_d) holds. Also, it is useful to note that if $s \leq t$, then $\delta_s \leq \delta_t$, i.e, RIP$(s,\delta_s) \implies$ RIP(t,δ_t).

3.3.1.1 The Good News

For both the vector and matrix cases, it has been proven that for some classes of random matrices (e.g., with i.i.d gaussian entries), the corresponding RIP can be proven to be satisfied with overwhelming probability. Details on these probabilistic analysis can be found for instance in [9, 14, 21]. However, not much is known for arbitrary *deterministic* matrices.

3.3.1.2 The Bad News

The RIP suffers from two major drawbacks:

1. The RIP is lacking scalability.
2. Finding/computing the isometry parameter δ_d can be as difficult as solving the original NP hard problems (CS) and (AR).

Both issues will be addressed in this paper within our general model and the proposed algorithm. The first issue is addressed next by introducing a natural modification of RIP.

3.3.2 *Scalable Restricted Isometry Property*

As just mentioned, an evident drawback of the RIP assumption is its lack of scalability. For example, if a linear operator $\mathscr{A}$ satisfies the RIP with some parameters (s,δ_s), then surely $2\mathscr{A}$ will *not* satisfy the RIP with the same parameters. This is not the case for the notion introduced below which remedies this drawback by considering a straightforward and natural generalization of the RIP for our general problem (NCF), and which we call SRIP.

Let $\varphi \in \mathscr{S}$, $d > 0$ and $\mathscr{A} : \mathbb{E} \to \mathbb{V}$. We write SRIP$(d,\alpha)$ if the following holds:

SRIP(d,α): There exist $\nu_d, \mu_d > 0$ satisfying $\frac{\mu_d}{\nu_d} < \alpha$ such that

$$\nu_d\|\mathbf{x}\| \le \|\mathscr{A}(\mathbf{x})\| \le \mu_d\|\mathbf{x}\| \quad \text{for every } \mathbf{x} \in \mathscr{C}_d.$$

By its definition, if SRIP(d,α) holds for some (d,α), then $\alpha > 1$. Of course, SRIP(d,α) might hold true for certain values of d,α and fail for others. The assumption is restrictive when d is "large" and α is "small" and loose when d is "small" and α is "large." This is reflected in the following lemma whose simple proof is omitted.

Lemma 3.8. *Suppose that $d_1 \le d_2$ and $\alpha_1 \ge \alpha_2$. If SRIP(d_1,α_1) is satisfied, then SRIP(d_2,α_2) is also satisfied.*

Plugging

$$\mu_d^2 = 1+\delta_d, \nu_d^2 = 1-\delta_d, \tag{3.8}$$

in SRIP, the relationship between RIP and SRIP (in the settings of Examples 3.2 and 3.3) is revealed through the following obvious result.

Lemma 3.9. *Let $\beta \in (0,1)$. If RIP(d,δ_d) is satisfied for $\delta_d < \beta$, then SRIP $\left(d,\sqrt{\frac{1+\beta}{1-\beta}}\right)$ holds true.*

We reemphasize that here we are concerned with solving the nonconvex model (NCF) directly rather than relaxing it. Much like the second drawback of the RIP alluded above (i.e., the necessity of knowing δ_{2s}), the determination of the unknown parameters (ν_d,μ_d) of SRIP appears as equally difficult. However, thanks to the proposed algorithmic framework which is developed next, we will show that finding/approximating these parameters is not an issue.

3.4 A Linearly Convergent Gradient Projection Method

3.4.1 *The Gradient Projection Method for Solving Problem (NCF)*

The gradient projection algorithm for minimizing a smooth function over some closed set is very well known and due to its simplicity is particularly adequate for solving large scale problems. However, even for convex problems, it suffers from a slow (e.g., sublinear) rate of convergence, see [4], and references therein.

We will prove that if SRIP($2s, \sqrt{2}$) holds, the gradient projection method actually converges linearly to the solution of the nonconvex problem (NCF), which is also shown to be unique.

Before proceeding, we recall the notion of orthogonal projection. For a nonempty closed possibly nonconvex set $C \subseteq \mathbb{E}$, the projection of $\mathbf{y} \in \mathbb{E}$ onto C, written $P_C(\mathbf{y})$ is a multi-valued map (as opposed to the convex case in which orthogonal projections are guaranteed to be single-valued operators) defined by

$$P_C(\mathbf{y}) := \operatorname{argmin}\{\|\mathbf{x}-\mathbf{y}\|^2 : \mathbf{x} \in C\}.$$

Consider the basic gradient projection method for solving problem (NC):

$$\min\{f(\mathbf{x}) : \mathbf{x} \in \mathscr{C}_s\}, \text{ where } f(\mathbf{x}) := \frac{1}{2}\|\mathscr{A}(\mathbf{x}) - \mathbf{b}\|^2.$$

The gradient of f is simply given by $\nabla f(\mathbf{x}) = \mathscr{A}^*(\mathscr{A}(\mathbf{x}) - \mathbf{b})$, where $\mathscr{A}^*$ stands for the adjoint map to $\mathscr{A}$. The gradient projection method generates a sequence $\mathbf{x}_k$ via:

$$\text{(GP)} \quad \mathbf{x}_{k+1} \in P_{\mathscr{C}_s}\left(\mathbf{x}_k - \frac{1}{T_k}\nabla f(\mathbf{x}_k)\right), k = 0, 1, 2, \ldots,$$

where T_k is an appropriately chosen (inverse) stepsize and $\mathbf{x}_0 \in \mathbb{E}$ is arbitrary.

Note that applying (GP) requires to compute an orthogonal projection onto the set $\mathscr{C}_s$ defined in (3.5). This set is nonempty and closed by the lower semi-continuity of φ. Finding an orthogonal projection onto a nonconvex set is by itself a nonconvex optimization problem, and as such is not necessarily an easy one. However, as seen below, it can be efficiently computed for the sets involved in sparse recovery problems. Note that in both cases below the resulting projections are in general not single valued, and when applying (GP) we can select any element of the resulting multivalued projection in an arbitrary fashion.

- *Case A*. Let $\mathbf{x} \in \mathbb{R}^n$ and $\varphi(\mathbf{x}) = \|\mathbf{x}\|_0$. In this case, the orthogonal projection $P_{C_s}(\mathbf{x})$ of $\mathbf{x} \in \mathbb{R}^n$ onto the set $\mathscr{C}_s$ is simply a vector consisting of the s components of $\mathbf{x}$ with the largest absolute values and zeros otherwise.
- *Case B*. Let $\mathbf{x} \in \mathbb{R}^{m \times n}$ and $\varphi(\mathbf{x}) = \operatorname{rank}(\mathbf{x})$. The set of orthogonal projections $P_{\mathscr{C}_s}(\mathbf{x})$ is computed via a truncated singular value decomposition [19] as

follows: if $\mathbf{x} = \mathbf{U}\Sigma\mathbf{V}^{\mathrm{T}}$ is a singular value decomposition of $\mathbf{x}$, then $P_{\mathscr{C}_s}(\mathbf{x})$ consists of matrices of the form $\mathbf{x} = \mathbf{U}\Sigma_s\mathbf{V}^{\mathrm{T}}$ where the diagonal Σ_s includes the s singular values with largest absolute value (otherwise zero).

3.4.2 Linear Rate of Convergence Analysis for GP

We assume that SRIP$(2s, \sqrt{2})$ holds. We will consider two versions of algorithm (GP). The first one is with a constant stepsize where we assume that

$$T_k = \bar{T} \in [\mu_{2s}^2, 2\nu_{2s}^2),$$

where μ_{2s}, ν_{2s} are as in the definition of SRIP. An evident drawback of the fixed stepsize setting is the requirement that at least μ_{2s} should be known. To avoid the need for knowing this parameter, we also introduce a variant of the method with a backtracking stepsize rule that *does not* require the knowledge of μ_{2s} for computational implementation, see Remark 3.10. This backtracking procedure requires that SRIP should hold with a parameter α which is smaller than $\sqrt{2}$ (but on the other hand, can be arbitrary close to $\sqrt{2}$).

Gradient Projection with backtracking:

Input: $\varphi \in \mathscr{S}$, $s > 0$, $\mathbf{x}_0 \in \mathscr{C}_s$ arbitrary,
$\eta > 1$ – backtracking parameter,
$T_0 \in (0, \mu_{2s})$ initial stepsize.
Step $k (k \geq 0)$**:**
(a) Compute $\mathbf{x}_{k+1} \in P_{\mathscr{C}_s}\left(\mathbf{x}_k - \frac{1}{T_k}\nabla f(\mathbf{x}_k)\right)$.
(b) If $\|\mathscr{A}(\mathbf{x}_{k+1} - \mathbf{x}_k)\| > \sqrt{T_k}\|\mathbf{x}_{k+1} - \mathbf{x}_k\|$, set $T_k \leftarrow \eta T_k$ and go back to (a).
(c) Set $T_{k+1} \leftarrow T_k$.
(d) Set $k \leftarrow k+1$.

Remark 3.10. It is very easy to find a $T_0 \in (0, \mu_{2s}^2)$ without actually knowing μ_{2s}. For example, by taking an arbitrary $\mathbf{v} \in \mathscr{C}_{2s}$, we get that $\frac{\|A(\mathbf{v})\|}{\|\mathbf{v}\|} \leq \mu_{2s}$, so we can pick $T_0 \in \left(0, \frac{\|A(\mathbf{v})\|^2}{\|\mathbf{v}\|^2}\right)$.

From the definition of the backtracking procedure, we first establish the following useful fact on the inverse step size T_k.

Proposition 3.11. *For all* $k \geq 0$,

$$T_k \leq \eta\mu_{2s}^2. \tag{3.9}$$

Proof. This is proved by induction on k. For $k = 0$ the claim is valid by the choice of T_0. Suppose that the claim is true for k and we will prove it for $k+1$. If no backtracking steps were done in step (b), then $T_{k+1} = T_k$ and the claim is correct by the induction assumption. Otherwise, if backtracking steps were performed during step (b), then, in particular, $\gamma = \frac{T_{k+1}}{\eta}$ satisfies $\|\mathscr{A}(\mathbf{x}_{k+2} - \mathbf{x}_{k+1})\| > \sqrt{\gamma}\|\mathbf{x}_{k+2} - \mathbf{x}_{k+1}\|$, which together with the fact that $\mathbf{x}_{k+2} - \mathbf{x}_{k+1} \in \mathscr{C}_{2s}$ and the SRIP assumption imply that $\sqrt{\gamma} \leq \mu_{2s}$ and hence $T_{k+1} \leq \eta \mu_{2s}^2$. ■

We are now ready to prove our main result which shows that if SRIP$(2s, \sqrt{2}/\eta)$ is satisfied, then the function values of the sequence generated by the above algorithm converges linearly to zero. For example, if $\eta = 1.1$, then SRIP$(2s, 1.285...)$ is required to hold true instead of SRIP$(2s, 1.414...)$.

Theorem 3.12. *Consider the GP method with either a constant stepsize $T_k = \bar{T} \in [\mu_{2s}^2, 2\nu_{2s}^2)$ or with a backtracking stepsize rule with parameter η and suppose that SRIP$(2s, \sqrt{2}/\xi)$ is satisfied where $\xi = 1$ for the constant stepsize setting and $\xi = \eta > 1$ for the backtracking scenario. Then*

$$f(\mathbf{x}_{k+1}) \leq (\rho - 1) f(\mathbf{x}_k), \ \forall k \geq 0$$

with $\rho < 2$ given by

$$\rho = \begin{cases} \frac{\bar{T}}{\nu_{2s}^2} & \text{constant stepsize} \\ \frac{\eta \mu_{2s}^2}{\nu_{2s}^2} & \text{backtracking.} \end{cases}$$

As a consequence,

$$f(\mathbf{x}_{k+1}) \leq (\rho - 1)^k f(\mathbf{x}_0), \ \textit{for every } k \geq 0$$

and $f(\mathbf{x}_k) \to 0$ as $k \to \infty$.

Proof. Let

$$q_k(\mathbf{x}, \mathbf{x}_k) := f(\mathbf{x}_k) + \langle \mathbf{x} - \mathbf{x}_k, \nabla f(\mathbf{x}_k) \rangle + \frac{T_k}{2} \|\mathbf{x} - \mathbf{x}_k\|^2. \tag{3.10}$$

Then the GP method can be equivalently rewritten as

$$\mathbf{x}_{k+1} \in \operatorname{argmin}\{q_k(\mathbf{x}, \mathbf{x}_k) : \mathbf{x} \in \mathscr{C}_s\},$$

and hence, in particular, for a solution $\bar{\mathbf{x}}$ of (NCF) it holds that

$$q_k(\mathbf{x}_{k+1}, \mathbf{x}_k) \leq q_k(\bar{\mathbf{x}}, \mathbf{x}_k). \tag{3.11}$$

Now, since $f(\mathbf{x}) = \frac{1}{2}\|\mathscr{A}(\mathbf{x}) - \mathbf{b}\|^2$, it follows that

$$\begin{aligned} f(\mathbf{x}_{k+1}) &= f(\mathbf{x}_k) + \langle \mathbf{x}_{k+1} - \mathbf{x}_k, \nabla f(\mathbf{x}_k)\rangle + \frac{1}{2}\|\mathscr{A}(\mathbf{x}_{k+1} - \mathbf{x}_k)\|^2 \\ &\leq f(\mathbf{x}_k) + \langle \mathbf{x}_{k+1} - \mathbf{x}_k, \nabla f(\mathbf{x}_k)\rangle + \frac{T_k}{2}\|\mathbf{x}_{k+1} - \mathbf{x}_k\|^2, \end{aligned}$$

where the last inequality follows from the fact that $\mathbf{x}_k - \mathbf{x}_{k+1} \in \mathscr{C}_{2s}$ (by the subadditivity and symmetry of the function $\varphi \in \mathscr{S}$) and from the fact that the definition of the stepsize (in the constant or backtracking settings) implies that $\|\mathscr{A}(\mathbf{x}_{k+1} - \mathbf{x}_k)\| \leq \sqrt{T_k}\|\mathbf{x}_{k+1} - \mathbf{x}_k\|$. Therefore, we have shown that $f(\mathbf{x}_{k+1}) \leq q_k(\mathbf{x}_{k+1}, \mathbf{x}_k)$ so that

$$f(\mathbf{x}_{k+1}) = q_k(\mathbf{x}_{k+1}, \mathbf{x}_k) \overset{(3.11)}{\leq} q_k(\bar{\mathbf{x}}, \mathbf{x}_k). \tag{3.12}$$

On the other hand,

$$\begin{aligned} q_k(\bar{\mathbf{x}}, \mathbf{x}_k) &= f(\mathbf{x}_k) + \langle \bar{\mathbf{x}} - \mathbf{x}_k, \nabla f(\mathbf{x}_k)\rangle + \frac{T_k}{2}\|\bar{\mathbf{x}} - \mathbf{x}_k\|^2 \\ &\leq f(\mathbf{x}_k) + \langle \bar{\mathbf{x}} - \mathbf{x}_k, \nabla f(\mathbf{x}_k)\rangle + \frac{T_k}{2\nu_{2s}^2}\|\mathscr{A}(\bar{\mathbf{x}} - \mathbf{x}_k)\|^2 \\ &\overset{\mathscr{A}(\bar{\mathbf{x}})=\mathbf{b}}{=} f(\mathbf{x}_k) + \langle \bar{\mathbf{x}} - \mathbf{x}_k, \nabla f(\mathbf{x}_k)\rangle + \frac{T_k}{2\nu_{2s}^2}\|\mathbf{b} - \mathscr{A}(\mathbf{x}_k)\|^2 \\ &= \left(1 + \frac{T_k}{\nu_{2s}^2}\right) f(\mathbf{x}_k) + \langle \bar{\mathbf{x}} - \mathbf{x}_k, \nabla f(\mathbf{x}_k)\rangle \\ &= \left(1 + \frac{T_k}{\nu_{2s}^2}\right) f(\mathbf{x}_k) - 2f(\mathbf{x}_k) \\ &= \left(\frac{T_k}{\nu_{2s}^2} - 1\right) f(\mathbf{x}_k), \end{aligned}$$

which along with (3.9) (in the backtracking setting), implies the result. ∎

Corollary 3.13. *Suppose that SRIP$(2s, \sqrt{2})$ holds true. Then the sequence $\{\mathbf{x}_k\}$ generated by GP converges to the unique optimal solution of (NC), and hence of NCF.*

Proof. Let $\{\mathbf{x}_k\}$ be the sequence generated by the GP method with a constant stepsize $\bar{T} = \mu_{2s}^2$ and let $\bar{\mathbf{x}}$ be a solution of NCF. Then by Theorem 3.12 we have that

$$f(\mathbf{x}_k) \leq (\rho - 1)^{k-1} f(\mathbf{x}_0)$$

On the other hand,

$$f(\mathbf{x}_k) = \frac{1}{2}\|\mathscr{A}(\mathbf{x}_k) - \mathbf{b}\|^2 = \frac{1}{2}\|\mathscr{A}(\mathbf{x}_k) - \mathscr{A}(\bar{\mathbf{x}})\|^2 \geq \frac{1}{2\nu_{2s}^2}\|\mathbf{x}_k - \bar{\mathbf{x}}\|^2.$$

Therefore, $\mathbf{x}_k \to \bar{\mathbf{x}}$ and since $\bar{\mathbf{x}}$ was chosen arbitrarily its uniqueness follows. ∎

The next corollary, follows immediately from Theorem 3.12, and bounds the number of iterations required to obtain an ε-optimal solution of (NC).

Corollary 3.14. *Consider the setting of Theorem 3.12. Then for* $k \geq 1 + \frac{\log(1/\varepsilon)+C}{D}$, *the (GP) algorithm produces an* $\mathbf{x}$ *such that*

$$\|\mathscr{A}(\mathbf{x}) - \mathbf{b}\|^2 \leq \varepsilon,$$

where $C := \log(2f(\mathbf{x}_0)), D := \log\left(\frac{1}{\rho-1}\right)$.

Remark 3.15. Recently, an algorithm called *the iterative M-sparse algorithm* was analyzed in [5] for solving the l_0 problem ($\mathbb{E} = \mathbb{R}^n, \mathbb{V} = \mathbb{R}^m, \varphi(\mathbf{x}) = \|\mathbf{x}\|_0$). This method is in fact nothing else but the gradient projection algorithm with a *constant step size fixed and equal to 1*. It was proved in [5] that if the columns of the matrix are normalized, and $\|\mathbf{A}\|_2 < 1$, this algorithm converges to a *local* minimum of (NC).

Our approach has focused on using SRIP to solve directly the NCF via a simple gradient projection method. On the other hand, RIP was used to determine conditions that warrant recovery of solutions for the nonconvex optimization problems such as (CS) and (AR) by solving their convex relaxations (ConvCS) and (ConvAR) respectively, namely, it is also needed to apply convex minimization schemes to solve these relaxed problems and achieve the same goals. While a direct comparison of these results is not fully transparent (e.g., in terms of complexity, the parameters involved etc.), it is nevertheless worthwhile to make the following remarks.

Remark 3.16. In the (CS) case, if RIP($2s, \delta_{2s}$) is satisfied with $\delta_{2s} < \sqrt{2} - 1$, then this implies (by Lemma 3.9) that SRIP ($2s, \alpha$) holds true with $\alpha = \sqrt{\frac{1+\sqrt{2}-1}{1-(\sqrt{2}-1)}} = \sqrt{\frac{1}{\sqrt{2}-1}} = 1.5538...$, which is less restrictive than the assumption $\alpha = \sqrt{2}$ used in Theorem 3.12. On the other hand, note that the later condition does not imply the condition on RIP of Theorem 3.6(ii).

Remark 3.17. In the (AR) case, we can be more precise. The condition in Theorem 3.7(ii) requires that RIP should hold with $\delta_{5s} < 0.1$. This condition is worse than the assumption of Theorem 3.12. Indeed, by Lemma 3.9 it implies that SRIP($5s, \alpha$) is satisfied with $\alpha = \sqrt{\frac{1.1}{0.9}} = 1.105...$, which is a more restrictive than the condition SRIP($2s, \sqrt{2}$), see Lemma 3.8.

We end by showing another interesting consequence of Theorem 3.12 which is particularly relevant to sparse recovery problems. Let us focus again on the setting of problem (CS) in Example 3.2, that is, $\mathbb{E} = \mathbb{R}^n, \mathbb{V} = \mathbb{R}^m$ and $\varphi(\mathbf{x}) = \|\mathbf{x}\|_0$. The support of a vector $\mathbf{x} \in \mathbb{R}^n$ is defined to be the set of indices of the nonzero components:

$$\text{supp}(\mathbf{x}) = \{i \in \{1, 2, \ldots, n\} : x_i \neq 0\}.$$

Our final result shows stabilization of the support in the sense that the support of $\mathbf{x}_k$ is contained in the support of the unique solution of NCF from a certain iteration of (GP).

Corollary 3.18. *Consider the setting of Theorem 3.12 and let* $\mathbb{E} = \mathbb{R}^n, \mathbb{V} = \mathbb{R}^m$ *and* $\varphi(\mathbf{x}) = \|\mathbf{x}\|_0$. *Let* $\bar{\mathbf{x}}$ *be the unique solution of NCF. Then there exists* $\bar{k}$ *such that for every* $k \geq \bar{k}$ *the inclusion*

$$\mathrm{supp}(\mathbf{x}_k) \subseteq \mathrm{supp}(\bar{\mathbf{x}})$$

holds true.

Proof. For every set $S \subseteq \{1,2,\ldots,n\}$, let us define:

$$f_S^* = \min\left\{\frac{1}{2}\|\mathbf{A}\mathbf{x} - \mathbf{b}\|^2 : x_i = 0, i \notin S\right\}. \tag{3.13}$$

If $|S| \leq s$ and $\mathrm{supp}(\bar{\mathbf{x}}) \not\subseteq S$ then $f_S^* > 0$ since otherwise, if $f_S^* = 0$, it would mean by the uniqueness of $\bar{\mathbf{x}}$ that the optimal solution of (3.13) is $\bar{\mathbf{x}}$ in contradiction to $\mathrm{supp}(\bar{\mathbf{x}}) \not\subseteq S$. Let us now define the number

$$g = \min_S \left\{f_S^* : |S| \leq s, S \subseteq \{1,2,\ldots,n\}, \mathrm{supp}(\bar{\mathbf{x}}) \not\subseteq S\right\}, \tag{3.14}$$

which is positive. Now, since $f(\mathbf{x}_k) \to 0$, it follows that there exists $\bar{k}$ such that

$$f(\mathbf{x}_k) < g \tag{3.15}$$

for all $k \geq \bar{k}$. Let $k \geq \bar{k}$ and let us assume in contradiction that $\mathrm{supp}(\mathbf{x}_k) \not\subseteq \mathrm{supp}(\bar{\mathbf{x}})$. Then

$$f(\mathbf{x}_k) \geq f^*_{\mathrm{supp}(\mathbf{x}_k)} \geq g,$$

where the last inequality follows from the definition of g, which is a contradiction to (3.15). ■

Appendix

We briefly summarize some of the first results providing sufficient conditions warranting recovery of sparse vectors for the compressed sensing l_0-minimization problem via the convex l_1-norm problem (ConvCS). These were obtained via the concept of *mutual coherence of a matrix*, see [6] for more details and references.

Definition 3.19. [11] Let $\mathbf{A} = [\mathbf{a}_1,\ldots,\mathbf{a}_n] \in \mathbb{R}^{m\times n}$ with $m \leq n$ and with normalized columns $\|\mathbf{a}_i\| = 1$ for all $i = 1,\ldots,n$. Then, the mutual coherence $M(\mathbf{A})$ of the matrix $\mathbf{A}$ is defined by

$$M(\mathbf{A}) := \max_{i\neq j} |\langle \mathbf{a}_i, \mathbf{a}_j\rangle| = \max_{i\neq j} |(\mathbf{A}^T\mathbf{A})_{ij}|.$$

Clearly, $0 \leq M(\mathbf{A}) \leq 1$. Furthermore, it has been shown that

$$M(\mathbf{A}) \geq \sqrt{\frac{n-m}{m(n-1)}}.$$

Note that the mutual coherence of a matrix is generally easy to compute even for large matrices.

Using the mutual coherence of a matrix given in Definition 3.19, the following sufficient condition relating (CS) to its convex relaxation (ConvCS) was proven in [15].

Theorem 3.20. *[15] Consider problem (CS) with $\mathscr{A}(\mathbf{x}) \equiv \mathbf{A}\mathbf{x}$. If a solution $\mathbf{x} \in \mathbb{R}^n$ of problem (CS) satisfies*

$$\|\mathbf{x}\|_0 < \frac{1}{2}\left(1 + \frac{1}{M(\mathbf{A})}\right),$$

then it is unique and coincides with the optimal solution of the convex problem (ConvCS).

We note that for a special class of matrices which are the concatenation of two orthogonal square matrices U, V, i.e., with $\mathbf{A} := [\mathbf{U}, \mathbf{V}]$, the above result has been improved in [16] by requiring the weaker condition:

$$\|\mathbf{x}\|_0 < \frac{\left(\sqrt{2} - \frac{1}{2}\right)}{M(\mathbf{A})}.$$

Further results in the same spirit have been derived for the noisy compressed sensing, see e.g., [8].

Finally, we note that the mutual coherence of a matrix $M(\mathbf{A})$ given in Definition 3.19 is closely related to RIP as shown in the following result.

Lemma 3.21. *Let $\mathbf{A} = [\mathbf{a}_1, \ldots, a_n] \in \mathbb{R}^{m \times n}$ with $m \leq n$ and with normalized columns $\|\mathbf{a}_i\| = 1$ for all $i = 1, \ldots, n$. Then, with $\delta_s \leq (s-1)M(\mathbf{A})$, the matrix $\mathbf{A}$ with mutual coherence $M(\mathbf{A})$ satisfies* RIP(s, δ_s).

Proof. This follows immediately from the definition of RIP and using the Gershgorin circles theorem (see e.g., [19]. ■

Acknowledgements We thank two anonymous referees for their useful comments and suggestions. This research was partially supported by the Israel Science Foundation under ISF Grant 489-06.

References

1. Bauschke, H.H., Borwein, J.M.: On projection algorithms for solving convex feasibility problems. SIAM Review **38**, 367–426 (1996)
2. Beck, A., Teboulle., M.: A fast iterative shrinkage-thresholding algorithm for linear inverse problems. SIAM J. Imaging Sciences **2**, 183–202 (2009)
3. Beck, A., Teboulle, M.: Gradient-based algorithms with applications to signal recovery problems. In: D. Palomar, Y. Eldar (eds.) Convex Optimization in Signal Processing and Communications, 42–88. Cambridge University Press (2010)
4. Bertsekas, D.: Non-Linear Programming, 2nd ed. Athena Scientific, Belmont, MA (1999)
5. Blumensath, T., Davies, M.E.: Iterative hard thresholding for Sparse Approximations. The Journal of Fourier Analysis and Applications **14**, 629–654 (2008)
6. Bruckstein, A. M., Donoho, D. L., Elad, M.: From sparse solutions of systems of equations to sparse modeling of signals and images. SIAM Review **51**, 34–81 (2009)
7. Cadzow, J.A.: Signal enhancement – a composite property mapping algorithm. IEEE Trans. Acoustics, Speech, Signal Process. **36**, 49–62 (1988)
8. Candès, E.J., Romberg, J., Tao, T.: Robust uncertainty principles: Exact signal reconstruction from highly incomplete frequency information. IEEE Transactions on Information Theory, 52:489–509 (2006)
9. Candès, E.J.: The restricted isometry property and its implications for compressed sensing. Compte Rendus de l'Academie des Sciences, Paris, Serie I **346**, 589–592 (2008)
10. Chartrand, R.: Exact reconstruction of sparse signals via nonconvex minimization. IEEE Signal Process. Letters **14**, 707–710 (2007)
11. Chen, S.S., Donoho, D.L., Saunders, M.A.: Atomic decomposition by basis pursuit. SIAM Review **43**, 129–159 (2001)
12. Combettes, P.L.: The foundations of set theoeretic estimation. Proc. IEEE **81**, 182–208 (1993)
13. Combettes, P.L., Trussell, H.J.: Method of successive projections for finding a common point of sets in metric spaces. J. Optim. Theory Appl. **67**, 487–507 (1990)
14. Donoho, D.L.: Compressed sensing. IEEE Transactions on Information Theory **52**, 1289–1306 (2006)
15. Donoho, D.L., Huo, X.: Uncertainty principles and ideal atomic decomposition. IEEE Trans. Inform. Theory **47**, 2845–2862 (2001)
16. Elad, M., Bruckstein, A.M.: A generalized uncertainty principle and sparse representation in pairs of bases. IEEE Trans. Inform. Theory **48**, 2558–2567 (2002)
17. Engl, H.W., Hanke, M., Neubauer, A.: Regularization of inverse problems, volume 375 of Mathematics and its Applications. Kluwer Academic Publishers Group, Dordrecht (1996)
18. Fazel, M., Hindi, H., Boyd, S.: Rank minimization and applications in system theory. In: American Control Conference, 3272–3278 (2004)
19. Golub, G., Loan, C.V.: Matrix computations, 3rd edn. Johns Hopkins University Press (1996)
20. Natarajan, B.K.: Sparse approximation solutions to linear systems. SIAM J. Computing **24**, 227–234 (1995)
21. Recht, B., Fazel, M., Parrilo, P.: Guaranteed minimum rank solutions of matrix equations via nuclear norm minimization. SIAM Review **52**, 471–501 (2010)
22. Santosa, F., Symes, W.W.: Linear inversion of band-limited reflection seismograms. SIAM J. Sci. Statist. Comput. **7**, 1307–1330 (1986)
23. Taylor, H.L., Banks, S.C., McCoy, J.F.: Deconvolution with the l_1 norm. Geophysics **44**, 39–52 (1979)

Chapter 4
The Newton Bracketing Method for Convex Minimization: Convergence Analysis

Adi Ben-Israel and Yuri Levin

Abstract Let f be a convex function bounded below with infimum $f_{\min}$ attained. A bracket is an interval $[L,U]$ containing $f_{\min}$. The Newton Bracketing (NB) method for minimizing f, introduced in [Levin and Ben-Israel, Comput. Optimiz. Appl. 21, 213–229 (2002)], is an iterative method that at each iteration transforms a bracket $[L,U]$ into a strictly smaller bracket $[L_+,U_+]$ with $L \leq L_+ < U_+ \leq U$. We show, under certain conditions on f, that an upper bound on the bracket ratio $(U_+ - L_+)/(U - L)$ can be guaranteed by the selection of the method parameters.

Keywords Newton Bracketing method · Directional Newton method · Convex functions · Unconstrained minimization · Fermat–Weber location problem

AMS 2010 Subject Classification: 52A41, 90C25, 49M15, 90B85

4.1 Introduction

The *Newton Bracketing* (NB) method, introduced in [8], is an iterative method for convex minimization. It was applied to location problems [9], semi-definite programming [4], and linearly-constrained convex programs [1].

Let $f : \mathbb{R}^n \to \mathbb{R}$ be a convex function of n variables with an attained infimum $f_{\min}$. An optimal solution $\mathbf{x}_{\min}$ (unique if f is strictly convex) can be approximated iteratively by a gradient method

$$\mathbf{x}_+ := \mathbf{x} - c\nabla f(\mathbf{x}), c > 0.$$

Gradient methods often suffer from slow convergence near $\mathbf{x}_{\min}$. They also lack a natural stopping rule.

A. Ben-Israel (✉)
RUTCOR – Rutgers Center for Operations Research, Rutgers University, 640 Bartholomew Road, Piscataway, NJ 08854-8003, USA
e-mail: adi.benisrael@gmail.com

H.H. Bauschke et al. (eds.), *Fixed-Point Algorithms for Inverse Problems in Science and Engineering*, Springer Optimization and Its Applications 49,
DOI 10.1007/978-1-4419-9569-8_4,

The problem can also be solved by approximating the optimal value $f_{\min}$. A *bracket* is a closed interval $[L,U]$ with

$$L \leq f_{\min} \leq U. \tag{4.1}$$

The length of the bracket $[L,U]$ is denoted $\Delta := U - L$. A *bracketing method* generates a sequence of nested brackets, shrinking to a point. The brackets are defined iteratively by

$$[L_+,U_+] := \Psi([L,U]),$$

where Ψ maps intervals to intervals,

$$L \leq L_+ \leq f_{\min} \leq U_+ \leq U, \text{ and } \Delta_+ := U_+ - L_+ < \Delta.$$

Using fixed point terminology, the optimal $\mathbf{x}_{\min}$ is a fixed point of the mapping

$$\Phi(\mathbf{x}) := \mathbf{x} - c\nabla f(\mathbf{x}), c > 0,$$

while the optimal value $f_{\min}$ (viewed as a degenerate interval) is a fixed point of Ψ.

The bracket size is a natural stopping criterion, stopping the iterations when

$$U - L < \varepsilon \tag{4.2}$$

for a given tolerance $\varepsilon > 0$. For fast convergence it is desirable to have large reductions of successive brackets, i.e., small values of the *bracket ratios*

$$\frac{\Delta_+}{\Delta} = \frac{U_+ - L_+}{U - L}, \tag{4.3}$$

and an upper bound on (4.3) translates to a guaranteed reduction. We study conditions that imply such upper bounds for the NB method.

First, a description of the method for $n = 1$. An iteration begins with a current solution x, where $f'(x) \neq 0$, and a bracket $[L, U := f(x)]$ containing $f_{\min}$ (an initial lower bound L on $f_{\min}$ is assumed given.) An intermediate value

$$M := \alpha U + (1 - \alpha)L \tag{4.4}$$

is selected for some $0 < \alpha < 1$, and one Newton iteration for solving

$$f(x) = M \tag{4.5}$$

is carried out, giving

$$x_+ := x - \frac{f(x) - M}{f'(x)}, \tag{4.6}$$

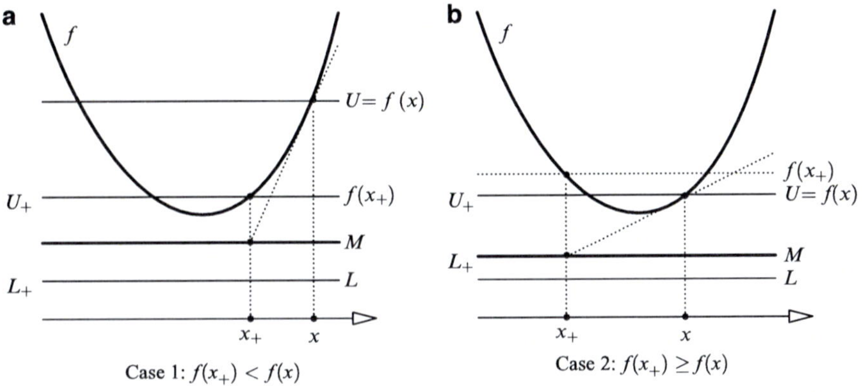

Fig. 4.1 Illustration of the 2 cases of the NB method

and two cases, illustrated in Fig. 4.1.

Case 1.

$$f(x_+) < f(x), \tag{4.7a}$$

and the bracket is updated,

$$U_+ := f(x_+), \; L_+ := L. \tag{4.7b}$$

Case 2.

$$f(x_+) \geq f(x), \tag{4.8a}$$

in which case the bracket is updated, keeping x,

$$U_+ := U, \; L_+ := M, \; x_+ := x. \tag{4.8b}$$

The iteration is summarized as follows:

(4.9)

1 Stopping rule. If $U - L < \varepsilon$, **stop** with x as solution.
2 Select a value $M := \alpha U + (1-\alpha)L$, for some $0 < \alpha < 1$.
3 Do one Newton iteration $x_+ := x - \dfrac{f(x) - M}{f'(x)}$.
4 Case 1: If $f(x_+) < f(x)$ then update U: $U_+ := f(x_+)$ and leave $L_+ := L$. Go to **1**.
5 Case 2: If $f(x_+) \geq f(x)$ then update L: $L_+ := M$ and leave $U_+ := U$, $x_+ := x$. Go to **1**.

For $n > 1$, the only change is in step 3:

$$\boxed{\begin{array}{l}\textbf{3 } \text{Do one directional Newton iteration} \\ \qquad\qquad \mathbf{x}_+ := \mathbf{x} - \dfrac{f(\mathbf{x}) - M}{\|\nabla f(\mathbf{x})\|^2} \nabla f(\mathbf{x}).\end{array}} \tag{4.10}$$

using the directional Newton method [7]. The iterate $\mathbf{x}_+$ satisfies

$$\begin{aligned} f(\mathbf{x}_+) &\geq f(\mathbf{x}) + \nabla f(\mathbf{x})^T (\mathbf{x}_+ - \mathbf{x}), \text{ since } f \text{ is convex,} \\ &= f(\mathbf{x}) + \nabla f(\mathbf{x})^T \left(-\frac{f(\mathbf{x}) - M}{\|\nabla f(\mathbf{x})\|^2} \nabla f(\mathbf{x}) \right) = f(\mathbf{x}) - (f(\mathbf{x}) - M). \end{aligned}$$

$$\therefore\ f(\mathbf{x}_+) \geq M, \text{ and} \tag{4.11a}$$

$$f(\mathbf{x}_+) > M, \text{ if } f \text{ is strictly convex and } f(\mathbf{x}) \neq M. \tag{4.11b}$$

The NB method is valid if (4.1) holds throughout the iterations, i.e., if the new interval $[L_+, U_+]$ also contains $f_{\min}$. This is clearly the case for $n = 1$ (the picture is the proof), but not in general for $n > 1$. Sufficient conditions for validity were given in [8], in particular, the method is valid for the quadratic function,

$$f(\mathbf{x}) = \frac{1}{2}\mathbf{x}^T Q\mathbf{x} + \mathbf{c}^T\mathbf{x} + \gamma, \ Q \text{ positive definite,} \tag{4.12}$$

if Q is well-conditioned,

$$\frac{\lambda_n}{\lambda_1} \geq 7 - \sqrt{48} \approx 0.071796768, \tag{4.13}$$

where λ_n and λ_1 are, respectively, the smallest and largest eigenvalues of Q.

In Case 2, the point $\mathbf{x}$ does not change (with the bonus that $\nabla f(\mathbf{x})$ need not be recomputed). Since the bracket decreases in every iteration of the NB method, a convergence analysis of the method must therefore be based on the brackets $[L, U]$ in $\mathbb{R}$, rather than on the iterates $\{\mathbf{x}\}$ in $\mathbb{R}^n$.

Example 4.1. One way the NB method may fail is illustrated by the function

$$f(x) = \frac{1}{2}x^2 + 1 \tag{4.14}$$

with initial $x := 1$, $U := f(1) = \frac{3}{2}$ and $L := 0$. For $\alpha = \frac{1}{3}$ we get $M = \frac{1}{2}$ by (4.4), and

$$x_+ := x - \frac{f(x) - M}{f'(x)} = 1 - \frac{\frac{3}{2} - \frac{1}{2}}{1} = 0,$$

which is the optimal solution. However, if $\varepsilon < 1$ the NB method does not stop since the bracket size is 1, but it also cannot continue since $f'(0) = 0$. This issue can be resolved by adding a derivative based stopping rule, but as a practical matter it can be ignored.

Example 4.2. To illustrate why the NB method requires the attainment of the infimum $f_{\min}$, consider the function

$$f(x) = e^{-x}$$

with initial $x := 0$, $U := f(0) = 1$, and $L := -1$. The NB method, for any choices of α, has only iterations of Case 1, the iterates $\to \infty$, and the bracket size remains ≥ 1.

Example 4.2 is special in that one case, Case 1, occurs in all iterations. In normal circumstances, Cases 1 and 2 alternate, with large [small] α making Case 1 [Case 2] more likely.

The bracket ratio (4.3) of the NB method is

$$\frac{\Delta_+}{\Delta} = \begin{cases} \dfrac{f(\mathbf{x}_+) - L}{U - L}, & \text{in case 1}, \\ \dfrac{U - M}{U - L} = 1 - \alpha, & \text{in case 2}. \end{cases} \tag{4.15}$$

We study here the convergence of the NB method in terms of the bracket ratio. These results are stated for $n = 1$ in Sect. 4.2, for $n > 1$ in Sect. 4.3, and are applied in Sect. 4.4 to the Fermat–Weber location problem.

Remark 4.3. For similar ideas and more general results, see Kim et al. [6], and Cegielski [2].

4.2 Bracket Reduction in the NB Method for $n = 1$

The results of this section, for the simple case $n = 1$, pave the way for the more interesting case, $n > 1$, in the next section.

Let $f : \mathbb{R} \to \mathbb{R}$ be convex and differentiable as needed, and consider the Newton method for solving the (4.5), namely

$$f(x) = M. \tag{4.16}$$

Let x be a point where $f'(x) \neq 0$, and let

$$x_+ := x - \frac{f(x) - M}{f'(x)} \tag{4.17}$$

be the next Newton iterate.

Remark 4.4. In what follows, there appear expressions like $|f(x_+) - M|$, where the absolute value sign is not necessary (because of (4.11)) but is given pro forma.

Let X_0 be the interval with endpoints x, x_+. If $f \in C_N^{1,1}(X_0)$ (i.e., $f'(x)$ is Lipschitz on X_0 with Lipschitz constant N) and $|f'(x)|$ is sufficiently large, then the value of $|f(x_+) - M|$ is bounded as follows.

Lemma 4.5. *Using the above notation, let $f \in C_N^{1,1}(X_0)$ and $f(x) > M$. If for some $\beta > 0$,*

$$|f'(x)|^2 \geq \beta N |f(x) - M|, \tag{4.18}$$

then

$$|f(x_+) - M| \leq \frac{1}{2\beta} |f(x) - M|. \tag{4.19}$$

Proof. Since $f \in C_N^{1,1}(X_0)$,

$$|f(x^+) - f(x) - f'(x)(x^+ - x)| \leq \frac{N}{2}(x^+ - x)^2,$$

by the descent lemma [11, 3.2.12, page 73]. Therefore,

$$\begin{aligned} |f(x^+) - M| &\leq \frac{N}{2} \frac{(f(x) - M)^2}{(f'(x))^2}, \quad \text{by (4.17)} \\ &\leq \frac{1}{2\beta} |f(x) - M|, \quad \text{by (4.18)}. \end{aligned}$$

■

Remark 4.6. To guarantee decrease in (4.19), it is required that

$$\beta > \frac{1}{2}. \tag{4.20}$$

Recall that $U := f(x)$. To apply Lemma 4.5 to an NB iteration, we note that

$$\begin{aligned} f(x) - M &= U - M \\ &= (1 - \alpha)(U - L), \quad \text{by (4.4)}, \\ &= (1 - \alpha)\Delta, \end{aligned} \tag{4.21}$$

and inequality (4.18) can be written as

$$(1 - \alpha)\beta \leq \frac{|f'(x)|^2}{N\Delta}, \tag{4.22}$$

relating α and β for any given x. We observe that β can be chosen arbitrarily, with α then constrained by (4.22).

The next theorem guarantees an upper bound on the bracket ratio (4.15).

Theorem 4.7. *Let $f : \mathbb{R} \to \mathbb{R}$ be convex, x a point where $f'(x) \neq 0$, X_0 the interval with endpoints x, x_+, and assume $f \in C_N^{1,1}(X_0)$. Let $[L,U]$ a bracket for $f_{\min}$ (i.e., $L \leq f_{\min} \leq U$), and let β satisfy*

$$\beta > \frac{|f'(x)|^2}{N\Delta}. \tag{4.23}$$

Then for

$$\alpha := 1 - \frac{|f'(x)|^2}{\beta N \Delta}, \tag{4.24}$$

the NB iteration results in a bracket ratio

$$\frac{\Delta_+}{\Delta} \leq \max\left\{\frac{1}{2\beta}(1-\alpha)+\alpha, 1-\alpha\right\}. \tag{4.25}$$

Proof. In Case 2, we get from (4.15)

$$\frac{\Delta_+}{\Delta} = 1 - \alpha,$$

and in Case 1,

$$\begin{aligned}
|f(x_+) - M| &\leq \frac{1}{2\beta}|U - M|, \text{ by Lemma 4.5, and } U = f(x), \\
&= \frac{1}{2\beta}(1-\alpha)|U - L|, \text{ by (4.4)}. \\
M - L &= \alpha(U - L), \text{ by (4.4)}. \\
\therefore\ U_+ - L_+ &= (f(x_+) - M) + (M - L), \text{ by (4.7b)}, \\
&\leq \left(\frac{1}{2\beta}(1-\alpha)+\alpha\right)(U - L),
\end{aligned}$$

completing the proof. ∎

Example 4.8. For

$$f(x) = \frac{1}{2}x^2 + 1$$

we have $f'(x) = x$ and $f'' = 1$, giving $N = 1$. Then (4.23) and (4.24) become

$$\beta > \frac{x^2}{\Delta}, \text{ and } \alpha = 1 - \frac{x^2}{\beta\Delta}.$$

Table 4.1 Some admissible βs and the corresponding bracket ratios

β	Case 1 $\frac{1}{2\beta}(1-\alpha)+\alpha$	Case 2 $(1-\alpha)$	Upper bound on bracket ratio, RHS (4.25)
0.8	0.6875	0.833333333	0.833333333
1	0.666666667	0.666666667	0.666666667
1.5	0.703703704	0.444444444	0.703703704
2	0.75	0.333333333	0.75
2.5	0.786666667	0.266666667	0.786666667

For the initial $x := 1$, $U := f(x) = \frac{3}{2}$, $L := 0$ and initial bracket size $\Delta = U - L = \frac{3}{2}$,

$$\beta > \frac{2}{3}, \text{ and } \alpha = 1 - \frac{2}{3\beta}.$$

Table 4.1 lists some admissible values of β, and the corresponding bracket ratios. The guaranteed ratios in the last column are pessimistic because of the high frequency of iterations of Case 2, with ratios given in the penultimate column.

Theorem 4.7 concerns a single iteration, and its application requires selecting an admissible β and recalculating α in each iteration.

To simplify matters, we fix the parameter β throughout the iterations (it may no longer be admissible in some iterations) and impose the following constraint on α,

$$\alpha_{\min} \le \alpha \le \alpha_{\max} \tag{4.26}$$

with given bounds $\{\alpha_{\min}, \alpha_{\max}\}$. The parameter α is computed in each iteration as the point in the interval $[\alpha_{\min}, \alpha_{\max}]$ that is closest to (4.24).

Example 4.9. Consider the quadratic (4.14)

$$f(x) = \frac{1}{2}x^2 + 1$$

with initial $x = 1$, $U := f(1) = \frac{3}{2}$, $L := 0$, and initial bracket size $\Delta = \frac{3}{2}$.

(a) Using $\beta = \frac{2}{3}$, $\alpha_{\min} = 0.2$, and $\alpha_{\max} = 0.9$, the NB method (with $\varepsilon = 10^{-6}$) stopped after 16 iterations, an average reduction per iteration

$$\frac{\Delta_+}{\Delta} = 0.41.$$

The reductions in each iterations are shown in Fig. 4.2a. Case 1 occurred in 11 iterations and Case 2 in 5.

(b) Using $\beta = 0.2$ (an inadmissible value), $\alpha_{\min} = 0.3$ and $\alpha_{\max} = 0.7$, the NB method (with $\varepsilon = 10^{-9}$) required 56 iterations, 15 of Case 1, and 41 of Case 2. The reductions in each iteration are shown in Fig. 4.2b. The average reduction per iteration is

$$\frac{\Delta_+}{\Delta} = 0.68.$$

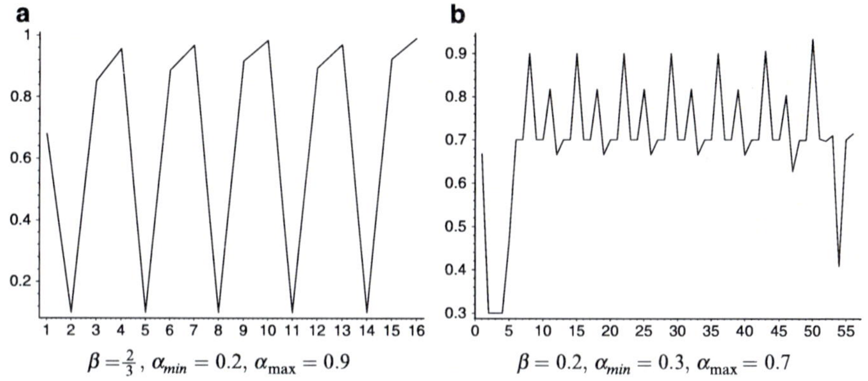

Fig. 4.2 Illustration of Example 4.9: Reduction per iteration for $f(x) = \frac{1}{2}x^2 + 1$, $L = 0$ and $x_0 = 1$

Remark 4.10. The periodic reductions in Fig. 4.2a, b are explained by self-similarity of the NB method for this particular function and the selected values of β.

4.3 Bracket Reduction in the NB Method for $n > 1$

Theorem 4.11. *Let $f : \mathbb{R}^n \to \mathbb{R}$ be convex and differentiable, let $\mathbf{x}$ be a point where $\nabla f(\mathbf{x}) \neq 0$, and let*

$$\mathbf{x}_+ := \mathbf{x} - \frac{f(\mathbf{x}) - M}{\|\nabla f(\mathbf{x})\|^2} \nabla f(\mathbf{x}), \tag{4.27}$$

be the next Newton iterate with step

$$\mathbf{h} := -\frac{f(\mathbf{x}) - M}{\|\nabla f(\mathbf{x})\|^2} \nabla f(\mathbf{x}). \tag{4.28}$$

Consider the ball

$$X_0 := \{\xi : \|\xi - \mathbf{x}_+\| \leq \|\mathbf{h}\|\},$$

and assume that $f \in C_N^{1,1}(X_0)$, i.e., ∇f is Lipschitz in X_0 with Lipschitz constant N. Finally, assume $\mathbf{x}$ satisfies,

$$\|\nabla f(\mathbf{x})\|^2 \geq \beta N |f(\mathbf{x}) - M|, \tag{4.29}$$

for some $\beta > 0$. Then:

(a) $\|\nabla f(\mathbf{x}_+)\| \geq \frac{\beta - 1}{\beta} \|\nabla f(\mathbf{x})\|$.

(b) $|f(\mathbf{x}_+) - M| \leq \frac{1}{2\beta} |f(\mathbf{x}) - M|$.

Proof. See Appendix A. ∎

Remark 4.12. For part (a) to be useful, it is required that $\beta > 1$.

As in (4.22), the inequality (4.29) can be written as,

$$(1-\alpha)\beta \le \frac{\|\nabla f(\mathbf{x})\|^2}{N\Delta}. \tag{4.30}$$

An admissible β can thus be selected at will, and α then determined by

$$\alpha = 1 - \frac{\|\nabla f(\mathbf{x})\|^2}{\beta N \Delta} \tag{4.31}$$

the analog of (4.24).

Example 4.13. Consider the quadratic function $f : \mathbb{R}^n \to \mathbb{R}$,

$$f(\mathbf{x}) = \frac{1}{2}\mathbf{x}^T Q\mathbf{x} + 1,\ Q \text{ positive definite,} \tag{4.32}$$

with $f_{\min} = 1$. Let Q have the eigenvalues

$$\lambda_1 \ge \lambda_2 \ge \cdots \ge \lambda_n > 0, \quad \text{and corresponding eigenvectors} \quad \mathbf{v}_1, \mathbf{v}_2, \cdots, \mathbf{v}_n.$$

Then $\nabla f(\mathbf{x}) = Q\mathbf{x}$, and $f''(\mathbf{x}) = Q$. The norm of Q corresponding to the Euclidean norm is $\|Q\| = \lambda_1$, which is taken as the Lipschitz constant N in Theorem 4.11. The inequality (4.30) becomes,

$$(1-\alpha)\beta \le \frac{\|Q\mathbf{x}\|^2}{\lambda_1 \Delta}, \tag{4.33}$$

giving for $\mathbf{x} = \mathbf{v}_n$,

$$(1-\alpha)\beta \le \frac{\lambda_n^2}{\lambda_1 \Delta}.$$

The following theorem is the analog of Theorem 4.7, giving an upper bound on the bracket ratio in a single iteration.

Theorem 4.14. *Let $f : \mathbb{R}^n \to \mathbb{R}$ be convex and differentiable, let $\mathbf{x}$ be a point where $\nabla f(\mathbf{x}) \ne \mathbf{0}$, and let $[L, U]$ be a bracket for $f_{\min}$. Let β satisfy,*

$$\beta > \frac{\|\nabla f(\mathbf{x})\|^2}{N\Delta}. \tag{4.34}$$

Then for α satisfying (4.31), the NB iteration results in a reduction

$$\frac{\Delta_+}{\Delta} \leq \max\left\{\frac{1}{2\beta}(1-\alpha)+\alpha, 1-\alpha\right\}. \tag{4.35}$$

Proof. The proof of Theorem 4.7 applies verbatim. ∎

4.4 Application to the Fermat–Weber Location Problem

The *Fermat–Weber Location Problem* is to find a point $\mathbf{x}$ minimizing the sum of Euclidean distances

$$f(\mathbf{x}) = \sum_{i=1}^{m} \|\mathbf{a}_i - \mathbf{x}\| \tag{4.36}$$

from m given points $\{\mathbf{a}_i : i = 1,\ldots,m\}$, see [3, 10, 17] and their references.

For large m, the contours of the function (4.36) are close to circular, see, e.g., Fig. 4.3b, and the problem is well-conditioned, so the NB method is valid by (4.13).

To apply Theorem 4.14, we need an estimate of $N = \sup f''$, which is problematic for the function (4.36). However, $f(\mathbf{x})$ can be approximated, near the optimal solution, by a quadratic

$$f(\mathbf{x}) \approx \frac{1}{2}\mathbf{x}^T Q\mathbf{x}$$

giving $f'' \approx Q$, $N \approx \lambda_1$ (the largest eigenvalue of Q), and therefore $N = O(m)$, say $N = \frac{m}{2}$.

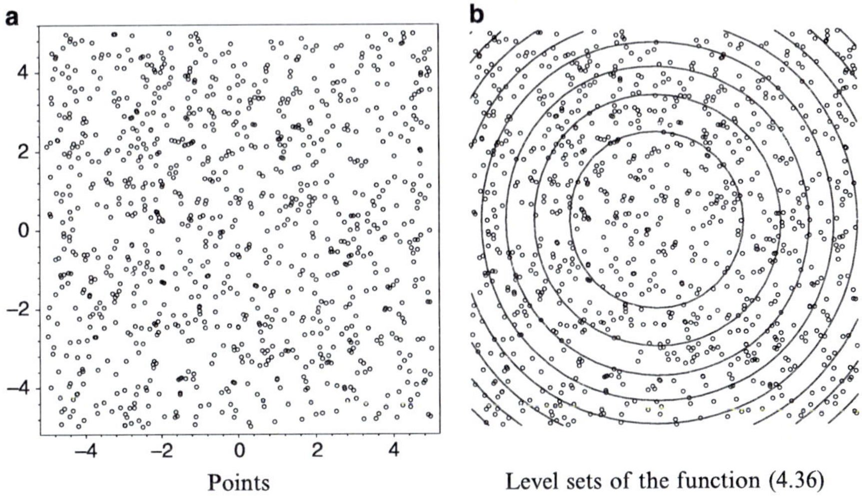

Fig. 4.3 Illustration of Example 4.16: 1,000 random points in $[-5,5]^2$

As in the case $n = 1$, we fix the parameter β throughout the iterations (it may no longer be admissible in some iterations) and impose the constraint (4.26),

$$\alpha_{\min} \le \alpha \le \alpha_{\max},$$

where the bounds $\{\alpha_{\min}, \alpha_{\max}\}$ are given. The parameter α is computed in each iteration as the point in the interval $[\alpha_{\min}, \alpha_{\max}]$ that is closest to (4.31).

The value of the parameter α depends on the bounds $\{\alpha_{\min}, \alpha_{\max}\}$. The following considerations apply to choosing these bounds (and indirectly α.)

(a) For large values of α, say $\alpha \ge 0.8$, the target M is close to U by (4.4), making Case 1 more likely, and the bracket ratio (see proof of Theorem 4.7),

$$\frac{\Delta_+}{\Delta} \le \frac{1-\alpha}{2\beta} + \alpha \tag{4.37}$$

is large. However, when Case 2 occurs, the ratio

$$\frac{\Delta_+}{\Delta} = 1 - \alpha \tag{4.38}$$

is small since α is large. We thus alternate between small and large reductions, see, e.g., Fig. 4.4.

(b) Small values of α (say $\alpha \le 0.2$) make Case 2 more likely, with a large ratio (4.38), i.e., a small reduction. However, in Case 2 the derivative need not be computed, so the overall time may be smaller.

Our numerical experience suggests that convergence is faster for higher values of α, as illustrated in Examples 4.16 and 4.17 below.

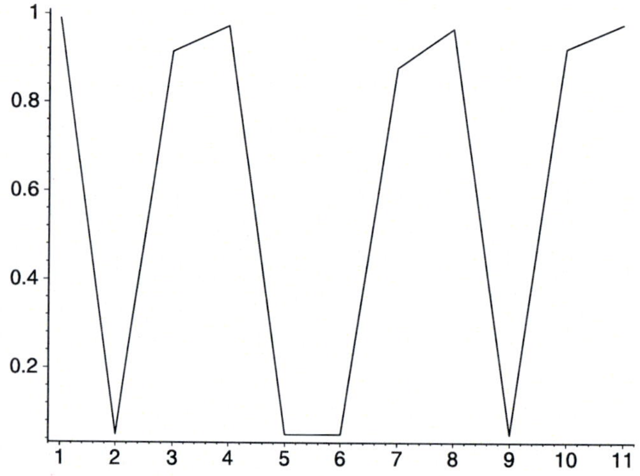

Fig. 4.4 Illustration of Example 4.16: Reduction per iteration for $\beta = \frac{2}{3}$, $\alpha_{\min} = 0.2$, $\alpha_{\max} = 0.95$

Remark 4.15. For large m, the function (4.36) is very flat near the optimal solution. Using a small ε as a stopping criterion, and stopping the computations with a final bracket $\Delta < \varepsilon$, does not guarantee that the final iterate $\mathbf{x}$ is close to the optimal solution, only that its value $f(\mathbf{x})$ is within ε of the optimal value.

Example 4.16. Consider a problem with $m = 1{,}000$ points, randomly generated in $[-5,5]^2$. The points are shown in Fig. 4.3a, and the level sets of the sum of distances (4.36) in Fig. 4.3b.

The initial lower bound was taken as $L = 0$ (a better lower bound would be the distance between any two points, but the method converges so fast that we do not save much by improving L). For the needed Lipschitz constant we substitute the value $N = \frac{m}{2} = 500$, for no good reason other than it works.

The problem was solved with an initial point chosen randomly in $[-5,5]^2$. Table 4.2 shows the results for the initial $\mathbf{x} = (-0.640353501, 0.937409957)$, with $U := f(\mathbf{x}) = 3809.901722$, and initial $\Delta = 3809.901722$. The value of the parameter β was fixed at $\frac{2}{3}$ and the bounds for α were $\alpha_{\min} = 0.2,\ \alpha_{\max} = 0.95$.

Using a tolerance $\varepsilon = 10^{-3}$, the method converged in 12 iterations. Case 1 occurred in 7 iterations, and Case 2 in 5 iterations (shown in bold numbers in Table 4.2). The lower bound $\alpha_{\min}$ was too low to be activated, but the upper bound $\alpha_{\max}$ applied in all but 3 iterations.

The last row of the table lists the bracket ratios, that are plotted in Fig. 4.4. The average reduction per iteration is

$$\left(\frac{0.0008}{3809.9}\right)^{1/12} = 0.277$$

Example 4.17. The results of Example 4.16 are typical: we solved 20 problems, each with 1,000 random points in $[-5,5]^2$, and a random initial solution, using the same parameters as above,

$$\varepsilon = 10^{-3}, L = 0, \beta = \frac{2}{3}, \alpha_{\min} = 0.2, \alpha_{\max} = 0.9.$$

Table 4.3 shows the total number of iterations (until a bracket with length $\leq 10^{-3}$ is reached), the number of iterations of Case 2 (5 in all but one problem), and the average reduction per iteration, that is around 30%.

Table 4.2 Results for Example 4.16 with $\beta = \frac{2}{3},\ \alpha_{\min} = 0.2,\ \alpha_{\max} = 0.95$

Iteration	0	1	**2**	3	4	**5**	**6**	7	8	**9**	10	11	**12**
α	0.95	0.95	**0.95**	0.88	0.95	**0.95**	**0.95**	0.83	0.95	**0.95**	0.89	0.95	**0.95**
Case	1	1	**2**	1	1	**2**	**2**	1	1	**2**	1	1	**2**
Δ	3809.9	3771.1	**188.5**	172.3	167.3	**8.37**	**0.418**	0.367	0.355	**0.018**	0.0163	0.016	**0.0008**
Reduction		0.989	**0.05**	0.914	0.971	**0.05**	**0.05**	0.881	0.945	**0.05**	0.905	0.981	**0.05**

Table 4.3 Results for Example 4.17, with 20 random problems, $\beta = \frac{2}{3}$, $\alpha_{\min} = 0.2$, $\alpha_{\max} = 0.95$

Problem	1	2	3	4	5	6	7	8	9	10	11	12	13	14	15	16	17	18	19	20
Iterations	14	12	16	15	12	13	12	11	13	15	13	14	11	14	13	14	13	13	13	11
Case 2	5	5	5	5	5	5	5	5	5	5	5	5	5	5	5	5	5	5	6	5
Reduction	0.33	0.28	0.37	0.35	0.27	0.30	0.27	0.24	0.30	0.35	0.30	0.32	0.25	0.33	0.30	0.32	0.30	0.30	0.24	0.25

Remark 4.18. The NB method for convex minimization is based on the fact that the graph of a convex function f is supported by its tangents, but differentiability of f, i.e., uniqueness of tangents, is not required. The NB method can therefore be translated for the minimization of nondifferentiable convex functions, using the subgradient methods of Shor [14, 15]; see also [5, 13, 16], to mention but a few.

Acknowledgements We thank the referees, and Professor A. Cegielski, for their constructive suggestions.

Appendix A: Proof of Theorem 4.11

Part (a). Proof that

$$\|\nabla f(\mathbf{x}_+)\| \geq \frac{\beta-1}{\beta}\|\nabla f(\mathbf{x})\|. \tag{A.1}$$

Since $f \in C_N^{1,1}(X_0)$,

$$\|\nabla f(\xi) - \nabla f(\mathbf{x})\| \leq N\|\xi - \mathbf{x}\| \tag{A.2}$$

for all $\xi \in X_0$. In particular, for $\mathbf{x}_+$,

$$\begin{aligned} \|\nabla f(\mathbf{x}_+) - \nabla f(\mathbf{x})\| &\leq N\|\mathbf{x}_+ - \mathbf{x}\| = N\frac{|f(\mathbf{x}) - M|}{\|\nabla f(\mathbf{x})\|} \\ &\leq \frac{1}{\beta}\|\nabla f(\mathbf{x})\|, \text{ by (4.29)}. \end{aligned} \tag{A.3}$$

$$\begin{aligned} \therefore\ \|\nabla f(\mathbf{x}_+)\| &\geq \|\nabla f(\mathbf{x})\| - \|\nabla f(\mathbf{x}_+) - \nabla f(\mathbf{x})\|, \\ &\geq \|\nabla f(\mathbf{x})\| - \frac{1}{\beta}\|\nabla f(\mathbf{x})\|, \text{ by (A.3)}, \\ &= \frac{\beta-1}{\beta}\|\nabla f(\mathbf{x})\|, \text{ proving (A.1)}. \end{aligned}$$

Part (b). The proof that

$$|f(\mathbf{x}_+) - M| \leq \frac{1}{2\beta}|f(\mathbf{x}) - M| \tag{A.4}$$

is analogous to that of Lemma 4.5.

Since $f \in C_N^{1,1}(X_0)$,

$$f(\mathbf{x}^+) - f(\mathbf{x}) \leq \nabla f(\mathbf{x})^T(\mathbf{x}^+ - \mathbf{x}) + \frac{N}{2}\|\mathbf{x}^+ - \mathbf{x}\|^2,$$

by the descent lemma, [11, 3.2.12, page 73]. Therefore,

$$
\begin{aligned}
|f(\mathbf{x}^+) - M| &\le \frac{N}{2} \frac{(f(\mathbf{x}) - M)^2}{||\nabla f(\mathbf{x})||^2} \text{ by (4.27)} \\
&\le \frac{1}{2\beta} |f(\mathbf{x}) - M| \text{ by (4.29), proving (A.4).}
\end{aligned}
$$

References

1. Ben-Israel, A., Levin, Y.: The Newton bracketing method for the minimization of convex functions subject to affine constraints. Discrete Appl. Math. **156**, 1977–1987 (2008)
2. Cegielski, A.: A method of projection onto an acute cone with level control in convex minimization. Math. Prog. **85**, 469–490 (1999)
3. Drezner, Z., Klamroth, K., Schöbel, A., Wesolowsky, G.: The Weber problem. In: Z. Dezner, H.W. Hamacher (eds.) Facility Location: Applications and Theory, Springer (2002)
4. Fortin, C., Wolkowicz, H.: The trust region subproblem and semidefinite programming. Optimization Methods and Software **19**, 41–67 (2004)
5. Held, M., Wolfe, P., Crowder, H.: Validation of subgradient optimization. Math. Prog. **6**, 62–88 (1974)
6. Kim, S., Ahn, H., Cho, S.-C.: Variable target value subgradient method. Math. Prog. **49**, 359–369 (1991)
7. Levin, Y., Ben-Israel, A.: Directional Newton methods in *n* variables. Math. of Comput. **71**, 251–262 (2001)
8. Levin, Y., Ben-Israel, A.: The Newton bracketing method for convex minimization. Comput. Optimiz. Appl. **21**, 213–229 (2002)
9. Levin, Y., Ben-Israel, A.: A heuristic method for large–scale multifacility location problems. Computers and Operations Research **31**, 257–272 (2004)
10. Love, R.F., Morris, J.G., Wesolowsky, G.O.: Facilities Location: Models and Methods. North-Holland (1988)
11. Ortega, J.M., Rheinboldt, W.C.: Iterative Solution of Nonlinear Equations in Several Variables. Academic, London (1970)
12. Ostrwoski, A.M.: Solutions of Equations in Euclidean and Banach Spaces, 3rd edn. Academic (1973)
13. Poljak, B.T.: Minimization of unsmooth functionals. Z. Vycis. Mat. i Mat. Fiz. **9**, 509–521 (1969)
14. Shor, N.Z.: Application of the Gradient Method for the solution of Network Transportation Problems. Notes, Scientific Seminar on Theory and Application of Cybernetics and Operations Research, Academy of Sciences, Kiev (1962)
15. Shor, N.Z.: Generalized gradient methods for non-smooth functions and their applications to mathematical programming problems. Ekonomika i Matematicheskie Metody **2**, 337–356 (1976)
16. Shor, N.Z., Kiwiel, K., Ruszcayǹski, A.: Minimization methods for non-differentiable functions. Springer (1985)
17. Wesolowsky, G.O.: The Weber problem: its history and perspectives. Location Science **1**, 5–23 (1993)

Chapter 5
Entropic Regularization of the ℓ_0 Function

Jonathan M. Borwein and D. Russell Luke

Abstract Many problems of interest where more than one solution is possible seek, among these, the one that is sparsest. The objective that most directly accounts for sparsity, the ℓ_0 metric, is usually avoided since this leads to a combinatorial optimization problem. The function $\|x\|_0$ is often viewed as the limit of the ℓ_p metrics. Naturally, there have been some attempts to use this as an objective for p small, though this is a nonconvex function for $p < 1$. We propose instead a scaled and shifted Fermi–Dirac entropy with two parameters, one controlling the smoothness of the approximation and the other the *steepness* of the metric. Our proposed metric is a convex relaxation for which a strong duality theory holds, yielding dual methods for metrics approaching the desired $\|\cdot\|_0$ function. Without smoothing, we propose a dynamically reweighted subdifferential descent method with "exact" line search that is finitely terminating for constraints that are well-separated. This algorithm is shown to recapture in a special case certain well-known "greedy" algorithms. Consequently we are able to provide an explicit algorithm whose fixed point, under the appropriate assumptions, is the sparsest possible solution. The variational perspective yields general strategies to make the algorithm more robust.

Keywords Convex optimization · Fenchel duality · Entropy · Regularization · Sparsity · Signal processing

AMS 2010 Subject Classification: 49M20, 65K10, 90C30

D.R. Luke (✉)
Institute for Numerical and Applied Mathematics, University of Goettingen,
Lotzestr. 16–18, 37073 Goettingen, Germany
e-mail: r.luke@math.uni-goettingen.de

H.H. Bauschke et al. (eds.), *Fixed-Point Algorithms for Inverse Problems in Science and Engineering*, Springer Optimization and Its Applications 49,
DOI 10.1007/978-1-4419-9569-8_5,

5.1 Introduction

Let $\mathbb{E}$ and $\mathbb{Y}$ be Euclidean spaces, and let $A : \mathbb{E} \to \mathbb{Y}$ be linear. We consider the problem

$$\begin{array}{ll} \underset{x\in\mathbb{E}}{\text{minimize}} & \varphi(x) \\ \text{subject to} & A(x) = b, \end{array} \tag{5.1}$$

where $\varphi(x) : \mathbb{E} \to \mathbb{R}$ is a lower semi-continuous (lsc), symmetric subadditive function that, in one way or another, counts the nonzero elements of x. This model has received a great deal of attention recently in applications where the number of constraints is much smaller than the dimension of the domain. Examples include the well-known compressed sensing [4], where $\mathbb{E} = \mathbb{R}^n$, $\mathbb{Y} = \mathbb{R}^m$ $(m \ll n)$ and $\varphi(x) \equiv \sum_j |\text{sign}(x_j)|$.

Another instance of importance is low-rank matrix reconstruction [5, 13]. Here $\mathbb{E} = \mathbb{R}^{m\times n}$, $\mathbb{Y} = \mathbb{R}^{m\times n}$ and $\varphi(x) \equiv \text{rank}(x)$. The goal in both of these applications is to find a "sparsest" solution x^* to $A(x) = b$. Both of the optimization problems associated with these examples are combinatorial and, in general, NP-hard [12]. At the expense of some generality, we will narrow our discussion to the case where $\mathbb{E} = \mathbb{R}^n$ and $\mathbb{Y} = \mathbb{R}^m$.

Before addressing the counting objective directly, we review some elementary observations about the most common relaxation of this problem, ℓ_1 optimization.

5.1.1 Elementary ℓ_1 Minimization

A natural first step toward solving such problems has been to solve convex relaxations instead, $\varphi(x) = \|x\|_1 \equiv \varphi_1(x)$. It has been known for some time that ℓ_1 optimization promotes sparsity in underdetermined systems [7, 15]. Later works established criteria under which the solution to (5.1) is unique and exactly matches the true signal x^* [6, 8, 9]. Sparsity of the true signal x^* and the structure of the matrix A are key requirements.

A qualitative geometric interpretation of these facts is obtained by considering the Fenchel dual [1] to Problem (5.1) when $\varphi = \varphi_1$:

$$\begin{array}{ll} \underset{y\in\mathbb{R}^m}{\text{maximize}} & b^{\mathrm{T}}y \\ \text{subject to} & \left(A^{\mathrm{T}}y\right)_j \in [-1, 1] \quad j = 1, 2, \ldots, n. \end{array} \tag{5.2}$$

By *strong Fenchel duality*, the optimal values of the primal and dual problems are equivalent, and a solution of the dual problem yields a solution to the primal. The dual problem yields valuable geometric insight. Elementary facts from linear programming guarantee that the solution includes a vertex of the polyhedron described by the constraints. The number of active constraints in the dual problem provides a crude upper bound on the number of nonzero elements of the sparsest solution to the primal problem. Unless the number of active constraints in the dual problem is less than or equal to the number of measurements m, there is no hope of uniquely recovering x^*. Supposing that the solution to (5.2) is indeed unique, a more vexing question is whether or not the corresponding primal solution is the sparsest solution to $Ax = b$. Here, it appears, convex analysis is at a loss to provide an answer.

5.1.2 ℓ_0 *Minimization*

We gain some insight into this breakdown by considering the dual of the original sparse optimization problem. For $\varphi(x) = \sum_j |\mathrm{sign}(x_j)| \equiv \varphi_0(x)$ in (5.1), the equivalence of the primal and dual problems is lost due to the nonconvexity of the objective. The theory of Fenchel duality still yields *weak duality*, but this is of limited use in this instance. The Fenchel dual to (5.1) when $\varphi = \varphi_0$ is

$$\begin{array}{ll} \underset{y \in \mathbb{R}^m}{\text{maximize}} & b^{\mathrm{T}} y \\ \text{subject to} & \left(A^{\mathrm{T}} y\right)_j = 0 \quad j = 1, 2, \ldots, n. \end{array} \tag{5.3}$$

If we denote the *values* of the primal (5.1) and dual problems (5.3) by p and d respectively, then these values satisfy the *weak duality inequality* $p \geq d$. The primal problem is a combinatorial optimization problem, and hence NP-hard; the dual problem, however, is a linear program, which is finitely terminating. Relatively elementary variational analysis provides a lower bound on the sparsity of signals x that satisfy the measurements. In this instance, however, the lower bound only reconfirms what we already know. Indeed, if A is full rank, then the only solution to the dual problem is $y = 0$. In other words, the minimal sparsity of the solution to the primal problem is greater than or equal to zero, which is obvious. The loss of information in passing from primal to dual formulations of nonconvex problems is a common phenomenon and at the heart of the difficulties in answering some very basic questions about sparse, and more generally nonconvex, optimization.

Our goal in this paper is twofold: first, to dig deeper into the convex analysis to see what can indeed be learned about the nonconvex problem from various convex relaxations, and second, to take what has been learned by other means and incorporate these advances into convex analysis and algorithms. As we showed with Example (5.3), the dual of the ℓ_0 problem is uninformative but trivial to solve. The conventional approach is to view ℓ_0 as a limit of the nonconvex p-metrics. However, the ℓ_p problems for $0 < p < 1$ are also NP hard and the duals to these optimization problems suffer the same loss of information that the dual to the ℓ_0 function suffers. The question that motivates our work is whether one can use convex relaxations approaching something related to the ℓ_0 function – something in the dual space – that are still informative with respect to the original ℓ_0 problem, but yield optimization problems that are solvable in polynomial time. The connection between the nonconvex and the convex that we explore is the Fenchel conjugate of the ℓ_0 function, which can be written as the limit of convex functions. We then study how well our proposed convex relaxations work for solving the sparse recovery problem.

5.1.3 Notation

Throughout this work, we use $\|\cdot\|$ without any subscript to denote the L^2-norm. When a different norm is meant, a subscript is added explicitly to the norm as with $\|\cdot\|_1$. We denote the *projection* of a point z onto the set C with respect to the L^2 norm by $P_C(z)$, where

$$P_C(z) \equiv \left\{ x \in C \mid \|x - z\| = \inf_{y \in C} \|z - y\| \right\}.$$

We denote the nonnegative orthant in $\mathbb{R}^n$ by $\mathbb{R}^n_+$ and the *extended reals* by $\overline{\mathbb{R}} \equiv \mathbb{R} \cup \{+\infty\}$. It is not uncommon to define the objective φ on the extended reals as a mapping from $\mathbb{R}^n$ to $\overline{\mathbb{R}}$. The *normal cone mapping* of a set $C \subset \mathbb{R}^n$ at a point x is defined by

$$N_C(x) \equiv \begin{cases} \{w \in \mathbb{R}^n \text{ with } (z-x)^{\mathrm{T}} w \le 0 \text{ for all } z \in C\} & \text{if } x \in C \\ \emptyset & \text{if } x \notin C. \end{cases}$$

We denote by $ri(C)$ the *relative interior* of C, that is the interior of C relative to its affine hull. The *indicator function* of a set C, ι_C is defined by

$$\iota_C(x) \equiv \begin{cases} 0 & \text{for } x \in C \\ +\infty & \text{for } x \notin C. \end{cases}$$

We use the indicator function to treat constraint sets as functions. For a function $f : \mathbb{R}^n \to \overline{\mathbb{R}}$ and a point $\overline{x}$ in the domain of f, the *subdifferential* of f at $\overline{x}$, denoted $\partial f(\overline{x})$ is defined by

$$\partial f(\overline{x}) \equiv \left\{ w \in \mathbb{R}^n \mid w^T (x - \overline{x}) \le f(x) - f(\overline{x}), \text{ for all } x \in \mathbb{R}^n \right\}. \tag{5.4}$$

When $\overline{x}$ is not in the domain of f we define $\partial f(\overline{x}) = \emptyset$. The *Fenchel conjugate* of a mapping $f : \mathbb{R}^n \to [-\infty, +\infty]$, denoted $f^* : \mathbb{R}^n \to [-\infty, +\infty]$, is defined by

$$f^*(y) = \sup_{x \in \mathbb{R}^n} \{ y^T x - f(x) \}. \tag{5.5}$$

The conjugate is always convex (as a supremum of affine functions) while $f = f^{**}$ exactly if f is convex, proper (not everywhere infinite) and lower semi-continuous (lsc) [1]. Finally, we make frequent reference to boxes in $\mathbb{R}^n$ centered at the origin with sides of length $2I_j$ ($j = 1,2,\dots,n$); these are denoted by $R_I \equiv [-I_1,I_1] \times [-I_2,I_2] \times \dots [-I_n,I_n]$ for $I = (I_1,I_2,\dots,I_n)$.

5.2 Entropic Regularization of the Zero Metric

The Fenchel conjugates of the functions $\varphi_1(x) \equiv \|x\|_1$ and $\varphi_0(x) \equiv \sum_j |\text{sign}(x_j)|$ are given respectively by

$$\varphi_1^*(y) \equiv \begin{cases} 0 & y \in [-1,1] \\ +\infty & \text{else} \end{cases} \qquad (\varphi_1(x) \equiv \|x\|_1) \tag{5.6}$$

$$\varphi_0^*(y) \equiv \begin{cases} 0 & y = 0 \\ +\infty & \text{else} \end{cases} \qquad (\varphi_0(x) \equiv \|x\|_0). \tag{5.7}$$

It is not uncommon to consider the function $\|\cdot\|_0$ as the limit of $\left(\sum_j |x_j|^p\right)^{1/p}$ as $p \to 0$. The notation is misleading since $\|\cdot\|_0$ is not a norm; the fact that

$$\|x\|_0 = \lim_{p \to 0^+} \sum_j |x_j|^p$$

shows that $d_0(x,y) := \|x-y\|_0$ still produces a metric since $\sum_j |x_j - y_j|^p$ does for $0 < p < 1$.

We propose an alternative strategy based on regularization of the conjugates. For $L \in \mathbb{R}^n_+$ and $\varepsilon > 0$ define the rectangle $R_L \equiv [-L_1,L_1] \times [-L_2,L_2] \times \cdots \times [-L_n,L_n]$ and

$$\phi_{\varepsilon,L}(y) \equiv \sum_{j=1}^{n} \psi_{\varepsilon,L_j}(y_j) \qquad (y = (y_1,y_2,\dots,y_n) \in \mathbb{R}^n), \tag{5.8}$$

where

$$\psi_{\varepsilon,L_j}(y_j) \equiv \begin{cases} \varepsilon\left(\frac{(L_j+y_j)\ln(L_j+y_j)+(L_j-y)\ln(L_j-y_j)}{2L_j\ln(2)} - \frac{\ln(L_j)}{\ln(2)}\right) & \text{for } |y_j| < L_j \\ \varepsilon & \text{for } |y_j| = L_j \\ +\infty & \text{for } |y_j| > L_j. \end{cases} \tag{5.9}$$

This is a scaled and shifted *Fermi–Dirac entropy* [1, 3]. The value at the endpoints $y_j = \pm L_j$ follows from defining $0\ln(0) = 0$, which is standard in the literature. The inclusion of the endpoints ($y_j = \pm L_j$) in the domain of definition of $\psi_{\varepsilon,L_j}(y_j)$ provides a type of continuity in the limiting cases, namely as the closed interval

$[-L_j, L_j]$ degenerates to the point $[0]$ and the relaxation parameter $\varepsilon \to 0$. This entropy is a smooth convex function on the interior of its domain and so elementary calculus can be used to calculate the Fenchel conjugate,

$$\phi^*_{\varepsilon,L}(x) = \sum_{j=1}^{n} \left(\frac{\varepsilon}{\ln(2)} \ln \left(4^{x_j L_j/\varepsilon} + 1 \right) - x_j L_j - \varepsilon \right). \tag{5.10}$$

(Calculate the gradient of the objective in the Fenchel problem (5.5), satisfy first order conditions for optimality and substitute the optimal solution back into (5.5) to get (5.10) for the optimal value parameterized by the dual variable.)

- For $\varepsilon > 0$ fixed we have

$$\lim_{L \to 0} \phi_{\varepsilon,L}(y) = \begin{cases} 0 & y = 0 \\ +\infty & \text{else} \end{cases} \quad \text{and} \quad \lim_{L \to 0} \phi^*_{\varepsilon,L}(x) = 0.$$

- For $L > 0$ fixed, in the limit as $\varepsilon \to 0$ we have

$$\lim_{\varepsilon \to 0} \phi_{\varepsilon,L}(y) = \begin{cases} 0 & y \in R_L \\ +\infty & \text{else} \end{cases} \quad \text{and} \quad \lim_{\varepsilon \to 0} \phi^*_{\varepsilon,L}(x) = L|x|.$$

We write $\phi_{0,L}$, $\phi^*_{0,L}$, $\phi_{\varepsilon,0}$ and $\phi^*_{\varepsilon,0}$ for the limits. In contrast to the limit of $\phi_{\varepsilon,L}(y)$ for $\varepsilon > 0$ fixed, if $y = L$ in the limiting process we have $\lim_{L \to 0} \phi_{\varepsilon,L}(L) = \varepsilon$. By $\phi_{\varepsilon,0}(0)$ we mean the former limit, so that $\phi_{\varepsilon,0}(0) = 0$. Note that $\|\cdot\|_0$ and $\phi^*_{\varepsilon 0}$ have the same conjugate, but unlike $\|\cdot\|_0$ the biconjugate of $\phi^*_{\varepsilon 0}$ is itself, that is, $\phi^{***} = \phi^*$. Also note that $\phi_{\varepsilon,L}$ and $\phi^*_{\varepsilon,L}$ are convex and smooth on the interior of their domains for all $\varepsilon, L > 0$. This is in contrast to metrics of the form $\left(\sum_j |x_j - y_j|^p\right)$ which are nonconvex for $p < 1$.

To maintain identification with φ in (5.1) we define

$$\varphi_{\varepsilon,L} \equiv \phi^*_{\varepsilon,L} \quad \text{and} \quad \varphi^*_{\varepsilon,L} \equiv \phi^{**}_{\varepsilon,L} = \phi_{\varepsilon,L},$$

where we have used the fact that the biconjugate of $\phi_{\varepsilon,L}$ is itself. We therefore consider the problem

$$\inf\{\varphi_{\varepsilon,L}(x) \mid x \in \mathbb{R}^n \text{ with } Ax = b\} \tag{5.11}$$

as a smooth convex relaxation of the conventional ℓ_p optimization for $0 \le p \le 1$. Our numerical approach to solve this problem will be to solve the dual.

Using Fenchel duality, the dual to this problem is the concave optimization problem

$$\sup\{y^{\mathrm{T}} b - \varphi^*_{\varepsilon,L}(A^{\mathrm{T}} y) \mid y \in \mathbb{R}^m\}, \tag{5.12}$$

where, again, $\varphi^*_{\varepsilon,L}(x) = \phi_{\varepsilon,L}(x)$ is given by (5.8). We reformulate this as a minimization problem

$$\underset{y\in\mathbb{R}^m}{\text{minimize}} \qquad \varphi^*_{\varepsilon,L}(A^{\mathrm{T}}y) - y^{\mathrm{T}}b, \tag{5.13}$$

which we will solve with the method described next.

The objective in the dual problem is smooth and convex, so we could in principle apply any number of efficient unconstrained optimization algorithms. Also, for this relaxation, the same numerical techniques can be used for all $L \to 0$.

5.3 Algorithms: Subgradient Descent

The central algorithm we explore in this note is simple (sub)gradient descent on the dual problem (5.13):

Algorithm 5.1 (Subgradient descent). Given $y^0 \in \mathbb{R}^m$, for $\nu = 0,1,2\ldots$ generate the sequence $\{y^\nu\}_{\nu=0}^\infty$ via

$$y^{\nu+1} = y^\nu + \lambda_\nu d^\nu,$$

where $d^\nu \in -\partial\left(\varphi^*_{\varepsilon,L}\left(A^{\mathrm{T}}y^\nu\right) - b^{\mathrm{T}}y^\nu\right)$ and λ_ν is an appropriate step length parameter.

For $\varepsilon > 0$, $\varphi^*_{\varepsilon,L}$ is continuously differentiable on its domain, and the algorithm amounts to the method of steepest descent.

5.3.1 *Nonsmooth Case:* $\varepsilon = 0$

In this section, we present and analyze a subgradient descent method with exact line search and variants thereof suitable for solving the dual problem above for the case $\varepsilon = 0$, that is, we do not smooth the problem.

Using the notation of indicator functions, we have

$$\varphi^*_{0,L}(x) = \iota_{R_L}(x) \equiv \begin{cases} 0 & \text{for } x \in R_L \\ +\infty & \text{otherwise.} \end{cases}$$

The specific instance of (5.13) that we address is

$$\min_{y\in\mathbb{R}^m} \iota_{R_L}(A^{\mathrm{T}}y) - y^{\mathrm{T}}b. \tag{5.14}$$

Since the set R_L is a rectangle, nonsmooth calculus yields the following simple expression for the subdifferential of the dual objective:

$$\partial\left(\iota_{R_L}\left(A^{\mathrm{T}}y^\nu\right) - b^{\mathrm{T}}y^\nu\right) = AN_{R_L}\left(A^T y^\nu\right) - b. \tag{5.15}$$

Here, we have used the fact that the subdifferential of the indicator function to the box R_L at the point x, denoted $\partial\iota_{R_L}(x)$ is equivalent to the normal cone mapping of R_L at the point x

$$\partial\iota_{R_L}(x) = N_{R_L}(x) \equiv \begin{cases} \{w \in \mathbb{R}^n \text{ with } (z-x)^{\mathrm{T}}w \le 0 \text{ for all } z \in R_L\} & \text{if } x \in R_L \\ \emptyset & \text{if } x \notin R_L. \end{cases}$$

Remark 5.2. It is important to note that we assume that we can perform exact arithmetic. This assumption is necessary due to the composition of the normal cone mapping of R_L with A^{T}: while we can determine the exact evaluation of the normal cone for a given $A^{\mathrm{T}}y^{\nu}$, we cannot guarantee exact evaluation of the matrix–vector product and, since the normal cone mapping is not Lipschitz continuous on R_L, this can lead to large computational errors.

Problem (5.14) is a linear programming problem. The algorithm we analyze below solves problem a *parametric* version of problem (5.14), where the parameter L changes dynamically at each iteration. To see how the parameter might be changed from one iteration to the next, we look to a trivial extension of the primal problem:

$$\begin{array}{ll} \underset{(x,L)\in\mathbb{R}^n\times\mathbb{R}^n_+}{\text{minimize}} & \sum_{j=1}^{n} L_j|x_j| \\ \text{subject to} & Ax = b. \end{array} \tag{5.16}$$

It is clear that $L = 0$ and any feasible x is an optimal solution to problem (5.16), and that the (global) optimal value is 0. However, this is not the only solution. Indeed, the sparsest solution x^* to $Ax = b$ and the weight L^* satisfying $L_j^* = 0$ only for those elements j on the support of x^* is also a solution. The algorithm we study below finds a weight *compatible* with the sparsest element x^*. A more satisfying reformulation would yield a weight that is in some sense *optimal* for the sparsest element x^*, but this is beyond the scope of this work.

5.3.2 Dynamically Rescaled Descent with Exact Line Search

There are three unresolved issues in our discussion to this point, namely the choice of elements from the subdifferential, the choice of the step length and the adjustment of the weights L_j. Our strategy is given in Algorithm 5.4 below. In the description of the algorithm, we use some geometric notions that we introduce first. It will be convenient to define the set C by

$$C \equiv \{y \in \mathbb{R}^m \mid A^{\mathrm{T}}y \in R_L\}.$$

This set is polyhedral as the domain of a linear mapping with box constraints.

Lemma 5.3 (Normal cone projection). *Let A be full rank and denote the normal cone to $C \equiv \{y \in \mathbb{R}^m \mid A^T y \in R_L\}$ at $\overline{y} \in C$ by $N_C(\overline{y})$. Then*

$$P_{N_C(\overline{y})} b = A\overline{w} \tag{5.17}$$

for

$$\overline{w} = \operatorname{argmin}\{\|Aw - b\|^2 \mid w \in N_{R_L}(A^T\overline{y})\}. \tag{5.18}$$

Proof. If A is full rank, then all points $y \in C$ satisfy the constraint qualification that A is injective on $N_{R_L}(A^T y)$, that is, the only vector $w \in N_{R_L}(A^T y)$ for which $Aw = 0$ is $w = 0$. Then, by convex or nonsmooth analysis (see e.g., [14, Theorem 6.14]) the set C is *regular* and

$$N_C(y) = AN_{R_L}(A^T y) = \left\{u = Aw \mid w \in N_{R_L}(A^T y)\right\}.$$

By the definition of the projection

$$P_{N_C(\overline{y})} b \equiv \operatorname{argmin}\left\{\|u - b\|^2 \mid u \in N_C(\overline{y})\right\}$$

hence,

$$\begin{aligned} P_{N_C(\overline{y})} b &= \operatorname{argmin}\left\{\|u - b\|^2 \mid u \in AN_{R_L}(A^T\overline{y})\right\} \\ &= A\operatorname{argmin}\left\{\|Aw - b\|^2 \mid w \in N_{R_L}(A^T\overline{y})\right\} = A\overline{w}. \end{aligned}$$

□

Algorithm 5.4 (Dynamically rescaled descent with exact line search).

Initialization: Set $\nu = 0$, $\tau > 0$, $L^0 = (\|a_1\|, \|a_2\|, \ldots, \|a_n\|)$, where a_j is the jth column of A, $y^0 = 0$ and the direction $d^0 = b$.
Main iteration: While $\|d^\nu\| > \tau$ do

- (Exact line search.) Calculate the step length $\lambda_\nu > 0$ according to

$$\lambda_\nu \equiv \operatorname*{argmin}_{\lambda > 0}\left\{\iota_{R_{L^\nu}}\left(A^T\left(y^\nu + \lambda d^\nu\right)\right) - b^T\left(y^\nu + \lambda d^\nu\right)\right\}. \tag{5.19}$$

Set $y' = y^\nu + \lambda_\nu d^\nu$.
- (Subgradient selection and preliminary rescaling.) Define

$$\mathbb{J}^{\nu+1} = \{j \mid |a_j^T y'| = L_j^\nu\}, \tag{5.20}$$

$$S(L, \mathbb{J}, \gamma) = (s_1(L, \mathbb{J}, \gamma), s_2(L, \mathbb{J}, \gamma), \ldots, s_n(L, \mathbb{J}, \gamma)),$$

$$\text{where} \quad s_j(L, \mathbb{J}, \gamma) = \begin{cases} \gamma L_j & \text{for all } j \in \mathbb{J} \\ L_j & \text{else,} \end{cases} \tag{5.21}$$

and

$$C(L,\mathbb{J},\gamma) = \{y \in \mathbb{R}^n \mid A^T y \in R_{S(L,\mathbb{J},\gamma)}\}, \quad \text{where}$$
$$R_{S(L,\mathbb{J},\gamma)} \equiv [-s_1(L,\mathbb{J},\gamma), s_1(L,\mathbb{J},\gamma)] \times \cdots \times [-s_n(L,\mathbb{J},\gamma), s_n(L,\mathbb{J},\gamma)] \tag{5.22}$$

Choose $\gamma' \geq 0$ small enough that $P_{N_{C(L^\nu,\mathbb{J}^{\nu+1},\gamma')}(y'')}b \in ri(N_{C(L^\nu,\mathbb{J}^{\nu+1},\gamma')}(y''))$ for $y'' = \gamma' y'$. Compute the direction

$$d^{\nu+1} \equiv b - P_{N_{C(L^\nu,\mathbb{J}^{\nu+1},\gamma')}(y'')}b \tag{5.23}$$

- (Rescaling.) Let

$$\mathbb{J}_+^{\nu+1} \equiv \{j \mid a_j^T d^{\nu+1} > 0\}, \qquad \mathbb{J}_-^{\nu+1} \equiv \{j \mid a_j^T d^{\nu+1} < 0\},$$

and define

$$\mathbb{I}^{\nu+1}(\gamma) \equiv \underset{j \in \mathbb{J}_+^{\nu+1} \cup \mathbb{J}_-^{\nu+1}}{\operatorname{argmin}} \left\{ \left\{ \frac{L_j^\nu - \gamma a_j^T y'}{a_j^T d^{\nu+1}} \;\middle|\; j \in \mathbb{J}_+^{\nu+1} \right\}, \left\{ \frac{-L_j^\nu - \gamma a_j^T y'}{a_j^T d^{\nu+1}} \;\middle|\; j \in \mathbb{J}_-^{\nu+1} \right\} \right\}. \tag{5.24}$$

Choose $\gamma^{\nu+1} \in [0, \gamma']$ to satisfy

$$\mathbb{I}^{\nu+1}(\gamma^{\nu+1}) \subset \mathbb{I}^{\nu+1}(0). \tag{5.25}$$

Set

$$L_j^{\nu+1} = \begin{cases} \gamma^{\nu+1} L_j^\nu & \text{for } j \in \mathbb{J}^{\nu+1} \\ L_j^\nu & \text{else} \end{cases} \tag{5.26}$$

and $y^{\nu+1} = \gamma^{\nu+1} y'$. Increment $\nu = \nu + 1$.

End do.

We begin with some observations. The next proposition shows that the directions chosen by (5.23) with $P_{N_{C(L^\nu,\mathbb{J}^{\nu+1},\gamma')}(y'')}b \in ri(N_{C(L^\nu,\mathbb{J}^{\nu+1},\gamma')}(y''))$ for $y'' = \gamma' y'$ are subgradient descent directions that are not only feasible, but orthogonal to the active constraints. We use orthogonality of the search directions to the active constraints to guarantee finite termination of the algorithm.

Proposition 5.5 (Feasible directions). *Let $C \equiv \{y \in \mathbb{R}^m \mid A^T y \in R_L\}$, $\bar{y} \in C$ and define the direction $\bar{d} \equiv b - P_{N_C(\bar{y})}b$. Then $-\bar{d} \in \partial\left(\iota_{R_L}(A^T\bar{y}) - b^T\bar{y}\right)$ and there exists a $\overline{\lambda} > 0$ such that $\bar{y} + \lambda\bar{d} \in C$ for all $\lambda \in [0, \overline{\lambda}]$.*

Moreover, if $P_{N_C(\bar{y})}b \in ri(N_C(\bar{y}))$, then the direction $\bar{d}$ is orthogonal to the jth column of A for all j such that $a_j^T\bar{y} = L_j$.

Proof. The inclusion $-\bar{d} \in \partial\left(\iota_{R_L}(A^{\mathrm{T}}\bar{y}) - b^{\mathrm{T}}\bar{y}\right)$ follows immediately from (5.15). The feasibility of this direction follows from Lemma 5.3 and the polyhedrality of C since the polar to the normal cone to C at a point $y \in C$ is therefore equivalent to the tangent cone, which consists only of *feasible directions* to C at y, defined as a direction d for which $\bar{y} + \lambda d \in C$ for all $\lambda > 0$ sufficiently small.

Indeed, let a_j denote the jth column of the matrix A and recall the definition of the *contingent cone* to C at $y \in C$:

$$K_C(y) \equiv \{w \in \mathbb{Y} \mid \text{ for all } \nu \; y + \lambda^\nu w^\nu \in C \text{ for some } w^\nu \to w, \; \lambda^\nu \searrow 0\}.$$

Since C is convex the contingent cone and the tangent cone are equivalent [2, Corollary 6.3.7] and since C is polyhedral the tangent cone can be written as

$$T_C(y) \equiv \{w \in \mathbb{Y} \mid \text{ for all } \nu \; y + \lambda^\nu w \in C \text{ for some } \lambda^\nu \searrow 0\},$$

that is, the tangent cone consists entirely of feasible directions. Now the tangent and normal cones to C are convex and polar to each other [14, Corollary 6.30], so, by Lemma 5.3, what remains to be shown is that $b - P_{N_C(\bar{y})}b$ lies in the polar to the normal cone to C. This follows since $N_C(y)$ is nonempty, closed, and convex. Hence for all $w \in N_C(\bar{y})$ and for any b

$$w^{\mathrm{T}}(b - P_{N_C(\bar{y})}b) \le 0,$$

that is, $b - P_{N_C(\bar{y})}b$ is in the polar to the normal cone.

To see the final statement of the proposition, denote by $\bar{\mathbb{J}}$ the set

$$\{j = 1, 2, \ldots, n \mid a_j^{\mathrm{T}}\bar{y} = L_j\}.$$

If the projection lies on the relative interior to $N_C(\bar{y})$, then the projection onto $N_C(\bar{y})$ is equivalent to the projection onto the subspace containing $N_C(\bar{y})$:

$$P_{N_C(\bar{y})}b = P_{D(\bar{y})}b,$$

where

$$D(\bar{y}) \equiv \left\{Aw \mid w \in \mathbb{R}^n \text{ with } w_j = 0 \text{ for } j \notin \bar{\mathbb{J}}\right\}.$$

Thus, $a_j^{\mathrm{T}}(b - P_{N_C(\bar{y})}b) = a_j^{\mathrm{T}}(b - P_{N_D(\bar{y})}b) = 0$ for $j \in \bar{\mathbb{J}}$ as claimed. □

Remark 5.6 (Detection of orthogonality of feasible directions). The interiority condition $P_{N_C(\bar{y})}b \in ri(N_C(\bar{y}))$ guaranteeing orthogonality of the directions can easily be checked. Let $\bar{w} \equiv \operatorname{argmin}\{\|Aw - b\|^2 \mid w \in N_{R_L}(A^{\mathrm{T}}\bar{y})\}$. By Lemma 5.3 $P_{N_C(\bar{y})}b = A\bar{w}$. Then $A\bar{w}$ and hence $P_{N_C(\bar{y})}b$ lies in $ri(N_C(\bar{y}))$ if and only if $\bar{w}_j \neq 0$ for all j such that $a_j^{\mathrm{T}}\bar{y} = L_j$.

Calculation of the direction in (5.23) of Algorithm 5.4 is suggested by Lemma 5.3, where it is shown that the projection is the mapping of the solution to a least squares problem over a cone. Also, by Proposition 5.5, the direction is the negative of a subgradient of the objective in (5.14) with the box $S(L^\nu, \mathbb{J}^{\nu+1}, \gamma')$, that is

$$-d^{\nu+1} \equiv -b + P_{N_{C(L^\nu,\mathbb{J}^{\nu+1},\gamma')}(y'')} b \in \partial\left(\iota_{R_{S(L^\nu,\mathbb{J}^{\nu+1},\gamma')}}(A^{\mathrm{T}}y'') - b^{\mathrm{T}}y''\right).$$

The description as a projection onto the normal cone of a polyhedron is perhaps less helpful than the explicit formulation of Lemma 5.3 for suggesting how this can be computed, but it provides greater geometrical insight. Moreover, the projection provides an elegant criterion for maintaining orthogonality of the search directions with the active constraints.

The exact line search step has an explicit formulation given in the next proposition.

Proposition 5.7 (exact line search). *Let $\bar{y} \in C$ and $\bar{d} = b - P_{N_C(\bar{y})} b$. Define the index sets*

$$\bar{\mathbb{J}}_+ \equiv \{j \mid a_j^{\mathrm{T}}\bar{d} > 0\}, \qquad \bar{\mathbb{J}}_- \equiv \{j \mid a_j^{\mathrm{T}}\bar{d} < 0\}.$$

The exact line search step length $\bar{\lambda}$ given by (5.19) has the explicit representation

$$\bar{\lambda} \equiv \min\left\{ \min_{j\in\bar{\mathbb{J}}_+}\left\{\frac{L_j - a_j^{\mathrm{T}}\bar{y}}{a_j^{\mathrm{T}}\bar{d}}\right\},\ \min_{j\in\bar{\mathbb{J}}_-}\left\{\frac{-L_j - a_j^{\mathrm{T}}\bar{y}}{a_j^{\mathrm{T}}\bar{d}}\right\}\right\} > 0. \tag{5.27}$$

Proof. Application of nonsmooth calculus provides a generalization to the fact from optimization of smooth objectives that the exact line search step extends to the tangent of a level set of the objective, from which we can extract (5.27). However, it is perhaps easiest to see the explicit formulation by direct inspection: the indicator function ι_{R_L} is zero at all points in R_L, so the step length is the largest λ such that $A^{\mathrm{T}}\left(\bar{y} + \lambda\bar{d}\right) \in R_L$, i.e., the largest λ such that

$$a_j^{\mathrm{T}}\left(\bar{y} + \lambda\bar{d}\right) \le L_j \quad \text{for all } j \in \bar{\mathbb{J}}_+.$$

and

$$a_j^{\mathrm{T}}\left(\bar{y} + \lambda\bar{d}\right) \ge -L_j \quad \text{for all } j \in \bar{\mathbb{J}}_-.$$

Note that by Proposition 5.5, it is not possible to have $a_j^{\mathrm{T}}\bar{d} > 0$ and $a_j^{\mathrm{T}}\bar{y} = L_j$ or, similarly $a_j^{\mathrm{T}}\bar{d} < 0$ and $a_j^{\mathrm{T}}\bar{y} = -L_j$, hence the step length is guaranteed to be positive, and we are done. □

5.4 Convergence to Sparse Solutions

We show in this section that for sufficiently sparse solutions x^* to $Ax = b$, the steepest subgradient descent algorithm with exact line search (Algorithm 5.4) recovers x^* exactly. Before we continue, however, we must specify precisely what is meant by "sufficiently sparse."

Definition 5.8 (Mutual coherence). Let a_j denote the jth column of A. The *mutual coherence* of A is defined as

$$\mu(A) \equiv \max_{1\le k,j\le n,\ k\neq j} \frac{|a_k^{\mathrm{T}} a_j|}{\|a_k\|\|a_j\|},$$

where $0/0 \equiv 1$.

The mutual coherence characterizes the dependence between columns of A. The mutual coherence of unitary matrices, for instance, is zero; for matrices with columns of zeros, the mutual coherence is 1.

Lemma 5.9 (Uniqueness of sparse representations [8]). *Let $A \in \mathbb{R}^{m\times n}$ $(m < n)$ be full rank. If there exists an element x^* such that $Ax^* = b$ and*

$$\|x^*\|_0 < \frac{1}{2}\left(1 + \frac{1}{\mu(A)}\right), \tag{5.28}$$

then it is unique and sparsest possible (has minimal support).

In the case of matrices that are not full rank – and thus unitarily equivalent to matrices with columns of zeros – only the trivial equation $Ax = 0$ has a unique sparsest possible solution. For unitary matrices $\mu(A) = 0$, we interpret $1/0 = +\infty$.

The sparsity condition of Lemma 5.9 yields a more direct representation that will be useful later.

Lemma 5.10 (sparsity conditions). *Let $A \in \mathbb{R}^{m\times n}$ $(m < n)$ be full rank. For $b \in \mathbb{R}^m \setminus \{0\}$ given and x^* a solution to $Ax = b$, define $\mathbb{J} = \{j \mid x_j^* \neq 0\}$ and denote by $J \in \mathbb{J}$ an element of x^* satisfying*

$$|x_J^*|\|a_J\| \geq |x_j^*|\|a_j\| \quad \text{for all} \quad j = 1,2,\ldots,n.$$

If the solution x^* satisfies condition (5.28) then there exists a $\bar{\gamma} > 0$ such that, for all $y \in \mathbb{B} \equiv \{y \in \mathbb{R}^m \mid \|y\| = 1\}$ and all $\gamma \in [0, \bar{\gamma}]$

$$\max_{k\notin\mathbb{J}} \frac{|a_k^{\mathrm{T}} b|}{\|a_k\| - \gamma|a_k^{\mathrm{T}} y|} < \frac{|a_J^{\mathrm{T}} b|}{\|a_J\| + \gamma|a_J^{\mathrm{T}} y|}. \tag{5.29}$$

Proof. We use continuity of the terms in (5.29) with respect to γ and y to simplify the operative inequality and prove the statement for the case $\gamma = 0$.

Reduction to the Case $\gamma = 0$

For all $\bar{\gamma}$ small enough, the function

$$g(y,\gamma) \equiv \max_{k\notin\mathbb{J}} \frac{|a_k^{\mathrm{T}} b|}{\|a_k\| - \gamma|a_k^{\mathrm{T}} y|}$$

is a continuous function on the compact domain $\mathbb{B} \times [0, \bar{\gamma}]$. Likewise, for *any* $\bar{\gamma} > 0$ the function

$$h(y, \gamma) \equiv \frac{|a_J^T b|}{\|a_J\| + \gamma |a_J^T y|}$$

is continuous. By continuity, the existence of $\bar{\gamma} > 0$ such that (5.29) holds for all $\gamma \in [0, \bar{\gamma}]$ and $y \in \mathbb{B}$ is then equivalent to

$$g(y, 0) = \max_{k \notin \mathbb{J}} \frac{|a_k^T b|}{\|a_k\|} < \frac{|a_J^T b|}{\|a_J\|} = h(y, 0). \tag{5.30}$$

We therefore limit our attention to (5.30).

Reformulation of (5.28)

Starting with (5.28), we have

$$\begin{aligned} \|x^*\|_0 = |\mathbb{J}| &< \frac{1}{2}\left(\frac{1}{\mu(A)} + 1\right) \\ &\Longleftrightarrow \\ \mu(A)|\mathbb{J}| &< \frac{1}{2}(1 + \mu(A)) \\ &\Longleftrightarrow \\ |x_J^*|\|a_J\||\mathbb{J}|\mu(A) &< \frac{1}{2}|x_J^*|\|a_J\|\,(1 + \mu(A)) \\ &\Longleftrightarrow \\ |x_J^*|\|a_J\||\mathbb{J}|\mu(A) &< |x_J^*|\|a_J\|\,(1 + \mu(A)(1 - |\mathbb{J}|)). \end{aligned} \tag{5.31}$$

Here, we have denoted the cardinality of $\mathbb{J}$ by $|\mathbb{J}|$.

Upper and Lower Bounds

It remains to show that the left hand side of (5.31) is an upper bound for the left hand side of (5.30) and, similarly, that the right hand side of (5.31) is a lower bound for the right hand side of (5.30).

Substituting Ax^* for b in (5.30) yields the equivalent statement

$$\frac{|\sum_{i \in \mathbb{J}} x_i^* a_k^T a_i|}{\|a_k\|} < \frac{|\sum_{i \in \mathbb{J}} x_i^* a_J^T a_i|}{\|a_J\|} \quad \text{for all } k \notin \mathbb{J}. \tag{5.32}$$

For the lower bound, we have

$$\begin{aligned}\frac{|\sum_{i\in\mathbb{J}} x_i^* a_J^\mathrm{T} a_i|}{\|a_J\|} &\geq |x_J^*|\|a_J\| - \sum_{i\in\mathbb{J}\setminus\{J\}} \frac{|x_i^*||a_J^\mathrm{T} a_i|}{\|a_J\|} \\ &\geq |x_J^*|\|a_J\| - \sum_{i\in\mathbb{J}\setminus\{J\}} |x_i^*|\|a_i\|\mu(A) \\ &\geq |x_J^*|\|a_J\|\left(1-(|\mathbb{J}|-1)\mu(A)\right).\end{aligned}$$

In summary

$$|x_J^*|\|a_J\|\left(1+(1-|\mathbb{J}|)\mu(A)\right) \leq \frac{|\sum_{i\in\mathbb{J}} x_i^* a_J^\mathrm{T} a_i|}{\|a_J\|}. \tag{5.33}$$

For the upper bound, we have

$$\begin{aligned}\frac{|\sum_{i\in\mathbb{J}} x_i^* a_k^\mathrm{T} a_i|}{\|a_k\|} &\leq \sum_{i\in\mathbb{J}} \frac{|x_i^*|\,|a_k^\mathrm{T} a_i|}{\|a_k\|} \\ &\leq \sum_{i\in\mathbb{J}} |x_i^*|\,\|a_i\|\mu(A) \\ &\leq |x_J^*|\,\|a_J\||\mathbb{J}|\mu(A)\end{aligned}$$

or

$$\frac{|\sum_{i\in\mathbb{J}} x_i^* a_k^\mathrm{T} a_i|}{\|a_k\|} \leq |x_J^*|\,\|a_J\||\mathbb{J}|\mu(A). \tag{5.34}$$

Inequality (5.31) together with (5.32), (5.33) and (5.34) yield (5.30). By the continuity argument at the beginning of the proof, we have thus shown that (5.28) implies (5.29) as claimed. □

The next lemma provides a sufficient condition for monotonicity of the cardinality of the set of active indices from one iteration of Algorithm 5.4. This is an important feature for the finite termination of Algorithm 5.4 proved in Theorem 5.12.

Lemma 5.11 (Step length). *For a given $L = (L_1, L_2, \ldots, L_n)$ and the corresponding sets R_L and $C \equiv \{y \in \mathbb{R}^m \mid A^\mathrm{T} y \in R_L\}$, let the point $\bar{y} \in C$ satisfy $P_{N_C(\bar{y})} b \in ri(N_C(\bar{y}))$. For this point define $\bar{d} \equiv b - P_{N_C(\bar{y})} b$ and the index sets $\mathbb{J} = \{j \mid a_j^\mathrm{T}\bar{y} = L_j\}$*

$$\mathbb{J}_+ \equiv \{j \mid a_j^\mathrm{T}\bar{d} > 0\}, \qquad \mathbb{J}_- \equiv \{j \mid a_j^\mathrm{T}\bar{d} < 0\}.$$

Then $(\mathbb{J}_+ \cup \mathbb{J}_-) \cap \mathbb{J} = \emptyset$ and for the step length given by (5.27) the set of active indices set is increasing, that is, $\mathbb{J} \subset \mathbb{J}' = \{j \mid a_j^\mathrm{T}(\bar{y} + \bar{\lambda}\bar{d}) = L_j\}$.

In the special case that $\bar{y} = 0$, then the step length $\bar{\lambda}$ is given by

$$\bar{\lambda} \equiv \min_{j\notin\mathbb{J}} \left\{ \frac{L_j}{|a_j^\mathrm{T}\bar{d}|} \right\}. \tag{5.35}$$

Proof. By Proposition 5.5 and Remark 5.6, if $P_{N_C(\bar{y})}b \in ri(N_C(\bar{y}))$ then $\bar{d}$ is orthogonal to the columns of A corresponding to the set of active indices $\mathbb{J}$. Thus $(\mathbb{J}_+ \cup \mathbb{J}_-) \cap \mathbb{J} = \emptyset$ as claimed. It follows immediately from (5.27) that $\mathbb{J} \subset \mathbb{J}' = \{j \mid a_j^T(\bar{y} + \bar{\lambda}\bar{d}) = L_j\}$ since $\bar{\lambda}$ is computed from the elements belonging to $\mathbb{J}_+ \cup \mathbb{J}_-$, and, again by Proposition 5.5, the active constraints corresponding to $\mathbb{J}$ remain unchanged in the direction $\bar{d}$.

When $\bar{y} = 0$, the step length given by (5.27) simplifies to

$$\bar{\lambda} = \min_{j \in \bar{\mathbb{J}}_+ \cup \bar{\mathbb{J}}_-} \left\{ \frac{L_j}{|a_j^{\mathrm{T}}\bar{d}|} \right\} > 0. \tag{5.36}$$

Hence, (5.36) is equivalent to (5.35). This completes the proof. □

We are now ready to state and prove the main result of this section, the convergence of Algorithm 5.4 for a particular choice of initial weights $L_j^0 = \|a_j\|$ for $j = 1, 2, \ldots, n$. Theorem 5.12 says that the algorithm finds a point y^* and a weight L^* for which $0 \in \partial\left(\iota_{R_{L^*}}(A^{\mathrm{T}}y^*) - (y^*)^{\mathrm{T}}b\right)$ exactly (tolerance $\tau = 0$), as opposed to finding a point where the chosen subgradient is smaller than some tolerance. Since the problem is convex, this is sufficient for optimality. Of course, this is only possible with exact arithmetic.

Theorem 5.12 (Exact recovery of sufficiently sparse solutions). *Let $A \in \mathbb{R}^{m\times n}$ $(m < n)$ be full rank and denote the jth column of A by a_j. Initialize Algorithm 5.4 with initial guess y^0 and weight L^0 such that $y_j^0 = 0$ and $L_j^0 = \|a_j\|$ for $j = 1, 2, \ldots, n$.*

If an element $x^* \in \mathbb{R}^n$ with $Ax^* = b$ satisfies (5.28), then with tolerance $\tau = 0$, Algorithm 5.4 converges in finitely many steps to a point y^* and a weight L^* where,

$$\operatorname{argmin}\{\|Aw - b\|^2 \mid w \in N_{R_{L^*}}(y^*)\} = x^*,$$

the unique sparsest solution to $Ax = b$.

Proof. The proof is by induction and follows a pattern similar to the convergence proof of the orthogonal matching pursuit algorithm [4, Theorem 6], though the details are more technical. (Indeed, we show in Sect. 5.5 that this is no coincidence.) To facilitate the proof, we will in fact prove convergence of a slightly more general procedure than Algorithm 5.4. The difference is in the initialization. Rather than initializing $y^0 = 0$, as any practical method would do, we will choose an arbitrary $y^0 = \gamma^0 y$ for any fixed vector y with $\gamma^0 \geq 0$ small enough. This allows us to establish the pattern for later iterations at the very beginning.

Let $C^0 \equiv \{y \in \mathbb{R}^m \mid A^{\mathrm{T}}y \in R_{L^0}\}$. The open unit ball lies in the (relative) interior of C^0 since, for any y with $\|y\| < 1$, we have $|(A^{\mathrm{T}}y)_j| \leq \|a_j\|\|y\| \leq \|a_j\| = L_j^0$ with the last inequality strict if $a_j \neq 0$. (Without loss of generality, we can assume that A has no zero columns.) Then $N_{C^0}(y^0) = \{0\}$, so that $P_{N_{C^0}(y^0)}b = 0$ and $d^0 = b$ is in fact a direction of (subgradient) descent according to Proposition 5.5 for any y^0 small enough.

Identifying the Active Constraints

Computing the step length, by (5.27) we have

$$\lambda_0 \equiv \min\left\{\min_{j\in\mathbb{J}_+^0}\left\{\frac{\|a_j\|-\gamma^0 a_j^{\mathrm{T}}y}{a_j^{\mathrm{T}}b}\right\},\ \min_{j\in\mathbb{J}_-^0}\left\{\frac{-\|a_j\|-\gamma^0 a_j^{\mathrm{T}}y}{a_j^{\mathrm{T}}b}\right\}\right\} > 0, \tag{5.37}$$

where, recall, $\gamma^0 y = y^0$, and

$$\mathbb{J}_+^0 = \{j \mid a_j^{\mathrm{T}}b > 0\} \quad \text{and} \quad \mathbb{J}_-^0 = \{j \mid a_j^{\mathrm{T}}b < 0\}.$$

Let j_0 be the index of a minimum element of the set above. We show that, for any choice of minimum element (in the case that there is more than one) $j_0 \in \mathbb{J}^* \equiv \{j \mid x_j^* \neq 0\}$. In other words, we show that

$$|a_k^{\mathrm{T}}y^0 + \lambda_0 a_k^{\mathrm{T}}b| < \|a_k\| \quad \text{for all } k \notin \mathbb{J}^*. \tag{5.38}$$

By the triangle inequality, (5.38) holds if

$$|a_k^{\mathrm{T}}y^0| + |\lambda_0 a_k^{\mathrm{T}}b| < \|a_k\| \quad \text{for all } k \notin \mathbb{J}^*. \tag{5.39}$$

Expanding λ_0 and rearranging terms in (5.39) yields, for γ^0 small enough,

$$\frac{|a_k^{\mathrm{T}}b|}{\|a_k\|-\gamma^0|a_k^{\mathrm{T}}y|} < \frac{1}{\lambda_0} = \begin{cases} \dfrac{|a_{j_0}^{\mathrm{T}}b|}{\left|\|a_{j_0}\|-\gamma^0 a_{j_0}^{\mathrm{T}}y\right|}, & \text{if } j_0 \in \mathbb{J}_+^0 \\[2ex] \dfrac{|a_{j_0}^{\mathrm{T}}b|}{\left|-\|a_{j_0}\|-\gamma^0 a_{j_0}^{\mathrm{T}}y\right|}, & \text{if } j_0 \in \mathbb{J}_-^0 \end{cases} \quad \text{for all } k \notin \mathbb{J}^*. \tag{5.40}$$

Let $J \in \mathbb{J}^*$ be the index of an element of x^* satisfying

$$|x_J^*|\|a_J\| \geq |x_j^*|\|a_j\| \quad \text{for all} \quad j = 1,2,\ldots,n.$$

By definition of λ_0,

$$\frac{|a_J^{\mathrm{T}}b|}{\|a_J\|+\gamma^0|a_J^{\mathrm{T}}y|} \leq \left\{\begin{array}{ll} \dfrac{|a_J^{\mathrm{T}}b|}{\left|\|a_J\|-\gamma^0 a_J^{\mathrm{T}}y\right|}, & \text{if } J \in \mathbb{J}_+^0 \\[2ex] \dfrac{|a_J^{\mathrm{T}}b|}{\left|-\|a_J\|-\gamma^0 a_J^{\mathrm{T}}y\right|}, & \text{if } J \in \mathbb{J}_-^0 \end{array}\right\} \leq \frac{1}{\lambda_0}.$$

By Lemma 5.10, the sparsity condition (5.28) implies (5.29) which immediately yields (5.40), and hence (5.38), for γ^0 small enough.

Letting $y' = \gamma^0 y + \lambda_0 b$, we conclude that, as (5.28) holds, then for γ^0 small enough (as it certainly would be for the initial guess of zero)

$$\mathbb{J}^1 \equiv \{j \mid |a_j^{\mathrm{T}} y'| = \|a_j\| = L_j^0\} \cap \mathbb{J}^* \neq \emptyset,$$

where $\mathbb{J}^\nu$ is defined by (5.20).

The question remains as to how small γ^0 need be. For this we refer to the index set $\mathbb{I}^0(\gamma)$ defined by (5.24) with $L^{-1} \equiv L^0$. Note that this is just the set of indices of active faces in $\mathbb{J}_+^0 \cup \mathbb{J}_+^0$ corresponding to the exact line search step length λ_0 computed by (5.37). Viewed as a function, λ^0 is the minimum of a finite collection of affine functions of γ^0 and is thus a continuous function of γ^0. Moreover, the set of indices corresponding to the affine functions at which the minimum is attained, $\mathbb{I}(\gamma^0)$, satisfies $\mathbb{I}(\gamma^0) \subset \mathbb{I}(0)$ on a neighborhood of 0. In other words, the index j_0 of the minimum element at which the exact step length λ_0 is attained belongs to $\mathbb{I}^0(0)$ for all γ^0 small enough. This yields an implementable strategy for determining the proper scaling in subsequent iterations by checking the coincidence of the set of active indices $\mathbb{I}^\nu(\gamma)$ with the set of faces reached from the origin, $\mathbb{I}^\nu(0)$.

Subgradient Selection

There always exists $\gamma' \geq 0$ with $y'' = \gamma' y'$ such that $P_{N_{C(L^0,\mathbb{J}^1,\gamma')}(y'')} b \in ri(N_{C(L^0,\mathbb{J}^1,\gamma')}(y''))$ since for $\gamma' = 0$ the normal cone to $C(L^0, \mathbb{J}^1, 0)$ at $y'' = 0$ defined by (5.22) is the subspace

$$N_{C(L^0,\mathbb{J}^1,0)}(0) = \left\{ Aw \;\middle|\; \begin{cases} w_j \in \mathbb{R} & \text{for } j \in \mathbb{J}^1 \\ w_j = 0 & \text{for } j \notin \mathbb{J}^1 \end{cases} \right\}.$$

This follows from the fact that the only active faces of the polyhedron $C(L^0, \mathbb{J}^1, 0)$ at the origin are the ones corresponding to the point [0] (the degenerated interval). Thus, at least for $\gamma' = 0$, the projection of b onto the subspace spanned by the columns of A corresponding to $\mathbb{J}^1$ is equivalent to $P_{N_{C(L^0,\mathbb{J}^1,0)}(0)} b$. By Proposition 5.5, then, for γ' small enough (possibly zero) the direction of descent $d^1 \equiv b - P_{N_{C(L^0,\mathbb{J}^1,\gamma')}(y'')} b$ is orthogonal to the columns of A corresponding to the index set $\mathbb{J}^1$.

Rescaling

For the choice of γ' above, we have $(\mathbb{J}_+^1 \cup \mathbb{J}_-^1) \cap \mathbb{J}^1 = \emptyset$, where

$$\mathbb{J}_+^1 \equiv \{j \mid a_j^{\mathrm{T}} d^1 > 0\}, \qquad \mathbb{J}_-^1 \equiv \{j \mid a_j^{\mathrm{T}} d^1 < 0\}.$$

There are two cases to consider: $\gamma' = 0$ and $\gamma' > 0$. If $\gamma' = 0$, then $\gamma^1 = 0$ and by Lemma 5.11

$$\mathbb{I}^1(0) = \operatorname*{argmin}_{j \notin \mathbb{J}_0^1} \left\{ \frac{L_j^0}{|a_j^\mathrm{T} d^1|} \right\},$$

so that

$$L_j^1 = \begin{cases} 0 & \text{for all } j \in \mathbb{I}^1(0) \\ L_j^0 & \text{else} \end{cases}$$

and $y^1 = 0$.

If, on the other hand, $\gamma' > 0$, the previous argument shows that there exists at least *some* $\gamma^1 \in [0, \gamma']$, such that $\mathbb{I}^1(\gamma^1) \subset \mathbb{I}^1(0)$, which is sufficient for our purposes.

With γ^1 in hand, we set the weights

$$L_j^1 = \begin{cases} \gamma^1 L_j^0 & \text{for all } j \in \mathbb{I}^1(\gamma^1) \\ L_j^0 & \text{else.} \end{cases}$$

and update the iterate $y^1 = \gamma^1 y'$ as prescribed.

Note that y^1 is feasible and the set of active faces $\mathbb{J}^1$ is unchanged since $a_j^\mathrm{T} y^1 = a_j^\mathrm{T} \gamma^1 y'$ with $a_j^\mathrm{T} \gamma^1 y' = \gamma^1 L_j^0 = L_j^1$ for all $j \in \mathbb{J}^1$, and $a_j^\mathrm{T} \gamma^1 y' < a_j^\mathrm{T} y' < L_j^1$ otherwise.

Induction. Proceeding now by induction, we suppose for $\nu \geq 0$ that $a_j^\mathrm{T} y^\nu = L_j^\nu$ for all $j \notin \mathbb{J}^\nu \subset \mathbb{J}^*$ and that $|a_j^\mathrm{T} y^\nu| < L_j^\nu = \|a_j\|$ for all $j \notin \mathbb{J}^\nu$, where $\nu \leq |\mathbb{J}^\nu| \leq |\mathbb{J}^*|$. We show that there are only two possibilities for the next iteration: either $d^{\nu+1} = 0$, in which case $\mathbb{J}^{\nu+1} = \mathbb{J}^*$ and $w^{\nu+1} = x^*$; or $d^{\nu+1} \neq 0$, in which case $\mathbb{J}^{\nu+1} \subset \mathbb{J}^*$ with $|\mathbb{J}^{\nu+1}| < |\mathbb{J}^{\nu+2}| \leq |\mathbb{J}^*|$ and $|a_j^\mathrm{T} y^{\nu+1}| = L_j^{\nu+1}$ for $j \in \mathbb{J}^{\nu+2}$ and $|a_j^\mathrm{T} y^\nu| < L_j^{\nu+1}$ for $j \notin \mathbb{J}^{\nu+2}$.

In either case, in a somewhat awkward consequence of our indexing, note that for γ^ν satisfying (5.25) and the induction hypothesis we have that $\mathbb{J}^{\nu+1} \subset \mathbb{J}^*$. Our task is to show that $\mathbb{J}^{\nu+2} \subset \mathbb{J}^*$.

Case 1. $d^{\nu+1} = 0$. In this case, we have

$$b = P_{N_{C(L^\nu, \mathbb{J}^{\nu+1}, \gamma')}(y'')} b \in ri(N_{C(L^\nu, \mathbb{J}^{\nu+1}, \gamma')}(y''))$$

for $y'' = \gamma' y'$ with $y' = y^\nu + \lambda^\nu d^\nu$ and, by assumption (5.28),

$$\mathbb{J}^{\nu+1} = \{ j \mid |a_j^\mathrm{T}(y^\nu + \lambda^\nu d^\nu)| = L_j^\nu \} \subset \mathbb{J}^*.$$

Also note that $\mathbb{J}_+^{\nu+1} = \emptyset$, $\mathbb{J}_-^{\nu+1} = \emptyset$ because $d^{\nu+1} = 0$ and hence $\mathbb{I}^{\nu+1}(\gamma) = \emptyset$ for all $\gamma \geq 0$. So without any calculation one can choose $\gamma^{\nu+1} = \gamma'$ and determine $L^{\nu+1}$ according to (5.26) and $y^{\nu+1} = \gamma^{\nu+1} y'$. Define $C^* \equiv \{ y \mid A^\mathrm{T} y \in R_{L^{\nu+1}} \}$. Then $d^{\nu+1} = 0 \in \partial \left(\iota_{C^*}(A^\mathrm{T} y^{\nu+1}) - b^\mathrm{T} y^{\nu+1} + \iota_{\mathbb{R}_+^n}(L^{\nu+1}) \right)$ and $y^{\nu+1}$ for $L^{\nu+1}$ defined by (5.26) is a fixed point of the iteration. By the definition of the subdifferential (5.4), $y^{\nu+1}$ is an optimal solution to (5.14). The corresponding subgradient

$$w^{\nu+1} \equiv \text{argmin}\left\{\|Aw-b\|^2 \mid w \in N_{R_{L^{\nu+1}}}(y^{\nu+1})\right\}$$

satisfies $Aw^{\nu+1} = b$ and is supported on $\mathbb{J}^{\nu+1} \subset \mathbb{J}^*$. Lemma 5.9 shows that x^* is the unique sparsest solution to $Ax = b$. Thus, $\mathbb{J}^{\nu+1} = \mathbb{J}^*$ and $w^{\nu+1} = x^*$ as claimed.

Case 2. $d^{\nu+1} \neq 0$. In this case $b \notin N_{C(L^\nu,\mathbb{J}^{\nu+1},\gamma')}(y'')$, and it must be that $|\mathbb{J}^{\nu+1}| < |\mathbb{J}^*|$. By the induction hypothesis $\mathbb{J}^{\nu+1} \subset \mathbb{J}^*$. By the choice of γ' we have

$$P_{N_{C(L^\nu,\mathbb{J}^{\nu+1},\gamma')}(y'')} b \in ri(N_{C(L^\nu,\mathbb{J}^{\nu+1},\gamma')}(y''))$$

and thus by Lemma 5.11 $\left(\mathbb{J}_+^{\nu+1} \cup \mathbb{J}_-^{\nu+1}\right) \cap \mathbb{J}^{\nu+1} = \emptyset$ and the active set is monotonically increasing, so we must show that $\mathbb{J}^{\nu+2} \subset \mathbb{J}^*$.

We continue to the rescaling step to find $\gamma^{\nu+1}$ satisfying (5.25). Since by construction $d^{\nu+1}$ is orthogonal to the columns a_j with $j \in \mathbb{J}^{\nu+1}$, we can deflate the matrix A to contain only those columns with indices not in $\mathbb{J}^{\nu+1}$. The weights corresponding to the remaining indices, denoted $\bar{L}^{\nu+1}$, are unchanged from the initialization, that is, $L_j^\nu = \|a_j\|$ for $j \notin \mathbb{J}^{\nu+1}$ and so the elements of $\bar{L}^{\nu+1}$ are just the norms of the remaining columns of the deflated matrix $A^{\nu+1}$. Repeating the argument for the first iteration with b replaced by $d^{\nu+1}$, condition (5.28) with $\gamma^{\nu+1}$ satisfying (5.25) guarantees that $|y^{\nu+1} + \lambda_{\nu+1}d^{\nu+1}| = \|a_j\| = \bar{L}_j^{\nu+1}$ for some j corresponding to an element of $\mathbb{J}^* \setminus \mathbb{J}^{\nu+1}$, while $|y^{\nu+1} + \lambda_{\nu+1}d^{\nu+1}| < \|a_j\| = L_j^\nu$ for j corresponding to the complement of $\mathbb{J}^*$. (Note that because of the deflation technique, the correspondence between these indices is not direct.) Defining $y' = y^{\nu+1} + \lambda_{\nu+1}d^{\nu+1}$ $\mathbb{J}^{\nu+2} \equiv \{j \mid |a_j^T y'| = L^{\nu+1}\}$, by orthogonality and rescaling of the previous weights we have that $\mathbb{J}^{\nu+2} \subset \mathbb{J}^*$ and $|\mathbb{J}^{\nu+1}| < |\mathbb{J}^{\nu+2}| \leq |\mathbb{J}|$, as claimed.

Since the cardinality of the active set increases strictly monotonically with each iteration, the algorithm is finitely terminating as asserted. □

The next corollary is an immediate consequence of Theorem 5.12. We will show in the next section that the corollary is actually a statement of finite termination of the orthogonal matching pursuit algorithm [4, Theorem 6].

Corollary 5.13 (Greedy rescaling). *Let $A \in \mathbb{R}^{m\times n}$ $(m < n)$ be full rank and denote the jth column of A by a_j. Initialize Algorithm 5.4 with initial guess $y_j^0 = 0$ and $L_j^0 = \|a_j\|$ for $j = 1,2,\ldots,n$, and at the rescaling step choose $L_j^{\nu+1} = \gamma^{\nu+1} = 0$ for all $j \in \mathbb{J}^{\nu+1}$.*

If a point x^* solves $Ax = b$ and satisfies (5.28), then with tolerance $\tau = 0$, Algorithm 5.4 converges in finitely many steps to $y^* = 0$ with the weight L^* where,

$$\text{argmin}\{\|Aw-b\|^2 \mid w \in N_{R_{L^*}}(0)\} = x^*,$$

the unique sparsest solution to $Ax = b$.

Remark 5.14. We have called the rescaling strategy of Corollary 5.13 *greedy* to conform with precedent, however, in light of the variational derivation that we have developed here, we would prefer to use the descriptor *dogmatic*. To see why we prefer this, note that when the scaling of the active indices is set to zero, these elements are forever "committed" to the active set, even if in later iterations it might be determined that this was an error for some elements. In our algorithm, the detection of a possible error would occur in the determination of the preliminary scaling stage. If $P_{N_{C(L^\nu,\mathbb{J}^{\nu+1},\gamma')}(y'')}b \in ri(N_{C(L^\nu,\mathbb{J}^{\nu+1},\gamma')}(y''))$ only for $\gamma' = 0$ this is an indication that the direction of descent will cause a sign change in one of the active elements.

If the scaling is bounded away from 0, then the orthogonality of the descent directions with the active columns of A, see Proposition 5.5, is no longer guaranteed and the strict monotonicity of the cardinality of the active set Lemma 5.11 is also lost. This reflects the fact that, in this case, the algorithm can "change its mind" about the active set, that is, it has *recourse*. The more general Algorithm 5.4 is, in fact, no less dogmatic than the greedy variant since we enforce orthogonality of the descent direction with the active columns of A. It can be modified to include recourse by simply not enforcing orthogonality of the descent direction with the active constraints. The analysis of this implementation, however, is beyond the scope of this work.

5.5 Greedy Algorithms

As promised above, we now show that the greedy rescaling of Algorithm 5.4 specified in Corollary 5.13, is equivalent to a well-known *greedy algorithm* (see [4] and references therein). The prototype greedy algorithm is formulated in [4] as follows:

Algorithm 5.15 (Orthogonal Matching Pursuit). Input the matrix A, the vector b and a solution tolerance $\tau > 0$.

Initialization: Let $\nu = 0$, $y^0 = 0$, $r^0 = b$, and the support set $\mathbb{J}^0 = \emptyset$.
Main iteration: For a given tolerance $\tau > 0$ do

- (Sweep.) For $j = 1, 2, \ldots, n$ compute the errors $\iota(j) = \min_{z_j} \|a_j z_j - r^{\nu-1}\|^2$, where a_j denotes the jth column of A.
- (Update support.) Compute $J^\nu \equiv \operatorname{argmin}\{\iota(j) \mid j \notin \mathbb{J}^{\nu-1}\}$ and update $\mathbb{J}^\nu \equiv \mathbb{J}^{\nu-1} \cup \{J^\nu\}$.
- (Compute provisional solution and residual.) Compute

$$x^\nu \equiv \operatorname{argmin}\{\|Ax-b\|^2 \mid \text{support}(x) = \mathbb{J}^\nu\} \quad \text{and} \quad r^\nu \equiv b - Ax^\nu. \tag{5.41}$$

- (Increment or stop.) If $\|r^\nu\| < \tau$, stop; otherwise set $\nu = \nu + 1$ and repeat.

Note that the calculation of the provisional solution (5.41) is almost the same as the calculation of the normal cone projection in Lemma 5.3, the only difference being that x^ν in (5.41) is the projection onto the *subspace* corresponding to the index set $\mathbb{J}^\nu$ while the subgradient w^ν in Lemma 5.3 is the projection onto the associated *normal cone* mapping.

Lemma 5.16 (Provisional solution/subgradient equivalence). *Let $\bar{\mathbb{J}} \subset \mathbb{J}$, where $\mathbb{J} \equiv \{j \mid x_j^* \neq 0\}$ for x^* a solution to (5.1) with the counting objective $\varphi(x) = \|x\|_0$. Let $L = (L_1, L_2, \ldots, L_n)$ and choose any $\bar{y} \in \mathbb{R}^m$ such that $|a_j^{\mathrm{T}}\bar{y}| \leq L_j$ with equality holding only for $j \in \bar{\mathbb{J}}$, and such that $\bar{w}_j \neq 0$ for any $j \in \bar{\mathbb{J}}$ where $\bar{w} = \mathrm{argmin}\{\|Aw - b\|_2^2 \mid w \in N_{R_L}(A^{\mathrm{T}}\bar{y})\}$. Then $\bar{w} = \bar{x} \equiv \mathrm{argmin}\{\|Ax - b\|^2 \mid x_j = 0 \quad \forall j \notin \bar{\mathbb{J}}\}$.*

Proof. If $\bar{w}_j \neq 0$ for all $j \in \bar{\mathbb{J}}$ then the minimizer of $\|Aw - b\|^2$ is in the relative interior to $N_{R_L}(A^{\mathrm{T}}\bar{y})$, an orthant of the subspace containing the support of $\bar{x}$. Hence, minimizers of $\|Aw - b\|^2$ over the orthant and the entire subspace are equivalent, that is $\bar{w} = \bar{x}$. □

Less obvious is the fact that the active index selection in Algorithm 5.15 is equivalent to an exact line search with a dynamically reweighted ℓ_1 norm.

Lemma 5.17 (Step length/active index selection). *Define $\bar{\mathbb{J}} \subset \{1, 2, \ldots, n\}$ and $\bar{L} = (\bar{L}_1, \bar{L}_2, \ldots, \bar{L}_n)$ with*

$$\bar{L}_j \equiv \begin{cases} \|a_j\| & \text{for } j \notin \bar{\mathbb{J}} \\ 0 & \text{for } j \in \bar{\mathbb{J}}. \end{cases}$$

and the sets $R_{\bar{L}}$ and $\bar{C} \equiv \{y \in \mathbb{R}^m \mid A^{\mathrm{T}}y \in R_{\bar{L}}\}$ accordingly. Let $\bar{d} = b - P_{N_{\bar{C}}(0)}b$. Then

$$\bar{J} \equiv \underset{j \notin \bar{\mathbb{J}}}{\mathrm{argmin}} \left\{ \min_{z_j} \|a_j z_j - \bar{d}\|^2 \right\} \tag{5.42}$$

is the index set corresponding to the step length $\bar{\lambda}$ given by (5.35), that is,

$$\min_{k \notin \bar{\mathbb{J}}} \left\{ \frac{\|a_k\|}{|a_k^{\mathrm{T}}\bar{d}|} \right\} = \frac{\|a_j\|}{|a_j^{\mathrm{T}}\bar{d}|} \quad \forall j \in \bar{J}.$$

Proof. We work forward from the definition of $\bar{J}$. Substituting

$$\frac{a_j^{\mathrm{T}}\bar{d}}{\|a_j\|^2} = \min_{z_j} \|a_j z_j - \bar{d}\|^2$$

into (5.42) yields

$$\begin{aligned} \bar{J} &= \underset{j \notin \bar{\mathbb{J}}}{\mathrm{argmin}} \left\{ \left\| \frac{a_j^{\mathrm{T}}\bar{d}}{\|a_j\|^2} a_j - \bar{d} \right\|^2 \right\} \\ &= \underset{j \notin \bar{\mathbb{J}}}{\mathrm{argmin}} \left\{ \frac{|a_j^{\mathrm{T}}\bar{d}|^2}{\|a_j\|^2} \left(\frac{\|a_j\|^2 \|\bar{d}\|^2}{|a_j^{\mathrm{T}}\bar{d}|^2} - 1 \right) \right\} \end{aligned}$$

$$= \underset{j \notin \mathbb{J}}{\operatorname{argmin}} \left\{ \|\bar{d}\|^2 - \frac{|a_j^{\mathrm{T}} \bar{d}|^2}{\|a_j\|^2} \right\}$$

$$= \underset{j \notin \mathbb{J}}{\operatorname{argmax}} \left\{ \frac{|a_j^{\mathrm{T}} \bar{d}|^2}{\|a_j\|^2} \right\}$$

$$= \underset{j \notin \mathbb{J}}{\operatorname{argmin}} \left\{ \frac{\|a_j\|}{|a_j^{\mathrm{T}} \bar{d}|} \right\}.$$

This completes the proof. □

We conclude that orthogonal matching pursuit is equivalent to the dynamically reweighted steepest subgradient descent method with exact line search.

Proposition 5.18. *Algorithm 5.15 is equivalent to Algorithm 5.4 initialized with* $y^0 = 0$ *and* $L^0 = (\|a_1\|, \|a_2\|, \ldots, \|a_n\|)$, *and with the rescaling* $\gamma^\nu = 0$ *for all* ν.

Proof. This follows immediately from Lemmas 5.16 and 5.17. □

5.6 Numerical Examples

The equivalence of Algorithm 5.4 with $\gamma^\nu = 0$ for all ν to the orthogonal matching pursuit Algorithm 5.15 makes the wealth of numerical experience with orthogonal matching pursuit immediately available to our more general algorithm. We only demonstrate in this section that the greedy version of the algorithm and the more general version behave similarly on sufficiently sparse problems.

Remark 5.19. Before presenting our numerical examples, a few comments about practical implementations are in order. As pointed out earlier, in the absence of exact arithmetic, practical implementations cannot directly apply the most general form of Algorithm 5.4. However, even without exact arithmetic, we can determine precisely the operative quantities as long as the numerical error is below the threshold needed to discriminate between certain discrete cases.

For example, suppose we have 14 digits of accuracy and $|a_j^{\mathrm{T}} y^\nu|$ is to within 10^{-15} of L_j^ν: would it be equal to L_j^ν if we had exact arithmetic? If $L_j^\nu = 0$, then it must be that $a_j^{\mathrm{T}} y^\nu = 0$ with exact arithmetic since it was proved in Propositions 5.5 and 5.7 that the iterates are generated from feasible directions with step length chosen so that the iterates are always feasible. If the dynamic reweighting were chosen so that $L_j^\nu > 0$, then it is impossible to determine whether $a_j^{\mathrm{T}} y^\nu$ *should* equal, say, $-L_j^\nu$, unless it is known that $a_j^{\mathrm{T}} d^\nu = 0$, in which case it *should* hold that $a_j^{\mathrm{T}} y^{\nu-1} = a_j^{\mathrm{T}} y^\nu$, where it has been determined from previous iterations that $a_j^{\mathrm{T}} y^{\nu-1} = -L_j^{\nu-1}$. Again, by Proposition 5.5, if $w_j^\nu \neq 0$ for j in the active set $\mathbb{J}^\nu$ and

$$w^\nu = \operatorname{argmin} \{ \|Aw - b\|^2 \mid w \in N_{R_{L^\nu}}(A^{\mathrm{T}} y^\nu) \}$$

then $a_j^T d^\nu = 0$. Let δ be the numerical accuracy of the computation. If $|w_j^\nu| > \delta$, then we are certain that $w_j^\nu \neq 0$, and thus $a_j^T d^\nu = 0$ so that $a_j^T y^\nu = a_j^T y^{\nu-1} = L_j^{\nu-1}$. If instead $|w_j^\nu| \leq \delta$, then we cannot be sure that $|w_j^\nu| \neq 0$ and consequently we cannot be certain that d^ν is orthogonal to the active columns of A.

This numerical uncertainty is related to the *ill-posedness* of the problem $Ax = b$: if the sparsest signal x^* has elements whose magnitude is below the numerical noise level, then the algorithm must be regularized. We will have more to say about this in the conclusion. For our numerical study we only take examples for which the signal is above the numerical noise level, and so our exact arithmetic algorithm is still implementable.

We turn to our numerical illustration:

5.6.1 Our "Toy" Problem

For our numerical example, we construct a real signal of length 128^2 ($n = 2 \times 16,384$ to account for real and imaginary parts) with 70 nonzero components ($|\mathbb{J}^*| = 70$), chosen at random, and randomly sample the discrete Fourier transform of this signal at a rate of about $1/8$. Since the true signal is real-valued, our effective sampling rate is about $1/4$ due to symmetry in the Fourier coefficients ($m = 2 \times 3,588$ for the real and imaginary parts). Since we are dealing with the Fourier transform, the scaling of columns of

$$A \in \mathbb{R}^{(2*16,384)\times(2*3,588)}$$

is just $\|a_j\| = 1/\sqrt{2*3,588}$.

5.6.2 Algorithm Illustrations

We illustrate the theory with two different implementations of Algorithm 5.4, the first with scaling parameter $\gamma^\nu > 0$ for each iteration ν (in fact, we need only take $\gamma^\nu = 1$ to satisfy the requirements of the algorithm) and the second with $\gamma^\nu = 0$ for all iterations corresponding to the "greedy" implementation. The complexity of the two implementations is identical. Both instances converge in 70 iterations and require the same work to compute the subgradient.

Although the normal equations provide an explicit closed-form expression for the calculation of the subgradient w in (5.18), this still involves the inversion of a matrix, albeit small relative to the overall problem size. As we are interested in applications for which the sparsity is on the order of 10^3–10^4 nonzero elements, instead, we solve (5.18) iteratively using the Relaxed Average Alternating Reflection (RAAR) algorithm [10, 11] for finding *best approximation pairs* between the sets $N_{R_L}(A^T y'')$ and $B \equiv \{x \mid Ax = b\}$. (It is important to note that we can only find best approximation pairs since for all but the last iteration $N_{R_L}(A^T y'') \cap B = \emptyset$.) Ordinary alternating projections would have also sufficed to solve this subproblem, however, we found that the RAAR algorithm required, on average, 33% fewer iterations with the proper choice of relaxation parameter.

Both of our implementations of Algorithm 5.4 require exactly the same number of iterations of the RAAR algorithm to compute (5.18) since they both solve the exact same subproblem at each iteration. The subproblems require, on average, 82.6 iterations to get to within the numerical tolerance (10^{-12}).

5.6.3 *Complexity*

Rather than explicitly forming the partial Fourier matrix A we take advantage of the fast Fourier transform. The FFT is the most complex computation in the algorithm. The RAAR algorithm requires 2 FFT computations per iteration on a complex-valued vector of length 128^2 and the main loop of Algorithm 5.4 requires 3 FFT computations of the same complexity. For the example reported here, over all the iterations, the algorithm required in total 821,871 FFT computations on complex-valued vectors of length 128^2, or on the order of 10^{11} floating point operations. On a 2.2 GHz Intel Core 2 Duo processor with 2GB 667 MHz memory this takes 32 s of CPU time.

If instead of using the FFT, we had used the normal equations to explicitly compute the subgradients we would have needed only 211 FFT computations, and the matrix inversions required in the normal equations would have required, at the worst, inversion of a 70×70 real-valued matrix. The computational complexity of this approach is estimated to be on the order of 10^7 floating point operations. For problems with sparsity <700, elements the normal equation approach will probably be faster; thereafter iterative methods, such as RAAR, using the FFT become competitive.

Figure 5.1a shows the error between the reconstructed signal and the true signal. The reconstruction for both implementations are identical. Figure 5.1b shows the weights corresponding to the implementation with scaling $\gamma^\nu = 0$ for all ν. The weights for the implementation with $\gamma^\nu = 1$ for all ν are not shown since these are all identical and unchanged from the initialization. Note that $\gamma^\nu = 1$ for all ν is then the behavior of the algorithm for solving the fixed, reweighted ℓ_1 optimization problem *for this problem*. These will not, in general, be the scalings chosen by the algorithm on different problems. Finally, in Fig. 5.1c we give a comparison of the step lengths at each iteration of the two implementations.

5.7 Comments and Conclusion

Our goals herein were to apply convex analysis to the nonconvex problem of sparse signal recovery and to take notions that have evolved from different approaches and incorporate these advances into convex analysis and algorithms. With this work, we have made a first step in this direction.

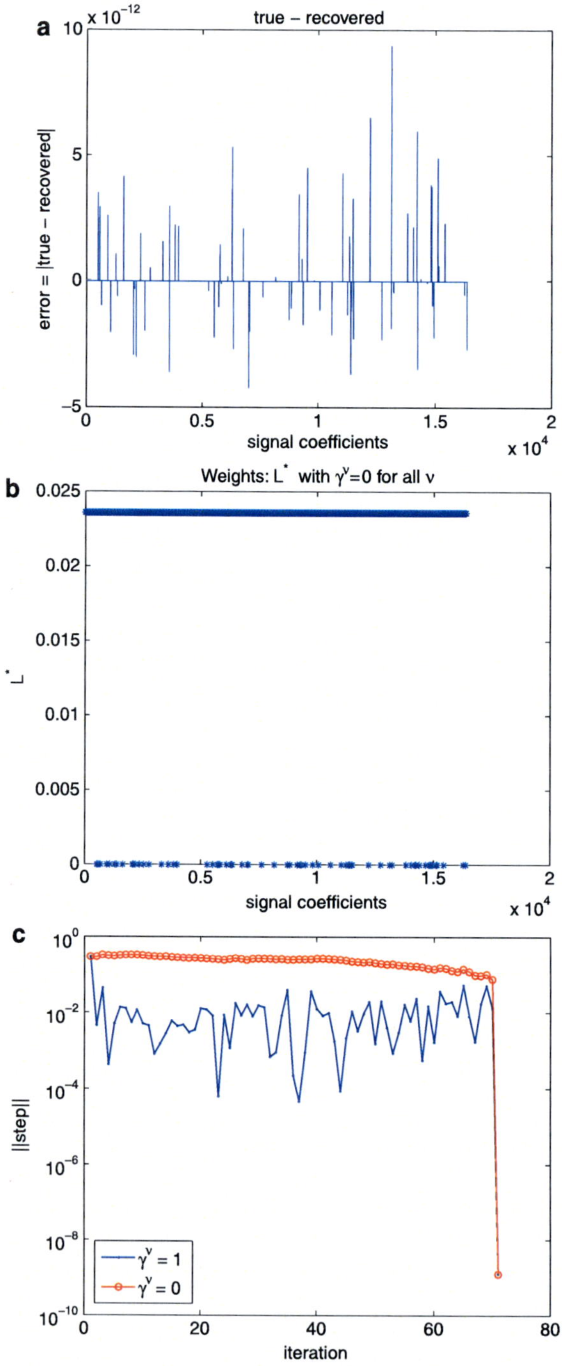

Fig. 5.1 (**a**) Pointwise reconstruction error. (**b**) Weights at the optimal solution for the implementation with $\gamma^\nu = 0$ for all ν. (**c**) Comparison of magnitude of steps between $\gamma^\nu = 0$ and $\gamma^\nu = 1$ implementations at each iteration

We proposed convex dual-space relaxations of the original nonconvex problem and have analyzed one extreme of possible relaxations. We have proved convergence in finitely many steps of a nonsmooth steepest descent method with exact line search and dynamically reweighted ℓ_1 norms when applied to problems satisfying the mutual coherence condition.

An instance of our algorithm is shown to be equivalent to orthogonal matching pursuit, which has been well-studied in the literature, though we are unaware of any identification of this method to dual-space linesearch methods as presented here. This explicit connection of orthogonal matching pursuit to reweighted ℓ_1 minimization in the dual opens the door to a greater synthesis of algorithms and a better understanding of the behavior of these algorithms.

Indeed, the proof of the coincidence of the solution to the ℓ_1 minimization problem to the solution of the corresponding minimization of the counting metric $\|\cdot\|_0$ is usually given indirectly. Here, under the assumption of mutual coherence and certain interiority qualifications on the projection of the data onto the normal cone associated with the active constraints, we have an explicit proof of the equivalence of the solutions to the ℓ_1 and $\|\cdot\|_0$ problems. An instance of this equivalence was demonstrated in the numerical example.

Our numerical examples do not extend to circumstances not covered by the theory developed here. There are two sources of failure of the algorithm, one due to the sparsity conditions not being met, and the other due to numerical error. We emphasized the importance of recognizing algorithms that implicitly rely on exact arithmetic and how implementations can succeed or fail without it. We are unaware of a numerical study that distinguishes between instances where the sparsity conditions are not met and instances where the numerical tolerance is not precise enough for a practical implementation. This is a topic worthy of greater attention than we have space for here.

The next step in this research will be to investigate the other relaxations, $\varepsilon > 0$ of (5.8). For this instance the objective is smooth (infinitely differentiable) in its domain R_L, and the gradient can be written in closed-form. We conjecture that the corresponding steepest descent, exact linesearch algorithm with dynamic reweighting will behave much like an interior point algorithm since the effect of the parameter ε is to keep the iterates on the interior of the feasible region.

Another direction that needs to be addressed is sparse *approximate* solutions to the model $Ax = b$. This is more appropriate for applications where the image b is corrupted by noise, or, as we have seen, numerical error. There has been a lot of very good work in this direction by other researchers. Our approach is appropriate for fast (finitely terminating), highly accurate exact solutions. It remains to be seen whether this basic program extends to fast (polynomial time), reasonably accurate approximate solutions.

Acknowledgements Jonathan M. Borwein, Research was supported by the Australian Research Council. D. Russell Luke, Research was supported by the US National Science Foundation grant DMS-0712796.

References

1. Borwein, J., Vanderwerff, J.: Convex Functions: Constructions, Characterizations and Counterexamples, *Encyclopedias in Mathematics*, vol. 109. Cambridge University Press, New York (2010)
2. Borwein, J.M., Lewis, A.S.: Convex analysis and nonlinear optimization: theory and examples, 2nd edn. Springer, New York (2006)
3. Borwein, J.M., Luke, D.R.: Duality and convex programming. In: O. Scherzer (ed.) Handbook of Mathematical Methods in Imaging, pp. 229–270. Springer-Verlag (2011)
4. Bruckstein, A.M., Donoho, D.L., Elad, M.: From sparse solutions of systems of equations to sparse modeling of signals and images. SIAM Rev. **51**, 34–81 (2009)
5. Candès, E., Recht, B.: Exact matrix completion via comvex optimization. Found. of Comput. Math. **9**, 717–772 (2009)
6. Candès, E., Tao, T.: Near-optimal signal recovery from random projections: Universal encoding strategies? IEEE Trans. Inform. Theory **52**, 5406–5425 (2006)
7. Chen, S.S., Donoho, D.L., Saunders, M.A.: Atomic decomposition by basis pursuit. SIAM J. Sci. Comput. **20**, 33–61 (1998)
8. Donoho, D.L., Elad, M.: Optimally sparse representation in general (non-orthogonal) dictionaries via l1 minimization,. Proc. Natl. Acad. Sci. **100**, 2197–2202 (2003)
9. Gribonval, R., Nielsen, M.: Sparse decompositions in unions of bases. IEEETrans. Inform. Theory **49**, 3320–3325 (2003)
10. Luke, D.R.: Relaxed averaged alternating reflections for diffraction imaging. Inverse Problems **21**, 37–50 (2005)
11. Luke, D.R.: Finding best approximation pairs relative to a convex and a prox-regular set in Hilbert space. SIAM J. Optim. **19**, 714–739 (2008)
12. Natarajan, B.K.: Sparse approximate solutions to linear systems. SIAM J. Comput. **24**, 227–234 (1995)
13. Recht, B., Fazel, M., Parrilo, P.: Guaranteed minimum rank solutions of matrix equations via nuclear norm minimization. SIAM Rev. **52**, 471–501 (2010)
14. Rockafellar, R.T., Wets, R.J.: Variational Analysis. Grundlehren der mathematischen Wissenschaften. Springer, Berlin (1998)
15. Santosa, F., Symes, W.W.: Linear inversion of band-limited reflection seismograms. SIAM J. Sci. Statist Comput. **7**, 1307–1330 (1986)

Chapter 6
The Douglas–Rachford Algorithm in the Absence of Convexity

Jonathan M. Borwein and Brailey Sims

Abstract The Douglas–Rachford iteration scheme, introduced half a century ago in connection with nonlinear heat flow problems, aims to find a point common to two or more closed constraint sets. Convergence of the scheme is ensured when the sets are convex subsets of a Hilbert space, however, despite the absence of satisfactory theoretical justification, the scheme has been routinely used to successfully solve a diversity of practical problems in which one or more of the constraints involved is non-convex. As a first step toward addressing this deficiency, we provide convergence results for a prototypical non-convex two-set scenario in which one of the sets is the Euclidean sphere.

Keywords Non-convex feasibility problem · Fixed point theory · Dynamical system · Iteration

AMS 2010 Subject Classification: 46B45, 47H10, 90C26

6.1 Introduction

In recent times variations of alternating projection algorithms have been applied in Hilbert space to various important applied problems – from optical aberration correction to three satisfiability, protein folding and construction of giant *Sudoku* puzzles [8]. While the theory of such methods is well understood in the convex case [3] and [4–6, 11], there is little corresponding theory when some of the sets involved are non-convex – and that is the case for the examples mentioned above [8, 9].

Our intention is to analyse the simplest non-convex prototype in Euclidean space: that of finding a point on the intersection of a sphere and a line or more

J.M. Borwein (✉)
CARMA, School of Mathematical and Physical Sciences, University of Newcastle, NSW 2308, Australia
e-mail: jonathan.borwein@newcastle.edu.au

H.H. Bauschke et al. (eds.), *Fixed-Point Algorithms for Inverse Problems in Science and Engineering*, Springer Optimization and Its Applications 49,
DOI 10.1007/978-1-4419-9569-8_6,

generally a proper affine subset. The sphere provides an accessible model of many reconstruction problems in which the magnitude, but not the phase, of a signal is measured.

6.2 Preliminaries

For any closed subset A of a Hilbert space $(X, \langle \cdot, \cdot \rangle)$, we say that a mapping $P_A : D_A \subseteq X \longrightarrow A$ is a *closest point projection* of D_A onto A if $A \subseteq D_A$, $P_A^2 = P_A$ and

$$\|x - P_A(x)\| = \operatorname{dist}(x, A) \ := \inf\{\|x - a\| : a \in A\},$$

for all $x \in D_A$.

For a given closest point projection, P_A, onto A we take the *reflection* of x in A (relative to P_A) to be,

$$R_A := 2P_A - I.$$

In this note, we will focus on the cases when the subset A is a sphere, which without loss of generality we take to be the unit sphere of the space; $S := \{x : \|x\| = 1\}$, or a line $L := \{x = \lambda a + \alpha b : \lambda \in \mathbf{R}\}$, where, without loss of generality, we take $\|a\| = \|b\| = 1$, $a \perp b$ and $\alpha > 0$.

The closest point projection of $x \neq 0$ onto the unit sphere S is,

$$P_S(x) := \frac{x}{\|x\|}$$

and so,

$$R_S(x) = \left(\frac{2}{\|x\|} - 1 \right) x.$$

Excluding $x = 0$ from the domain of P_S, and hence also R_S, avoids the problem of non-unique closest points and hence the need to make a selection. The closest point projection of $x \in X$ onto L is the orthogonal projection,

$$P_L(x) := \langle x, a \rangle\, a \, + \, \alpha b$$

and so,

$$R_L(x) = 2\langle x, a \rangle\, a \, + \, 2\alpha b \, - \, x.$$

Given two closed sets A and B together with closest point projections P_A and P_B, starting from an arbitrary initial point $x_0 \in D_A$ the *Douglas–Rachford iteration scheme* (reflect-reflect-average), introduced in [7] for numerical solution of partial differential equations, is a method for finding a point in the intersection of the two sets. That is, it aims to find a feasible point for the possibly non-convex constraint $x \in A \cap B$. Explicitly it is the iterative scheme,

$$x_{n+1} := T_{A,B}(x_n),$$

where $T_{A,B}$ is the operator $T_{A,B} := \frac{1}{2}(R_B R_A + I)$. This method also goes under many other names, see [4].

When either of the sets is non-convex various compatibility restrictions between the domains and ranges of the mappings involved are required to ensure all iterates are defined. For instance, $R_A(D_A) \subseteq D_B$ and $\frac{1}{2}(R_B R_A + I)(D_A) \subseteq D_A$.

With our particular S and L we have for $x \neq 0$ that,

$$T_{S,L}(x) = \left(1 - \frac{1}{\|x\|}\right) x + \left(\frac{2}{\|x\|} - 1\right) \langle x, a\rangle\, a + \alpha b.$$

Thus, if X is N-dimensional and $(x(1), x(2), x(3), \ldots, x(N))$ denotes the coordinates of x relative to an orthonormal basis B whose first two elements are respectively a and b we have,

$$T_{S,L}(x) = \left(\frac{x(1)}{\rho}, \left(1 - \frac{1}{\rho}\right) x(2) + \alpha, \left(1 - \frac{1}{\rho}\right) x(3), \ldots, \left(1 - \frac{1}{\rho}\right) x(N)\right),$$

where $\rho := \|x\| = \sqrt{x(1)^2 + \cdots + x(N)^2}$.

Let us note that the only fixed points of $T_{S,L}$ are $\pm\sqrt{1-\alpha^2}a + \alpha b$, the two points of intersection of S with L.

In this case the *Douglas–Rachford* scheme becomes,

$$x_{n+1}(1) = x_n(1)/\rho_n, \tag{6.1}$$

$$x_{n+1}(2) = \alpha + (1 - 1/\rho_n) x_n(2), \quad \text{and} \tag{6.2}$$

$$x_{n+1}(k) = (1 - 1/\rho_n) x_n(k), \quad \text{for } k = 3, \ldots, N, \tag{6.3}$$

where $\rho_n := \|x_n\| = \sqrt{x_n(1)^2 + \cdots + x_n(N)^2}$.

From this it is clear that if the initial point x_0 lies in the hyperplane $\langle x, a\rangle = 0$; that is $x_0(1) = 0$, then all of the iterates remain in that hyperplane, which we will refer to as a *singular manifold* for the problem. We will analyse this case in greater detail in a subsequent section. Similarly, if the initial point lies in either of the two open half-spaces $\langle x, a\rangle > 0$ or $\langle x, a\rangle < 0$; that is, $x_0(1) > 0$ or $x_0(1) < 0$ respectively, then all subsequent iterates will remain in the same open half space. Further, by symmetry, it suffices to only consider initial points lying in the positive open half-space $x_0(1) > 0$.

Figure 6.1 shows two steps of the underlying geometric construction: the smaller (green) points are the intermediate reflections in the sphere. Most figures were constructed in *Cinderella*, a software geometry package [www.cinderella.de]. A web applet version of the underlying Cinderella construction is available at

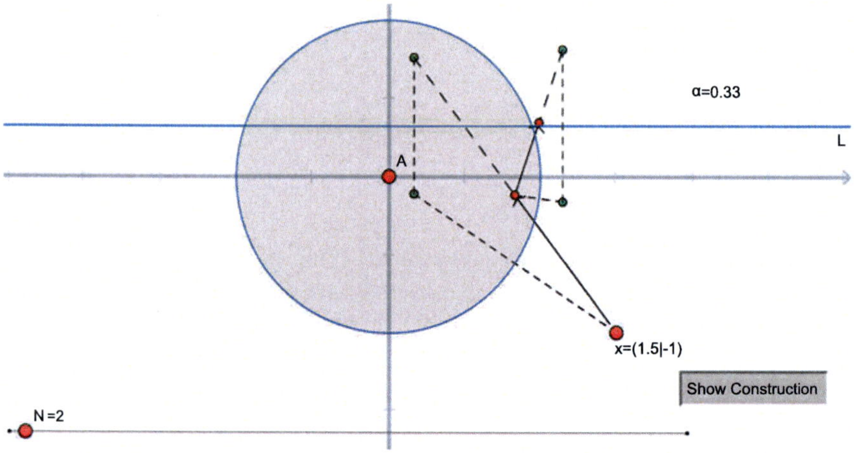

Fig. 6.1 Two steps showing the construction

http://www.carma.newcastle.edu.au/~jb616/reflection.html. Indeed, many of the insights for the proofs below came from examining the constructions. The number of iterations N, the height of the line (α), and the initial point are all dynamic – changing one changes the entire visible trajectory.

Success of the Douglas–Rachford scheme relies on convergence of the (*Picard*) iterates, $x_n = T^n_{A,B}(x_0)$, to a fixed point of the generally nonlinear operator $T_{A,B}$ in $A \cap B$, as $n \longrightarrow \infty$. When both A and B are closed convex sets convergence of the scheme (in the weak topology) from any initial point in X to some point in $A \cap B$ was established by Lions and Mercier [11].

However, as noted, many practical situations yield feasibility problems in which one or more of the constraint sets is non-convex. That the Douglas–Rachford scheme works well in many of these situations has been observed and exploited for some years, despite the absence of any really satisfactory theoretical underpinning.

Remark 6.1 (Divide-and-concur). If one wishes to find a point in the intersection of M sets $A_1, A_2, \ldots A_k, \ldots, A_M$ in X, we can instead consider the subset $A := \prod_{k=1}^{M} A_k$ and the linear subset

$$B := \{x = (x_1, x_2, \ldots, x_M) \colon x_1 = x_2 = \cdots = x_M\}$$

of the Hilbert space product $\prod_{k=1}^{M} X$. Then we observe that

$$R_A(x) = \prod_{k=1}^{M} R_{A_k}(x_k),$$

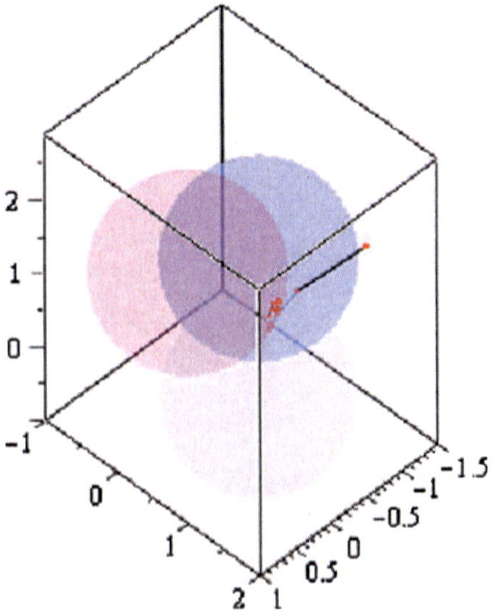

Fig. 6.2 Douglas–Rachford for three spheres in three-space

so that the reflections may be 'divided' up and

$$P_B(x) = \left(\frac{x_1 + x_2 + \cdots + x_M}{M}, \ldots, \frac{x_1 + x_2 + \cdots + x_M}{M} \right),$$

so that the projection and hence reflection on B are averaging ('concurrences'); thence comes the name. In this form the algorithm is particularly suited to parallelization [12].

We can also compose more reflections in serial as illustrated for reflect-reflect-reflect-average with spheres in Figure 6.2, where we observe iterates spiralling to a feasible point.

Example 6.2 (Linear equations). For the hyperplane $H := \{x\colon \langle b, x\rangle = \alpha\}$, where without loss of generality we take $\|b\| = 1$, the projection is

$$x \mapsto x + (\alpha - \langle b, x\rangle)\, b.$$

The consequent averaged-reflection version of the Douglas–Rachford recursion for a point in the intersection of M distinct hyperplanes is:

$$x \mapsto x + \frac{2}{M} \sum_{k=1}^{M} (\alpha_k - \langle b_k, x\rangle)\, b_k, \tag{6.4}$$

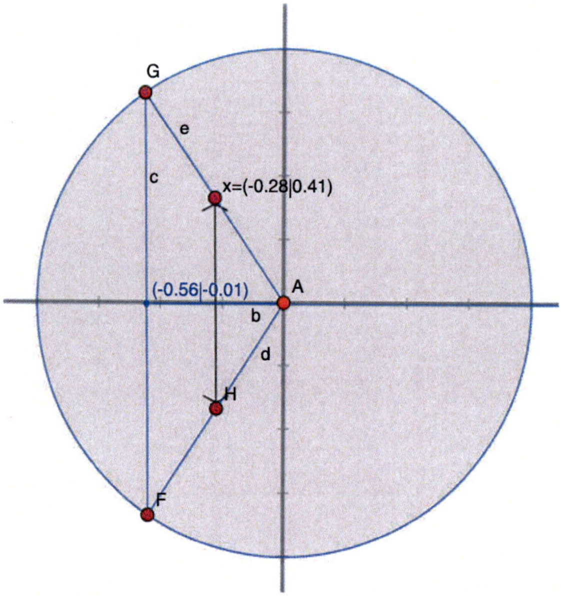

Fig. 6.3 Iterated reflection with a ray

while the corresponding-averaged projection algorithm is:

$$x \mapsto x + \frac{1}{M}\sum_{k=1}^{M}(\alpha_k - \langle b_k, x\rangle)\, b_k \tag{6.5}$$

In more general situations, the difference between projection and reflection algorithms is even greater.

Remark 6.3 (The case of a half-line or segment). Note, even in two dimensions, alternating projections, alternating reflections, project-project and average, and reflect-reflect and average will all often converge to (locally nearest) infeasible points even when A is simply the ray $R := \{(x,0)\colon x \geqslant -1/2\}$ and B is the circle as before. They can also behave quite 'chaotically'. (See Fig. 6.3 for a periodic illustration in *Cinderella* and Fig. 6.4 for more complex behaviour.) So the affine nature of the convex set seems quite important.

For any two closed sets A and B and feasible point $p \in A \cap B$ we say that the Douglas–Rachford scheme is *locally convergent* at p if there is a neighbourhood, N_p of p such that starting from any point x_0 in N_p the iterates $T^n_{A,B}(x_0)$ converge to p. The set comprising all initial points x_0 for which the iterates converge to p is the *basin of attraction* of p.

As a first step toward an understanding of the Douglas–Rachford scheme in the absence of convexity, we analyse its behaviour in the indicative situation when one of the sets is the non-convex sphere S and the other is the affine line L. We begin by establishing local convergence of the scheme when $0 \leq \alpha < 1$.

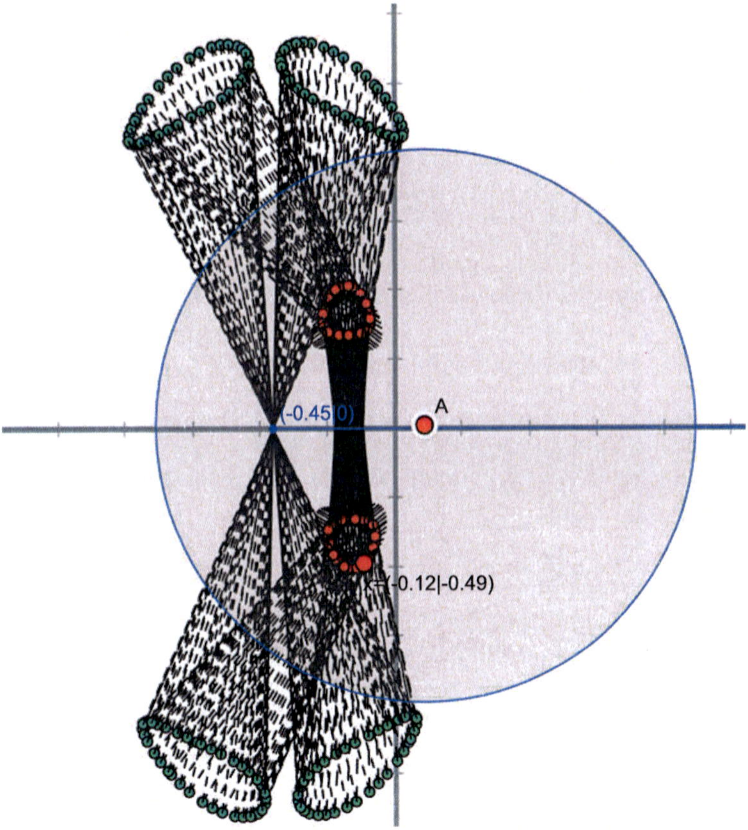

Fig. 6.4 More complex behaviour for a ray and circle

6.3 Local Convergence When $0 \le \alpha < 1$

In this section we show, at least when X is finite dimensional, that for $0 \le \alpha < 1$ local convergence at each of the feasible points is a consequence of the following theorem from the stability theory of difference equations.

Theorem 6.4 (Perron [10], Corollary 4.7.2, page 104). *If $f : \boldsymbol{N} \times \boldsymbol{R}^m \longrightarrow \boldsymbol{R}^m$ satisfies,*

$$\lim_{x \to 0} \frac{\|f(n,x)\|}{\|x\|} = 0,$$

uniformly in n and M is a constant $m \times m$ matrix all of whose eigenvalues lie inside the unit disk, then the zero solution (provided it is an isolated solution; that is, there is a neighbourhood of 0 containing no other solution) of the difference equation,

$$x_{n+1} = Mx_n + f(n, x_n),$$

is exponentially asymptotically stable*; that is, there exists* $\delta > 0$, $K > 0$ *and* $\zeta \in (0,1)$ *such that if* $\|x_0\| < \delta$ *then* $\|x_n\| \le K\|x_0\|\zeta^n$.

To apply this in our context, we begin by noting that the operator $T := T_{S,L}$ is differentiable at any non-zero point y with derivative the linear operator,

$$T_y'(x) = \left\langle \left(\frac{2}{\|y\|} - 1\right)x - 2\frac{\langle x,y\rangle}{\|y\|^3}y,\ a\right\rangle a + \left(1 - \frac{1}{\|y\|}\right)x + \frac{\langle x,y\rangle}{\|y\|^3}y.$$

By symmetry it suffices to consider local convergence at the unique fixed point of $T_{S,L}$ lying in the positive open half-space $\langle x, a\rangle > 0$; namely, $p := \sqrt{1-\alpha^2}a + \alpha b$. Observing that, p is an isolated fixed point of $T_{S,L}$ (see the discussion before (6.1)) and, using $\|p\| = 1$ and $\langle p, a\rangle = \sqrt{1-\alpha^2}$, we obtain,

$$T_p'(x) = \left\langle x,\ \alpha^2 a - \alpha\sqrt{1-\alpha^2}b\right\rangle a \ +\ \left\langle x, \alpha\sqrt{1-\alpha^2}a + \alpha^2 b\right\rangle b,$$

which, relative to the basis B, corresponds to the $n \times n$ matrix,

$$\begin{pmatrix} \alpha^2 & -\alpha\sqrt{1-\alpha^2} & 0 & \cdots & 0 \\ \alpha\sqrt{1-\alpha^2} & \alpha^2 & 0 & \cdots & 0 \\ 0 & 0 & 0 & \cdots & 0 \\ \cdot & \cdot & \cdot & & \cdot \\ \cdot & \cdot & \cdot & & \cdot \\ \cdot & \cdot & \cdot & & \cdot \\ 0 & 0 & 0 & \cdots & 0 \end{pmatrix}.$$

From this, we immediately deduce that the only points in the spectrum of T_p' are the eigenvalues 0, and $\alpha^2 \pm i\alpha\sqrt{1-\alpha^2}$.

Introducing the change of variable $\xi := x - p$ and defining f by,

$$f(\xi) := T_{S,L}(p+\xi) \ - \ T_{S,L}(p) \ - \ T_p'(\xi),$$

we see that the Douglas–Rachford scheme becomes,

$$\xi_{n+1} = T_{S,L}(p+\xi_n) - p = T_{S,L}(p+\xi_n) - T_{S,L}(p) = T_p'(\xi_n) + f(\xi_n).$$

Further, by the very definition of the derivative we have,

$$\lim_{\xi\to 0}\frac{\|f(\xi)\|}{\|\xi\|} = \lim_{\xi\to 0}\frac{\|T_{S,L}(p+\xi) - T_{S,L}(p) \ - \ T_p'(\xi)\|}{\|\xi\|} = 0.$$

Thus, all the conditions of Perron's theorem are satisfied, provided T_p' has its spectrum contained in the open unit disk. But, this follows immediately since both non-zero eigenvalues have modulus equal to $\alpha < 1$, establishing that locally the Douglas–Rachford scheme converges exponentially to $\xi = 0$; that is, to $x = p$. Thus, we have proved,

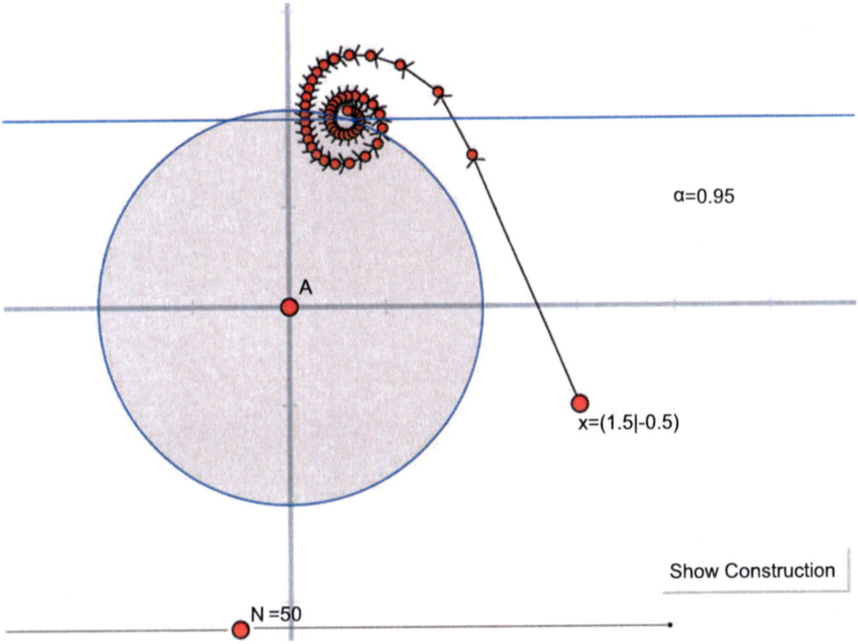

Fig. 6.5 Case with $\alpha = 0.95$

Theorem 6.5. *If* $0 \leq \alpha < 1$ *then the Douglas–Rachford scheme is locally convergent at each of the points* $\pm\sqrt{1-\alpha^2}a + \alpha b$.

Remark 6.6 (Explaining the spiral). It is also worthy of note that the non-zero eigenvalues both have arguments whose cosines have absolute value α, so 'spiraling', as illustrated in Fig. 6.5, should be less rapid the larger the value of α, an observation born out by experiment. It should also be noted that when $\alpha = 1$; that is, the line L is tangential to the sphere S, Perron's theorem fails to apply, as in this case T'_p has eigenvalues lying on the unit circle. Indeed, the conclusion of Theorem 6.5 is false as we will show in the following sections.

6.4 Convergence When $\alpha = 0$

We show that starting from any initial point with $x_0(1) > 0$ the Douglas–Rachford scheme converges to the feasible point $a = (1, 0, 0, \cdots, 0)$, as illustrated in Fig. 6.6. In this case the scheme (6.1)–(6.3) reduces to,

$$\begin{aligned} x_{n+1}(1) &= x_n(1)/\rho_n, \quad \text{and} \\ x_{n+1}(k) &= (1 - 1/\rho_n)x_n(k), \quad \text{for } k = 2, \ldots, N, \end{aligned}$$

with $\rho_n = \|x_n\| = \sqrt{x_n(1)^2 + \cdots + x_n(N)^2} \geq x_n(1) > 0$.

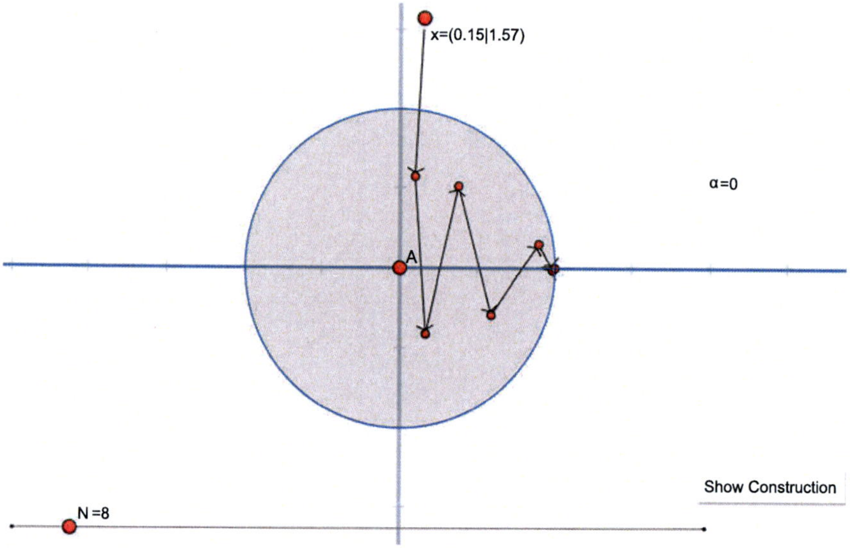

Fig. 6.6 Case with $\alpha = 0$

Proposition 6.7. *If* $\rho_n > 1$ *then* $\rho_{n+1}^2 < \rho_n^2$.

Proof. We may estimate as follows.

$$\begin{aligned}
\rho_{n+1}^2 &= \frac{x_n(1)^2}{\rho_n^2} + \left(1 - \frac{1}{\rho_n}\right)^2 \sum_{k=2}^{N} x_n(k)^2 \\
&= \frac{x_n(1)^2 + x_n(2)^2 + \cdots + x_n(N)^2}{\rho_n^2} + \left(1 - \frac{2}{\rho_n}\right) \sum_{k=2}^{N} x_n(k)^2 \\
&= 1 + \left(1 - \frac{2}{\rho_n}\right) \sum_{k=2}^{N} x_n(k)^2 \\
&\le 1 + \left(1 - \frac{2}{\rho_n} + \frac{1}{\rho_n^2}\right) \sum_{k=2}^{N} x_n(k)^2 \\
&= 1 + \left(1 - \frac{1}{\rho_n}\right)^2 \sum_{k=2}^{N} x_n(k)^2 \\
&\le 1 + \left(1 - \frac{1}{\rho_n}\right)^2 \rho_n^2 \\
&= 1 + (\rho_n - 1)^2 \\
&= \rho_n^2 + 2(1 - \rho_n) \\
&< \rho_n^2, \quad \text{as } \rho_n > 1.
\end{aligned}$$

■

Corollary 6.8. *If $\rho_n > 1$ for all n then $\rho_n \longrightarrow 1$.*

Proof. By the above proposition, the ρ_n are decreasing and so converge to some limit $\rho \geq 1$. But then, taking limits in $\rho_{n+1}^2 \leq \rho_n^2 + 2(1-\rho_n)$ leads to $\rho \leq 1$, so $\rho = 1$. ∎

Proposition 6.9. *If $\rho_n \leq 1$ then so too is $\rho_{n+1} \leq 1$.*

Proof. From the first three lines in the proof of the above proposition, we have

$$\begin{aligned}\rho_{n+1}^2 &= 1 + \left(1 - \frac{2}{\rho_n}\right)\sum_{k=2}^{N} x_n(k)^2 \\ &\leq 1 - \sum_{k=2}^{N} x_n(k)^2, \quad \text{provided } \rho_n \leq 1 \\ &\leq 1.\end{aligned}$$

∎

Theorem 6.10. *If $\alpha = 0$ and the initial point has $x_0(1) > 0$ then the Douglas–Rachford scheme converges to the feasible point $(1,0,0,\ldots,0)$.*

Proof. In case $\rho_n > 1$ for all n then, by the above corollary, $\rho_n \to 1$, so by the recurrence $x_n(k) \to 0$ for $k = 2, \cdots, N$ and $x_n \to (1,0,0,\ldots,0)$.

On the other hand, if this is not the case then there is a smallest n_0 with $\rho_{n_0} \leq 1$ and then either $\rho_{n'} = 1$ for some $n' \geq n_0$, in which case we have $x_{n'+1}(k) = 0$ for $k = 2,\ldots,N$, so $x_{n'+1} = (1,0,\ldots,0)$ and we have arrived at the feasible point after a finite number of steps, or alternatively from the last proposition $\rho_n < 1$ for all $n \geq n_0$. Consequently, the sequence $(x_n(1))_{n=n_0}^{\infty}$ is strictly increasing (hence convergent to some $x(1) \leq 1$) and so for $n \geq n_0$ we have $\rho_n \geq x_n(1) \geq x_{n_0} > 0$. But then, for each integer $k \geq 2$ and $n \geq n_0$ we see from the recurrence that,

$$\begin{aligned}\left|\frac{x_{n+1}(k)}{x_{n+1}(1)}\right| &= (1-\rho_n)\left|\frac{x_n(k)}{x_n(1)}\right| \\ &\leq (1 - x_{n_0}(1))\left|\frac{x_n(k)}{x_n(1)}\right|.\end{aligned}$$

Hence, $\dfrac{x_n(k)}{x_n(1)}$ converges to 0 and we conclude that $x_n \longrightarrow (1,0,\ldots,0)$. ∎

6.5 The Tangential Case When $\alpha = 1$

When $\alpha = 1$ the only feasible point is $b = (0,1,0,\ldots,0)$, however, we show that starting from an initial point with $x_0(1) > 0$ the Douglas–Rachford scheme converges to a point $\hat{y}b := (0,\hat{y},0,\ldots,0)$ with $\hat{y} > 1$, whose projection onto either S or L is the feasible point (Fig. 6.7). The following result will be needed.

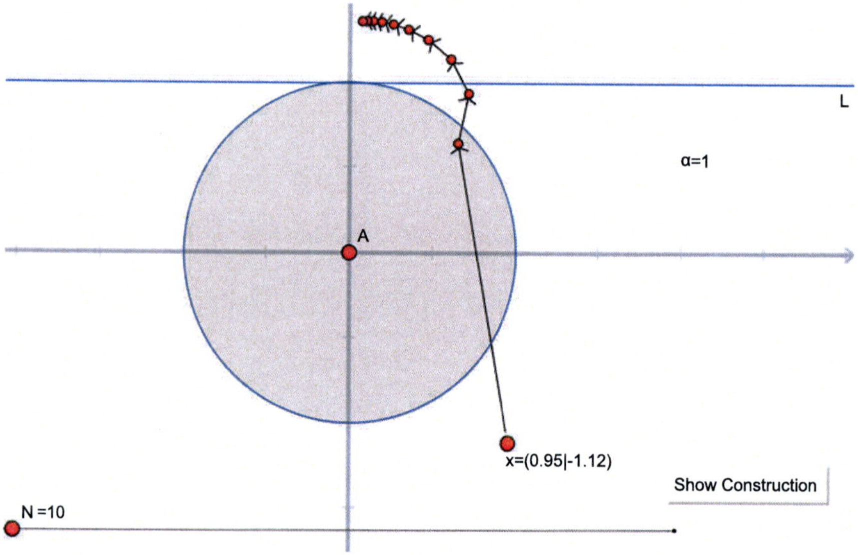

Fig. 6.7 Case with $\alpha = 1$

Proposition 6.11. *If* $\rho_n > 2$ *then* $\rho_{n+1} \le \rho_n$.

Proof. The proof is similar to that of Proposition 6.7. We may estimate as follows.

$$\begin{aligned}
\rho_{n+1}^2 &= \frac{x_n(1)^2}{\rho_n^2} + \left(\left(1-\frac{1}{\rho_n}\right)x_n(2)+1\right)^2 + \left(1-\frac{1}{\rho_n}\right)^2 \sum_{k=3}^{N} x_n(k)^2 \\
&= \frac{x_n(1)^2 + x_n(2)^2 + \cdots + x_n(N)^2}{\rho_n^2} \\
&\quad + \left(1-\frac{2}{\rho_n}\right)\sum_{k=2}^{N} x_n(k)^2 + 2\left(1-\frac{1}{\rho_n}\right)x_n(2)+1 \\
&= 2 + \left(1-\frac{2}{\rho_n}\right)\sum_{k=2}^{N} x_n(k)^2 + 2\left(1-\frac{1}{\rho_n}\right)x_n(2) \\
&\le 2 + \left(1-\frac{2}{\rho_n}\right)\rho_n^2 + 2\left(1-\frac{1}{\rho_n}\right)\rho_n, \quad \text{as } \rho_n > 2 \\
&= \rho_n^2.
\end{aligned}$$

■

To show the asserted behaviour, we begin by noting that from the recurrence,

$$x_{n+1}(2) = x_n(2) + 1 - \frac{x_n(2)}{\rho_n} \ge x_n(2), \tag{6.6}$$

since $\frac{x_n(2)}{\rho_n} \leq 1$. Thus, the $x_n(2)$ are increasing and so either they converge to a finite limit, $\hat{y}$ say, or they diverge to $+\infty$.

In the first case, taking limits in the above equation (6.6) yields $\hat{y} = \lim_n x_n(2) = \lim_n \rho_n \geq 0$ and so $x_n \longrightarrow (0, \hat{y}, 0, \cdots, 0)$. To see that $\hat{y} > 1$ we argue as follows. We have $x_n(1) \to 0$. But (6.1) shows $x_{n+1}(1) = x_n(1)/\rho_n$ so we must have $\lim_n \rho_n > 1$.

To show that the second, divergent, case is impossible we appeal to Proposition 6.11. to deduce that if the $x_n(2)$ diverges to $+\infty$, we must have for all sufficiently large n that $2 < x_n(2) \leq \rho_n$ and so eventually the ρ_n are decreasing and hence convergent to a finite limit which is necessarily greater than or equal to $\limsup_n x_n(2)$ which cannot therefore be infinite; a contradiction.

Consequently, we have proved,

Theorem 6.12. *When L is tangential to S at b (that is, when $\alpha = 1$), starting from any initial point with $x_0(1) \neq 0$, the Douglas–Rachford scheme converges to a point $\hat{y}b$ with $\hat{y} > 1$.*

This is consistent with the behaviour in the convex case [4, 11].

6.6 Behaviour in the Infeasible Case When $\alpha > 1$

Satisfyingly, when there are no feasible solutions, starting from any point off the singular manifold, the Douglas–Rachford scheme diverges. More precisely,

Theorem 6.13. *If there are no feasible solutions (that is, when $\alpha > 1$), then starting from any initial point with $x_0(1) \neq 0$, we have that $x_n(2)$ and hence ρ_n diverge to $+\infty$ at a linear or faster rate in the sense that* $\liminf_n x_{n+1}(2) - x_n(2) \geqslant \alpha - 1$.

Proof. From the recursion we have,

$$\begin{aligned} x_{n+1}(2) - x_n(2) &= \alpha - \frac{x_n(2)}{\rho_n} \\ &> \alpha - 1, \quad \text{as } x_n(2) < \rho_n \\ &> 0, \end{aligned}$$

from which the result follows. ∎

It is also worth noting that, as a consequence of the above theorem and the recurrence, $x_n(1) \to 0$ and so asymptotically the iterates approach the hyperplane $\langle x, a \rangle = 0$.

6.7 Behaviour on the Singular Manifold, $\langle \mathbf{x}, \mathbf{a} \rangle = 0$

Here, we consider the iterates of a non-zero initial point with $x_0(1) = 0$ and so $x_n(1) = 0$ for all n.

We again distinguish the cases; $\alpha = 0, 0 < \alpha < 1, \alpha = 1$. The case $\alpha > 1$ having already been dealt with in the previous section.

When $\alpha = 0$ it is readily seen that for any non-zero point x in the singular manifold we have $T_{S,L}(x) = \left(1 - \frac{1}{\|x\|}\right)x$. If $\|x\| = 1$ then the first iteration yields $x_1 = 0 \notin D_{T_{S,L}}$, so subsequent iterates are not defined. At points with $\|x\| < 1$ we see that $T_{S,L}$ has period two (that is, $T_{S,L}^2(x) = x$), while for $\|x\| > 1$ we have $T_{S,L}^2(x) = \left(1 - \frac{2}{\|x\|}\right)x$, so again the scheme breaks down as above, but after two iterations if $\|x\| = 2$.

We observe that the iterates of any non-zero point on the line $\{x : x = \lambda b,\ \lambda \in \mathbf{R}\}$ remain on this line and that when $\alpha = 1$ (that is, L is tangential to S at b) all points on the open half line corresponding to $\lambda > 0$ remain fixed under $T_{S,L}$.

In the other cases the scheme exhibits periodic behaviour when rational commensurability is present, while in the absence of such commensurability the behaviour may be quite chaotic. To make this precise we need to consider interval-valued mappings to deal with the jump at the origin. Luckily, the work in [1, 2] shows that various interval mapping analogues of Sharkovskii's theorem –'period three implies chaos' –are applicable. The interval mapping is needed to deal with the multivalued nature of the projection P_S at zero.

Remark 6.14 (Hilbert space analogues). It is not essential that X be finite dimensional for any of the arguments in Sects. 6.3–6.5, since the iterates are tracked by a finite number of coordinates. However, since convergence (to zero) in the other dimensions is only coordinate wise, we can in general only guarantee weak convergence of the iterates.

6.8 Some Final Remarks

A wealth of experimental evidence, using both *Maple* and the dynamic geometry package *Cinderella*, leads to the conclusion that the basin of attraction for $p = \sqrt{1-\alpha^2}a + hb$ is the open half space $\{x : \langle x, a\rangle > 0\}$ – the largest region possible. See also http://www.carma.newcastle.edu.au/~jb616/expansion.html.

Moreover, we found that for stable computation in Cinderella it was necessary to have access to precision beyond *Cinderella's* built-in double precision. This was achieved by taking input directly from *Maple*. We illustrate in Fig. 6.8 which show various spurious red points on the left and accurate data on the right. The figures show the effect of roughly ten steps of the Douglas–Rachford iteration for 400 different starting points – where the points are coloured by their original distance from the vertical axis with red closest.

However, we are as yet unable to furnish a proof of this, leaving open the following conjecture:

Conjecture 6.1. In the simple example of a sphere and a line with two intersection points, the basins of attraction are the two open half-spaces forming the complement of the singular manifold.

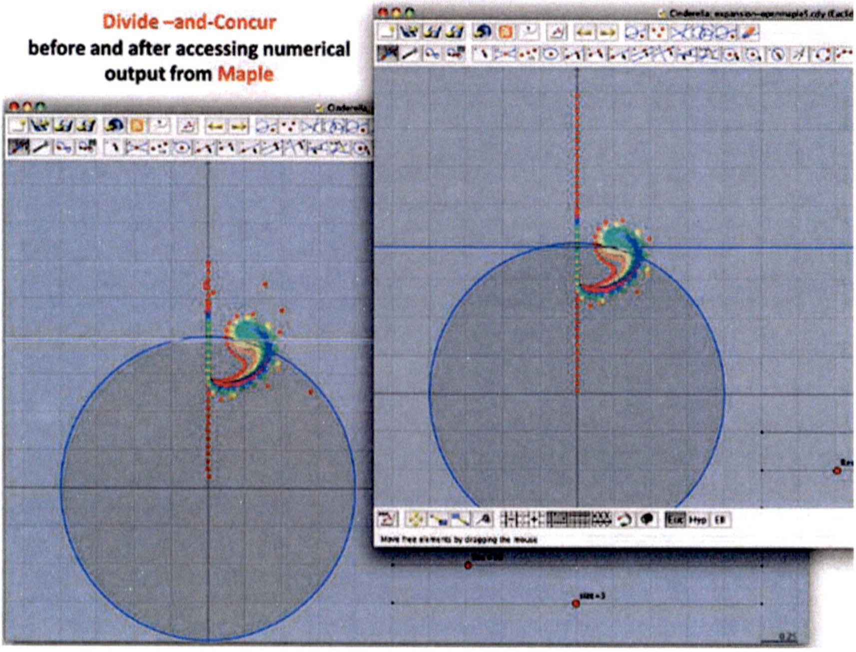

Fig. 6.8 Multiple iterations in Cinderella

Remark 6.15 (The case of a sphere and a proper affine subset of X). If we replace the line L by a proper affine subset, say $A := \{\lambda_1 a_1 + \lambda_2 a_2 + \cdots \lambda_K a_K + \alpha b : \lambda_1, \cdots, \lambda_K \in \mathbf{R}\}$, where $0 \leq \alpha$, $1 < K < N$, and $a_1, a_2, \cdots, a_K, b$ are mutually orthogonal norm one elements, then when $\alpha < 1$ the feasible points are no longer isolated, so Theorem 6.4 no longer applies, indeed local convergence in the sense described above is impossible. Nonetheless, *all our results appropriately viewed continue to hold* and we shall sketch the argument. Details will be given elsewhere.

Indeed, if for any non-feasible point $q \neq 0$ we let $Q := A_0^{\perp} + \mathbf{R}q$, where $A_0^{\perp}$ is the orthogonal complement of the subspace $A_0 := A - \alpha b$, then we see that for any initial point $x_0 \in Q$ the sequence of iterates, $x_n = T_{S,A}^n(x_0)$ remains confined to the subspace Q. So, if the Douglas–Rachford scheme converges it will converge to a point in $S \cap A \cap Q$. Further, the fixed points of $T_{S,A}|_Q$ consists of two isolated points comprising $S \cap A \cap Q$; namely, $p = (kq(1), kq(2), \cdots, kq(K), \alpha, 0, \cdots, 0)$, where,

$$k := \pm\sqrt{\frac{1-\alpha^2}{q(1)^2 + \cdots + q(K)^2}}.$$

And so we have 'local convergence' in the following sense. For either feasible point $p \in S \cap A \cap Q$ there is a neighbourhood, N_p of p in the subspace Q such that starting from any point x_0 in N_p the iterates converge to p.

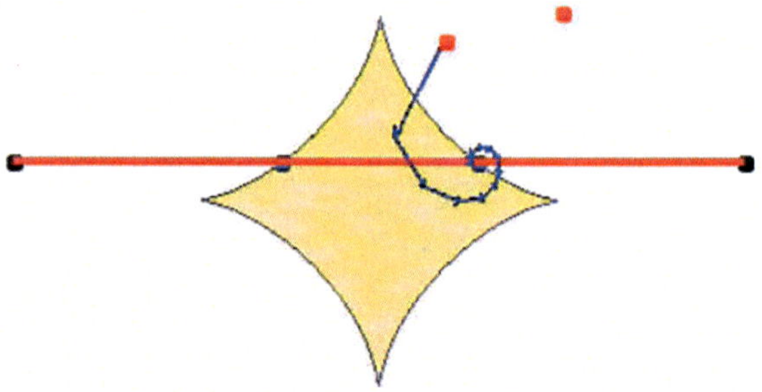

Fig. 6.9 Spiralling with the $1/2$-sphere

Additionally, we may derive similar conclusions to those obtained above in the cases when $\alpha = 0$, $\alpha = 1$ and $\alpha > 1$. Further, in this case the singular manifold is the subspace $A_0^{\perp}$.

In conclusion, our analysis sheds some significant light on the behaviour of non-convex Douglas–Rachford schemes but much remains to be studied.

Example 6.16 (Other regions). For example, we observe that neither convexity nor so much symmetry is essential to the behaviour exhibited in Theorem 6.4. Figure 6.9 shows the situation for a line and a non-convex p-sphere, where $S(p) := \{(x,y)\colon |x|^p + |y|^p = 1\}$, in the plane. The details of such analysis remain to be performed.

Acknowledgements This research was supported by the Australian Research Council. We also express our thanks to Chris Maitland, Matt Skerritt and Ulli Kortenkamp for helping us exploit the full resources of *Cinderella*.

References

1. Andres, J., Pastor, K., Šnyrychová: A multivalued version of Sharkovskii's theorem holds with at most two exceptions. J. Fixed Point Theory Appl. **2**, 153–170 (2007)
2. Andres, J., Fürst, T., Pastor, K.: Full analogy of Sharkovskii's theorem for lower semicontinuous maps. J. Math. Anal. Appl. **340**, 1132–1144 (2008)
3. Bauschke, H.H., Borwein, J.M.: On projection algorithms for solving convex feasibility problems. SIAM Review **38**, 367–426 (1996)
4. Bauschke, H.H., Combettes, P.L., Luke, D.R.: Phase retrieval, error reduction algorithm, and Fienup variants: a view from convex optimization. J. Opt. Soc. Amer. A **19**, 1334–1345 (2002)
5. Bauschke, H.H., Combettes, P.L., Luke, D.R..: Finding best approximation pairs relative to two closed convex sets in Hilbert spaces. J. Approx. Theory **127**, 178–192 (2004)
6. Bauschke, H.H., Combettes, P.L., Luke, D.R.: A strongly convergent reflection method for finding the projection onto the intersection of two closed convex sets in a Hilbert space. J. Approx. Theory **141**, 63–69 (2006)
7. Douglas, J., Rachford, H.H.: On the numerical solution of heat conduction problems in two or three space variables. Trans. Amer. Math. Soc. **82**, 421–439 (1956)

8. Elser, V., Rankenburg, I., Thibault, P.: Searching with iterated maps. Proceedings of the National Academy of Sciences **104**, 418–423 (2007)
9. Gravel, S., Elser, V.: Divide and concur: A general approach constraint satisfaction. Phys. Rev. E **78** 036706, pp. 5 (2008), http://link.aps.org/doi/10.1103/PhysRevE.78.036706
10. Lakshmikantham, V., Trigiante, D.: Theory of Difference Equations – Numerical Methods and Applications. Marcel Dekker (2002)
11. Lions, P.-L., Mercier, B.: Splitting algorithms for the sum of two nonlinear operators. SIAM J. Numer. Anal. **16**, 964–979 (1979)
12. Pierra, G.: Eclatement de contraintes en parallèle pour la miniminisation d'une forme quadratique. Lecture Notes in Computer Science, Springer, **41** 200–218 (1976)

Chapter 7
A Comparison of Some Recent Regularity Conditions for Fenchel Duality

Radu Ioan Boţ and Ernö Robert Csetnek

Abstract This article provides an overview on regularity conditions for Fenchel duality in convex optimization. Our attention is focused, on the one hand, on three generalized interior-point regularity conditions expressed by means of the quasi interior and of the quasi-relative interior and, on the other hand, on two closedness-type conditions that have been recently introduced in the literature. We discuss how they do relate to each other, but also to several other classical ones and illustrate these investigations by numerous examples.

Keywords Convex optimization · Fenchel duality · Quasi interior · Quasi-relative interior · Generalized Interior-point Regularity conditions · Closedness-type regularity conditions

AMS 2010 Subject Classification: 46N10, 42A50

7.1 Introduction

The primal problem we investigate in this section is an unrestricted optimization problem having as objective function the sum of two proper and convex functions defined on a separated locally convex space. To it we attach the *Fenchel dual* problem and f urther we concentrate ourselves on providing regularity conditions for *strong duality* for this primal-dual pair, which is the situation when the optimal objective values of the two problems coincide and the dual has an optimal solution. First of all, we bring into the discussion several conditions of this kind that one can find in the literature, where along the one which asks for the *continuity* of one of the two functions at a point from the intersection of the effective domains, we enumerate some classical generalized interior-point ones. Here, we refer to the regularity

R.I. Boţ (✉)
Faculty of Mathematics, Chemnitz University of Technology, 09107 Chemnitz, Germany
e-mail: radu.bot@mathematik.tu-chemnitz.de

H.H. Bauschke et al. (eds.), *Fixed-Point Algorithms for Inverse Problems in Science and Engineering*, Springer Optimization and Its Applications 49,
DOI 10.1007/978-1-4419-9569-8_7, © Springer Science+Business Media, LLC 2011

conditions employing not only the *interior*, but also the *algebraic interior* (cf. [21]), the *intrinsic core* (cf. [17]) and the *strong-quasi relative interior* (cf. [1, 24]) of the difference of the domains of the two functions. The latter conditions guarantee strong duality if we suppose additionally that the two functions are lower semicontinuous and the space we work within is a Fréchet one. A general scheme containing the relations between these sufficient conditions is also furnished.

The central role in the paper is played by some regularity conditions for Fenchel duality recently introduced in the literature. First of all, we consider some regularity conditions expressed via the *quasi interior* and *quasi-relative interior* (cf. [8,9]), which presents the advantage that they do not ask for any topological assumption regarding the functions involved and work in general separated locally convex spaces. We consider three conditions of this kind, relate them to each other, but also to the classical ones mentioned above. By means of some examples we are able to underline their wider applicability, by providing optimization problems where these are fulfilled, while the consecrated ones fail.

The second class of recently introduced regularity conditions we discuss here is the one of the so-called *closedness-type regularity conditions*, which additionally ask for lower semicontinuity for the two functions, but work in general separated locally convex spaces, too. We discuss here two closedness-type conditions (cf. [7, 10]), we relate them to each other, to the classical interior-point ones, but also, more important, to the ones expressed via the quasi interior and quasi-relative interior. More precisely, we show that, unlike in finite-dimensional spaces, in the infinite-dimensional setting these two classes of regularity conditions for Fenchel duality are not comparable. In this way we give a negative answer to an open problem stated in [19, Remark 4.3].

The paper is organized as follows. In Sect. 7.2, we introduce some elements of convex analysis, whereby the accent is put on different generalized interiority notions. The notions quasi interior and quasi-relative interior are also introduced and some of their important properties are mentioned. The third section starts with the definition of the Fenchel dual problem, followed by a subsection dedicated to the classical interior-point regularity conditions. The second subsection of Sect. 7.3 deals with the new conditions expressed via the quasi interior and quasi-relative interior, while in the third one the closedness-type conditions are studied.

7.2 Preliminary Notions and Results

Consider X a (real) separated locally convex space and X^* its topological dual space. We denote by $w(X^*,X)$ the weak* topology on X^* induced by X. For a nonempty set $U \subseteq X$, we denote by $\operatorname{co}(U), \operatorname{cone}(U), \operatorname{coneco}(U), \operatorname{aff}(U), \operatorname{lin}(U), \operatorname{int}(U), \operatorname{cl}(U)$, its *convex hull, conic hull, convex conic hull, affine hull, linear hull, interior* and *closure*, respectively. In case U is a linear subspace of X we denote by $U^\perp$ the *annihilator* of U. Let us mention the following property: if U is convex then

$$\operatorname{coneco}(U \cup \{0\}) = \operatorname{cone}(U). \tag{7.1}$$

If $U \subseteq \mathbb{R}^n$ ($n \in \mathbb{N}$) we denote by $\mathrm{ri}(U)$ the *relative interior* of U, that is the interior of U with respect to its affine hull. We denote by $\langle x^*, x \rangle$ the value of the linear continuous functional $x^* \in X^*$ at $x \in X$ and by $\ker x^*$ the *kernel* of x^*. The *indicator function* of U, $\delta_U : X \to \overline{\mathbb{R}}$, is defined as

$$\delta_U(x) = \begin{cases} 0, & \text{if } x \in U, \\ +\infty, & \text{otherwise,} \end{cases}$$

where $\overline{\mathbb{R}} = \mathbb{R} \cup \{\pm\infty\}$ is the extended real line. We make the following conventions: $(+\infty) + (-\infty) = +\infty$, $0 \cdot (+\infty) = +\infty$ and $0 \cdot (-\infty) = 0$. For a function $f : X \to \overline{\mathbb{R}}$ we denote by $\mathrm{dom} f = \{x \in X : f(x) < +\infty\}$ the *domain* of f and by $\mathrm{epi} f = \{(x,r) \in X \times \mathbb{R} : f(x) \le r\}$ its *epigraph*. Moreover, we denote by $\widehat{\mathrm{epi}}(f) = \{(x,r) \in X \times \mathbb{R} : (x,-r) \in \mathrm{epi} f\}$, the symmetric of $\mathrm{epi} f$ with respect to the x-axis. For a given real number α, $f - \alpha : X \to \overline{\mathbb{R}}$ is, as usual, the function defined by $(f-\alpha)(x) = f(x) - \alpha$ for all $x \in X$. We call f *proper* if $\mathrm{dom} f \neq \emptyset$ and $f(x) > -\infty$ for all $x \in X$. The *normal cone* of U at $x \in U$ is $N_U(x) = \{x^* \in X^* : \langle x^*, y - x \rangle \le 0 \ \forall y \in U\}$.

The *Fenchel–Moreau conjugate* of f is the function $f^* : X^* \to \overline{\mathbb{R}}$ defined by

$$f^*(x^*) = \sup_{x \in X} \{\langle x^*, x \rangle - f(x)\} \ \forall x^* \in X^*.$$

We have the so-called *Young–Fenchel inequality*

$$f^*(x^*) + f(x) \ge \langle x^*, x \rangle \ \forall x \in X \ \forall x^* \in X^*.$$

Having $f, g : X \to \overline{\mathbb{R}}$ two functions we denote by $f \Box g : X \to \overline{\mathbb{R}}$ their *infimal convolution*, defined by $f \Box g(x) = \inf_{u \in X} \{f(u) + g(x-u)\}$ for all $x \in X$. We say that the infimal convolution is *exact at* $x \in X$ if the infimum in its definition is attained. Moreover, $f \Box g$ is said to be *exact* if it is exact at every $x \in X$.

Let us recall in the following the most important generalized interiority notions introduced in the literature. The set $U \subseteq X$ is supposed to be nonempty and convex. We have:

- $\mathrm{core}(U) := \{x \in U : \mathrm{cone}(U - x) = X\}$, the *algebraic interior* (the *core*) of U (cf. [21, 26]);
- $\mathrm{icr}(U) := \{x \in U : \mathrm{cone}(U - x)$ is a linear subspace of $X\}$, the *relative algebraic interior* (*intrinsic core*) of U (cf. [2, 18, 26]);
- $\mathrm{sqri}(U) := \{x \in U : \mathrm{cone}(U - x)$ is a closed linear subspace of $X\}$ the *strong quasi-relative interior* (*intrinsic relative algebraic interior*) of U (cf. [4, 26]).

We mention the following characterization of the strong quasi-relative interior (cf. [17, 26]): $x \in \mathrm{sqri}(U) \Leftrightarrow x \in \mathrm{icr}(U)$ and $\mathrm{aff}(U - x)$ is a closed linear subspace.

The *quasi-relative interior* of U is the set (cf. [3])

$$\mathrm{qri}(U) = \{x \in U : \mathrm{cl}\left(\mathrm{cone}(U - x)\right) \text{ is a linear subspace of } X\}.$$

The quasi-relative interior of a convex set is characterized by means of the normal cone as follows.

Proposition 7.1 (cf. [3]). *Let U be a nonempty convex subset of X and $x \in U$. Then $x \in \operatorname{qri}(U)$ if and only if $N_U(x)$ is a linear subspace of X^*.*

Next we consider another generalized interiority notion introduced in connection with a convex set, which is close to the quasi-relative interior. The *quasi interior* of U is the set

$$\operatorname{qi}(U) = \{x \in U : \operatorname{cl}\big(\operatorname{cone}(U - x)\big) = X\}.$$

It can be characterized as follows.

Proposition 7.2 (cf. [8, Proposition 2.4]). *Let U be a nonempty convex subset of X and $x \in U$. Then $x \in \operatorname{qi}(U)$ if and only if $N_U(x) = \{0\}$.*

Remark 7.3. The above characterization of the quasi interior of a convex set was given in [16], where the authors supposed that X is a reflexive Banach space. It is proved in [8, Proposition 2.4] that this property holds in a more general context, namely in separated locally convex spaces.

We have the following relations between the different generalized interiority notions considered above

$$\operatorname{int}(U) \subseteq \operatorname{core}(U) \subseteq \begin{array}{c} \operatorname{sqri}(U) \subseteq \operatorname{icr}(U) \\ \\ \operatorname{qi}(U) \end{array} \subseteq \operatorname{qri}(U) \subseteq U, \tag{7.2}$$

all the inclusions being in general strict. As one can also deduce from some of the examples which follows in this paper in general between $\operatorname{sqri}(U)$ and $\operatorname{icr}(U)$, on the one hand, and $\operatorname{qi}(U)$, on the other hand, no relation of inclusion can be provided. In case $\operatorname{int}(U) \neq \emptyset$ all the generalized interior-notions considered in (7.2) collapse into $\operatorname{int}(U)$ (cf. [3, Corollary 2.14]).

It follows from the definition of the quasi-relative interior that $\operatorname{qri}(\{x\}) = \{x\}$ for all $x \in X$. Moreover, if $\operatorname{qi}(U) \neq \emptyset$, then $\operatorname{qi}(U) = \operatorname{qri}(U)$. Although this property is given in [20] in the case of real normed spaces, it holds also in separated locally convex spaces, as it easily follows from the properties given above. For U, V two convex subsets of X such that $U \subseteq V$, we have $\operatorname{qi}(U) \subseteq \operatorname{qi}(V)$, a property which is no longer true for the quasi-relative interior (however this holds whenever $\operatorname{aff}(U) = \operatorname{aff}(V)$, see [13, Proposition 1.12]). If X if finite-dimensional then $\operatorname{qri}(U) = \operatorname{sqri}(U) = \operatorname{icr}(U) = \operatorname{ri}(U)$ (cf. [3, 17]) and $\operatorname{core}(U) = \operatorname{qi}(U) = \operatorname{int}(U)$ (cf. [20, 21]). We refer the reader to [2, 3, 17, 18, 20, 21, 23, 26] and to the references therein for more properties and examples regarding the above considered generalized interiority notions.

Example 7.4. Take an arbitrary $p \in [1, +\infty)$ and consider the real Banach space $\ell^p = \ell^p(\mathbb{N})$ of real sequences $(x_n)_{n \in \mathbb{N}}$ such that $\sum_{n=1}^{\infty} |x_n|^p < +\infty$, equipped with the norm $\|\cdot\| : \ell^p \to \mathbb{R}$, $\|x\| = \left(\sum_{n=1}^{\infty} |x_n|^p\right)^{1/p}$ for all $x = (x_n)_{n\in\mathbb{N}} \in \ell^p$. Then (cf. [3])

$$\operatorname{qri}(\ell^p_+) = \{(x_n)_{n\in\mathbb{N}} \in \ell^p : x_n > 0 \ \forall n \in \mathbb{N}\},$$

where $\ell^p_+ = \{(x_n)_{n\in\mathbb{N}} \in \ell^p : x_n \geq 0 \ \forall n \in \mathbb{N}\}$ is the positive cone of ℓ^p. Moreover, one can prove that

$$\operatorname{int}(\ell^p_+) = \operatorname{core}(\ell^p_+) = \operatorname{sqri}(\ell^p_+) = \operatorname{icr}(\ell^p_+) = \emptyset.$$

In the setting of separable Banach spaces, every nonempty closed convex set has a nonempty quasi-relative interior (cf. [3, Theorem 2.19], see also [2, Theorem 2.8] and [26, Proposition 1.2.9]) and every nonempty convex set which is not contained in a hyperplane possesses a nonempty quasi interior (cf. [20]). This result may fail if the condition X is separable is removed, as the following example shows.

Example 7.5. For $p \in [1, +\infty)$ consider the real Banach space

$$\ell^p(\mathbb{R}) = \Big\{s : \mathbb{R} \to \mathbb{R} \ \Big| \ \sum_{r\in\mathbb{R}} |s(r)|^p < \infty\Big\},$$

equipped with the norm $\|\cdot\| : \ell^p(\mathbb{R}) \to \mathbb{R}$, $\|s\| = \left(\sum_{r\in\mathbb{R}} |s(r)|^p\right)^{1/p}$ for all $s \in \ell^p(\mathbb{R})$, where

$$\sum_{r\in\mathbb{R}} |s(r)|^p = \sup_{F\subseteq\mathbb{R}, F \text{ finite}} \sum_{r\in F} |s(r)|^p.$$

Considering the positive cone $\ell^p_+(\mathbb{R}) = \{s \in \ell^p(\mathbb{R}) : s(r) \geq 0 \ \forall r \in \mathbb{R}\}$, we have (cf. [3, Example 3.11(iii)], see also [5, Remark 2.20]) that $\operatorname{qri}\left(\ell^p_+(\mathbb{R})\right) = \emptyset$.

Let us mention some properties of the quasi-relative interior. For the proof of (i) and (ii) we refer to [2,3], while property (iii) was proved in [8, Proposition 2.5] (see also [9, Proposition 2.3]).

Proposition 7.6. *Consider U a nonempty convex subset of X. Then:*

(i) $t\operatorname{qri}(U) + (1-t)U \subseteq \operatorname{qri}(U)\ \forall t \in (0,1]$; *hence* $\operatorname{qri}(U)$ *is a convex set. If, additionally,* $\operatorname{qri}(U) \neq \emptyset$ *then:*
(ii) $\operatorname{cl}\left(\operatorname{qri}(U)\right) = \operatorname{cl}(U)$;
(iii) $\operatorname{cl}\Big(\operatorname{cone}\left(\operatorname{qri}(U)\right)\Big) = \operatorname{cl}\left(\operatorname{cone}(U)\right)$.

The first part of the next lemma was proved in [8, Lemma 2.6] (see also [9, Lemma 2.1]).

Lemma 7.7. *Let U and V be nonempty convex subsets of X and $x \in X$. Then:*

(i) if $\operatorname{qri}(U) \cap V \neq \emptyset$ *and* $0 \in \operatorname{qi}(U - U)$, *then* $0 \in \operatorname{qi}(U - V)$;
(ii) $x \in \operatorname{qi}(U)$ *if and only if* $x \in \operatorname{qri}(U)$ *and* $0 \in \operatorname{qi}(U - U)$.

Proof. (ii) Suppose that $x \in \mathrm{qi}(U)$. Then $x \in \mathrm{qri}(U)$ and since $U - x \subseteq U - U$ and $0 \in \mathrm{qi}(U - x)$, the direct implication follows. The reverse one follows as a direct consequence of (i) by taking $V := \{x\}$. □

Remark 7.8. Consider the setting of Example 7.4. By applying the previous result, we get (since $\ell^p_+ - \ell^p_+ = \ell^p$) that

$$\mathrm{qi}(\ell^p_+) = \mathrm{qri}(\ell^p_+) = \{(x_n)_{n\in\mathbb{N}} \in \ell^p : x_n > 0 \; \forall n \in \mathbb{N}\}.$$

The proof of the duality theorem presented in the next section is based on the following separation theorem.

Theorem 7.9 (cf. [8, Theorem 2.7]). *Let U be a nonempty convex subset of X and $x \in U$. If $x \notin \mathrm{qri}(U)$, then there exists $x^* \in X^*, x^* \neq 0$, such that*

$$\langle x^*, y\rangle \leq \langle x^*, x\rangle \; \forall y \in U.$$

Viceversa, if there exists $x^* \in X^*$, $x^* \neq 0$, such that

$$\langle x^*, y\rangle \leq \langle x^*, x\rangle \; \forall y \in U$$

and

$$0 \in \mathrm{qi}(U - U),$$

then $x \notin \mathrm{qri}(U)$.

Remark 7.10. (a) The above separation theorem is a generalization to separated locally convex spaces of a result stated in [15, 16] in the framework of real normed spaces (cf. [8, Remark 2.8]).
(b) The condition $x \in U$ in Theorem 7.9 is essential (see [16, Remark 2]). However, if x is an arbitrary element of X, an alternative separation theorem has been given by Cammaroto and Di Bella in [12, Theorem 2.1]. Let us mention that some strict separation theorems involving the quasi-relative interior can be found in [13].

7.3 Fenchel Duality

Let us briefly recall some considerations regarding Fenchel duality. We deal in the following with the following optimization problem

$$(P_{\mathrm{F}}) \quad \inf_{x\in X}\{f(x) + g(x)\},$$

where X is a separated locally convex space and $f, g : X \to \overline{\mathbb{R}}$ are proper functions such that $\mathrm{dom}\, f \cap \mathrm{dom}\, g \neq \emptyset$.

The classical *Fenchel dual problem* to (P_F) has the following form

$$(D_F) \quad \sup_{y^* \in X^*} \{-f^*(-y^*) - g^*(y^*)\}.$$

We denote by $v(P_F)$ and $v(D_F)$ the optimal objective values of the primal and dual problems, respectively. Weak duality always holds, that is $v(P_F) \geq v(D_F)$. To guarantee strong duality, the situation when $v(P_F) = v(D_F)$ and (D_F) has an optimal solution, several regularity conditions were introduced in the literature.

7.3.1 Classical Interior-Point Regularity Conditions

In this subsection, we deal with generalized interior-point regularity conditions, by enumerating the classical ones existing in the literature and by studying the relations between them. Let us start by recalling the most known conditions of this type:

(RC_1^F) | $\exists x' \in \operatorname{dom} f \cap \operatorname{dom} g$ such that f (or g) is continuous at x';

(RC_2^F) | X is a Fréchet space, f and g are lower semicontinuous and $0 \in \operatorname{int}(\operatorname{dom} f - \operatorname{dom} g)$;

(RC_3^F) | X is a Fréchet space, f and g are lower semicontinuous and $0 \in \operatorname{core}(\operatorname{dom} f - \operatorname{dom} g)$;

(RC_4^F) | X is a Fréchet space, f and g are lower semicontinuous, $\operatorname{aff}(\operatorname{dom} f - \operatorname{dom} g)$ is a closed linear subspace of X and $0 \in \operatorname{icr}(\operatorname{dom} f - \operatorname{dom} g)$

and

(RC_5^F) | X is a Fréchet space, f and g are lower semicontinuous and $0 \in \operatorname{sqri}(\operatorname{dom} f - \operatorname{dom} g)$.

The condition (RC_3^F) was considered by Rockafellar (cf. [21]), (RC_5^F) by Attouch and Brézis (cf. [1]) and Zălinescu (cf. [24]), while Gowda and Teboulle proved that (RC_4^F) and (RC_5^F) are equivalent (cf. [17]).

Theorem 7.11. *Let $f, g : X \to \overline{\mathbb{R}}$ be proper and convex functions. If one of the regularity conditions (RC_i^F), $i \in \{1,2,3,4,5\}$, is fulfilled, then $v(P_F) = v(D_F)$ and (D_F) has an optimal solution.*

Remark 7.12. In case X is a Fréchet space and f, g are proper, convex and lower semicontinuous functions we have the following relations between the above regularity conditions (see also [17, 25] and [26, Theorem 2.8.7])

$$(RC_1^F) \Rightarrow (RC_2^F) \Leftrightarrow (RC_3^F) \Rightarrow (RC_4^F) \Leftrightarrow (RC_5^F).$$

Let us notice that the regularity conditions (RC_2^F) and (RC_3^F) are equivalent. Indeed, assume that X is a Fréchet space, f,g are proper, convex and lower semicontinuous functions such that $\operatorname{dom} f \cap \operatorname{dom} g \neq \emptyset$ and consider the *infimal value function* $h : X \to \overline{\mathbb{R}}$, defined by $h(y) = \inf_{x\in X}\{f(x)+g(x-y)\}$ for all $y \in X$. The function h is convex and not necessarily lower semicontinuous, while one has that $\operatorname{dom} h = \operatorname{dom} f - \operatorname{dom} g$. Nevertheless, the function $(x,y) \mapsto f(x)+g(x-y)$ is ideally convex (being convex and lower semicontinuous), hence h is li-convex (cf. [26, Proposition 2.2.18]). Now by [26, Theorem 2.2.20] it follows that $\operatorname{core}(\operatorname{dom} h) = \operatorname{int}(\operatorname{dom} h)$, which has as consequence the equivalence of the regularity conditions (RC_2^F) and (RC_3^F). Let us mention that this fact has been noticed in the setting of Banach spaces by Simons in [22, Corollary 14.3].

7.3.2 *Interior-Point Regularity Conditions Expressed via Quasi Interior and Quasi-Relative Interior*

Taking into account the relations that exist between the generalized interiority notions presented in Sect. 7.2 a natural question arises: is the condition $0 \in \operatorname{qri}(\operatorname{dom} f - \operatorname{dom} g)$ sufficient for strong duality? The following example (which can be found in [17]) shows that even if we impose a stronger condition, namely $0 \in \operatorname{qi}(\operatorname{dom} f - \operatorname{dom} g)$, the above question has a negative answer and this means that we need to look for additional assumptions in order to guarantee Fenchel duality.

Example 7.13. Consider the Hilbert space $X = \ell^2(\mathbb{N})$ and the sets

$$C = \{(x_n)_{n\in\mathbb{N}} \in \ell^2 : x_{2n-1} + x_{2n} = 0 \ \forall n \in \mathbb{N}\}$$

and

$$S = \{(x_n)_{n\in\mathbb{N}} \in \ell^2 : x_{2n} + x_{2n+1} = 0 \ \forall n \in \mathbb{N}\},$$

which are closed linear subspaces of ℓ^2 and satisfy $C \cap S = \{0\}$. Define the functions $f,g : \ell^2 \to \overline{\mathbb{R}}$ by $f = \delta_C$ and $g(x) = x_1 + \delta_S(x)$, respectively, for all $x = (x_n)_{n\in\mathbb{N}} \in \ell^2$. One can see that f and g are proper, convex and lower semicontinuous functions with $\operatorname{dom} f = C$ and $\operatorname{dom} g = S$. As $v(P_F) = 0$ and $v(D_F) = -\infty$ (cf. [17, Example 3.3]), there is a duality gap between the optimal objective values of the primal problem and its Fenchel dual problem. Moreover, $S - C$ is dense in ℓ^2 (cf. [17]), thus $\operatorname{cl}\big(\operatorname{cone}(\operatorname{dom} f - \operatorname{dom} g)\big) = \operatorname{cl}(C - S) = \ell^2$. The last relation implies $0 \in \operatorname{qi}(\operatorname{dom} f - \operatorname{dom} g)$, hence $0 \in \operatorname{qri}(\operatorname{dom} f - \operatorname{dom} g)$.

We notice that if $v(P_F) = -\infty$, by the weak duality result follows that for the primal-dual pair $(P_F)-(D_F)$ strong duality holds. This is the reason why we suppose in what follows that $v(P_F) \in \mathbb{R}$.

Consider now the following regularity conditions expressed by means of the quasi interior and quasi-relative interior:

$$(RC_6^{\mathrm{F}}) \left| \begin{array}{l} \operatorname{dom} f \cap \operatorname{qri}(\operatorname{dom} g) \neq \emptyset,\ 0 \in \operatorname{qi}(\operatorname{dom} g - \operatorname{dom} g) \text{ and} \\ (0,0) \notin \operatorname{qri}\left[\operatorname{co}\left((\operatorname{epi} f - \widehat{\operatorname{epi}}(g - v(P_{\mathrm{F}}))) \cup \{(0,0)\}\right)\right]; \end{array} \right.$$

$$(RC_7^{\mathrm{F}}) \left| \begin{array}{l} 0 \in \operatorname{qi}(\operatorname{dom} f - \operatorname{dom} g) \text{ and} \\ (0,0) \notin \operatorname{qri}\left[\operatorname{co}\left((\operatorname{epi} f - \widehat{\operatorname{epi}}(g - v(P_{\mathrm{F}}))) \cup \{(0,0)\}\right)\right] \end{array} \right.$$

and

$$(RC_8^{\mathrm{F}}) \left| \begin{array}{l} 0 \in \operatorname{qi}\left[(\operatorname{dom} f - \operatorname{dom} g) - (\operatorname{dom} f - \operatorname{dom} g)\right],\ 0 \in \operatorname{qri}(\operatorname{dom} f - \operatorname{dom} g) \text{ and} \\ (0,0) \notin \operatorname{qri}\left[\operatorname{co}\left((\operatorname{epi} f - \widehat{\operatorname{epi}}(g - v(P_{\mathrm{F}}))) \cup \{(0,0)\}\right)\right]. \end{array} \right.$$

Let us notice that these three regularity conditions were first introduced in [8]. We study in the following the relations between these conditions. We remark that

$$\begin{aligned} &\operatorname{epi} f - \widehat{\operatorname{epi}}(g - v(P_{\mathrm{F}})) \\ &\quad = \{(x - y, f(x) + g(y) - v(P_{\mathrm{F}}) + \varepsilon) : x \in \operatorname{dom} f, y \in \operatorname{dom} g, \varepsilon \geq 0\}, \end{aligned}$$

thus if the set $\operatorname{epi} f - \widehat{\operatorname{epi}}(g - v(P_{\mathrm{F}})$ is convex, then $\operatorname{dom} f - \operatorname{dom} g$ is convex, too.

Proposition 7.14. *Let $f, g : X \to \overline{\mathbb{R}}$ be proper functions such that $v(P_{\mathrm{F}}) \in \mathbb{R}$ and $\operatorname{epi} f - \widehat{\operatorname{epi}}(g - v(P_{\mathrm{F}}))$ is a convex subset of $X \times \mathbb{R}$ (the latter is the case if for instance f and g are convex functions). The following statements are true:*

(i) $(RC_7^{\mathrm{F}}) \Leftrightarrow (RC_8^{\mathrm{F}})$; if, moreover, f and g are convex, then $(RC_6^{\mathrm{F}}) \Rightarrow (RC_7^{\mathrm{F}}) \Leftrightarrow (RC_8^{\mathrm{F}})$;

(ii) if (P_{F}) has an optimal solution, then $(0,0) \notin \operatorname{qri}\left[\operatorname{co}\left((\operatorname{epi} f - \widehat{\operatorname{epi}}(g - v(P_{\mathrm{F}}))) \cup \{(0,0)\}\right)\right]$ can be equivalently written as $(0,0) \notin \operatorname{qri}\left(\operatorname{epi} f - \widehat{\operatorname{epi}}(g - v(P_{\mathrm{F}}))\right)$;

(iii) if $0 \in \operatorname{qi}\left[(\operatorname{dom} f - \operatorname{dom} g) - (\operatorname{dom} f - \operatorname{dom} g)\right]$, then $(0,0) \notin \operatorname{qri}\left[\operatorname{co}\left((\operatorname{epi} f - \widehat{\operatorname{epi}}(g - v(P_{\mathrm{F}}))) \cup \{(0,0)\}\right)\right]$ is equivalent to $(0,0) \notin \operatorname{qi}\left[\operatorname{co}\left((\operatorname{epi} f - \widehat{\operatorname{epi}}(g - v(P_{\mathrm{F}}))) \cup \{(0,0)\}\right)\right]$.

Proof. (i) That (RC_7^{F}) is equivalent to (RC_8^{F}) is a direct consequence of Lemma 7.7 (ii). Let us suppose that f and g are convex and (RC_6^{F}) is fulfilled. By applying Lemma 7.7(i) with $U := \operatorname{dom} g$ and $V := \operatorname{dom} f$ we get $0 \in \operatorname{qi}(\operatorname{dom} g - \operatorname{dom} f)$ or, equivalently, $0 \in \operatorname{qi}(\operatorname{dom} f - \operatorname{dom} g)$. This means that (RC_7^{F}) holds.

(ii) One can prove that the primal problem (P_{F}) has an optimal solution if and only if $(0,0) \in \operatorname{epi} f - \widehat{\operatorname{epi}}(g - v(P_{\mathrm{F}})))$ and the conclusion follows.

(iii) See [8, Remark 3.4 (a)]. □

Remark 7.15. (a) The condition $0 \in \mathrm{qi}(\mathrm{dom} f - \mathrm{dom} g)$ implies relation

$$0 \in \mathrm{qi}\left[(\mathrm{dom} f - \mathrm{dom} g) - (\mathrm{dom} f - \mathrm{dom} g)\right]$$

in Proposition 7.14(iii). This is a direct consequence of the inclusion $\mathrm{dom} f - \mathrm{dom} g \subseteq (\mathrm{dom} f - \mathrm{dom} g) - (\mathrm{dom} f - \mathrm{dom} g)$.

(b) We have the following implication

$$(0,0) \in \mathrm{qi}\left[\mathrm{co}\left(\left(\mathrm{epi} f - \widehat{\mathrm{epi}}(g - v(P_{\mathrm{F}}))\right) \cup \{(0,0)\}\right)\right] \Rightarrow 0 \in \mathrm{qi}(\mathrm{dom} f - \mathrm{dom} g).$$

Indeed, suppose that $(0,0) \in \mathrm{qi}\left[\mathrm{co}\left(\left(\mathrm{epi} f - \widehat{\mathrm{epi}}(g - v(P_{\mathrm{F}}))\right) \cup \{(0,0)\}\right)\right]$. Then $\mathrm{cl}\left[\mathrm{coneco}\left(\left(\mathrm{epi} f - \widehat{\mathrm{epi}}(g - v(P_F))\right) \cup \{(0,0)\}\right)\right] = X \times \mathbb{R}$, hence (cf. (7.1))

$$\mathrm{cl}\left[\mathrm{cone}\left(\mathrm{epi} f - \widehat{\mathrm{epi}}(g - v(P_{\mathrm{F}}))\right)\right] = X \times \mathbb{R}.$$

As the inclusion

$$\mathrm{cl}\left[\mathrm{cone}\left(\mathrm{epi} f - \widehat{\mathrm{epi}}(g - v(P_{\mathrm{F}}))\right)\right] \subseteq \mathrm{cl}\left(\mathrm{cone}(\mathrm{dom} f - \mathrm{dom} g)\right) \times \mathbb{R}$$

trivially holds, we have $\mathrm{cl}\left(\mathrm{cone}(\mathrm{dom} f - \mathrm{dom} g)\right) = X$, that is

$$0 \in \mathrm{qi}(\mathrm{dom} f - \mathrm{dom} g).$$

Hence, the following implication is true

$$0 \notin \mathrm{qi}(\mathrm{dom} f - \mathrm{dom} g) \Rightarrow (0,0) \notin \mathrm{qi}\left[\mathrm{co}\left(\left(\mathrm{epi} f - \widehat{\mathrm{epi}}(g - v(P_{\mathrm{F}}))\right) \cup \{(0,0)\}\right)\right].$$

Nevertheless, in the regularity conditions given above one cannot substitute the condition $(0,0) \notin \mathrm{qi}\left[\mathrm{co}\left(\left(\mathrm{epi} f - \widehat{\mathrm{epi}}(g - v(P_{\mathrm{F}}))\right) \cup \{(0,0)\}\right)\right]$ by the stronger, but more handleable one $0 \notin \mathrm{qi}(\mathrm{dom} f - \mathrm{dom} g)$, since in all the regularity conditions (RC_i^{F}), $i \in \{6,7,8\}$, the other hypotheses imply $0 \in \mathrm{qi}(\mathrm{dom} f - \mathrm{dom} g)$ (cf. Proposition 7.14(i)).

We give now the following strong duality result concerning the primal-dual pair $(P_{\mathrm{F}}) - (D_{\mathrm{F}})$. It was first stated in [8] under convexity assumptions for the functions involved.

Theorem 7.16. *Let $f, g : X \to \overline{\mathbb{R}}$ be proper functions such that $v(P_{\mathrm{F}}) \in \mathbb{R}$ and* epi $f - \widehat{\mathrm{epi}}(g - v(P_{\mathrm{F}}))$ *is a convex subset of $X \times \mathbb{R}$ (the latter is the case if, for instance, f and g are convex functions). Suppose that either f and g are convex and (RC_6^{F}) is fulfilled, or one of the regularity conditions (RC_i^{F}), $i \in \{7,8\}$, holds. Then $v(P_{\mathrm{F}}) = v(D_{\mathrm{F}})$ and (D_{F}) has an optimal solution.*

Proof. One has to use the techniques employed in the proof of [8, Theorem 3.5]. □

When the condition $(0,0) \notin \operatorname{qri}\left[\operatorname{co}\left((\operatorname{epi} f - \widehat{\operatorname{epi}}(g - v(P_{\mathrm{F}}))) \cup \{(0,0)\}\right)\right]$ is removed, the duality result given above may fail. In the setting of Example 7.13, strong duality does not hold. Moreover, it has been proved in [8, Example 3.12(b)] that $(0,0) \in \operatorname{qri}\left[\operatorname{co}\left((\operatorname{epi} f - \widehat{\operatorname{epi}}(g - v(P_{\mathrm{F}}))) \cup \{(0,0)\}\right)\right]$.

The following example (given in [8, Example 3.13]) justifies the study of the regularity conditions expressed by means of the quasi interior and quasi-relative interior.

Example 7.17. Consider the real Hilbert space $\ell^2 = \ell^2(\mathbb{N})$. We define the functions $f, g : \ell^2 \to \overline{\mathbb{R}}$ by

$$f(x) = \begin{cases} \|x\|, \text{ if } x \in x^0 - \ell^2_+, \\ +\infty, \text{ otherwise} \end{cases}$$

and

$$g(x) = \begin{cases} \langle c, x \rangle, \text{ if } x \in \ell^2_+, \\ +\infty, \quad \text{otherwise}, \end{cases}$$

respectively, where $x^0, c \in \ell^2_+$ are arbitrarily chosen such that $x^0_n > 0$ for all $n \in \mathbb{N}$. Note that

$$v(P_{\mathrm{F}}) = \inf_{x \in \ell^2_+ \cap (x^0 - \ell^2_+)} \{\|x\| + \langle c, x \rangle\} = 0$$

and the infimum is attained at $x = 0$. We have $\operatorname{dom} f = x^0 - \ell^2_+ = \{(x_n)_{n\in\mathbb{N}} \in \ell^2 : x_n \leq x^0_n \ \forall n \in \mathbb{N}\}$ and $\operatorname{dom} g = \ell^2_+$. By using Example 7.4, we get

$$\operatorname{dom} f \cap \operatorname{qri}(\operatorname{dom} g) = \{(x_n)_{n\in\mathbb{N}} \in \ell^2 : 0 < x_n \leq x^0_n \ \forall n \in \mathbb{N}\} \neq \emptyset.$$

Also, $\operatorname{cl}\big(\operatorname{cone}(\operatorname{dom} g - \operatorname{dom} g)\big) = \ell^2$ and so $0 \in \operatorname{qi}(\operatorname{dom} g - \operatorname{dom} g)$. Further,

$$\operatorname{epi} f - \widehat{\operatorname{epi}}(g - v(P_{\mathrm{F}})) = \{(x - y, \|x\| + \langle c, y \rangle + \varepsilon) : x \in x^0 - \ell^2_+, y \in \ell^2_+, \varepsilon \geq 0\}.$$

In the following, we prove that $(0,0) \notin \operatorname{qri}\left(\operatorname{epi} f - \widehat{\operatorname{epi}}(g - v(P_{\mathrm{F}}))\right)$. Assuming the contrary, one would have that the set $\operatorname{cl}\left[\operatorname{cone}\left(\operatorname{epi} f - \widehat{\operatorname{epi}}(g - v(P_{\mathrm{F}}))\right)\right]$ is a linear subspace of $\ell^2 \times \mathbb{R}$. Since $(0,1) \in \operatorname{cl}\left[\operatorname{cone}\left(\operatorname{epi} f - \widehat{\operatorname{epi}}(g - v(P_{\mathrm{F}}))\right)\right]$ (take $x = y = 0$ and $\varepsilon = 1$), $(0,-1)$ must belong to this set, too. On the other hand, one can easily see that for all (x,r) belonging to $\operatorname{cl}\left[\operatorname{cone}\left(\operatorname{epi} f - \widehat{\operatorname{epi}}(g - v(P_{\mathrm{F}}))\right)\right]$ it holds $r \geq 0$. This leads to the desired contradiction.

Hence, the regularity condition (RC^{F}_6) is fulfilled, thus strong duality holds (cf. Theorem 7.16). On the other hand, ℓ^2 is a Fréchet space (being a Hilbert space), the functions f and g are proper, convex and lower semicontinuous and, as $\operatorname{sqri}(\operatorname{dom} f - \operatorname{dom} g) = \operatorname{sqri}(x^0 - \ell^2_+) = \emptyset$, none of the conditions (RC^{F}_i), $i \in \{1,2,3,4,5\}$, listed at the beginning of this section, can be applied for this optimization problem.

As for all $x^* \in \ell^2$ it holds $G^*(x^*) = \delta_{c-\ell^2_+}(x^*)$ and (cf. [26, Theorem 2.8.7])

$$f^*(-x^*) = \inf_{x_1^*+x_2^*=-x^*} \{\|\cdot\|^*(x_1^*) + \delta^*_{x^0-l^2_+}(x_2^*)\} = \inf_{\substack{x_1^*+x_2^*=-x^*,\\ \|x_1^*\|\leq 1, x_2^*\in\ell^2_+}} \langle x_2^*, x^0\rangle,$$

the optimal objective value of the Fenchel dual problem is

$$v(D_{\mathrm{F}}) = \sup_{\substack{x_2^*\in\ell^2_+-c-x_1^*,\\ \|x_1^*\|\leq 1, x_2^*\in\ell^2_+}} \langle -x_2^*, x^0\rangle = \sup_{x_2^*\in\ell^2_+} \langle -x_2^*, x^0\rangle = 0$$

and $x_2^* = 0$ is the optimal solution of the dual.

The following example (see also [14, Example 2.5]) underlines the fact that in general the regularity condition (RC_7^{F}) (and automatically also (RC_8^{F})) is weaker than (RC_6^{F}) (see also Example 7.28 below).

Example 7.18. Consider the real Hilbert space $\ell^2(\mathbb{R})$ and the functions f, g: $\ell^2(\mathbb{R}) \to \overline{\mathbb{R}}$ defined for all $s \in \ell^2(\mathbb{R})$ by

$$f(s) = \begin{cases} s(1), & \text{if } s \in \ell^2_+(\mathbb{R}),\\ +\infty, & \text{otherwise} \end{cases}$$

and

$$g(s) = \begin{cases} s(2), & \text{if } s \in \ell^2_+(\mathbb{R}),\\ +\infty, & \text{otherwise,} \end{cases}$$

respectively. The optimal objective value of the primal problem is

$$v(P_{\mathrm{F}}) = \inf_{s\in\ell^2_+(\mathbb{R})} \{s(1) + s(2)\} = 0$$

and $s = 0$ is an optimal solution (let us notice that (P_{F}) has infinitely many optimal solutions). We have $\mathrm{qri}(\mathrm{dom}\, g) = \mathrm{qri}(\ell^2_+(\mathbb{R})) = \emptyset$ (cf. Example 7.5), hence the condition (RC_6^{F}) fails. In the following, we show that (RC_7^{F}) is fulfilled. One can prove that $\mathrm{dom}\, f - \mathrm{dom}\, g = \ell^2_+(\mathbb{R}) - \ell^2_+(\mathbb{R}) = \ell^2(\mathbb{R})$, thus $0 \in \mathrm{qi}(\mathrm{dom}\, f - \mathrm{dom}\, g)$. Like in the previous example, we have

$$\mathrm{epi}\, f - \widehat{\mathrm{epi}}(g - v(P_{\mathrm{F}})) = \{(s - s', s(1) + s'(2) + \varepsilon) : s, s' \in \ell^2_+(\mathbb{R}), \varepsilon \geq 0\}$$

and with the same technique one can show that $(0,0) \notin \mathrm{qri}\left(\mathrm{epi}\, f - \widehat{\mathrm{epi}}(g - v(P_{\mathrm{F}}))\right)$, hence the condition (RC_7^{F}) holds.

Let us take a look at the formulation of the dual problem. To this end we have to calculate the conjugates of f and g. Let us recall that the scalar product on $\ell^2(\mathbb{R})$, $\langle\cdot,\cdot\rangle : \ell^2(\mathbb{R}) \times \ell^2(\mathbb{R}) \to \mathbb{R}$ is defined by $\langle s, s'\rangle = \sup_{F\subseteq\mathbb{R}, F \text{ finite}} \sum_{r\in F} s(r)s'(r)$, for $s, s' \in \ell^2(\mathbb{R})$ and that the dual space $\left(\ell^2(\mathbb{R})\right)^*$ is identified with $\ell^2(\mathbb{R})$. For an arbitrary $u \in \ell^2(\mathbb{R})$, we have

$$f^*(u) = \sup_{s\in\ell^2_+(\mathbb{R})} \{\langle u,s\rangle - s(1)\} = \sup_{s\in\ell^2_+(\mathbb{R})} \left\{ \sup_{\mathrm{F}\subseteq\mathbb{R},\mathrm{F}\text{finite}} \sum_{r\in\mathrm{F}} u(r)s(r) - s(1) \right\}$$

$$= \sup_{\mathrm{F}\subseteq\mathbb{R},F\text{ finite}} \left\{ \sup_{s\in\ell^2_+(\mathbb{R})} \left\{ \sum_{r\in\mathrm{F}} u(r)s(r) - s(1) \right\} \right\}.$$

If there exists $r\in\mathbb{R}\setminus\{1\}$ with $u(r)>0$ or if $u(1)>1$, then one has $f^*(u)=+\infty$. Assuming the contrary, for every finite subset F of $\mathbb{R}$, independently from the fact that 1 belongs to F or not, it holds $\sup_{s\in\ell^2_+(\mathbb{R})}\{\sum_{r\in\mathrm{F}}u(r)s(r)-s(1)\}=0$. Consequently,

$$f^*(u) = \begin{cases} 0, & \text{if } u(r)\le 0\ \forall r\in\mathbb{R}\setminus\{1\} \text{ and } u(1)\le 1, \\ +\infty, & \text{otherwise.} \end{cases}$$

Similarly, one can provide a formula for g^* and in this way we obtain that $v(D_{\mathrm{F}})=0$ and that $u=0$ is an optimal solution of the dual ((D_{F}) has actually infinitely many optimal solutions).

Let us compare in the following the regularity conditions expressed by means of the quasi interior and quasi relative interior with the classical ones form the literature, mentioned at the beginning of the section. To this end, we need an auxiliary result.

Proposition 7.19. *Suppose that for the primal-dual pair* $(P_{\mathrm{F}})-(D_{\mathrm{F}})$ *strong duality holds. Then* $(0,0)\notin \mathrm{qi}\left[\mathrm{co}\left((\mathrm{epi}\, f - \widehat{\mathrm{epi}}(g-v(P_{\mathrm{F}})))\cup\{(0,0)\}\right)\right]$.

Proof. By the assumptions we made, there exists $x^*\in X^*$ such that $v(P_{\mathrm{F}}) = -f^*(-x^*)-g^*(x^*) = \inf_{x\in X}\{\langle x^*,x\rangle + f(x)\} + \inf_{x\in X}\{\langle -x^*,x\rangle + g(x)\}$, hence

$$v(P_{\mathrm{F}}) \le \langle x^*,x\rangle + f(x) + \langle -x^*,y\rangle + g(y)\ \forall (x,y)\in X\times Y,$$

that is

$$\langle -x^*, x-y\rangle - (f(x)+g(y)-v(P_{\mathrm{F}})) \le 0\ \forall (x,y)\in \mathrm{dom}\, f\times \mathrm{dom}\, g.$$

We obtain

$$\langle (-x^*,-1),(z,r)\rangle \le 0\ \forall (z,r)\in \mathrm{epi}\, f - \widehat{\mathrm{epi}}(g-v(P_{\mathrm{F}})),$$

hence

$$\langle (-x^*,-1),(z,r)\rangle \le 0\ \forall (z,r)\in \mathrm{co}\left((\mathrm{epi}\, f - \widehat{\mathrm{epi}}(g-v(P_{\mathrm{F}})))\cup\{(0,0)\}\right).$$

The last relation ensures $(-x^*,-1) \in N_{[\mathrm{co}((\mathrm{epi} f-\widehat{\mathrm{epi}}(g-v(P_{\mathrm{F}})))\cup\{(0,0)\})]}(0,0)$ and Proposition 7.2 implies that $(0,0) \notin \mathrm{qi}\left[\mathrm{co}\left((\mathrm{epi} f - \widehat{\mathrm{epi}}(g-v(P_{\mathrm{F}}))) \cup \{(0,0)\}\right)\right]$. □

A comparison of the above regularity conditions is provided in the following.

Proposition 7.20. *Suppose that X is a Fréchet space and $f,g : X \to \overline{\mathbb{R}}$ are proper, convex and lower semicontinuous functions. The following relations hold*

$$(RC_1^{\mathrm{F}}) \Rightarrow (RC_2^{\mathrm{F}}) \Leftrightarrow (RC_3^{\mathrm{F}}) \Rightarrow (RC_7^{\mathrm{F}}) \Leftrightarrow (RC_8^{\mathrm{F}}).$$

Proof. In view of Remark 7.12 and Proposition 7.14(i) we have to prove only the implication $(RC_3^{\mathrm{F}}) \Rightarrow (RC_7^{\mathrm{F}})$. Let us suppose that (RC_3^{F}) is fulfilled. We apply (7.2) and obtain $0 \in \mathrm{qi}\left(\mathrm{dom} f - \mathrm{dom}\, g\right)$. Moreover, the regularity condition (RC_3^{F}) ensures strong duality for the pair $(P_{\mathrm{F}}) - (D_{\mathrm{F}})$ (cf. Theorem 7.11), hence $(0,0) \notin \mathrm{qi}\left[\mathrm{co}\left((\mathrm{epi} f - \widehat{\mathrm{epi}}(g-v(P_{\mathrm{F}}))) \cup \{(0,0)\}\right)\right]$ (cf. Proposition 7.19). Applying Proposition 7.14(iii) (see also Remark 7.15(a)) we get that the condition (RC_7^{F}) holds and the proof is complete. □

Remark 7.21. One can notice that the implications

$$(RC_1^{\mathrm{F}}) \Rightarrow (RC_7^{\mathrm{F}}) \Leftrightarrow (RC_8^{\mathrm{F}})$$

hold in the framework of separated locally convex spaces and for $f,g : X \to \overline{\mathbb{R}}$ proper and convex functions (nor completeness for the space neither lower semicontinuity for the functions is needed here).

Next we show that, in general, the conditions (RC_i^{F}), $i \in \{4,5\}$, cannot be compared with (RC_i^{F}), $i \in \{6,7,8\}$. Example 7.17 provides a situation for which (RC_i^{F}), $i \in \{6,7,8\}$, are fulfilled, unlike (RC_i^{F}), $i \in \{4,5\}$. In the following example, the conditions (RC_i^{F}), $i \in \{4,5\}$, are fulfilled, while (RC_i^{F}), $i \in \{6,7,8\}$, fail.

Example 7.22. Consider $(X, \|\cdot\|)$ a nonzero real Banach space, $x_0^* \in X^* \setminus \{0\}$ and the functions $f,g : X \to \overline{\mathbb{R}}$ defined by $f = \delta_{\ker x_0^*}$ and $g = \|\cdot\| + \delta_{\ker x_0^*}$, respectively. The optimal objective value of the primal problem is

$$v(P_{\mathrm{F}}) = \inf_{x \in \ker x_0^*} \|x\| = 0$$

and $\bar{x} = 0$ is the unique optimal solution of (P_{F}). The functions f and g are proper, convex and lower semicontinuous. Further, $\mathrm{dom} f - \mathrm{dom}\, g = \ker x_0^*$, which is a closed linear subspace of X, hence (RC_i^{F}), $i \in \{4,5\}$, are fulfilled. Moreover, $\mathrm{dom}\, g - \mathrm{dom}\, g = \mathrm{dom} f - \mathrm{dom}\, g = \ker x_0^*$ and it holds $\mathrm{cl}(\ker x_0^*) = \ker x_0^* \neq X$. Thus, $0 \notin \mathrm{qi}(\mathrm{dom}\, g - \mathrm{dom}\, g)$ and $0 \notin \mathrm{qi}(\mathrm{dom} f - \mathrm{dom}\, g)$ and this means that all the three regularity conditions (RC_i^{F}), $i \in \{6,7,8\}$, fail.

The conjugate functions of f and g are

$$f^* = \delta_{(\ker x_0^*)^\perp} = \delta_{\mathbb{R}x_0^*} \text{ and } g^* = \delta_{B_*(0,1)} \Box \delta_{\mathbb{R}x_0^*} = \delta_{B_*(0,1)+\mathbb{R}x_0^*},$$

respectively (cf. [26, Theorem 2.8.7]), where $B_*(0,1)$ is the closed unit ball of the dual space X^*. Hence, $v(D_{\mathrm{F}}) = 0$ and the set of optimal solutions of (D_{F}) coincides with $\mathbb{R}x_0^*$. Finally, let us notice that instead of $\ker x_0^*$ one can consider any closed linear subspace S of X such that $S \neq X$.

7.3.3 Closedness-Type Regularity Conditions

Besides the generalized interior-point regularity conditions, there exist in the literature so-called *closedness-type regularity conditions* for conjugate duality. In the following, we will recall two sufficient conditions of this type for Fenchel duality and we will relate them to the ones investigate in the previous subsection. Let these two conditions be:

$$(RC_9^{\mathrm{F}}) \left| \begin{array}{l} f \text{ and } g \text{ are lower semicontinuous and} \\ \operatorname{epi} f^* + \operatorname{epi} g^* \text{ is closed in } (X^*, w(X^*,X)) \times \mathbb{R} \end{array} \right.$$

and

$$(RC_{10}^{\mathrm{F}}) \left| \begin{array}{l} f \text{ and } g \text{ are lower semicontinuous, } f^* \Box g^* \text{ is } w(X^*,X)\text{-lower} \\ \text{semicontinuous on } X^* \text{ and exact at } 0. \end{array} \right.$$

The condition (RC_9^{F}) has been first considered by Burachik and Jeyakumar in Banach spaces (cf. [10]) and by Boţ and Wanka in separated locally convex spaces (cf. [7]), while the second one, (RC_{10}^{F}), has been introduced in [7]. We have the following duality results (cf. [7]).

Theorem 7.23. *Let $f,g : X \to \overline{\mathbb{R}}$ be proper and convex functions such that* $\operatorname{dom} f \cap \operatorname{dom} g \neq \emptyset$. *If (RC_9^{F}) is fulfilled, then*

$$(f+g)^*(x^*) = \min\{f^*(x^*-y^*) + g^*(y^*) : y^* \in X^*\} \ \forall x^* \in X^*. \tag{7.3}$$

Theorem 7.24. *Let $f,g : X \to \overline{\mathbb{R}}$ be proper and convex functions such that* $\operatorname{dom} f \cap \operatorname{dom} g \neq \emptyset$. *If (RC_{10}^{F}) is fulfilled, then $v(P_{\mathrm{F}}) = v(D_{\mathrm{F}})$ and (D_{F}) has an optimal solution.*

Remark 7.25. (a) Let us notice that condition (7.3) is referred in the literature as *stable strong duality* (see [6, 11, 22] for more details) and obviously guarantees strong duality for $(P_{\mathrm{F}}) - (D_{\mathrm{F}})$. When $f,g : X \to \overline{\mathbb{R}}$ are proper, convex and lower semicontinuous functions with $\operatorname{dom} f \cap \operatorname{dom} g \neq \emptyset$ one has in fact that (RC_9^{F}) is fulfilled if and only if (7.3) holds (cf. [7, Theorem 3.2]).

(b) If f,g are proper, convex and lower semicontinuous such that $\operatorname{dom} f \cap \operatorname{dom} g \neq \emptyset$, then $(RC_9^{\mathrm{F}}) \Rightarrow (RC_{10}^{\mathrm{F}})$ (cf. [7, Sect. 4]). Moreover, there are examples showing that in general (RC_{10}^{F}) is weaker than (RC_9^{F}) (see[7]). Finally, let us mention that (under the same hypotheses) $f^*\Box g^*$ is a $w(X^*,X)$-lower semicontinuous function on X^* if and only if $(f+g)^* = f^*\Box g^*$. This is a direct consequence of the equality $(f+g)^* = \operatorname{cl}(f^*\Box g^*)$, where the closure is considered with respect to the weak* topology on X^* (cf. [7, Thorem 2.1]).

(c) In case X is a Fréchet space and f,g are proper, convex and lower semicontinuous functions, we have the following relations between the regularity conditions considered for the primal-dual pair $(P_{\mathrm{F}}) - (D_{\mathrm{F}})$ (cf. [7], see also [17] and [26, Theorem 2.8.7])

$$(RC_1^{\mathrm{F}}) \Rightarrow (RC_2^{\mathrm{F}}) \Leftrightarrow (RC_3^{\mathrm{F}}) \Rightarrow (RC_4^{\mathrm{F}}) \Leftrightarrow (RC_5^{\mathrm{F}}) \Rightarrow (RC_9^{\mathrm{F}}) \Rightarrow (RC_{10}^{\mathrm{F}}).$$

We refer to [6, 7, 10, 22] for several examples showing that in general the implications above are strict. The implication $(RC_1^{\mathrm{F}}) \Rightarrow (RC_9^{\mathrm{F}}) \Rightarrow (RC_{10}^{\mathrm{F}})$ holds in the general setting of separated locally convex spaces (in the hypotheses that f,g are proper, convex and lower semicontinuous).

We observe that if X is a finite-dimensional space and f,g are proper, convex and lower semicontinuous, then $(RC_6^{\mathrm{F}}) \Rightarrow (RC_7^{\mathrm{F}}) \Leftrightarrow (RC_8^{\mathrm{F}}) \Rightarrow (RC_9^{\mathrm{F}}) \Rightarrow (RC_{10}^{\mathrm{F}})$. However, in the infinite-dimensional setting this is no longer true. In the following two examples, the conditions (RC_9^{F}) and (RC_{10}^{F}) are fulfilled, unlike (RC_i^{F}), $i \in \{6,7,8\}$ (we refer to [6, 7, 10, 19, 22] for examples in the finite-dimensional setting).

Example 7.26. Consider the same setting as in Example 7.22. We know that (RC_5^{F}) is fulfilled, hence also (RC_9^{F}) and (RC_{10}^{F}) (cf. Remark 7.25(c)). This is not surprising, since $\operatorname{epi} f^* + \operatorname{epi} g^* = (B_*(0,1) + \mathbb{R}x_0^*) \times [0,\infty)$, which is closed in $(X^*, w(X^*,X)) \times \mathbb{R}$ (note that by the Banach–Alaoglu Theorem, the unit ball $B_*(0,1)$ is compact in $(X^*, w(X^*,X))$). As shown in Example 7.22, none of the regularity conditions (RC_i^{F}), $i \in \{6,7,8\}$, is fulfilled.

Example 7.27. Consider the real Hilbert space $\ell^2(\mathbb{R})$ and the functions $f,g : \ell^2(\mathbb{R}) \to \overline{\mathbb{R}}$ defined by $f = \delta_{\ell^2_+(\mathbb{R})}$ and $g = \delta_{-\ell^2_+(\mathbb{R})}$, respectively. We have $\operatorname{qri}(\operatorname{dom} f - \operatorname{dom} g) = \operatorname{qri}\left(\ell^2_+(\mathbb{R})\right) = \emptyset$ (cf. Example 7.5), hence all the generalized interior-point regularity conditions (RC_i^{F}), $i \in \{1,2,3,4,5,6,7,8\}$, fail (see also Proposition 7.14(i)). The conjugate functions of f and g are $f^* = \delta_{-\ell^2_+(\mathbb{R})}$ and $g^* = \delta_{\ell^2_+(\mathbb{R})}$, respectively, hence $\operatorname{epi} f^* + \operatorname{epi} g^* = \ell^2(\mathbb{R}) \times [0,\infty)$, that is the condition (RC_9^{F}) holds (hence also (RC_{10}^{F}), cf. Remark 7.25(b)). One can see that $v(P_{\mathrm{F}}) = v(D_{\mathrm{F}}) = 0$ and $y^* = 0$ is an optimal solution of the dual problem.

The next issue we investigate concerns the relation between the generalized interior-point conditions (RC_i^{F}), $i \in \{6,7,8\}$ and the closedness-type ones (RC_9^{F}) and (RC_{10}^{F}). In the last two examples the conditions (RC_9^{F}) and (RC_{10}^{F}) are fulfilled, while (RC_i^{F}), $i \in \{6,7,8\}$, fail. In the following we provide an example for which (RC_7^{F}) is fulfilled, unlike (RC_i^{F}), $i \in \{9,10\}$. In this way we give a negative answer to an open problem stated in [19, Remark 4.3], concomitantly proving that in general (RC_7^{F}) (and automatically also (RC_8^{F})) and (RC_9^{F}) are not comparable.

Example 7.28. (See also [14, Example 2.7]) Like in Example 7.13, consider the real Hilbert space $X = \ell^2(\mathbb{N})$ and the sets

$$C = \{(x_n)_{n\in\mathbb{N}} \in \ell^2 : x_{2n-1} + x_{2n} = 0 \ \forall n \in \mathbb{N}\}$$

and

$$S = \{(x_n)_{n\in\mathbb{N}} \in \ell^2 : x_{2n} + x_{2n+1} = 0 \ \forall n \in \mathbb{N}\},$$

which are closed linear subspaces of ℓ^2 and satisfy $C \cap S = \{0\}$. Define the functions $f, g : \ell^2 \to \overline{\mathbb{R}}$ by $f = \delta_C$ and $g = \delta_S$, respectively, which are proper, convex and lower semicontinuous. The optimal objective value of the primal problem is $v(P_{\mathrm{F}}) = 0$ and $\overline{x} = 0$ is the unique optimal solution of $v(P_{\mathrm{F}})$. Moreover, S–C is dense in ℓ^2 (cf. [17, Example 3.3]), thus $\mathrm{cl}\big(\mathrm{cone}(\mathrm{dom}\, f - \mathrm{dom}\, g)\big) = \mathrm{cl}(C - S) = \ell^2$. This implies $0 \in \mathrm{qi}(\mathrm{dom}\, f - \mathrm{dom}\, g)$. Further, one has

$$\mathrm{epi}\, f - \widehat{\mathrm{epi}}(g - v(P_{\mathrm{F}})) = \{(x - y, \varepsilon) : x \in C, y \in S, \varepsilon \geq 0\} = (C - S) \times [0, +\infty)$$

and $\mathrm{cl}\Big[\mathrm{cone}\Big(\mathrm{epi}\, f - \widehat{\mathrm{epi}}(g - v(P_{\mathrm{F}}))\Big)\Big] = \ell^2 \times [0, +\infty)$, which is not a linear subspace of $\ell^2 \times \mathbb{R}$, hence $(0,0) \notin \mathrm{qri}\Big(\mathrm{epi}\, f - \widehat{\mathrm{epi}}(g - v(P_{\mathrm{F}}))\Big)$. All together, we get that the condition (RC_7^{F}) is fulfilled, hence strong duality holds (cf. Theorem 7.16). One can prove that $f^* = \delta_{C^\perp}$ and $g^* = \delta_{S^\perp}$, where

$$C^\perp = \{(x_n)_{n\in\mathbb{N}} \in \ell^2 : x_{2n-1} = x_{2n} \ \forall n \in \mathbb{N}\}$$

and

$$S^\perp = \{(x_n)_{n\in\mathbb{N}} \in \ell^2 : x_1 = 0, x_{2n} = x_{2n+1} \ \forall n \in \mathbb{N}\}.$$

Further, $v(D_{\mathrm{F}}) = 0$ and the set of optimal solutions of the dual problem is exactly $C^\perp \cap S^\perp = \{0\}$.

We show that (RC_{10}^{F}) is not fulfilled (hence (RC_9^{F}) fails too, cf. Remark 7.25(b)). Let us consider the element $e^1 \in \ell^2$, defined by $e_1^1 = 1$ and $e_k^1 = 0$ for all $k \in \mathbb{N} \setminus \{1\}$. We compute $(f + g)^*(e^1) = \sup_{x\in\ell^2}\{\langle e^1, x\rangle - f(x) - g(x)\} = 0$ and $(f^* \Box g^*)(e^1) = \delta_{C^\perp + S^\perp}(e^1)$. If we suppose that $e^1 \in C^\perp + S^\perp$, then we would have $(e^1 + S^\perp) \cap C^\perp \neq \emptyset$. However, it has been proved in [17, Example 3.3] that $(e^1 + S^\perp) \cap C^\perp = \emptyset$. This shows that $(f^* \Box g^*)(e^1) = +\infty > 0 = (f + g)^*(e^1)$. Via Remark 7.25(b) it follows that the condition (RC_{10}^{F}) is not fulfilled and, consequently, (RC_i^{F}), $i \in \{1,2,3,4,5,9\}$, fail, too (cf. Remark 7.25(c)), unlike condition (RC_7^{F}). Concerning (RC_6^{F}), one can see that this condition is not fulfilled, since $0 \in \mathrm{qi}(\mathrm{dom}\, g - \mathrm{dom}\, g)$ does not hold.

In the next example, the conditions (RC_i^{F}), $i \in \{6,7,8\}$, are fulfilled and (RC_9^{F}) fails.

Example 7.29. The example we consider in the following is inspired by [22, Example 11.3]. Consider X an arbitrary Banach space, C a convex and closed subset of X and x_0 an extreme point of C which is not a support point of C. Taking for instance $X = \ell^2$, $1 < p < 2$ and $C := \left\{x \in \ell^2 : \sum_{n=1}^{\infty} |x_n|^p \leq 1\right\}$ one can find extreme points in C that are not support points (see [22]). Consider the functions $f,g : X \to \overline{\mathbb{R}}$ defined as $f = \delta_{x_0 - C}$ and $g = \delta_{C - x_0}$, respectively. They are both proper, convex and lower semicontinuous and fulfill, as x_0 is an extreme point of C, $f + g = \delta_{\{0\}}$. Thus $v(P_{\mathrm{F}}) = 0$ and $\overline{x} = 0$ is the unique optimal solution of (P_{F}). We show that, different to the previous example, (RC_6^{F}) is fulfilled and this will guarantee that both (RC_7^{F}) and (RC_8^{F}) are valid, too (cf. Proposition 7.14(i)). To this end, we notice first that $x_0 \in \mathrm{qi}(C)$. Assuming the contrary, one would have that there exists $x^* \in X^* \setminus \{0\}$ such that $\langle x^*, x_0\rangle = \sup_{x \in C}\langle x^*, x\rangle$ (cf. Proposition 7.2), contradicting the hypothesis that x_0 is not a support point of C. This means that $x_0 \in \mathrm{qri}(C)$, too, and so $0 \in \mathrm{dom} f \cap \mathrm{qri}(\mathrm{dom} g)$. Further, since it holds $\mathrm{cl}(\mathrm{cone}(C - x_0)) \subseteq \mathrm{cl}(\mathrm{cone}(C - C))$, we have $\mathrm{cl}(\mathrm{cone}(C - C)) = X$ and from here $0 \in \mathrm{qi}(C - C) = \mathrm{qi}(\mathrm{dom} g - \mathrm{dom} g)$. Noticing that

$$\mathrm{epi} f - \widehat{\mathrm{epi}}(g - v(P_{\mathrm{F}})) = \{(x - y, \varepsilon) : x, y \in C, \varepsilon \geq 0\} = (C - C) \times [0, +\infty),$$

it follows that $\mathrm{cl}\left[\mathrm{cone}\left(\mathrm{epi} f - \widehat{\mathrm{epi}}(g - v(P_{\mathrm{F}}))\right)\right] = X \times [0, +\infty)$, which is not a linear subspace of $X \times \mathbb{R}$. Thus, $(0,0) \notin \mathrm{qri}\left(\mathrm{epi} f - \widehat{\mathrm{epi}}(g - v(P_{\mathrm{F}}))\right)$ and this has as consequence the fact that (RC_6^{F}) is fulfilled. Hence strong duality holds (cf. Theorem 7.16), $v(D_{\mathrm{F}}) = 0$ and 0 is an optimal solution of the dual problem.

We show that (RC_9^{F}) is not fulfilled. Assuming the contrary, one would have that the equality in (7.3) holds for all $x^* \in X^*$. On the other hand, in [22, Example 11.3] it is proven that this is the case only when $x^* = 0$ and this provides the desired contradiction.

Remark 7.30. Consider the following optimization problem

$$(P_{\mathrm{F}}^A) \ \inf_{x \in X}\{f(x) + (g \circ A)(x)\},$$

where X and Y are separated locally convex spaces with topological dual spaces X^* and Y^*, respectively, $A : X \to Y$ is a linear continuous mapping, $f : X \to \overline{\mathbb{R}}$ and $g : Y \to \overline{\mathbb{R}}$ are proper functions such that $A(\mathrm{dom} f) \cap \mathrm{dom} g \neq \emptyset$. The Fenchel dual problem to (P_{F}^A) is

$$(D_{\mathrm{F}}^A) \ \sup_{y^* \in Y^*}\{-f^*(-A^* y^*) - g^*(y^*)\},$$

where $A^* : Y^* \to X^*$ is the *adjoint operator*, defined by $\langle A^* y^*, x\rangle = \langle y^*, Ax\rangle$ for all $y^* \in Y^*$ and $x \in X$. We denote by $v(P_{\mathrm{F}}^A)$ and $v(D_{\mathrm{F}}^A)$ the optimal objective values of the primal and the dual problem, respectively, and suppose that $v(P_{\mathrm{F}}^A) \in \mathbb{R}$. We consider the set

$$A \times \mathrm{id}_{\mathbb{R}}(\mathrm{epi} f) = \{(Ax, r) \in Y \times \mathbb{R} : f(x) \leq r\}.$$

By using the approach presented in the previous section one can provide similar discussions regarding strong duality for the primal-dual pair $(P_{\mathrm{F}}^{A})-(D_{\mathrm{F}}^{A})$. To this end, we introduce the following functions: $F,G : X\times Y\to\overline{\mathbb{R}}$, $F(x,y)=f(x)+\delta_{\{u\in X:Au=y\}}(x)$ and $G(x,y)=g(y)$ for all $(x,y)\in X\times Y$. The functions F and G are proper and their domains fulfill the relation

$$\operatorname{dom}F-\operatorname{dom}G=X\times\big(A(\operatorname{dom}f)-\operatorname{dom}g\big).$$

Since $\operatorname{epi}F=\{(x,Ax,r):f(x)\le r\}$ and $\widehat{\operatorname{epi}}(G-v(P_{\mathrm{F}}^{A}))=\{(x,y,r):r\le -G(x,y)+v(P_{\mathrm{F}}^{A})\}=X\times\widehat{\operatorname{epi}}(g-v(P_{\mathrm{F}}^{A}))$, we obtain

$$\operatorname{epi}F-\widehat{\operatorname{epi}}(G-v(P_{\mathrm{F}}^{A}))=X\times\Big(A\times\mathrm{id}_{\mathbb{R}}(\operatorname{epi}f)-\widehat{\operatorname{epi}}(g-v(P_{\mathrm{F}}^{A}))\Big).$$

Moreover,

$$\inf_{(x,y)\in X\times Y}\{F(x,y)+G(x,y)\}=\inf_{x\in X}\{f(x)+(g\circ A)(x)\}=v(P_{\mathrm{F}}^{A}).$$

On the other hand, for all $(x^*,y^*)\in X^*\times Y^*$ we have $F^*(x^*,y^*)=f^*(x^*+A^*y^*)$ and

$$G^*(x^*,y^*)=\begin{cases} g^*(y^*), & \text{if } x^*=0,\\ +\infty, & \text{otherwise.}\end{cases}$$

Therefore,

$$\sup_{\substack{x^*\in X^*\\ y^*\in Y^*}}\{-F^*(-x^*,-y^*)-G^*(x^*,y^*)\}=\sup_{y^*\in Y^*}\{-f^*(-A^*y^*)-g^*(y^*)\}=v(D_{\mathrm{F}}^{A}).$$

For more details concerning this approach, we refer to [8, 14].

We remark that Borwein and Lewis gave in [3] some regularity conditions by means of the quasi-relative interior, in order to guarantee strong duality for (P_{F}^{A}) and (D_{F}^{A}). However, they considered a more restrictive case, namely when the codomain of the operator A is finite-dimensional. Here we have considered the more general case, when both spaces X and Y are infinite-dimensional.

Finally, let us notice that several regularity conditions by means of the quasi interior and quasi-relative interior were introduced in the literature in order to guarantee strong duality between a primal optimization problem with geometric and cone constraints and its Lagrange dual problem. However, they have either contradictory assumptions, like in [12], or superfluous conditions, like in [16]. For a detailed argumentation of these considerations and also for correct alternative strong duality results in the case of Lagrange duality we refer to [8, 9].

Acknowledgements The research of the first author was partially supported by DFG (German Research Foundation), project WA 922/1-3.

References

1. Attouch, H., Brézis, H.: Duality for the sum of convex functions in general Banach spaces. In: J.A. Barroso (ed.) Aspects of Mathematics and Its Applications, North-Holland Publishing Company, Amsterdam, pp. 125–133 (1986)
2. Borwein, J.M., Goebel, R.: Notions of relative interior in Banach spaces. J. Math. Sci. (New York) **115**, 2542–2553 (2003)
3. Borwein, J.M., Lewis, A.S.: Partially finite convex programming, part I: Quasi relative interiors and duality theory. Math. Programming **57**, 15–48 (1992)
4. Borwein, J.M., Jeyakumar, V., Lewis, A.S., Wolkowicz, H.: Constrained approximation via convex programming (1988). Preprint, University of Waterloo
5. Borwein, J.M., Lucet, Y., Mordukhovich, B.: Compactly epi-Lipschitzian convex sets and functions in normed spaces. J. Convex Anal. **7**, 375–393 (2000)
6. Boţ, R.I.: Conjugate Duality in Convex Optimization, *Lecture Notes in Economics and Mathematical Systems*, vol. 637. Springer, Berlin (2010)
7. Boţ, R.I., Wanka, G.: A weaker regularity condition for subdifferential calculus and Fenchel duality in infinite dimensional spaces. Nonlinear Anal. **64**, 2787–2804 (2006)
8. Boţ, R.I., Csetnek, E.R., Wanka, G.: Regularity conditions via quasi-relative interior in convex programming. SIAM J. Optim. **19**, 217–233 (2008)
9. Boţ, R.I., Csetnek, E.R., Moldovan, A.: Revisiting some duality theorems via the quasirelative interior in convex optimization. J. Optim. Theory Appl. **139**, 67–84 (2008)
10. Burachik, R.S., Jeyakumar, V.: A new geometric condition for Fenchel's duality in infinite dimensional spaces. Math. Programming **104**, 229–233 (2005)
11. Burachik, R.S., Jeyakumar, V., Wu, Z.-Y.: Necessary and sufficient conditions for stable conjugate duality. Nonlinear Anal. **64**, 1998–2006 (2006)
12. Cammaroto, F., Di Bella, B.: Separation theorem based on the quasirelative interior and application to duality theory. J. Optim. Theory Appl. **125**, 223–229 (2005)
13. Cammaroto, F., Di Bella, B.: On a separation theorem involving the quasi-relative interior. Proc. Edinburgh Math. Soc. (2) **50**, 605–610 (2007)
14. Csetnek, E.R.: Overcoming the failure of the classical generalized interior-point regularity conditions in convex optimization. Applications of the duality theory to enlargements of maximal monotone operators. Ph.D. Thesis, Chemnitz University of Technology, Germany (2009)
15. Daniele, P., Giuffrè, S.: General infinite dimensional duality and applications to evolutionary network equilibrium problems. Optim. Lett. **1**, 227–243 (2007)
16. Daniele, P., Giuffrè, S., Idone, G., Maugeri, A.: Infinite dimensional duality and applications. Math. Ann. **339**, 221–239 (2007)
17. Gowda, M.S., Teboulle, M.: A comparison of constraint qualifications in infinite-dimensional convex programming. SIAM J. Control Optim. **28**, 925–935 (1990)
18. Holmes, R.B.: Geometric Functional Analysis and its Applications. Springer, Berlin (1975)
19. Li, C., Fang, D., López, G., López, M.A.: Stable and total Fenchel duality for convex optimization problems in locally convex spaces. SIAM J. Optim. **20**, 1032–1051 (2009)
20. Limber, M.A., Goodrich, R.K.: Quasi interiors, Lagrange multipliers, and L^p spectral estimation with lattice bounds. J. Optim. Theory Appl. **78**, 143–161 (1993)
21. Rockafellar, R.T.: Conjugate duality and optimization. Conference Board of the Mathematical Sciences Regional Conference Series in Applied Mathematics **16**, Society for Industrial and Applied Mathematics, Philadelphia (1974)
22. Simons, S.: From Hahn-Banach to Monotonicity, *Lecture Notes in Mathematics*, vol. 1693. Springer, New York (2008)
23. Tanaka, T., Kuroiwa, D.: The convexity of A and B assures $\text{int}A + B = \text{int}(A+B)$. Appl. Math. Lett. **6**, 83–86 (1993)
24. Zălinescu, C.: Solvability results for sublinear functions and operators. Math. Methods Oper. Res. **31**, A79–A101 (1987)
25. Zălinescu, C.: A comparison of constraint qualifications in infinite-dimensional convex programming revisited. J. Austral. Math. Soc. Ser. B **40**, 353–378 (1999)
26. Zălinescu, C.: Convex Analysis in General Vector Spaces. World Scientific, New Jersey (2002)

Chapter 8
Non-Local Functionals for Imaging

Jérôme Boulanger, Peter Elbau, Carsten Pontow, and Otmar Scherzer

Abstract Non-local functionals have been successfully applied in a variety of applications, such as spectroscopy or in general filtering of time-dependent data. We mention the patch-based denoising of image sequences [Boulanger et al. IEEE Transactions on Medical Imaging (2010)]. Another family of non-local functionals considered in these notes approximates total variation denoising. Thereby we rely on fundamental characteristics of Sobolev spaces and the space of functions of finite total variation (see [Bourgain et al. Journal d'Analyse Mathématique 87, 77–101 (2002)] and several follow up papers). Standard results of the calculus of variations, like for instance the relation between lower semi-continuity of the functional and convexity of the integrand, do not apply, in general, for the non-local functionals. In this paper we address the questions of the calculus of variations for non-local functionals and derive relations between lower semi-continuity of the functionals and separate convexity of the integrand. Moreover, we use the new characteristics of Sobolev spaces to derive novel approximations of the total variation energy regularisation. All the functionals are well-posed and reveal a unique minimising point. Even more, existing numerical schemes can be recovered in this general framework.

Keywords Non-local functionals · Derivative free model · Total variation regularisation · Neighbourhood filter · Patch-based filter

AMS 2010 Subject Classification: 49J05, 49J45, 49M25

8.1 Introduction

A standard way in image analysis to regularise a given image u_0 is to define for all images u an energy $E(u)$, which compromises the similarity of u and u_0 with the regularity of u. The regularised image is then the minimising point of E. A typical

P. Elbau (✉)
Johann Radon Institute for Computational and Applied Mathematics,
Austrian Academy of Sciences, Altenbergerstraße 69, 4040 Linz, Austria
e-mail: peter.elbau@ricam.oeaw.ac.at

H.H. Bauschke et al. (eds.), *Fixed-Point Algorithms for Inverse Problems in Science and Engineering*, Springer Optimization and Its Applications 49,
DOI 10.1007/978-1-4419-9569-8_8,

choice of E is the integral of an energy density $f(x,u(x),|\nabla u(x)|;u_0(x))$, which only depends on the image values of u and u_0 in an infinitesimal small neighbourhood around a point x of the image domain.

But such energy functionals are not suitable for regularisations that aim for taking into account multiple structures in an image. For that purpose, filtering techniques are used which compare the value $u(x)$ with values $u(y)$, which are similar to $u(x)$ or in a region similar to the one around $u(x)$. This idea leads to neighbourhood filters, as introduced in [23], and more generally to patch-based filters as for instance the non-local means filter, see [6]. Recently, it was indicated in [12] that these filters can be also formulated as non-local energy minimisation problems where the energy is (in the simplest case) an integral of a density of the form $f(x,y,u(x),u(y);u_0(x),u_0(y))$ over all pairs of points (x,y) in the image.

On the other hand, using in the classical energy formulation finite difference quotients $\frac{|u(x)-u(y)|}{|x-y|}$ as approximations of the norm $|\nabla u(x)|$ of the gradient results in the same class of non-local energy minimisation problems. This approach was studied in particular for total variation minimisation in [1] and was reviewed in [19]. A similar approach was considered in [11], where the concept of non-local operators was introduced.

Our aim is now to provide a general theory about the existence of minimising points of such non-local energy functionals, i.e. of functionals $\mathcal{J}$ of the form

$$\mathcal{J}: L^p(X;\mathbb{R}^n) \to \mathbb{R}\cup\{\infty\}, \quad \mathcal{J}(u) = \int_X \int_X f(x,y,u(x),u(y))\,\mathrm{d}x\,\mathrm{d}y,$$

where the density f now depends on pairs of points and their image values. In the case of patch-based filters, we even have to consider densities depending on the image values at all points close to x and y.

In the first section, we will review how the mentioned examples can be cast as a minimisation problem for a non-local functional. Then we will provide in the next section an existence result for minimising points, which we are going to apply in the last section to the three types of functionals introduced in the first section.

8.2 Examples of Non-Local Functionals

In these notes, we are going to analyse the behaviour of non-local functionals on some Lebesgue space. Let us first clarify the terminology non-local functional. Throughout the text, X shall denote a bounded, open subset of $\mathbb{R}^m$ for some $m \in \mathbb{N}$ endowed with the Lebesgue measure on the σ-algebra $\mathscr{A}$ of all Lebesgue measurable subsets of X. Moreover, we define $\mathscr{B}$ to be the Borel σ-algebra of $\mathbb{R}$.

Definition 8.1. Let $n \in \mathbb{N}$ and $p \in [1,\infty)$. We call a map $\mathcal{J}: L^p(X;\mathbb{R}^n) \to \mathbb{R}\cup\{\infty\}$, which is not constantly equal to infinity, a non-local functional on $L^p(X;\mathbb{R}^n)$ if there exists a function $f: X \times X \times \mathbb{R}^n \times \mathbb{R}^n \to \mathbb{R}$ such that

$$\mathscr{J}(u) = \int_X \int_X f(x,y,u(x),u(y))\,\mathrm{d}x\,\mathrm{d}y \quad \text{for all} \quad u \in L^p(X;\mathbb{R}^n) \tag{8.1}$$

and such that

(i) f is measurable with respect to the σ-algebras $\mathscr{A} \times \mathscr{A} \times \mathscr{B}^n \times \mathscr{B}^n$ and $\mathscr{B}$,
(ii) f has the symmetry

$$f(x,y,w,z) = f(y,x,z,w) \quad \text{for all} \quad x,y \in X,\ w,z \in \mathbb{R}^n, \tag{8.2}$$

(iii) The negative part f^- of f fulfils

$$\int_X \int_X f^-(x,y,u(x),u(y))\,\mathrm{d}x\,\mathrm{d}y < \infty \quad \text{for all} \quad u \in L^p(X;\mathbb{R}^n)\,.$$

We then say that $\mathscr{J}$ is the non-local functional defined by f. Sometimes it is convenient to express the dependency of the functional on p and f and then we write $\mathscr{J}_f^p$ instead of $\mathscr{J}$.

The symmetry condition (8.2) in this definition is just introduced for convenience, since a function $\tilde{f}$ and its symmetrisation f, given by

$$f(x,y,w,z) = \frac{1}{2}(\tilde{f}(x,y,w,z) + \tilde{f}(y,x,z,w)) \quad \text{for all} \quad x,y \in X,\ w,z \in \mathbb{R}^n,$$

would define the same non-local functional anyway.

Before giving a criterion for non-local functionals to possess a minimising point, we first list a few examples where such functionals have been documented in the literature recently.

8.2.1 A Derivative Free Model for the Sobolev and Total Variation Seminorm

Let $q \in [1,\infty)$ and $(\varphi_k)_{k\in\mathbb{N}}$ be a sequence of non-negative, radially symmetric and radially decreasing functions in $L^1(\mathbb{R}^m)$ such that we have for all $k \in \mathbb{N}$

$$\int_{\mathbb{R}^m} \varphi_k(x)\,\mathrm{d}x = 1 \quad \text{and} \quad \lim_{k\to\infty} \int_{\{y\in\mathbb{R}^m \,:\, |y|>\delta\}} \varphi_k(x)\,\mathrm{d}x = 0$$

for every $\delta \in (0,\infty)$. It is shown in [4] that there exist constants $K_{q,m} \in (0,\infty)$ such that the functionals

$$\mathscr{R}_k^q(u) = \int_X \int_X \frac{|u(x)-u(y)|^q}{|x-y|^q} \varphi_k(x-y)\,\mathrm{d}x\,\mathrm{d}y, \quad k \in \mathbb{N},$$

fulfil for every measurable real function u that

$$\lim_{k\to\infty} K_{q,m}\mathscr{R}_k^q(u) = \begin{cases} \int_X |\nabla u(x)|^q \,\mathrm{d}x & \text{if } q>1 \text{ and } u \in W^{1,q}(X), \\ |Du|(X) & \text{if } q=1 \text{ and } u \in BV(X), \\ \infty & \text{otherwise.} \end{cases}$$

So, we could think of the functional $K_{q,m}\mathscr{R}_k^q$ as an approximation for the qth power of the Sobolev seminorm of u if $q > 1$ and for the total variation seminorm of u if $q = 1$.

Let $p \in (1,\infty)$. The energy minimisation problem then consists in minimisation of the energy functional

$$\mathscr{J} : L^p(X) \to \mathbb{R} \cup \{\infty\}, \quad \mathscr{J}(u) = \alpha \mathscr{R}_k^q(u) + \frac{1}{p}\int_X |u(x) - u_0(x)|^p \,\mathrm{d}x$$

for some regularisation parameter $\alpha \in (0,\infty)$ and some $k \in \mathbb{N}$, which compromises smoothness for the regularised solution with respect to the approximation of the Sobolev seminorm and closeness of the data with respect to the L^p norm. These minimisation problems have been considered in [1]. In [19], numerical realisations have been derived which reveal the relations to various filtering techniques such as bilateral filtering.

8.2.2 Neighbourhood Filters and Non-Local Functionals

Neighbourhood filters approximate a noisy image $u_0 : X \to \mathbb{R}$ at a point $x \in X$ by an average over the domain with similar intensity to $u_0(x)$. The filtered image $u : X \to \mathbb{R}$ is thus calculated by

$$u(x) = \frac{1}{C}\int_X K(x,y,u_0)u_0(y)\,\mathrm{d}y, \quad C = \int_X K(x,y,u_0)\,\mathrm{d}y, \tag{8.3}$$

for some kernel function K providing a measure for the distance between the points $(x,u_0(x))$ and $(y,u_0(y))$ (assuming that the integrals are well-defined). A typical choice of K is

$$K(x,y,u_0) = g(|u_0(x) - u_0(y)|^2)k(x,y) \tag{8.4}$$

for some positive, bounded, continuous function $g \in C([0,\infty))$ and some non-negative, symmetric function $k \in L^\infty(X \times X)$, i.e. $k(x,y) = k(y,x)$ for all $x,y \in X$, with $\|k\|_\infty \neq 0$. The probably best known examples of this method are the Yaroslavsky filter [23], the SUSAN filter [20], and the bilateral filter [21].

In [12], it was shown that this neighbourhood filter can be written as the first step of a fixed point iteration starting with the initial data $u_0 \in L^2(X)$ for the minimisation of the functional

$$\mathscr{R} : L^2(X) \to \mathbb{R}, \quad \mathscr{R}(u) = \int_X \int_X G(|u(x) - u(y)|^2) k(x,y) \, \mathrm{d}x \, \mathrm{d}y, \tag{8.5}$$

where G is a primitive function for g. Indeed, if we calculate the derivative of $\mathscr{R}$ in the direction of a function $v \in L^2(X)$, we find (using the symmetry of the integrand with respect to the interchange of the variables x and y)

$$\begin{aligned} \delta\mathscr{R}(u;v) &= \lim_{t \to 0} \frac{\mathscr{R}(u+tv) - \mathscr{R}(u)}{t} \\ &= 4 \int_X \int_X g(|u(x) - u(y)|^2) k(x,y)(u(x) - u(y)) v(x) \, \mathrm{d}x \, \mathrm{d}y. \end{aligned}$$

So, $\mathscr{R}$ is Gâteaux differentiable and every minimising point $u \in L^2(X)$ fulfils that $\delta R(u;v) = 0$ for all $v \in L^2(X)$. This condition can be written in the form

$$u(x) = \frac{\int_X g(|u(x) - u(y)|^2) k(x,y) u(y) \, \mathrm{d}y}{\int_X g(|u(x) - u(y)|^2) k(x,y) \, \mathrm{d}y}$$

for almost all $x \in X$. The first iteration u_1 of a fixed point iteration

$$u_\ell(x) = \frac{\int_X g(|u_{\ell-1}(x) - u_{\ell-1}(y)|^2) k(x,y) u_{\ell-1}(y) \, \mathrm{d}y}{\int_X g(|u_{\ell-1}(x) - u_{\ell-1}(y)|^2) k(x,y) \, \mathrm{d}y}, \quad \ell \in \mathbb{N}, \tag{8.6}$$

with the initial value u_0 is then equivalent to the neighbourhood filter (8.3) with $K(x,y,u_0) = g(|u_0(x) - u_0(y)|^2) k(x,y)$.

Now, we may not stop after the first iteration step, but try to reach out for a minimising point of $\mathscr{R}$. Nevertheless, we do not want to go too far away from the original data u_0, which by the way do not appear in the functional $\mathscr{R}$ at all. Therefore, it is advisable to add to the functional $\mathscr{R}$ a term penalising a large distance to u_0. On this way, we would for instance end up with the problem of finding a minimising point of a functional $\mathscr{J} : L^p(X) \to \mathbb{R}$, $p \in [2, \infty)$, of the form

$$\mathscr{J}(u) = \alpha\mathscr{R}(u) + \int_X |u(x) - u_0(x)|^p \, \mathrm{d}x, \tag{8.7}$$

with some regularisation parameter $\alpha \in (0, \infty)$, as was suggested in [12].

8.2.3 *Patch-Based Filtering with a Non-Local Functional*

If we choose the weighting function K in (8.3) not only to depend on the difference between the intensities at the points x and y in X, but on the difference between two domains, the so-called patches, around these two points, then we get kernel functions of the form

$$K(x,y,u_0) = g(H_{u_0}(x,y))k(x,y), \tag{8.8}$$

where

$$H_u(x,y) = \int_X h(t)|u(x-t) - u(y-t)|^2 \mathrm{d}t$$

measures the distance in an intensity image $u \in L^2(X)$ between the patches around the points $x, y \in X$ with some non-negative weighting function $h \in L^\infty(X)$, $\|h\|_\infty \neq 0$. Since the term $H_u(x,y)$ involves also values of u at points outside of the domain X, we will assume that X is a rectangular domain in $\mathbb{R}^m$ and consider every function $u \in L^2(X)$ to be just periodically extended outside that domain. Moreover, we choose the function $g : [0,\infty) \to \mathbb{R}$ to be positive, bounded, and continuous and we assume for simplicity that the non-negative function $k \in L^\infty(X \times X)$ with $\|k\|_\infty \neq 0$ only depends on the distance between the two patches, so $k(x,y) = k_1(|x-y|)$ for some function $k_1 \in L^\infty([0,\infty))$.

We will call such a neighbourhood filter patch-based. The prime example for this method is the non-local means filter [6]. Other applications can be for instance found in [3] and [13].

Proceeding as before, we write the neighbourhood filter in the form of a fixed point iteration for minimising the functional

$$\mathscr{R} : L^2(X) \to \mathbb{R}, \quad \mathscr{R}(u) = \int_X \int_X G(H_u(x,y))k(x,y)\,\mathrm{d}x\,\mathrm{d}y \tag{8.9}$$

with some $G \in C^1(X)$ with positive and bounded derivative. Using that we have $k(x+s, y+s) = k(x,y)$ for all $x, y, s \in X$, we find for the directional derivative of $\mathscr{R}$ in the direction of a function $v \in L^2(X)$ the expression

$$\delta\mathscr{R}(u;v) = 4 \int_X \int_X \tilde{g}(x,y,u)k(x,y)(u(x) - u(y))v(x)\,\mathrm{d}x\,\mathrm{d}y,$$

where the function $\tilde{g}$ is given by

$$\tilde{g}(x,y,u) = \int_X h(s)G'(H_u(x+s,y+s))\,\mathrm{d}s. \tag{8.10}$$

So, $\mathscr{R}$ is a Gâteaux differentiable functional and every minimising point $u \in L^2(X)$ of $\mathscr{R}$ satisfies $\delta\mathscr{R}(u;v) = 0$ for all $v \in L^2(X)$. This can be written as

$$u(x) = \frac{\int_X \tilde{g}(x,y,u)k(x,y)u(y)\,\mathrm{d}y}{\int_X \tilde{g}(x,y,u)k(x,y)\,\mathrm{d}y} \tag{8.11}$$

for almost all $x \in X$.

To establish the relation between the functional (8.9) and the filter defined by the kernel (8.8) in analogy to the connection between the functional (8.5) and the filter defined by the kernel (8.4), we should thus choose the function G such that

$$g(H_{u_0}(x,y)) = \lambda_{u_0} \int_X h(s) G'(H_{u_0}(x+s,y+s))\,\mathrm{d}s \tag{8.12}$$

for almost all $(x,y) \in X \times X$, all initial data $u_0 \in L^2(X)$, and some constant $\lambda_{u_0} \in (0,\infty)$ possibly depending on u_0. In general, however, it is not necessarily true that there exists for a given function g a solution G to this equation.[1]

It was therefore suggested in [12] to choose instead for G a primitive function of g, define $\tilde{g}$ as before by relation (8.10), and consider the neighbourhood filter

$$u(x) = \frac{\int_X \tilde{g}(x,y,u_0) k(x,y) u_0(y)\,\mathrm{d}y}{\int_X \tilde{g}(x,y,u_0) k(x,y)\,\mathrm{d}y}, \tag{8.13}$$

which is the first step of a fixed point iteration for (8.11) with initial data u_0. This neighbourhood filter can now be seen as an approximation of the patch-based filter with kernel $K(x,y,u_0) = g(H_{u_0}(x,y))k(x,y)$. Indeed, the only difference is that the kernel K is replaced by the averaged kernel

$$\tilde{K}(x,y,u_0) = \int_X h(s) K(x+s,y+s,u_0)\,\mathrm{d}s.$$

Following the lines of the previous section to derive an energy functional related to the filter defined by the kernel K, we thus end up with a functional of the form

$$\mathscr{J} : L^p(X) \to \mathbb{R}, \quad \mathscr{J}(u) = \alpha \mathscr{R}(u) + \int_X |u(x) - u_0(x)|^p\,\mathrm{d}x, \tag{8.14}$$

with $\mathscr{R}$ given by (8.9), G chosen as primitive function of g, some regularisation parameter $\alpha \in (0,\infty)$, and $p \in [2,\infty)$.

8.3 Existence of Minimising Points for Non-Local Functionals

Let X be again a bounded, open subset of $\mathbb{R}^m$, $m \in \mathbb{N}$, endowed with the Lebesgue measure and $n \in \mathbb{N}$. To prove that a functional on $L^p(X;\mathbb{R}^n)$ has a minimising point, we use below that coercivity and sequential lower semi-continuity with respect to

[1] Take, for example, $X = [0,1]$, $h = \chi_{[0,\varepsilon]}$ for some $\varepsilon \in (0,1/8)$, and $u_0 = \chi_{[0,1/4]} + \chi_{[3/4,1]}$. Considering now the points $(x_0,y_0) = (1/4, 1/4+\varepsilon)$ and $(x_1,y_1) = (3/4+\varepsilon, 3/4)$, we see that $H_{u_0}(x_0+t, y_0+t) = H_{u_0}(x_1-t, y_1-t)$ for all $t \in [0,\varepsilon]$. Thus, the condition (8.12) would imply that $g(\varepsilon) = g(H_{u_0}(x_0,y_0)) = g(H_{u_0}(x_1-\varepsilon, y_1-\varepsilon)) = g(0)$.

the weak topology on $L^p(X;\mathbb{R}^n)$ if $p \in (1,\infty)$ and with respect to the weak-star topology if $p = \infty$ of the functionals are sufficient for the existence of a minimising point. Let us briefly recall this classical result.

Definition 8.2. Let V be a topological space. Then a function $\mathcal{J} : V \to \mathbb{R} \cup \{\infty\}$ is called sequentially lower semi-continuous if we have that for every sequence $(u_k)_{k\in\mathbb{N}} \subset V$ converging to $u \in V$

$$\liminf_{k\to\infty} \mathcal{J}(u_k) \geq \mathcal{J}(u).$$

Definition 8.3. Let $(V, \|\cdot\|)$ be a normed space. Then a function $\mathcal{J} : V \to \mathbb{R} \cup \{\infty\}$ is called coercive if

$$\lim_{k\to\infty} \mathcal{J}(u_k) = \infty$$

for all sequences $(u_k)_{k\in\mathbb{N}} \subset V$ with $\lim_{k\to\infty} \|u_k\| = \infty$.

Proposition 8.4. *Let $\mathcal{J} : L^p(X;\mathbb{R}^n) \to \mathbb{R} \cup \{\infty\}$ be a coercive functional which is not constantly equal to infinity and sequentially lower semi-continuous with respect to the weak topology on $L^p(X;\mathbb{R}^n)$ if $p \in (1,\infty)$ and with respect to the weak-star topology if $p = \infty$. Then $\mathcal{J}$ has a minimising point.*

8.3.1 Well-Posedness of Non-Local Functionals

To assure that a function $f : X \times X \times \mathbb{R}^n \times \mathbb{R}^n \to \mathbb{R}$ defines a non-local functional on $L^p(X;\mathbb{R}^n)$, we need to impose some regularity conditions on the function f. The measurability of the function f requested in Definition 8.1 guarantees that also the composition $f \circ g_u$ of f with the measurable function

$$g_u : X \times X \to X \times X \times \mathbb{R}^n \times \mathbb{R}^n, \quad g_u(x,y) = (x,y,u(x),u(y)),$$

(choosing the σ-algebra $\mathcal{A}$ of all Lebesgue measurable subsets on X and the Borel σ-algebra $\mathcal{B}^n$ on $\mathbb{R}^n$) is for all measurable functions $u : X \to \mathbb{R}^n$ again measurable.

To satisfy the third assumption of a non-local functional (see Definition 8.1), we have to guarantee that the integral over the negative part of $f \circ g_u$ is for all $u \in L^p(X;\mathbb{R}^n)$ finite. We confine ourselves here with a sufficient condition for f which is easy to deal with. In all our examples, the function f will be non-negative anyway, so that this condition will not be a restriction at all.

Proposition 8.5. *Let $p \in [1,\infty)$ and let $f : X \times X \times \mathbb{R}^n \times \mathbb{R}^n \to \mathbb{R}$ be a measurable function satisfying the symmetry condition (8.2). If there exist a constant $C \in (0,\infty)$ and non-negative functions $\gamma \in L^1(X \times X)$ and $\lambda \in L^1(X)$ such that*

$$f(x,y,w,z) \geq -(\gamma(x,y) + \lambda(x)|z|^p + \lambda(y)|w|^p + C|w|^p|z|^p) \tag{8.15}$$

for almost all $(x,y) \in X \times X$ *and all* $w,z \in \mathbb{R}^n$*, then* f *defines a non-local functional* $\mathscr{J}_f^p$ *on* $L^p(X;\mathbb{R}^n)$*.*

Proof. A direct estimate shows that

$$\int_X \int_X f^-(x,y,u(x),u(y))\,\mathrm{d}x\,\mathrm{d}y \le \|\gamma\|_1 + 2\|\lambda\|_1\|u\|_p^p + C\|u\|_p^{2p} < \infty$$

for every function $u \in L^p(X;\mathbb{R}^n)$. The function f thus fulfils all three assumption in the Definition 8.1 and therefore defines the non-local functional $\mathscr{J}_f^p$ on $L^p(X;\mathbb{R}^n)$. ∎

For $p = \infty$, we get a similar result.

Proposition 8.6. *Let* $f : X \times X \times \mathbb{R}^n \times \mathbb{R}^n \to \mathbb{R}$ *be a measurable function satisfying the symmetry condition (8.2). If for every* $M \in (0,\infty)$ *there exists a non-negative function* $\gamma_M \in L^1(X \times X)$ *such that*

$$f(x,y,w,z) \ge -\gamma_M(x,y) \tag{8.16}$$

for almost all $(x,y) \in X \times X$ *and all* $w,z \in \mathbb{R}^n$ *with* $|w| \le M$ *and* $|z| \le M$*, then* f *defines a non-local functional* $\mathscr{J}_f^\infty$ *on* $L^\infty(X;\mathbb{R}^n)$*.*

Proof. For an arbitrary function $u \in L^\infty(X;\mathbb{R}^n)$, we choose $M = \|u\|_\infty$ and find

$$\int_X \int_X f^-(x,y,u(x),u(y))\,\mathrm{d}x\,\mathrm{d}y \le \|\gamma_M\|_1 < \infty,$$

as desired. ∎

8.3.2 Sequential Lower Semi-Continuity of a Non-Local Functional

In the following, we investigate conditions on a function f such that it defines a non-local functional on $L^p(X;\mathbb{R}^n)$, which is sequentially lower semi-continuous with respect to the weak topology if $p \in [1,\infty)$ and with respect to the weak-star topology if $p = \infty$.

In the first step, we consider only lower semi-continuity with respect to the strong topology, which can be guaranteed under very mild conditions. Indeed, a sufficient criterion is already that f is lower semi-continuous with respect to the last two variables, see [9].

Proposition 8.7. *Let* $\mathscr{J}_f^p$ *be a non-local functional on* $L^p(X;\mathbb{R}^n)$*,* $p \in [1,\infty]$*, with* f *additionally satisfying (8.15) if* $p \in [1,\infty)$ *and (8.16) if* $p = \infty$*.*

Then the functional $\mathscr{I}_f^p$ is lower semi-continuous with respect to the strong topology on $L^p(X;\mathbb{R}^n)$ if the map

$$f_{(x,y)} : \mathbb{R}^n \times \mathbb{R}^n \to \mathbb{R}, \quad f_{(x,y)}(w,z) = f(x,y,w,z)$$

is for almost all $(x,y) \in X \times X$ lower semi-continuous.

Proof. Let $(u_k)_{k\in\mathbb{N}} \subset L^p(X;\mathbb{R}^n)$ be a sequence converging to $u \in L^p(X;\mathbb{R}^n)$. We choose a subsequence $(u_{k_\ell})_{\ell\in\mathbb{N}}$ of $(u_k)_{k\in\mathbb{N}}$ such that

$$\lim_{\ell\to\infty} \mathscr{I}_f^p(u_{k_\ell}) = \liminf_{k\to\infty} \mathscr{I}_f^p(u_k)$$

holds and such that

$$\lim_{\ell\to\infty} u_{k_\ell}(x) = u(x) \quad \text{for almost all} \quad x \in X.$$

Let now first $p \in [1,\infty)$. To be able to apply Fatou's Lemma, we use the lower bound (8.15) for f and consider instead of f the function

$$(x,y,w,z) \mapsto f(x,y,w,z) + \gamma(x,y) + \lambda(x)|z|^p + \lambda(y)|w|^p + C|w|^p|z|^p,$$

which is for almost all $(x,y) \in X \times X$ and all $w,z \in \mathbb{R}^n$ non-negative. Then we find with the lower semi-continuity of $f_{(x,y)}$ that

$$\begin{aligned}
&\liminf_{k\to\infty} \mathscr{I}_f^p(u_k) + \|\gamma\|_1 + 2\|\lambda\|_1\|u\|_p^p + C\|u\|_p^{2p} \\
&\quad\geq \int_X\int_X \liminf_{\ell\to\infty} \big(f(x,y,u_{k_\ell}(x),u_{k_\ell}(y)) + \gamma(x,y) + \lambda(x)|u_{k_\ell}(y)|^p \\
&\qquad + \lambda(y)|u_{k_\ell}(x)|^p + C|u_{k_\ell}(x)|^p|u_{k_\ell}(y)|^p\big)\,\mathrm{d}x\,\mathrm{d}y \\
&\quad\geq \int_X\int_X f(x,y,u(x),u(y))\,\mathrm{d}x\,\mathrm{d}y + \|\gamma\|_1 + 2\|\lambda\|_1\|u\|_p^p + C\|u\|_p^{2p}.
\end{aligned}$$

Thus, $\liminf_{k\to\infty} \mathscr{I}_f^p(u_k) \geq \mathscr{I}_f^p(u)$, and we conclude that $\mathscr{I}_f^p$ is sequentially lower semi-continuous.

If $p = \infty$, we choose $M = \sup_{\ell\in\mathbb{N}} \|u_{k_\ell}\|_\infty$ and get from condition (8.16) a non-negative function $\gamma_M \in L^1(X \times X)$ such that the function

$$(x,y) \mapsto f(x,y,u_{k_\ell}(x),u_{k_\ell}(y)) + \gamma_M(x,y)$$

is for almost all $(x,y) \in X \times X$ and all $\ell \in \mathbb{N}$ non-negative. Therefore, Fatou's Lemma implies as before that

$$\begin{aligned}\liminf_{k\to\infty} \mathscr{J}_f^\infty(u_k) + \|\gamma_M\|_1 &\geq \int_X\int_X \liminf_{\ell\to\infty}(f(x,y,u_{k_\ell}(x),u_{k_\ell}(y)) + \gamma_M(x,y))\,\mathrm{d}x\,\mathrm{d}y \\ &\geq \int_X\int_X f(x,y,u(x),u(y))\,\mathrm{d}x\,\mathrm{d}y + \|\gamma_M\|_1.\end{aligned}$$

So, $\liminf_{k\to\infty} \mathscr{J}_f^\infty(u_k) \geq \mathscr{J}_f^\infty(u)$, and we conclude that $\mathscr{J}_f^\infty$ is sequentially lower semi-continuous. ∎

However, we need sequential lower semi-continuity of the functional $\mathscr{J}_f^p$ with respect to a weaker topology, since coercivity only guarantees that a minimising sequence is bounded, but in the strong topology a bounded sequence does not need to have a convergent subsequence and the existence of such a subsequence is one of the essential steps in the proof of Proposition 8.4.

Similar to the classical case of local functionals, it is the convexity of the function f which provides us with the sequential lower semi-continuity of $\mathscr{J}_f^p$ with respect to the weak or weak-star topology, respectively, see [9]. Similar results already appeared in [2, 14, 16, 17], from where we also took the idea of this proof.

Proposition 8.8. *Let $\mathscr{J}_f^p$ be a non-local functional on $L^p(X;\mathbb{R}^n)$, $p \in [1,\infty]$, with $f_{(x,y)}$ being continuous for almost all $(x,y) \in X \times X$. We further assume that there exist a constant $C \in \mathbb{R}$ and functions $\gamma \in L^1(X \times X)$ and $\lambda \in L^1(X)$ such that*

$$|f(x,y,w,z)| \leq \gamma(x,y) + \lambda(x)|z|^p + \lambda(y)|w|^p + C|w|^p|z|^p \qquad (8.17)$$

for almost all $(x,y) \in X \times X$ and all $w,z \in \mathbb{R}^n$ if $p \in [1,\infty)$ and that there exists for every $M \in (0,\infty)$ a function $\gamma_M \in L^1(X \times X)$ such that

$$|f(x,y,w,z)| \leq \gamma_M(x,y) \qquad (8.18)$$

for almost all $(x,y) \in X \times X$ and all $w,z \in \mathbb{R}^n$ with $|w| \leq M$ and $|z| \leq M$ if $p = \infty$.

Then, $\mathscr{J}_f^p$ is sequentially lower semi-continuous with respect to the weak topology on $L^p(X;\mathbb{R}^n)$ if $p \in [1,\infty)$ and with respect to the weak-star topology if $p = \infty$ if the function

$$\Phi_{x,\psi} : \mathbb{R}^n \to \mathbb{R}, \quad \Phi_{x,\psi}(w) = \int_X f(x,y,w,\psi(y))\,\mathrm{d}y \qquad (8.19)$$

is for every $\psi \in L^p(X;\mathbb{R}^n)$ for almost all $x \in X$ convex.

Proof. For the proof of this proposition, we will use the notion of Young measures. For a short introduction to the theory of Young measures, we refer to Chap. 8 in [10].

Let $(u_k)_{k\in\mathbb{N}} \subset L^p(X;\mathbb{R}^n)$ be a sequence converging to $u \in L^p(X;\mathbb{R}^n)$ with respect to the weak topology on $L^p(X;\mathbb{R}^n)$ if $p \in [1,\infty)$ and with respect to the weak-star topology if $p = \infty$.

In particular, the sequence $(u_k)_{k\in\mathbb{N}}$ is bounded in $L^p(X;\mathbb{R}^n)$ and therefore, there exists a subsequence $(u_{k_\ell})_{\ell\in\mathbb{N}}$ of $(u_k)_{k\in\mathbb{N}}$ generating a Young measure ν. That is, we have a map $\nu: X \to \mathscr{M}(\mathbb{R}^n;\mathbb{R})$, $x \mapsto \nu_x$, where $\mathscr{M}(\mathbb{R}^n;\mathbb{R})$ denotes the set of all signed Radon measures on $\mathbb{R}^n$, fulfilling that ν_x is a probability measure for almost all $x \in X$, and that for all $\phi \in C_0(\mathbb{R}^n)$ the function $X \to \mathbb{R}$, $x \mapsto \int_{\mathbb{R}^n} \phi(w)\,\mathrm{d}\nu_x(w)$ is measurable and

$$\lim_{\ell\to\infty} \int_X h(x)\phi(u_{k_\ell}(x))\,\mathrm{d}x = \int_X h(x) \int_{\mathbb{R}^n} \phi(w)\,\mathrm{d}\nu_x(w)\,\mathrm{d}x \tag{8.20}$$

for every $h \in L^1(X)$.

Since $f_{(x,y)}$ is lower semi-continuous for almost all $(x,y) \in X \times X$ and the functions

$$X \times X \to [0,\infty), \quad (x,y) \mapsto f^-(x,y,u_{k_\ell}(x),u_{k_\ell}(y)), \quad \ell \in \mathbb{N},$$

are uniformly integrable because of the conditions (8.17) and (8.18), we get from the fundamental theorem for Young measures (see e.g. Theorem 8.6 in [10]) that

$$\liminf_{\ell\to\infty} \mathscr{J}_f^p(u_{k_\ell}) \geq \int_X \int_X \int_{\mathbb{R}^n} \int_{\mathbb{R}^n} f(x,y,w,z)\,\mathrm{d}\nu_y(z)\,\mathrm{d}\nu_x(w)\,\mathrm{d}y\,\mathrm{d}x. \tag{8.21}$$

Moreover, since $f_{(x,y)}$ is even continuous for almost all $(x,y) \in X \times X$ and the functions

$$X \to \mathbb{R}, \quad y \mapsto f(x,y,w,u_{k_\ell}(y)), \quad \ell \in \mathbb{N},$$

are again due to the conditions (8.17) and (8.18) for almost all $x \in X$ and all $w \in \mathbb{R}^n$ uniformly integrable, the fundamental theorem for Young measures additionally shows that

$$\lim_{\ell\to\infty} \Phi_{x,u_{k_\ell}}(w) = \lim_{\ell\to\infty} \int_X f(x,y,w,u_{k_\ell}(y))\,\mathrm{d}y = \int_X \int_{\mathbb{R}^n} f(x,y,w,z)\,\mathrm{d}\nu_y(z)\,\mathrm{d}y$$

for almost all $x \in X$ and all $w \in \mathbb{R}^n$. Therefore, the convexity of $\Phi_{x,u_{k_\ell}}$ for almost all $x \in X$ and all $\ell \in \mathbb{N}$ implies that the function

$$\tilde{\Phi}_{x,\nu} : \mathbb{R}^n \to \mathbb{R}, \quad \tilde{\Phi}_{x,\nu}(w) = \int_X \int_{\mathbb{R}^n} f(x,y,w,z)\,\mathrm{d}\nu_y(z)\,\mathrm{d}y \tag{8.22}$$

is convex for almost all $x \in X$.

Using that ν_x is by definition of a Young measure for almost all $x \in X$ a probability measure, we thus find with Jensen's inequality that

$$\begin{aligned}
&\int_X \int_{\mathbb{R}^n} \int_X \int_{\mathbb{R}^n} f(x,y,w,z)\,\mathrm{d}\nu_y(z)\,\mathrm{d}y\,\mathrm{d}\nu_x(w)\,\mathrm{d}x \\
&\quad \geq \int_X \int_X \int_{\mathbb{R}^n} f\left(x,y,\textstyle\int_{\mathbb{R}^n} w\,\mathrm{d}\nu_x(w),z\right)\mathrm{d}\nu_y(z)\,\mathrm{d}y\,\mathrm{d}x.
\end{aligned} \tag{8.23}$$

Since $(u_{k_\ell})_{\ell\in\mathbb{N}}$ converges weakly in $L^p(X;\mathbb{R}^n)$ if $p\in[1,\infty)$ and weakly-star in $L^\infty(X;\mathbb{R}^n)$ if $p=\infty$ to u, we get from (8.20) that

$$\int_{\mathbb{R}^n} w\,\mathrm{d}\nu_x(w) = u(x) \quad \text{for almost all} \quad x\in X.$$

Exploiting the symmetry of f and then using the convexity of $\Phi_{y,u}$ for almost all $y\in X$, we can again apply Jensen's inequality and get

$$\begin{aligned}&\int_X\int_{\mathbb{R}^n}\int_X f(x,y,u(x),z)\,\mathrm{d}x\,\mathrm{d}\nu_y(z)\,\mathrm{d}y\\ &\geq \int_X\int_X f(x,y,u(x),\textstyle\int_{\mathbb{R}^n} z\,\mathrm{d}\nu_y(z))\,\mathrm{d}x\,\mathrm{d}y = \mathscr{J}_f^p(u).\end{aligned} \tag{8.24}$$

Putting the inequalities (8.21), (8.23) and (8.24) together, we finally find that

$$\liminf_{\ell\to\infty} \mathscr{J}_f^p(u_{k_\ell}) \geq \mathscr{J}_f^p(u),$$

proving the sequential lower semi-continuity of $\mathscr{J}_f^p$. ∎

In fact, the convexity of the function $\Phi_{x,\psi}$ for every $\psi\in L^p(X;\mathbb{R}^n)$ for almost all $x\in X$ is even necessary for the sequential lower semi-continuity of the functional $\mathscr{J}_f^p$. For a proof of this fact, we refer to [9].

We remark that the only argument in the proof of Proposition 8.8, where we need the upper bound on f, is to show that the function $\tilde{\Phi}_{x,\nu}$ defined in (8.22) is for every generated Young measure $\nu: X\to\mathscr{M}(\mathbb{R}^n;\mathbb{R})$ for almost all $x\in X$ convex. If we thus guarantee the convexity of $\tilde{\Phi}_{x,\nu}$ by directly imposing convexity on the function $f_{(x,y)}$, we can neglect the upper bound, see also [17].

Corollary 8.9. *Let $\mathscr{J}_f^p$ be a non-local functional on $L^p(X;\mathbb{R}^n)$, $p\in[1,\infty]$, with f additionally fulfilling (8.15) if $p\in[1,\infty)$ and (8.16) if $p=\infty$.*

Then, $\mathscr{J}_f^p$ is sequentially lower semi-continuous with respect to the weak topology on $L^p(X;\mathbb{R}^n)$ if $p\in[1,\infty)$ and with respect to the weak-star topology if $p=\infty$ if the function $f_{(x,y)}$ is separately convex, i.e. if the function $\mathbb{R}^n\to\mathbb{R}$, $w\mapsto f_{(x,y)}(w,z)$ is for all $z\in\mathbb{R}^n$ convex, for almost all $(x,y)\in X\times X$.

In general, however, the separate convexity of $f_{(x,y)}$ is not necessary for the sequential lower semi-continuity of the functional $\mathscr{J}_f^p$. Though in the case $n=1$ (under some additional regularity assumptions on f), there exists for every sequentially lower semi-continuous non-local functional $\mathscr{J}_f^p$ on $L^p(X)$ a function $\tilde{f}$ defining the same non-local functional $\mathscr{J}_f^p$ as f such that $\tilde{f}_{(x,y)}$ is for almost all $(x,y)\in X\times X$ separately convex, see again [9].

8.4 Application of the Theory

We are now going to apply the results of the previous section to our examples of non-local functionals and show under which conditions we can guarantee a minimising point. For the derivative free model, we will additionally provide a new method to numerically determine the minimising point.

8.4.1 *A Derivative Free Model for the Sobolev and Total Variation Seminorm*

Let $p \in (1,\infty)$, $q \in [1,\infty)$, and $X \subset \mathbb{R}^m$, $m \in \mathbb{N}$, be a bounded, open set endowed with the Lebesgue measure. We consider the functional

$$\mathscr{J}^{p,q}(u) = \alpha \int_X \int_X \frac{|u(x)-u(y)|^q}{|x-y|^q} \varphi(x-y)\,\mathrm{d}x\,\mathrm{d}y + \frac{1}{p}\int_X |u(x)-u_0(x)|^p\,\mathrm{d}x \quad (8.25)$$

on $L^p(X)$ for some initial function $u_0 \in L^p(X)$, some positive parameter α, and some non-negative, radially symmetric, and radially decreasing function $\varphi \in L^1(\mathbb{R}^m)$.

This functional can be written as the non-local functional $\mathscr{J}_f^p$ on $L^p(X)$ defined by the function

$$f : X \times X \times \mathbb{R} \times \mathbb{R} \to \mathbb{R},$$

$$f(x,y,w,z) = \begin{cases} \alpha \frac{|w-z|^q}{|x-y|^q} \varphi(x-y) + \frac{|w-u_0(x)|^p + |z-u_0(y)|^p}{2p|X|} & \text{if} \quad x \neq y, \\ 0 & \text{if} \quad x = y, \end{cases}$$

where $|X|$ denotes the Lebesgue measure of the set X. The function $w \mapsto f(x,y,w,z)$ is for almost all $(x,y) \in X \times X$ and all $z \in \mathbb{R}$ convex. Thus, by Corollary 8.9, the functional $\mathscr{J}^{p,q}$ is sequentially lower semi-continuous with respect to the weak topology on $L^p(X)$.

Moreover, we have that

$$\mathscr{J}^{p,q}(u) \geq \frac{1}{p}\|u-u_0\|_p^p \geq \frac{1}{p}\big|\|u\|_p - \|u_0\|_p\big|^p$$

and therefore, $\mathscr{J}^{p,q}$ is also coercive. So, by Proposition 8.4, $\mathscr{J}^{p,q}$ has a minimising point which is unique due to the strict convexity of the second summand of $\mathscr{J}^{p,q}$.

In the following, we assume without loss of generality that u_0 has mean zero. Otherwise, we exchange u_0 with $u_0 - \frac{1}{|X|}\int_X u_0$ in $\mathscr{J}^{p,q}$ and add the mean value of u_0 to any minimising point of the changed functional to get the respective minimising point of the original functional. Now we consider a sequence $(\varphi_k)_{k\in\mathbb{N}}$ of non-negative, radially symmetric and radially decreasing functions from $L^1(\mathbb{R}^m)$ with integral one and

$$\lim_{k\to\infty}\int_{\{y\in\mathbb{R}^m\,:\,|y|>\delta\}}\varphi_k(x)\;\mathrm{d}x=0$$

for all $\delta\in(0,\infty)$. This sequence induces a sequence of functionals $(\mathscr{J}_k^{p,q})_{k\in\mathbb{N}}$, where $\mathscr{J}_k^{p,q}$ results from exchanging φ with φ_k in (8.25). The result from above supplies us with a sequence of unique minimising points $u_k\in L^p(X)$ of $\mathscr{J}_k^{p,q}$. In [1] (see also [19], where a very detailed proof is given), it is shown for the case $p=2$ and $q\in[1,\infty)$ that all members of the sequence $(u_k)_{k\in\mathbb{N}}$ lie also in $L^q(X)$ and that $(u_k)_{k\in\mathbb{N}}$ has a subsequence converging in the L^q norm to a limit function $u_*\in W^{1,q}(X)$ if $q>1$ and to a limit function $u_*\in BV(X)$ if $q=1$. This is done using some general compactness results provided in [4]. These results carry easily over to the case $p\in(1,\infty)$ treated here. Further, it can be shown that all functions u_k are also minimising points of $\mathscr{J}_k^{p,q}$ over the space $L^1(X)$.

Let us denote the converging subsequence of $(u_k)_{k\in\mathbb{N}}$ again with $(u_k)_{k\in\mathbb{N}}$. By applying a result of A. Ponce [18], it can be shown that the sequence of functionals $\mathscr{J}_k^{p,q}$ converges in the sense of Γ-convergence with respect to the $L^1(X)$ topology to the functional

$$\mathscr{J}^{p,q}:L^1(X)\to\mathbb{R}\cup\{\infty\},\quad \mathscr{J}^{p,q}(u)=\alpha K_{q,m}\mathscr{R}^q(u)+\frac{1}{p}\int_X|u(x)-u_0(x)|^p\,\mathrm{d}x$$

with some constants $K_{q,m}\in(0,\infty)$, where the functional $\mathscr{R}^q:L^1(X)\to\mathbb{R}\cup\{\infty\}$, $q\in[1,\infty)$, is defined by

$$\mathscr{R}^q(u)=\begin{cases}\int_X|\nabla u(x)|^q\,\mathrm{d}x & \text{if } q>1 \text{ and } u\in W^{1,q}(X),\\ |Du|(X) & \text{if } q=1 \text{ and } u\in BV(X),\\ \infty & \text{otherwise.}\end{cases}$$

It follows that the limit function u_* of $(u_k)_{k\in\mathbb{N}}$ is a minimising point of the limit functional $\mathscr{J}^{p,q}$ over $L^1(X)$ and that u_* also belongs to $L^p(X)\cap W^{1,q}(X)$ if $q>1$ and to $L^p(X)\cap BV(X)$ if $q=1$.

The results above show in particular that minimising points of the functionals

$$u\mapsto\alpha\int_X\int_X\frac{|u(x)-u(y)|}{|x-y|}\varphi_k(x-y)\,\mathrm{d}x\,\mathrm{d}y+\frac{1}{p}\int_X|u(x)-u_0(x)|^p\,\mathrm{d}x$$

are a suitable approximation to the minimising points of $\mathscr{J}^{p,q}$. We used this fact in order to generate new numerical schemes for total variation regularisation.

8.4.1.1 Numerical Implementation

Let $X=[0,1]$ and let $(\varphi_k^{(1)})_{k\in\mathbb{N}}$ be the sequence of kernel functions defined by

$$\varphi_k^{(1)}=\frac{k}{2}\chi_{[-\frac{1}{k},\frac{1}{k}]}.$$

Let further

$$\mathscr{R}_k^{(1)}(u) = \int_X \int_X \frac{|u(x) - u(y)|}{|x - y|} \varphi_k^{(1)}(x - y)\, \mathrm{d}x \mathrm{d}y.$$

For the transition to the discrete setting, we approximate functions $u \in L^1(X)$ by piecewise constant functions u_k of the form

$$u_k = \sum_{i=0}^{k} \upsilon_i \chi_{[\frac{i}{k+1}, \frac{i+1}{k+1})} \tag{8.26}$$

(using some averaging process), where $\upsilon_0, \ldots, \upsilon_k \in \mathbb{R}$.

Evaluating u_k with $\mathscr{R}_k^{(1)}$ yields the standard total variation seminorm of u_k (note that $K_{1,1} = 1$):

$$\mathscr{R}_k^{(1)}(u_k) = \sum_{i=1}^{k} |\upsilon_i - \upsilon_{i-1}| = |Du_k|.$$

Using instead of $(\varphi_k^{(1)})_{k \in \mathbb{N}}$ the family of kernels $(\varphi_k^{(2)})_{k \in \mathbb{N}}$ defined by

$$\varphi_k^{(2)} = \frac{k}{4} \chi_{[-\frac{2}{k}, \frac{2}{k}]}$$

yields

$$\mathscr{R}_k^{(2)}(u_k) = \log(2) \sum_{i=1}^{k} |\upsilon_i - \upsilon_{i-1}| + \frac{1 - \log(2)}{2} \sum_{i=1}^{k-1} |\upsilon_{i+1} - \upsilon_{i-1}|.$$

We compare the two schemes from above by using them for the computation of the total variation seminorm in an implementation of a standard steepest gradient algorithm for one-dimensional total variation regularisation minimisation. The latter is applied to signal denoising. Following the work of Vogel [22], we discretise the functional $\mathscr{I}^{2,1}$ using the two different approximations of the total variation seminorm from above.

Thereby, we identify the functions u_k from above with the $(k+1)$-tuples $\mathbf{u} = (\upsilon_0, \ldots, \upsilon_k) \in \mathbb{R}^{k+1}$ via $\upsilon_i = u_k(\frac{i}{k+1})$. The two discretisations of the functional $\mathscr{I}^{2,1}$ read

$$T_s(\mathbf{u}) = \frac{1}{2k} (\mathbf{u} - \mathbf{u}_0)^T (\mathbf{u} - \mathbf{u}_0) + \alpha J_s(\mathbf{u})$$

for $s = 1, 2$, where $J_s(\mathbf{u}) = \mathscr{R}_k^{(s)}(u_k)$ with u_k as in (8.26) and where $\mathbf{u_0}$ is the $(k+1)$-tuple corresponding to the function $u_0 \in L^p(X)$. For the numerical calculations, the absolute value is everywhere replaced by the smooth approximation $t \mapsto \sqrt{t^2 + \beta^2}$ with some small positive constant β.

The minimisation algorithm is as follows for $s \in \{1, 2\}$ (see e.g., [22]): Choose a stopping parameter $\varepsilon > 0$. Set $\mu = 1$. Choose a starting point $\mathbf{u}^1 \in \mathbb{R}^{k+1}$.

Compute $T^\mu = T_s(\mathbf{u}^\mu)$ and $\mathbf{p}^\mu = \nabla T_s(\mathbf{u}^\mu)$. Conduct an (inexact) line search: $\alpha^\mu = \min_{\tilde{\alpha}>0} T_s(\mathbf{u}^\mu + \tilde{\alpha}\mathbf{p}^\mu)$. Update $\mathbf{u}^\mu$ by $\mathbf{u}^{\mu+1} = \mathbf{u}^\mu + \alpha^\mu \mathbf{p}^\mu$. If $|T^{\mu+1} - T^\mu| < \varepsilon$ stop. Otherwise jump to the computation of $T^{\mu+1} = T_s\left(\mathbf{u}^{\mu+1}\right)$ and do another iteration.

We denote the algorithm involving J_1 as approximation to the total variation seminorm by Algorithm 1 and the other one involving J_2 as approximation to the total variation seminorm by Algorithm 2.

For our first test, we take as data function $u_0 = \chi_{(\frac{1}{2},1]}$ and parameters $k = 100$, $\alpha = 0.01$ and $\beta = 0.1$. To the signal u_0 we add uniformly distributed random noise with amplitudes between -0.25 and 0.25. We start the algorithms with the functions $u^1(x) = \frac{1}{2}$ for all $x \in [0,1]$. Our tests show that Algorithm 2 needs only 55% of the amount of iterations that Algorithm 1 needs to terminate when ε is set to 10^{-6} but is only 13% faster (the choice of ε guarantees a good reconstruction from the visual point of view in both cases). Another test shows that given an amount of 500 iterations the second algorithm gives a 35% better reconstruction than the first one but is on the other hand slower by the same factor. If we allow μ to become very large, both algorithms compute the same minimum in about the same amount of time. However, Algorithm 2 needs significantly less iterations. If we double the noise, the behaviour of the algorithms does not change. Figure 8.1 shows reconstructions after 500 iterations after the noise has been doubled.

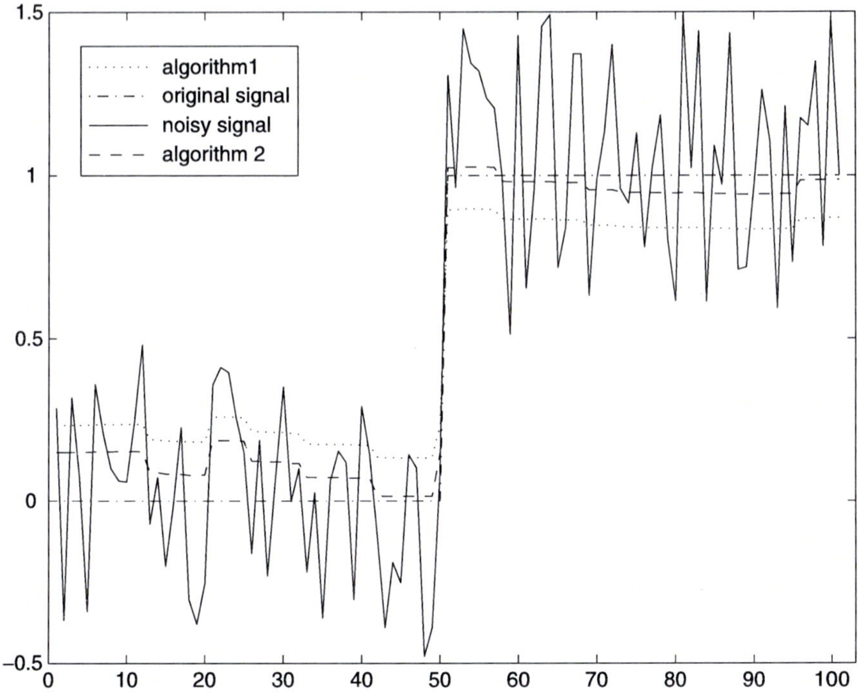

Fig. 8.1 First example after 500 iterations

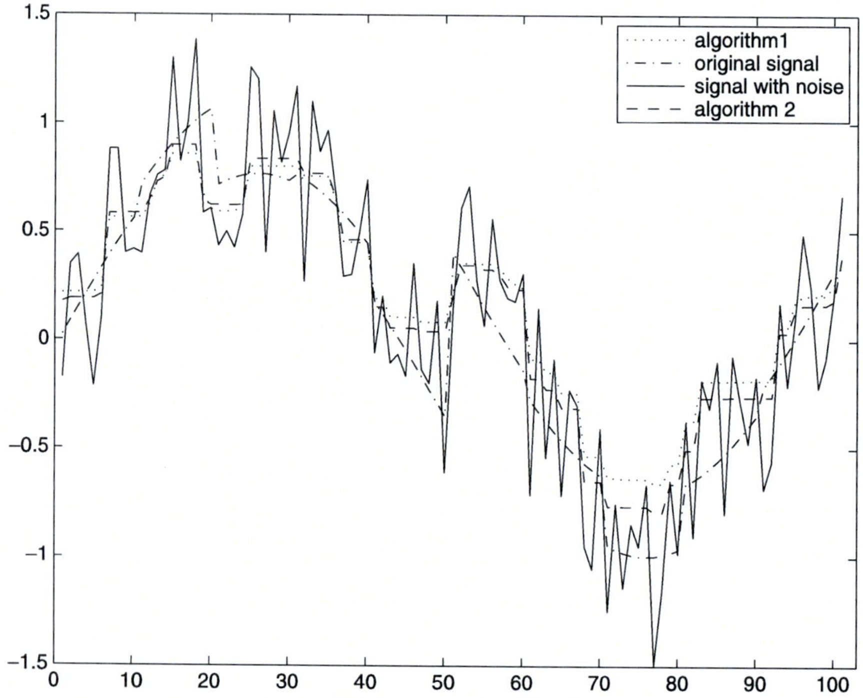

Fig. 8.2 Second example after 100 iterations

In the second test, our original signal u_0 is a sine function on $[0,1]$ that is degraded twice: first, for each $i \in [0,9] \cap \mathbb{N}$ a (uniformly distributed) random number k_i between -0.5 and 0.5 is added to u_0 on the interval $[\frac{i}{10}, \frac{i+1}{10})$ and second, uniformly distributed noise is added to the whole signal as in the example above. For $x \in [\frac{i}{10}, \frac{i+1}{10})$ we thus have $u_0(x) = \sin(2\pi x) + k_i + n(x)$ where $n(x)$ is the noise. All parameters are chosen as in the first example. Here Algorithm 2 needs about 10% less iterations to reach the stopping conditions than Algorithm 1 and provides slightly better results but is a little bit slower, too. After 500 iterations Algorithm 2 delivers a reconstruction that is only slightly better than that of Algorithm 1 and needs a little bit more time as well. In the long run, Algorithm 2 seems to deliver slightly superior reconstructions in comparison to Algorithm 1. Figure 8.2 shows reconstructions after 100 iterations.

8.4.2 Neighbourhood Filters

Let us consider a non-local functional $\mathcal{J}_f^p$ on $L^p(X)$, where X is a bounded, open subset of $\mathbb{R}^m$ endowed with the Lebesgue measure and $p \in (1,\infty)$, defined by a function f of the form

$$f(x,y,w,z) = G(|w-z|^2)k(x,y) + |w-u_0(x)|^p + |z-u_0(y)|^p$$

with some initial data $u_0 \in L^p(X)$, a Borel measurable function $G: [0,\infty) \to [0,\infty)$, and a measurable function $k: X \times X \to [0,\infty)$, where we additionally assume that k has the symmetry $k(x,y) = k(y,x)$ for all $x,y \in X$. So,

$$\mathscr{J}_f^p(u) = \int_X \int_X G(|u(x)-u(y)|^2)k(x,y)\,\mathrm{d}x\,\mathrm{d}y + 2|X| \int_X |u(x)-u_0(x)|^p\,\mathrm{d}x.$$

In particular, the regularisation functional (8.7) for a neighbourhood filter has such a form.

As in the case of the derivative free model, the last term in the functional $\mathscr{J}_f^p$ enforces that $\mathscr{J}_f^p$ is coercive, since

$$\mathscr{J}_f^p(u) \geq 2|X|\,\|u-u_0\|_p^p \geq 2|X|\,\big|\|u\|_p - \|u_0\|_p\big|^p.$$

To guarantee the existence of a minimising point of $\mathscr{J}_f^p$, it therefore suffices according to Corollary 8.9 to choose the functions G and k such that $f_{(x,y)}$ is for almost all $(x,y) \in X \times X$ separately convex. Let us for simplicity assume that $G \in C^2(\mathbb{R})$. Then we get the following criterion for the separate convexity of $f_{(x,y)}$.

Lemma 8.10. *Let $G \in C^2(\mathbb{R})$, $k: X \times X \to [0,\infty)$ be a measurable function which is not almost everywhere zero, and $p \in (1,\infty)$. Then the function $f_{(x,y)}: \mathbb{R}^2 \to \mathbb{R}$ given by*

$$f_{(x,y)}(w,z) = G(|w-z|^2)k(x,y) + |w-u_0(x)|^p + |z-u_0(y)|^p$$

is for every measurable function $u_0: X \to \mathbb{R}$ for almost every $(x,y) \in X \times X$ separately convex if and only if G fulfils

$$2\xi G''(\xi) + G'(\xi) \geq 0 \quad \textit{for all} \quad \xi \in [0,\infty) \tag{8.27}$$

if $p \neq 2$ or $\operatorname{ess\,sup} k = \infty$, and

$$2\xi G''(\xi) + G'(\xi) + \frac{1}{\operatorname{ess\,sup} k} \geq 0 \quad \textit{for all} \quad \xi \in [0,\infty) \tag{8.28}$$

if $p = 2$ and $\operatorname{ess\,sup} k < \infty$.

Proof. Taking two times the derivative of $f_{(x,y)}$ with respect to the variable w, we get the condition

$$2k(x,y)(2|w-z|^2 G''(|w-z|^2) + G'(|w-z|^2)) + p(p-1)|w-u_0(x)|^{p-2} \geq 0 \tag{8.29}$$

for all $w,z \in \mathbb{R}$ (and $w \neq u_0(x)$ if $p \in (1,2)$) and every measurable function $u_0: X \to \mathbb{R}$ for the convexity of $f_{(x,y)}$ in the first variable. Since the last term of the sum on the left hand side is equal to two if $p = 2$, the conditions (8.27) and (8.28) are sufficient for $f_{(x,y)}$ to be separately convex for almost all $(x,y) \in X \times X$.

To show that the separate convexity of $f_{(x,y)}$ also implies the conditions (8.27) and (8.28), we first choose for every $C \in (0, \operatorname{ess\,sup} k)$ a set $A_C \subset X \times X$ with positive measure such that $k(x,y) \geq C$ for all $(x,y) \in A_C$. Then the inequality (8.29) implies for all $(x,y) \in A_C$, all $w \in \mathbb{R}$, and every measurable function $u_0 : X \to \mathbb{R}$ that

$$2\xi G''(\xi) + G'(\xi) \geq -\frac{p(p-1)|w - u_0(x)|^{p-2}}{2C} \quad \text{for all} \quad \xi \in [0,\infty). \tag{8.30}$$

Now, if $p \neq 2$, we find for arbitrary $w \in \mathbb{R}$ and $\varepsilon \in (0,\infty)$ a constant function $u_0 : X \to \mathbb{R}$ such that $|w - u_0(x)|^{p-2} < \varepsilon$ for all $x \in X$. So, in the limit $\varepsilon \to 0$, inequality (8.30) gives us condition (8.27).

In the case $p = 2$, the inequality (8.30) simply reads

$$2\xi G''(\xi) + G'(\xi) \geq -\tfrac{1}{C} \quad \text{for all} \quad \xi \in [0,\infty).$$

Thus, if $\operatorname{ess\,sup} k = \infty$, we let C tend to infinity and find again (8.27). If $\operatorname{ess\,sup} k < \infty$, we let C tend to $\operatorname{ess\,sup} k$ and end up with condition (8.28). ■

8.4.2.1 Numerical Implementation

We consider the numerical minimisation of the non-local functional $\mathcal{J} : L^2(X) \to \mathbb{R}$ defined in (8.7) with $G(\xi) = 1 - e^{-\xi/\lambda}$ and $k(x,y) = \chi_{[-\sigma_k,\sigma_k]}(|x-y|)$. This functional has thus three parameters λ, α and σ_k.

Though the function G is not convex, the functions αG and k fulfil for sufficiently small $\alpha \in (0,\infty)$ the requirements of Lemma 8.10. Thus we are able to guarantee a minimising point of the functional $\mathcal{J}$ if the regularisation parameter $\alpha \in (0,\infty)$ is small enough.

Proceeding as in the derivation of the fixed point iteration (8.6), we get the fixed point iteration

$$u_\ell(x) = \frac{u_0(x) + \alpha \int_X g(|u_{\ell-1}(x) - u_{\ell-1}(y)|^2) k(x,y) u_{\ell-1}(y)\,\mathrm{d}y}{1 + \alpha \int_X g(|u_{\ell-1}(x) - u_{\ell-1}(y)|^2) k(x,y)\,\mathrm{d}y}, \quad \ell \in \mathbb{N},$$

for the minimisation of the functional $\mathcal{J}$ where $g(\xi) = G'(\xi) = \frac{1}{\lambda} e^{-\xi/\lambda}$. Initialising with u_0, we iterate until $\|u_\ell - u_{\ell-1}\|_2 < \varepsilon$, where $\varepsilon \in \mathbb{R}$ is a sufficiently small parameter. Figure 8.3b, c illustrates the result of this procedure for two different values of the parameter σ_k which defines the local character of the functional $\mathcal{J}$.

8.4.3 Patch-Based Filtering

Finally, we investigate the more complicated functionals (8.14) introduced for the variational description of the patch-based filtering method. Here, let X be a rectangular domain in $\mathbb{R}^m$, $p \in [2,\infty)$, and $\mathscr{J} : L^p(X) \to \mathbb{R} \cup \{\infty\}$ be a functional of the form

$$\mathscr{J}(u) = \alpha \int_X \int_X G(H_u(x,y))k(|x-y|)\,\mathrm{d}x\,\mathrm{d}y + \int_X |u(x) - u_0(x)|^p\,\mathrm{d}x,$$

where $G : [0,\infty) \to [0,\infty)$ is a Borel measurable function, $k : [0,\infty) \to [0,\infty)$ is a measurable function, $u_0 \in L^p(X)$, $\alpha \in (0,\infty)$, and

$$H_u(x,y) = \int_X h(t)|u(x-t) - u(y-t)|^2\,\mathrm{d}t$$

with some non-negative function $h \in L^\infty(X)$. Moreover, we consider $u \in L^p(X)$ to be periodically continued outside the domain X.

This type of functionals does not fit into our Definition 8.1 of a non-local functional since the integrand does not only depend on two values of the function u, but rather on all values of u in some neighbourhoods of two points. Nevertheless, we may try to find a sufficient condition for the functions G, k, and h such that $\mathscr{J}$ has a minimising point by requiring as before that $\mathscr{J}$ shall be coercive and sequentially lower semi-continuous with respect to the weak topology on $L^p(X)$.

The coercivity of $\mathscr{J}$ follows directly from the fact that $\mathscr{J}(u) \geq \|u - u_0\|_p^p$. And for the sequential lower semi-continuity of $\mathscr{J}$ we get the following result.

Proposition 8.11. *Let $p \in [2,\infty)$, $h \in L^\infty(X)$ and $k \in L^\infty(X \times X)$ be non-negative functions, and $G : [0,\infty) \to [0,\infty)$ be a monotonically increasing, convex function. Then the functional*

$$\mathscr{R} : L^p(X) \to \mathbb{R} \cup \{\infty\}, \quad \mathscr{R}(u) = \int_X \int_X G(H_u(x,y))k(x,y)\,\mathrm{d}x\,\mathrm{d}y$$

is sequentially lower semi-continuous with respect to the weak topology on $L^p(X)$.

Proof. First, we remark that $\mathscr{R}$ is a convex functional. Indeed, the map $L^p(X) \to \mathbb{R}$, $u \mapsto H_u(x,y)$ is for all $x,y \in X$ convex and so is the map $L^p(X) \to \mathbb{R}$, $u \mapsto G(H_u(x,y))$, since G is monotonically increasing and convex. Thus, by the monotonicity and linearity of the integral, $\mathscr{R}$ is convex.

Moreover, the functional $\mathscr{R}$ is sequentially lower semi-continuous with respect to the strong topology. To see this, let $(u_\ell)_{\ell \in \mathbb{N}} \subset L^p(X)$ be a sequence converging strongly to $u \in L^p(X)$. Then for every point $(x,y) \in X \times X$, the functions $X \to \mathbb{R}$, $t \mapsto u_\ell(x-t) - u_\ell(y-t)$ converge with $\ell \to \infty$ strongly to the function $X \to \mathbb{R}$, $t \mapsto u(x-t) - u(y-t)$. Since $p \geq 2$, we therefore have that

$$\lim_{\ell \to \infty} H_{u_\ell}(x,y) = H_u(x,y)$$

for all $x, y \in X$. So, we get with Fatou's Lemma

$$\liminf_{\ell\to\infty} \mathscr{R}(u_\ell) \geq \int_X \int_X \liminf_{\ell\to\infty} G(H_{u_\ell}(x,y))k(x,y)\,\mathrm{d}x\,\mathrm{d}y = \mathscr{R}(u).$$

Since a convex functional on $L^p(X)$ which is lower semi-continuous with respect to the strong topology on $L^p(X)$ is also sequentially lower semi-continuous with respect to the weak topology, we can conclude that $\mathscr{R}$ is sequentially lower semi-continuous with respect to the weak topology on $L^p(X)$. ■

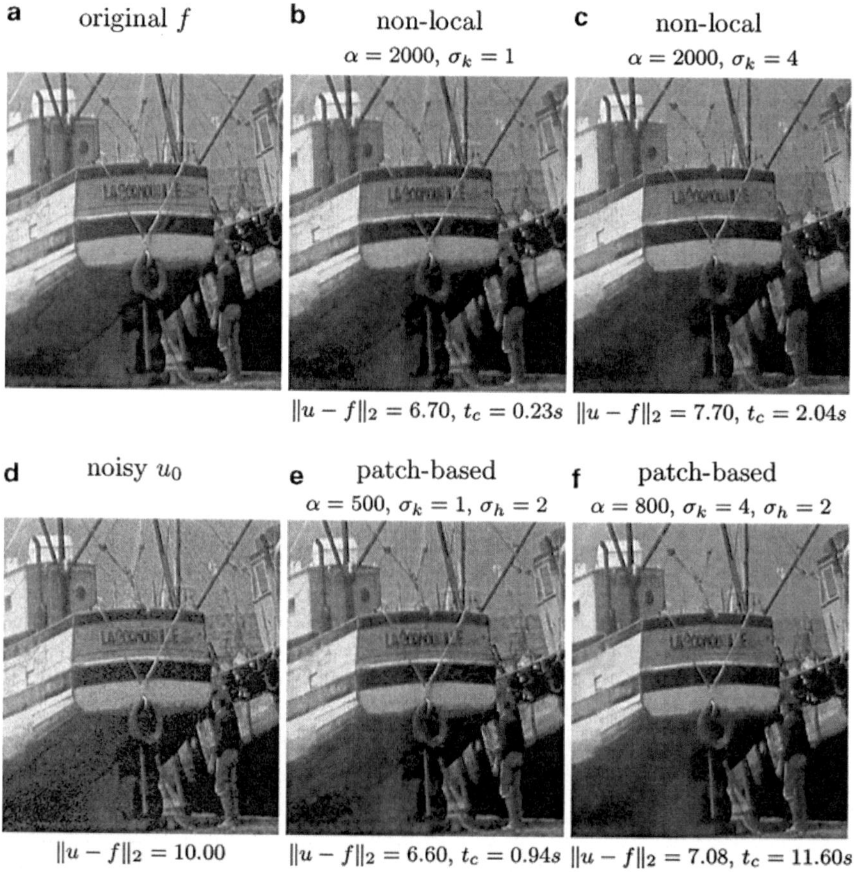

Fig. 8.3 Results obtained with the fixed point iterations of the non-local functional and the patch-based functional. The kernel used in the functionals are defined by $G(\xi) = 1 - e^{-\xi/\lambda}$, $k = \chi_{[-\sigma_k,\sigma_k]}$ and $h = \chi_{[-\sigma_h,\sigma_h]}$. The balance α between the regularisation term and the data term has been selected a posteriori. The initial image is a 243×270 fraction of the famous "boat" image whose intensities range from 0 to 255. The noisy version has been obtained by adding a Gaussian noise of standard deviation $\|u_0 - f\|_2 = 10$. Finally, the computation times t_c obtained on a $8{\times}3.40$ GHz computer are indicated in seconds

8.4.3.1 Numerical Implementation

We consider the numerical minimisation of the non-local functional $\mathcal{J} : L^2(X) \to \mathbb{R}$ defined in (8.14) with $G(\xi) = 1 - e^{-\xi/\lambda}$, $k(x,y) = \chi_{[-\sigma_k,\sigma_k]}(|x-y|)$, and $h = \chi_{[-\sigma_h,\sigma_h]}$. This functional has thus four parameters $\lambda, \alpha, \sigma_h$ and σ_k.

Since the function G is not convex, we cannot apply Proposition 8.11 to guarantee the existence of a minimising point of the functional $\mathcal{J}$ in this case. Nevertheless, we are trying to minimise the functional numerically.

Similarly to the derivation of (8.13), we find the fixed point iteration

$$u_\ell(x) = \frac{u_0(x) + \alpha \int_X \tilde{g}(x,y,u_{\ell-1})k(x,y)u_{\ell-1}(y)\,\mathrm{d}y}{1 + \alpha \int_X \tilde{g}(x,y,u_{\ell-1})k(x,y)\,\mathrm{d}y}, \quad \ell \in \mathbb{N},$$

for the minimisation of $\mathcal{J}$ where $\tilde{g}$ is given by (8.10).

We remark that this expression corresponds to the block implementation of the non-local means filter described in [6] if we consider a single iteration and let $\alpha \to \infty$. As in the previous section, we initialise with u_0 and iterate the fixed point equation until $\|u_\ell - u_{\ell-1}\|_2 < \varepsilon$ where $\varepsilon \in \mathbb{R}$ is a sufficiently small parameter. Figure 8.3e, f illustrates the result of this procedure for two different values of the parameter σ_k which defines the local character of the functional.

Acknowledgements The work of CP and OS has been supported by the Austrian Science Fund (FWF) within the research networks NFNs *Industrial Geometry*, Project S09203, and Photoacoustic Imaging in Biology and Medicine, Project S10505-N20.

References

1. Aubert, G., Kornprobst, P.: Can the nonlocal characterization of Sobolev spaces by Bourgain et al. be useful for solving variational problems? SIAM Journal on Numerical Analysis **47**, 844 (2009)
2. Bevan, J., Pedregal, P.: A necessary and sufficient condition for the weak lower semicontinuity of one-dimensional non-local variational integrals. Proc. Roy. Soc. Edinburgh Sect. A **136**, 701–708 (2006)
3. Boulanger, J., Kervrann, C., Salamero, J., Sibarita, J.-B., Elbau, P., Bouthemy, P.: Patch-based non-local functional for denoising fluorescence microscopy image sequences. IEEE Transactions on Medical Imaging (2010)
4. Bourgain, J., Brézis, H., Mironescu, P.: Another look at Sobolev spaces. In: Optimal Control and Partial Differential Equations, IOS Press, Amsterdam (2001)
5. Bourgain, J., Brézis, H., Mironescu, P.: Limiting embedding theorems for $W^{s,p}$ when $s \uparrow 1$ and applications. Journal d'Analyse Mathématique **87**, 77–101 (2002)
6. Buades, A., Coll, B., Morel, J.-M.: A review of image denoising algorithms, with a new one. Multiscale Modeling and Simulation **4**, 490–530 (2006)
7. Buades, A., Coll, B., Morel, J.-M.: Neighborhood filters and PDEs. Numerische Mathematik **105**, 1–34 (2006)
8. Efros, A.A., Leung, T.K.: Texture synthesis by non-parametric sampling. International Conference on Computer Vision **2**, 1033–1038 (1999)
9. Elbau, P.: Sequential lower semi-continuity of non-local functionals. Arxiv preprint (2011)

10. Fonseca, I., Leoni, G.: Modern methods in the calculus of variations: Lp spaces. Springer (2007)
11. Gilboa, G., Osher, S.: Nonlocal operators with applications to image processing. Multiscale Modeling and Simulation **7**, 1005–1028 (2008)
12. Kindermann, S., Osher, S., Jones, P.W.: Deblurring and denoising of images by nonlocal functionals. Multiscale Modeling and Simulation **4**, 1091–1115 (2006)
13. Mahmoudi, M., Sapiro, G.: Fast image and video denoising via nonlocal means of similar neighborhoods. IEEE Signal Processing Letters **12**, 839 (2005)
14. Muñoz, J.: On some necessary conditions of optimality for a nonlocal variational principle. SIAM J. Control Optim. **38**, 1521–1533 (2000)
15. Muñoz, J.: Extended variational analysis for a class of nonlocal minimization principles. Nonlinear Anal. **47**, 1413–1418 (2001)
16. Muñoz, J.: Characterisation of the weak lower semicontinuity for a type of nonlocal integral functional: the n-dimensional scalar case. J. Math. Anal. Appl. **360**, 495–502 (2009)
17. Pedregal, P.: Nonlocal variational principles. Nonlinear Analysis **29**, 1379–1392 (1997)
18. Ponce, A.C.: A new approach to Sobolev spaces and connections to Γ-convergence. Calculus of Variations and Partial Differential Equations **19**, 229–255 (2004)
19. Pontow, C., Scherzer, O.: A derivative-free approach to total variation regularization. Arxiv preprint arXiv:0911.1293 (2009)
20. Smith, S.M., Brady, J.M.: SUSAN – A new approach to low level image processing. International Journal of Computer Vision **23**, 45–78 (1997)
21. Tomasi, C., Manduchi, R.: Bilateral filtering for gray and color images. In: Proceedings of the Sixth International Conference on Computer Vision, Volume 846 (1998)
22. Vogel, C.R.: Computational methods for inverse problems. Frontiers in Mathematics, Volume 23. SIAM (2002)
23. Yaroslavsky, L.P., Yaroslavskij, L.P.: Digital picture processing: An introduction. Springer (1985)

Chapter 9
Opial-Type Theorems and the Common Fixed Point Problem

Andrzej Cegielski and Yair Censor

Abstract The well-known Opial theorem says that an orbit of a nonexpansive and asymptotically regular operator T having a fixed point and defined on a Hilbert space converges weakly to a fixed point of T. In this paper, we consider recurrences generated by a sequence of quasi-nonexpansive operators having a common fixed point or by a sequence of extrapolations of an operator satisfying Opial's demi-closedness principle and having a fixed point. We give sufficient conditions for the weak convergence of sequences defined by these recurrences to a fixed point of an operator which is closely related to the sequence of operators. These results generalize in a natural way the classical Opial theorem. We give applications of these generalizations to the common fixed point problem.

Keywords Common fixed point · Opial theorem · Cutter operators · Dos Santos method · Quasi-nonexpansive operators

AMS 2010 Subject Classification: 46B45, 37C25, 65K15, 90C25

9.1 Introduction

Iterative methods for convex optimization problems in a Hilbert space $\mathcal{H}$ have usually the form of the recurrence $x^{k+1} = U_k x^k$, where $x^0 \in X$, $X \subset \mathcal{H}$ is closed and convex, and $U_k : X \to X$ are operators related to the optimization problem at hand. Some of the methods employ the same operator $U_k = U$ in all iterations. If we suppose that U is a nonexpansive and asymptotically regular operator having a fixed point then it follows from the Opial theorem that a so generated sequence $\{x^k\}_{k=0}^{\infty}$ converges weakly to a fixed point of U (see [30, Theorem 1]). Many iterative methods employ, however, different operators U_k in successive iterations,

A. Cegielski (✉)
Faculty of Mathematics, Computer Science and Econometrics, University of Zielona Góra, ul. Szafrana 4a, 65-514 Zielona Góra, Poland

H.H. Bauschke et al. (eds.), *Fixed-Point Algorithms for Inverse Problems in Science and Engineering*, Springer Optimization and Its Applications 49,
DOI 10.1007/978-1-4419-9569-8_9,

usually assuming that all operators U_k have a common fixed point. Examples of such methods for solving the common fixed point problem include methods of successive projections (with various control sequences such as the almost cyclic control, the repetitive control, etc.), methods of simultaneous projections (also known as Cimmino-type methods), where the weights depend on the iteration index, surrogate projection methods, etc. Our main aim here is to give, in a unified manner, sufficient conditions for weak convergence of sequences generated by the recurrence $x^{k+1} = U_k x^k$ and to apply the results to the common fixed point problem.

An interesting point related to our current investigation is a *local acceleration* technique of Cimmino's [18] well-known simultaneous projection method for linear equations. This technique is referred to in the literature as the *Dos Santos* (DS) method, see Dos Santos [24] and Bauschke and Borwein [4, Sect. 7], although Dos Santos attributes it, in the linear case, to De Pierro's Ph.D. Thesis [23]. The method essentially uses the line through each pair of consecutive Cimmino iterates and chooses the point on this line which is closest to the solution x^* of the linear system $Ax = b$. The nice thing about it is that existence of the solution of the linear system must be assumed, but the method does not need the solution point x^* in order to proceed with the locally accelerated DS iterative process. This approach was also used by Appleby and Smolarski [3]. On the other hand, while trying to be as close as possible to the solution point x^* in each iteration, the method is not known to guarantee overall acceleration of the process. Therefore, we call it a *local acceleration* technique. In all the above references the DS method works for *convex feasibility problems* and one of our questions was whether it can also be extended to handle common fixed point problems. If so, for which classes of operators?

Here, we answer this question by focusing on the class of operators $T : \mathscr{H} \to \mathscr{H}$ that have the property that, for any $x \in \mathscr{H}$, the hyperplane through Tx whose normal is $x - Tx$ always "cuts" the space into two half-spaces one of which contains the point x while the other contains the (assumed nonempty) fixed points set of T. This explains the name *cutter operators* or *cutters* that we introduce here. These operators themselves, introduced and investigated by Bauschke and Combettes [5, Definition 2.2] and by Combettes [21], play an important role in optimization and feasibility theory since many commonly used operators are actually cutters. We define generalized relaxations and extrapolation of cutter operators and construct *extrapolated simultaneous cutter operators.* For these simultaneous extrapolated cutters we present convergence results of successive iteration processes for common fixed point problems which generalize the locally accelerated DS iterative processes, thus, cover some of the earlier results about such methods and present some new ones.

The paper is organized as follows. In Sect. 9.2 we give the definition of cutter operators and bring some of their properties that will be used here. Section 9.3 contains the Opial theorem and its generalization. Opial-type theorems for cutters are presented in Sect. 9.4 and applications to the common fixed point problem, including the connection to the DS method (Example 9.38), are studied in Sect. 9.5.

9.2 Preliminaries

Let $\mathscr{H}$ be a real Hilbert space with an inner product $\langle\cdot,\cdot\rangle$ and with the norm $\|\cdot\|$. Given $x, y \in \mathscr{H}$ we denote

$$H(x,y) := \{u \in \mathscr{H} \mid \langle u-y, x-y\rangle \le 0\}. \tag{9.1}$$

Definition 9.1. An operator $T : \mathscr{H} \to \mathscr{H}$ is called a *cutter operator* or, in short, a *cutter* iff

$$\operatorname{Fix} T \subseteq H(x, Tx) \text{ for all } x \in \mathscr{H}, \tag{9.2}$$

where $\operatorname{Fix} T$ is the fixed points set of T, equivalently,

$$q \in \operatorname{Fix} T \text{ implies that } \langle Tx - x, Tx - q\rangle \le 0 \text{ for all } x \in \mathscr{H}. \tag{9.3}$$

The class of cutter operators is denoted by $\mathscr{T}$, i.e.,

$$\mathscr{T} := \{T : \mathscr{H} \to \mathscr{H} \mid \operatorname{Fix} T \subseteq H(x, Tx) \text{ for all } x \in \mathscr{H}\}. \tag{9.4}$$

The class $\mathscr{T}$ of operators was introduced and investigated by Bauschke and Combettes in [5, Definition 2.2] and by Combettes in [21]. Operators in this class were named *directed operators* by Zaknoon [33] and further employed under this name by Segal [32] and Censor and Segal [14–16]. Cegielski [12, Definition 2.1] named and studied these operators as *separating operators*. Since both *directed* and *separating* are key words of other, widely-used, mathematical entities we decide to use from now on the term *cutter operators*. This name can be justified by the fact that the bounding hyperplane of $H(x, Tx)$ "cuts" the space into two half-spaces, one which contains the point x while the other contains the set $\operatorname{Fix} T$. We recall definitions and results on cutter operators and their properties as they appear in [5, Proposition 2.6] and [21], which are also sources for further references.

Bauschke and Combettes [5] showed the following:

(i) The set of all fixed points of a cutter operator assumed to be nonempty is closed and convex because $\operatorname{Fix} T = \cap_{x\in\mathscr{H}} H(x, Tx)$.
(ii) Denoting by Id the identity operator,

$$\text{if } T \in \mathscr{T} \text{ then } \mathrm{Id} + \lambda(T - \mathrm{Id}) \in \mathscr{T} \text{ for all } \lambda \in [0,1]. \tag{9.5}$$

This class of operators is fundamental because many common types of operators arising in convex optimization belong to the class and because it allows a complete characterization of Fejér-monotonicity [5, Proposition 2.7]. The localization of fixed points is discussed by Goebel and Reich in [26, pp. 43–44]. In particular, it is shown there that a firmly nonexpansive (FNE) operator, namely, an operator $T : \mathscr{H} \to \mathscr{H}$ that fulfills

$$\|Tx - Ty\|^2 \le \langle Tx - Ty, x - y\rangle \text{ for all } x, y \in \mathscr{H}, \tag{9.6}$$

which has a fixed point, satisfies (9.3) and is, therefore, a cutter operator. The class of cutter operators, includes additionally, according to [5, Proposition 2.3], among others, the resolvent of a maximal monotone operator, the orthogonal projections and the subgradient projectors. Another family of cutters appeared recently in Censor and Segal [15, Definition 2.7]. Note that every cutter operator belongs to the class of operators $\mathscr{F}^0$, defined by Crombez [22, p. 161],

$$\mathscr{F}^0 := \{T : \mathscr{H} \to \mathscr{H} \mid \|Tx - q\| \leq \|x - q\| \text{ for all } q \in \operatorname{Fix} T \text{ and } x \in \mathscr{H}\}, \tag{9.7}$$

whose elements are called elsewhere quasi-nonexpansive or paracontracting operators.

Definition 9.2. Let $T : \mathscr{H} \to \mathscr{H}$ and let $\lambda \in (0,2)$. We call the operator $T_\lambda := \operatorname{Id} + \lambda(T - \operatorname{Id})$ *a relaxation of* T.

Definition 9.3. We say that an operator $T : \mathscr{H} \to \mathscr{H}$ with $\operatorname{Fix} T \neq \emptyset$ is *strictly quasi-nonexpansive* if

$$\|Tx - z\| < \|x - z\| \tag{9.8}$$

for all $x \notin \operatorname{Fix} T$ and for all $z \in \operatorname{Fix} T$. We say that T is *α-strongly quasi-nonexpansive*, where $\alpha > 0$, or, in short, strongly quasi-nonexpansive if

$$\|Tx - z\|^2 \leq \|x - z\|^2 - \alpha\|Tx - x\|^2 \tag{9.9}$$

for all $x \in \mathscr{H}$ and for all $z \in \operatorname{Fix} T$.

We have the following result from [21, Proposition 2.3 (i) and (ii)].

Lemma 9.4. *Let $X \subset \mathscr{H}$ be a closed and convex set and $U : X \to X$ be an operator having a fixed point.*

(i) U is a cutter if and only if

$$\langle z - x, Ux - x\rangle \geq \|Ux - x\|^2 \tag{9.10}$$

for all $x \in X$ and for all $z \in \operatorname{Fix} U$.

(ii) Let $\lambda \in (0,2)$. If U is a cutter, then its relaxation U_λ is $\frac{2-\lambda}{\lambda}$-strongly quasi-nonexpansive.

One can show that the implication converse to (ii) is also true.

Definition 9.5. We say that an operator $T : \mathscr{H} \to \mathscr{H}$ is *demiclosed* at 0 if for any weakly converging sequence $\{x^k\}_{k=0}^\infty$, $x^k \rightharpoonup y \in \mathscr{H}$ as $k \to \infty$, with $Tx^k \to 0$ as $k \to \infty$, we have $Ty = 0$.

It is well-known that for a nonexpansive operator $T : \mathscr{H} \to \mathscr{H}$, the operator $T - \operatorname{Id}$ is demiclosed at 0, see Opial [30, Lemma 2].

Definition 9.6. We say that an operator $T : \mathscr{H} \to \mathscr{H}$ is *asymptotically regular* if

$$\|T^{k+1}x - T^k x\| \to 0, \text{ as } k \to \infty, \tag{9.11}$$

for all $x \in \mathscr{H}$.

9.3 The Opial Theorem and Its Generalization

Opial proved the following theorem [30, Theorem 1] which is widely applied in processes described by the recurrence

$$x^{k+1} = Ux^k, \tag{9.12}$$

where $x^0 \in X$ is arbitrary, $U : X \to X$ is a nonexpansive operator and $X \subset \mathscr{H}$ is a closed and convex subset of a Hilbert space $\mathscr{H}$. Many iterative methods for convex optimization problems have the form (9.12), where the operator U is defined in a natural way by the problem under consideration.

Theorem 9.7. *Let $X \subset \mathscr{H}$ be a nonempty, closed and convex subset of a Hilbert space $\mathscr{H}$ and let $U : X \to X$ be a nonexpansive and asymptotically regular operator with* $\operatorname{Fix} U \neq \emptyset$*. Then, for any arbitrary $x \in X$, the sequence $\{U^k x\}_{k=0}^{\infty}$ converges weakly to a fixed point z^* of U.*

An example of a nonexpansive and asymptotically regular operator is a strict relaxation of a firmly nonexpansive operator or, equivalently, an averaged operator. Therefore, the Krasnoselskii–Mann theorem (see, e.g., [9, Theorem 2.1]) follows from the Opial theorem.

Several optimization methods for convex optimization problems have, however, the form

$$x^{k+1} = U_k x^k, \tag{9.13}$$

where $x^0 \in X$ is arbitrary and $\{U_k\}_{k=0}^{\infty}$, $U_k : X \to X$, is a sequence of operators. The Opial theorem cannot be applied to such methods, even if we suppose that U_k are averaged operators having a common fixed point. Our aim is to give sufficient conditions for the weak convergence of sequences generated by the recurrence (9.13) to a common fixed point of the operators $\{U_k\}_{k=0}^{\infty}$. Before formulating our main results we extend the definition of an asymptotically regular operator to a sequence of operators.

Definition 9.8. We say that a sequence of operators $\{U_k\}_{k=0}^{\infty}$, $U_k : X \to X$, is *asymptotically regular*, if for any $x \in X$

$$\lim_{k \to \infty} \|U_k U_{k-1} \dots U_0 x - U_{k-1} \dots U_0 x\| = 0, \tag{9.14}$$

or, equivalently,

$$\lim_{k\to\infty} \|U_k x^k - x^k\| = 0, \tag{9.15}$$

where the sequence $\{x^k\}_{k=0}^{\infty}$ is generated by the recurrence (9.13) with $x^0 = x$.

It is clear that an operator $U : X \to X$ is asymptotically regular, if the constant sequence of operators $U_k = U$ is asymptotically regular. A weaker version of the following theorem was proved in [11, Theorem 1].

Theorem 9.9. *Let $X \subset \mathcal{H}$ be nonempty, closed and convex, let $S : X \to \mathcal{H}$ be an operator having a fixed point and such that $S - \mathrm{Id}$ is demiclosed at 0. Let $\{U_k\}_{k=0}^{\infty}$ be an asymptotically regular sequence of quasi-nonexpansive operators $U_k : X \to X$ such that $\bigcap_{k=0}^{\infty} \mathrm{Fix}\, U_k \supset \mathrm{Fix}\, S$. Let $\{x^k\}_{k=0}^{\infty}$ be any sequence generated by the recurrence (9.13). Under these conditions it is true that:*

(i) If the sequence of operators $\{U_k\}_{k=0}^{\infty}$ has the property

$$\lim_{k\to\infty} \|U_k x^k - x^k\| = 0 \Longrightarrow \lim_{k\to\infty} \|Sx^k - x^k\| = 0 \tag{9.16}$$

then $\{x^k\}_{k=0}^{\infty}$ converges weakly to a point $z^* \in \mathrm{Fix}\, S$.

(ii) If $\mathcal{H}$ is finite-dimensional and the sequence of operators $\{U_k\}_{k=0}^{\infty}$ has the property

$$\lim_{k\to\infty} \|U_k x^k - x^k\| = 0 \Longrightarrow \liminf_{k\to\infty} \|Sx^k - x^k\| = 0 \tag{9.17}$$

then $\{x^k\}_{k=0}^{\infty}$ converges to a point $z^* \in \mathrm{Fix}\, S$.

Proof. Let $x \in X$, $z \in \mathrm{Fix}\, S$ and let the sequence $\{x^k\}_{k=0}^{\infty}$ be generated by the recurrence (9.13). Since U_k is quasi-nonexpansive and $\mathrm{Fix}\, U_k \supset \mathrm{Fix}\, S$, we have

$$\|x^{k+1} - z\| = \|U_k x^k - z\| \le \|x^k - z\|, \text{ for all } k \ge 0. \tag{9.18}$$

Therefore, $\{x^k\}_{k=0}^{\infty}$ is Fejér-monotone with respect to $\mathrm{Fix}\, S$, thus bounded.

(i) Suppose that condition (9.16) is satisfied. By the asymptotic regularity of the sequence $\{U_k\}_{k=0}^{\infty}$ we have $\lim_{k\to\infty} \|U_k x^k - x^k\| = 0$, consequently, $\lim_{k\to\infty} \|Sx^k - x^k\| = 0$. Let $x^* \in X$ be a weak cluster point of $\{x^k\}_{k=0}^{\infty}$ and let $\{x^{n_k}\}_{k=0}^{\infty} \subset \{x^k\}_{k=0}^{\infty}$ be a subsequence converging weakly to x^*. Then $\lim_{k\to\infty} \|Sx^{n_k} - x^{n_k}\| = 0$ and $x^* \in \mathrm{Fix}\, S$, by the demiclosedness of $S - \mathrm{Id}$ at 0. Since x^* is an arbitrary weak cluster point of $\{x^k\}_{k=0}^{\infty}$ and $\{x^k\}_{k=0}^{\infty}$ is Fejér-monotone with respect to $\mathrm{Fix}\, S$, the weak convergence of the whole sequence $\{x^k\}_{k=0}^{\infty}$ to x^* follows from [7, Lemma 6] (see also [4, Theorem 2.16 (ii)]).

(ii) Let $\mathcal{H}$ be finite-dimensional and suppose that condition (9.17) is satisfied. By the asymptotic regularity of $\{U_k\}_{k=0}^{\infty}$, we have $\lim_{k\to\infty} \|U_k x^k - x^k\| = 0$, consequently, $\lim_{k\to\infty} \|Sx^{n_k} - x^{n_k}\| = 0$ for a subsequence $\{x^{n_k}\}_{k=0}^{\infty} \subset \{x^k\}_{k=0}^{\infty}$. Since $\{x^{n_k}\}_{k=0}^{\infty}$ is bounded, a subsequence $\{x^{m_{n_k}}\}_{k=0}^{\infty} \subset \{x^{n_k}\}_{k=0}^{\infty}$ which converges to a point $x^* \in X$ exists. Since $S - \mathrm{Id}$ is closed at 0, we have $x^* \in \operatorname{Fix} S$. The convergence of the whole sequence $\{x^k\}_{k=0}^{\infty}$ to x^* follows now from [4, Theorem 2.16 (v)]. ∎

Note that if $U : X \to X$ is a nonexpansive operator having a fixed point, then U is quasi-nonexpansive and $U - \mathrm{Id}$ is demiclosed at 0 (see [30, Lemma 2]). Therefore, Theorem 9.9 (i) indeed generalizes the Opial theorem.

Remark 9.10. It follows from the proof that Theorem 9.9 remains true if we replace the assumption that $\{U_k\}_{k=0}^{\infty}$ is asymptotically regular and the assumption (9.16) in case (i) or (9.17) in case (ii) by a weaker assumption $\lim_{k\to\infty} \|Sx^k - x^k\| = 0$ in case (i) or $\liminf_{k\to\infty} \|Sx^k - x^k\| = 0$ in case (ii), respectively. The formulation presented in Theorem 9.9 is preferred, because in applications, the operators U_k are often relaxed cutters with relaxation parameters guaranteeing the asymptotic regularity of $\{U_k\}_{k=0}^{\infty}$. Furthermore, various practical algorithms which apply relaxed cutters have properties which yield (9.16), (9.17) or some related conditions (see the examples presented in Sect. 9.5).

9.4 Opial-Type Theorems for Cutters

In this section, we focus our attention on cutters. We first recall some properties of sequences of real numbers. Let $\alpha_k, \beta_k \geq 0$, for all $k \geq 0$, and let $\sum_{k=0}^{\infty} \alpha_k \beta_k < +\infty$. Then

$$\liminf_{k\to\infty} \alpha_k > 0 \Longrightarrow \sum_{k=0}^{\infty} \beta_k < +\infty \tag{9.19}$$

or, equivalently,

$$\sum_{k=0}^{\infty} \beta_k = +\infty \Longrightarrow \liminf_{k\to\infty} \alpha_k = 0. \tag{9.20}$$

If $\lambda_k \in [0,2]$ then the following equivalence holds

$$\liminf_{k\to\infty} \lambda_k(2-\lambda_k) > 0 \Longleftrightarrow \left(\liminf_{k\to\infty} \lambda_k > 0 \text{ and } \limsup_{k\to\infty} \lambda_k < 2 \right). \tag{9.21}$$

Lemma 9.11. *Let the sequence $\{x^k\}_{k=0}^{\infty} \subset X$ be generated by the recurrence*

$$x^{k+1} = P_X \left(x^k + \lambda_k \left(T_k x^k - x^k \right) \right), \tag{9.22}$$

where P_X is the metric projection onto X, $\lambda_k \in [0,2]$ and $\{T_k\}_{k=0}^{\infty}$ is a sequence of cutters, $T_k : X \to \mathcal{H}$, with

$$\bigcap_{k=0}^{\infty} \operatorname{Fix} T_k \neq \emptyset. \tag{9.23}$$

Then

$$\|x^{k+1}-z\|^2 \le \|x^k-z\|^2 - \lambda_k(2-\lambda_k)\|T_kx^k-x^k\|^2 \tag{9.24}$$

for all $z \in \bigcap_{k=0}^{\infty} \operatorname{Fix} T_k$. *Consequently,*

$$\|x^{k+1}-z\|^2 \le \|x^0-z\|^2 - \sum_{l=0}^{k} \lambda_l(2-\lambda_l)\|T_lx^l-x^l\|^2 \tag{9.25}$$

and

$$\sum_{k=0}^{\infty} \lambda_k(2-\lambda_k)\|T_kx^k-x^k\|^2 \le d^2\left(x^0, \bigcap_{k=0}^{\infty} \operatorname{Fix} T_k\right). \tag{9.26}$$

Moreover,

(i) If $\liminf_{k\to\infty} \lambda_k(2-\lambda_k) > 0$ *then* $\sum_{k=0}^{\infty} \|T_kx^k-x^k\|^2 < +\infty$,
(ii) If $\sum_{k=0}^{\infty} \lambda_k(2-\lambda_k) = +\infty$ *then* $\liminf_{k\to\infty} \|T_kx^k-x^k\| = 0$.

Proof. Let $z \in \bigcap_{k=0}^{\infty} \operatorname{Fix} T_k$. It is clear that $z \in X$, so that $P_X z = z$. By the nonexpansivity of the metric projection P_X and by Lemma 9.4 (i), we have

$$\begin{aligned}
\|x^{k+1}-z\|^2 &= \|P_X\left(x^k+\lambda_k\left(T_kx^k-x^k\right)\right)-z\|^2 \\
&= \|P_X\left(x^k+\lambda_k\left(T_kx^k-x^k\right)\right)-P_Xz\|^2 \\
&\le \|x^k+\lambda_k\left(T_kx^k-x^k\right)-z\|^2 \\
&= \|x^k-z\|^2+\lambda_k^2\|T_kx^k-x^k\|^2-2\lambda_k\langle z-x, T_kx^k-x^k\rangle \\
&\le \|x^k-z\|^2+\lambda_k^2\|T_kx^k-x^k\|^2-2\lambda_k\|T_kx^k-x^k\|^2,
\end{aligned} \tag{9.27}$$

which yields (9.24). Iterating this inequality k times we obtain (9.25). Since $\|x^{k+1}-z\|^2 \ge 0$, we obtain (9.26).

(i) Suppose that $\liminf_{k\to\infty} \lambda_k(2-\lambda_k) > 0$. If we set $\alpha_k = \lambda_k(2-\lambda_k)$ and $\beta_k = \|T_kx^k-x^k\|^2$ in (9.19) we obtain $\sum_{k=0}^{\infty} \|T_kx^k-x^k\|^2 < +\infty$.
(ii) Suppose that $\sum_{k=0}^{\infty} \lambda_k(2-\lambda_k) = +\infty$. If we set $\beta_k = \lambda_k(2-\lambda_k)$ and $\alpha_k = \|T_kx^k-x^k\|^2$ in (9.20) we obtain $\liminf_{k\to\infty} \|T_kx^k-x^k\| = 0$. ∎

Proposition 9.12. *Let* $S: X \to \mathcal{H}$ *be an operator having a fixed point and such that* $S - \operatorname{Id}$ *is demiclosed at* 0, *let* $x^0 \in X$ *and let the sequence* $\{x^k\}_{k=0}^{\infty} \subset X$ *be generated by the recurrence (9.22), where* $\lambda_k \in [0,2]$ *for all* $k \ge 0$, *and* $\{T_k\}_{k=0}^{\infty}$, $T_k : X \to \mathcal{H}$, *is a sequence of cutters with* $\bigcap_{k=0}^{\infty} \operatorname{Fix} T_k \supset \operatorname{Fix} S$.

(i) If $\liminf_{k\to\infty} \lambda_k(2-\lambda_k) > 0$ *and*

$$\sum_{k=0}^{\infty} \|T_k x^k - x^k\|^2 < +\infty \Longrightarrow \lim_{k\to\infty} \|Sx^k - x^k\| = 0, \tag{9.28}$$

then $\{x^k\}_{k=0}^{\infty}$ *converges weakly to a fixed point of* S.

(ii) If $\liminf_{k\to\infty} \lambda_k(2-\lambda_k) > 0$, $\mathscr{H}$ *is finite-dimensional and*

$$\sum_{k=0}^{\infty} \|T_k x^k - x^k\|^2 < +\infty \Longrightarrow \liminf_{k\to\infty} \|Sx^k - x^k\| = 0, \tag{9.29}$$

then $\{x^k\}_{k=0}^{\infty}$ *converges to a fixed point of* S.

(iii) If $\sum_{k=0}^{\infty} \lambda_k(2-\lambda_k) = +\infty$ *and*

$$\liminf_{k\to\infty} \|T_k x^k - x^k\| = 0 \Longrightarrow \lim_{k\to\infty} \|Sx^k - x^k\| = 0 \tag{9.30}$$

then $\{x^k\}_{k=0}^{\infty}$ *converges weakly to a fixed point of* S.

(iv) If $\sum_{k=0}^{\infty} \lambda_k(2-\lambda_k) = +\infty$, $\mathscr{H}$ *is finite-dimensional and*

$$\liminf_{k\to\infty} \|T_k x^k - x^k\| = 0 \Longrightarrow \liminf_{k\to\infty} \|Sx^k - x^k\| = 0 \tag{9.31}$$

then $\{x^k\}_{k=0}^{\infty}$ *converges to a fixed point of* S.

Proof. Let $C = \bigcap_{k=0}^{\infty} \operatorname{Fix} T_k$ and $z \in C$. Denote $U_k = P_X(\mathrm{Id} + \lambda_k(T_k - \mathrm{Id}))$. By Lemma 9.11 the sequence $\{x^k\}_{k=0}^{\infty}$ is Fejér-monotone with respect to C, thus bounded. Suppose that $\liminf_{k\to\infty} \lambda_k(2-\lambda_k) > 0$.

(i) Lemma 9.11 (i) and (9.28) yield $\lim_{k\to\infty} \|Sx^k - x^k\| = 0$. Let $x^* \in X$ be a weak cluster point of $\{x^k\}_{k=0}^{\infty}$. By the demiclosedness of $S - \mathrm{Id}$ we have $x^* \in \operatorname{Fix} S$. The weak convergence of $\{x^k\}_{k=0}^{\infty}$ to x^* follows now from [4, Theorem 2.16 (ii)].

(ii) Suppose that $\mathscr{H}$ is finite-dimensional. Lemma 9.11 (i) and (9.29) yield $\lim_{k\to\infty} \|Sx^{n_k} - x^{n_k}\| = 0$ for a subsequence $\{x^{n_k}\}_{k=0}^{\infty} \subset \{x^k\}_{k=0}^{\infty}$. Let $\{x^{m_{n_k}}\}_{k=0}^{\infty} \subset \{x^{n_k}\}_{k=0}^{\infty}$ be a subsequence which converges to a point $x^* \in X$. By the closedness of $S - \mathrm{Id}$ we have $x^* \in \operatorname{Fix} S$. The convergence of $\{x^k\}_{k=0}^{\infty}$ to x^* follows now from [4, Theorem 2.16 (v)].

If $\sum_{k=0}^{\infty} \lambda_k(2-\lambda_k) = +\infty$ then (iii) and (iv) can be proved similarly to (i) and (ii) by application of Lemma 9.11 (ii) and (9.30), (9.31), respectively. ■

Special cases of Proposition 9.12 were proved in [10, Corollary 3.4.F], where $X = \mathbb{R}^n$ and $S = P_{C_i}$, $i = 1, 2, \ldots, m$, with $\bigcap_{i=1}^{m} C_i \subset \bigcap_{k=0}^{\infty} \operatorname{Fix} T_k$. Other results which are closely related to Proposition 9.12 can be found in [31, Theorem 2], where, instead of assumptions (9.28)–(9.31), there appears

$$\liminf_{k\to\infty} \|T_k x^k - x^k\| = 0 \Longrightarrow \liminf_{k\to\infty} \|x^k - P_F x^k\| = 0, \tag{9.32}$$

where $F = \bigcap_{k=0}^{\infty} \operatorname{Fix} T_k$. As shown in the next section, the assumptions (9.28)–(9.31) are easier to verify than (9.32).

Remark 9.13. (a) If $\{a_k\}_{k=0}^{\infty} \subset \mathbb{R}_+$ then $\sum_{k=0}^{\infty} a_k^2 < +\infty$ implies $\lim_{k\to\infty} a_k = 0$. Therefore, if we replace $\sum_{k=0}^{\infty} \|T_k x^k - x^k\|^2 < +\infty$ by $\lim_{k\to\infty} \|T_k x^k - x^k\| = 0$ in Proposition 9.12 (i), we obtain the following weaker result:

(i') If $\liminf_k \lambda_k(2-\lambda_k) > 0$ and

$$\lim_{k\to\infty} \|T_k x^k - x^k\| = 0 \Longrightarrow \lim_{k\to\infty} \|Sx^k - x^k\| = 0 \tag{9.33}$$

then $\{x^k\}_{k=0}^{\infty}$ converges weakly to a fixed point of S.

(b) Since relaxed cutters are quasi-nonexpansive (see, [5, equivalence (v)⇔(vi) in Proposition 2.3]), iteration (9.22) with $X = \mathcal{H}$ is a special case of (9.13), where $U_k = \operatorname{Id} + \lambda_k(T_k - \operatorname{Id})$, $k \geq 0$. Then inequality (9.24) for $\lambda_k \in (0,2]$ can be written as

$$\|x^{k+1} - z\|^2 \leq \|x^k - z\|^2 - \frac{2-\lambda_k}{\lambda_k}\|U_k x^k - x^k\|^2. \tag{9.34}$$

This shows that result (i') also follows from Theorem 9.9 (i). Indeed, by (9.34) $\{U_k\}_{k=0}^{\infty}$ is asymptotically regular. If (9.33) holds then (9.16) holds, because of the equivalence $\lim_{k\to\infty} \|T_k x^k - x^k\| = 0 \Longleftrightarrow \lim_{k\to\infty} \|U_k x^k - x^k\| = 0$ which is valid if $\liminf_k \lambda_k > 0$. Now Theorem 9.9 (i) yields the weak convergence of $\{x^k\}_{k=0}^{\infty}$ to a fixed point of S.

(c) We also see that (9.29) is weaker than (9.28), and (9.31) is weaker than (9.30), i.e., in the finite-dimensional case convergence holds under weaker assumptions than in the infinite-dimensional one.

Corollary 9.14. *Let $T : X \to \mathcal{H}$ be a nonexpansive cutter (e.g., a firmly nonexpansive operator having a fixed point), let $x^0 \in X$ and let a sequence $\{x^k\}_{k=0}^{\infty}$ be generated by the recurrence*

$$x^{k+1} = P_X\left(x^k + \lambda_k\left(Tx^k - x^k\right)\right), \tag{9.35}$$

where $\lambda_k \in [0,2]$.

(i) If $\liminf_{k\to\infty} \lambda_k(2-\lambda_k) > 0$, then $\{x^k\}_{k=0}^{\infty}$ converges weakly to a fixed point of T.

(ii) If $\mathcal{H}$ is finite-dimensional and $\sum_{k=0}^{\infty} \lambda_k(2-\lambda_k) = +\infty$, then $\{x^k\}_{k=0}^{\infty}$ converges to a fixed point of T.

Proof. Denote $T_k = T$, for all $k \geq 0$, and $S = T$. Since S is nonexpansive, $S - \operatorname{Id}$ is demiclosed at 0 (see [30, Lemma 2]). Implications (9.28) and (9.31) are obvious. Therefore, (i) follows from Proposition 9.12 (i), while (ii) follows from Proposition 9.12 (iv). ∎

Remark 9.15. Since A firmly nonexpansive operator having a fixed point is a cutter and an averaged operator is relaxed firmly nonexpansive, the Krasnoselskii–Mann theorem (see, e.g., [9, Theorem 2.1]) follows from Corollary 9.14 (i) by setting $X = \mathcal{H}$ and $\lambda_k = \lambda \in (0,2)$ for $k \geq 0$.

Before formulating our next result, we introduce the notion of a generalized relaxation of an operator (compare [12, Sect. 1]).

Definition 9.16. Let $T : X \to \mathcal{H}$, $\lambda \in [0,2]$ and let $\sigma : X \to (0,+\infty)$. The operator $T_{\sigma,\lambda} : X \to \mathcal{H}$,

$$T_{\sigma,\lambda}x := x + \lambda\sigma(x)(Tx - x) \tag{9.36}$$

is called the *generalized relaxation* of T, the value λ is called the *relaxation parameter* and σ is called the *step-size function*. If $\sigma(x) \geq 1$ for all $x \in X$, then the operator $T_{\sigma,\lambda}$ is called an *extrapolation* of T_λ.

Definition 9.17. We say that an operator $T : X \to \mathcal{H}$ having a fixed point is *oriented* if, for all $x \notin \operatorname{Fix} T$,

$$\delta(x) := \inf\left\{ \frac{\langle z - x, Tx - x\rangle}{\|Tx - x\|^2} \mid z \in \operatorname{Fix} T \right\} > 0. \tag{9.37}$$

If $\delta(x) \geq \alpha > 0$ for all $x \notin \operatorname{Fix} T$, then we call the operator T *α-strongly oriented* or *strongly oriented.*

Lemma 9.4 (i) means that a cutter is 1-strongly oriented. Denoting $T_\sigma = T_{\sigma,1}$ for an operator $T : X \to \mathcal{H}$ and a step-size function $\sigma : X \to (0,+\infty)$, it is clear that $T_{\sigma,\lambda}$ is a λ-relaxation of T_σ, i.e., $T_{\sigma,\lambda} = (T_\sigma)_\lambda$ for any $\lambda \in [0,2]$.

Lemma 9.18. *Let $T : X \to \mathcal{H}$ be an oriented operator with* $\operatorname{Fix} T \neq \emptyset$. *If a step-size function $\sigma : X \to (0,+\infty)$ satisfies the inequality*

$$\sigma(x) \leq \frac{\langle z - x, Tx - x\rangle}{\|Tx - x\|^2} \tag{9.38}$$

for all $x \notin$ Fix T and for all $z \in$ Fix T, then T_σ is a cutter.

Proof. Let $x \notin \operatorname{Fix} T$ and $z \in \operatorname{Fix} T$. Let $\sigma : X \to (0,+\infty)$ be a step-size function satisfying (9.38). The existence of σ follows from the assumption that T is oriented. By inequality (9.38) we have

$$\begin{aligned} \langle z - T_\sigma x, x - T_\sigma x\rangle &= \langle z - x, x - T_\sigma x\rangle + \|x - T_\sigma x\|^2 \\ &= -\langle z - x, \sigma(x)(Tx - x)\rangle + \|x - T_\sigma x\|^2 \\ &\leq -\|\sigma(x)(Tx - x)\|^2 + \|x - T_\sigma x\|^2 = 0, \end{aligned} \tag{9.39}$$

i.e., T_σ is a cutter. ∎

Corollary 9.19. *Let $U : X \to \mathscr{H}$ be a strongly oriented operator having a fixed point and such that $U - \mathrm{Id}$ is demiclosed at 0, and let the sequence $\{x^k\}_{k=0}^{\infty} \subset X$ be generated by the recurrence*

$$x^{k+1} = P_X U_{\sigma_k,\lambda_k}(x^k), \tag{9.40}$$

where $x^0 \in X$, $\liminf_{k\to\infty} \lambda_k(2-\lambda_k) > 0$ and let the step-size functions $\sigma_k : X \to (0,+\infty)$ satisfy the condition

$$\alpha \le \sigma_k(x) \le \frac{\langle z - x, Ux - x\rangle}{\|Ux - x\|^2} \tag{9.41}$$

for all $x \notin \operatorname{Fix} U$, for all $z \in \operatorname{Fix} U$ and for some $\alpha > 0$. Then $\{x^k\}_{k=0}^{\infty}$ converges weakly to a fixed point of U.

Proof. Let $z \in \operatorname{Fix} U$. The existence of step-size functions $\sigma_k : X \to (0,+\infty)$ satisfying (9.41) for all $x \notin \operatorname{Fix} U$ and for some $\alpha > 0$, follows from the assumption that U is strongly oriented. It is clear that the recurrence (9.40) is a special case of (9.22) with $T_k = U_{\sigma_k} = U_{\sigma_k,1}$. By Lemma 9.18 the operator T_k is a cutter. We have

$$\|T_k x^k - x^k\| = \|U_{\sigma_k} x^k - x^k\| = \sigma_k(x^k)\|Ux^k - x^k\| \ge \alpha\|Ux^k - x^k\|. \tag{9.42}$$

Therefore,

$$\lim_{k\to\infty} \|T_k x^k - x^k\| = 0 \Longrightarrow \lim_{k\to\infty} \|Ux^k - x^k\|, \tag{9.43}$$

which is stronger than condition (9.28) with $S = U$ (see Remark 9.13). The weak convergence of $\{x^k\}_{k=0}^{\infty}$ to a fixed point of U follows now from Proposition 9.12 (i), because $\operatorname{Fix} U_{\sigma_k} = \operatorname{Fix} U$ for all $k \ge 0$. ■

9.5 Applications to the Common Fixed Point Problem

Let $\mathscr{U} = \{U_i\}_{i\in I}$, where $I := \{1,2,...,m\}$, be a finite family of cutters $U_i : \mathscr{H} \to \mathscr{H}$, having a common fixed point. The *common fixed point problem* is to find $x^* \in \bigcap_{i\in I} \operatorname{Fix} U_i$. In this section, we study the convergence properties of sequences generated by the recurrence

$$x^{k+1} = x^k + \lambda_k \sigma_k(x^k)\left(\sum_{i\in J_k} w_i^k(x^k) V_i^k x^k - x^k\right), \tag{9.44}$$

where $\lambda_k \in [0,2]$, $\sigma_k : \mathscr{H} \to (0,+\infty)$ are step-size functions, $\mathscr{V}^k = \{V_i^k\}_{i\in J_k}$ is a family of cutters $V_i^k : \mathscr{H} \to \mathscr{H}$, $i \in J_k = \{1,2,...,m_k\}$ with the property $\bigcap_{i\in J_k} \operatorname{Fix} V_i^k \supset \bigcap_{i\in I} \operatorname{Fix} U_i$ and $w^k : \mathscr{H} \to \Delta_{m_k}$ are weight functions

$$w^k(x) = (w_1^k(x), w_2^k(x), ..., w_{m_k}^k(x)) \tag{9.45}$$

(the subset Δ_m denotes here the standard simplex, i.e., $\Delta_m = \{u \in \mathbb{R}^m : u_i \geq 0,\ i = 1,2,...,m, \text{ and } \sum_{i=1}^m u_i = 1\}$). If $\sigma_k(x) = 1$ for all $x \in \mathscr{H}$ and for all $k \geq 0$, then the method defined by the recurrence (9.44) takes the form

$$x^{k+1} = x^k + \lambda_k \left(\sum_{i \in J_k} w_i^k(x^k) V_i^k x^k - x^k \right), \tag{9.46}$$

and is called the *simultaneous cutter method.* If $\sigma_k(x) \geq 1$ for all $x \in \mathscr{H}$ and for all $k \geq 0$, then method (9.44) is called the *extrapolated simultaneous cutter method.* The recurrence (9.44) can be written in the form

$$x^{k+1} = x^k + \lambda_k \sigma_k(x^k) \left(V^k x^k - x^k \right), \tag{9.47}$$

where $V^k = \sum_{i \in J_k} w_i^k V_i^k$, or in the form

$$x^{k+1} = V^k_{\sigma_k, \lambda_k} x^k. \tag{9.48}$$

Remark 9.20. The sequence of weight functions $\{w^k\}_{k=0}^\infty$ induces a *control sequence*. This notion is usually applied in the literature if the values of w^k are extremal points of a standard simplex (see, e.g., [13, Definition 3.2] or [17, Definition 5.1.1]). One can recognize special cases of a sequence of weight functions w^k as known control sequences. In particular, the weight functions $\{w^k\}_{k=0}^\infty$ can be constant, i.e., $w^k(x) = (w_1^k, w_2^k, ..., w_{m_k}^k) \in \Delta_{m_k}$ for all $x \in \mathscr{H}$, $k \geq 0$. A simple example of such a control sequence is the *cyclic control* (see [27, Equality (2)], [13, (3.3)] or [17, Definition 5.1.1]) The sequence $\{w^k\}_{k=0}^\infty$ can also be a constant sequence, i.e., $J_k = J$ and $w^k = w : \mathscr{H} \to \Delta_m$ for all $k \geq 0$. A simple example of such a control is the *remotest set control* (see [27, Equality (3')] or [13, (3.5)] or [17, Definition 5.1.1]). Sequences of weights depending on $x \in \mathscr{H}$ enable, however, a more general model and demonstrate the importance of assumptions on the weight functions control.

Definition 9.21. Let $V_i : \mathscr{H} \to \mathscr{H}$, $i \in J = \{1,2,...,l\}$. We say that a weight function $w : \mathscr{H} \to \Delta_l$ is *appropriate with respect to the family* $\mathscr{V} = \{V_i\}_{i \in J}$ or, shortly, *appropriate* if for any $x \notin \bigcap_{i \in J} \operatorname{Fix} V_i$ there exists a $j \in J$ such that

$$w_j(x) \|V_j x - x\| \neq 0. \tag{9.49}$$

Lemma 9.22. *Let $V_i : \mathscr{H} \to \mathscr{H}$, $i \in J = \{1,2,...,l\}$, be cutters having a common fixed point and let $V = \sum_{i \in J} w_i V_i$, where $w : \mathscr{H} \to \Delta_l$ is appropriate with respect to the family $\mathscr{V} = \{V_i\}_{i \in J}$. Then*

(i) $\operatorname{Fix} V = \bigcap_{i \in J} \operatorname{Fix} V_i$,

(ii) V is a cutter, consequently, for all $\lambda \in (0,2)$, the operator V_λ is $\frac{2-\lambda}{\lambda}$-strongly quasi-nonexpansive,

(iii) The following inequalities hold

$$\|V_\lambda x - z\|^2 \le \|x-z\|^2 - \lambda(2-\lambda)\sum_{i\in J} w_i(x)\|V_i x - x\|^2 \tag{9.50}$$

$$\le \|x-z\|^2 - \lambda(2-\lambda)\|Vx-x\|^2 \tag{9.51}$$

for all $\lambda \in [0,2]$, $x \in \mathcal{H}$ *and* $z \in \operatorname{Fix} V$.

Proof. (i) The inclusion $\bigcap_{i\in J} \operatorname{Fix} V_i \subset \operatorname{Fix} V$ is obvious. We show that $\operatorname{Fix} V \subset \bigcap_{i\in J} \operatorname{Fix} V_i$. If $\bigcap_{i\in J} \operatorname{Fix} V_i = \mathcal{H}$ then the inclusion is clear. Otherwise, suppose that $x \in \operatorname{Fix} V$, $x \notin \bigcap_{i\in J} \operatorname{Fix} V_i$ and that $z \in \bigcap_{i\in J} \operatorname{Fix} V_i$. Since a cutter is strongly quasi-nonexpansive (see Lemma 9.4 (ii)) we have $\|V_i x - z\| < \|x-z\|$ for any $i \in J$ such that $x \notin \operatorname{Fix} V_i$. The convexity of the norm, the strict quasi-nonexpansivity of V_i and the fact that the weight function w is appropriate yield

$$\|Vx - z\| = \|\sum_{i\in J} w_i(x)(V_i x - z)\| \le \sum_{i\in J} w_i(x)\|V_i x - z\|$$

$$< \sum_{i\in J} w_i(x)\|x-z\| = \|x-z\|. \tag{9.52}$$

We get a contradiction, which shows that $\operatorname{Fix} V \subset \bigcap_{i\in J} \operatorname{Fix} V_i$.

(ii) Let $x \in \mathcal{H}$ and $z \in \operatorname{Fix} V$. It follows from (i) that $z \in \bigcap_{i\in J} \operatorname{Fix} V_i$. By Lemma 9.4 (i) and by the convexity of $\|\cdot\|^2$, we have

$$\langle Vx - x, z - x\rangle = \sum_{i\in J} w_i(x)\langle V_i x - x, z - x\rangle$$

$$\ge \sum_{i\in J} w_i(x)\|V_i x - x\|^2$$

$$\ge \|\sum_{i\in J} w_i(x) V_i x - x\|^2$$

$$= \|Vx - x\|^2. \tag{9.53}$$

Applying again Lemma 9.4 (i) we deduce that V is a cutter. By Lemma 9.4 (ii) the operator V_λ is $\frac{2-\lambda}{\lambda}$-strongly quasi-nonexpansive for any $\lambda \in (0,2)$.

(iii) Let $\lambda \in [0,2]$, $x \in \mathcal{H}$ and $z \in \operatorname{Fix} V$. The convexity of $\|\cdot\|^2$ and Lemma 9.4 (i) yield

$$\|V_\lambda x - z\|^2 = \|x + \lambda \sum_{i\in J} w_i(x)(V_i x - x) - z\|^2$$

$$= \|x-z\|^2 + \lambda^2 \|\sum_{i\in J} w_i(x)(V_i x - x)\|^2 - 2\lambda \sum_{i\in J} w_i(x)\langle z - x, V_i x - x\rangle$$

$$\leq \|x-z\|^2+\lambda^2\sum_{i\in J}w_i(x)\|V_ix-x\|^2-2\lambda\sum_{i\in J}w_i(x)\|V_ix-x\|^2$$
$$=\|x-z\|^2-\lambda(2-\lambda)\sum_{i\in J}w_i(x)\|V_ix-x\|^2, \tag{9.54}$$

i.e., the inequality (9.50) holds. Inequality (9.51) follows from the convexity of the function $\|\cdot\|^2$. ■

Definition 9.23. Let $\mathscr{V}=\{V_i\}_{i\in J}$ be a finite family of operators $V_i:\mathscr{H}\to\mathscr{H}$, $i\in J$, and let $\beta\in(0,1]$ be a constant. We say that a weight function $w:\mathscr{H}\to\Delta_{|J|}$ is *β-regular with respect to the family of cutters $\mathscr{U}=\{U_i\}_{i\in I}$*, or, shortly, *regular* if for any $x\in\mathscr{H}$ there exists a $j\in J$ such that

$$w_j(x)\|V_jx-x\|^2\geq\beta\max\left\{\|U_ix-x\|^2\mid i\in I\right\}. \tag{9.55}$$

If $\bigcap_{i\in J}\operatorname{Fix}V_i\supset\bigcap_{i\in I}\operatorname{Fix}U_i$ then a weight function which is regular with respect to the family $\mathscr{U}=\{U_i\}_{i\in I}$ is appropriate with respect to the family $\mathscr{V}=\{V_i\}_{i\in J}$.

Example 9.24. Let $\mathscr{V}=\mathscr{U}$ and let $I(x)=\{i\in I\mid x\notin\operatorname{Fix}U_i\}$ and let $m(x)=|I(x)|$ be the cardinality of $I(x)$, for $x\in\mathscr{H}$. The following weight functions $w:\mathscr{H}\to\Delta_m$, where $w(x)=(w_1(x),\ldots,w_m(x))$, are regular:

(a) Positive constant weights, i.e.,

$$w(x)=w\in\operatorname{ri}\Delta_m \tag{9.56}$$

for all $x\in\mathscr{H}$, where $\operatorname{ri}\Delta_m=\{w\in\mathbb{R}^m\mid w>0\text{ and }\langle e,w\rangle=1\}$ is the relative interior of Δ_m. A specific example is furnished by equal weights, i.e., $w_i(x)=1/m$, $i\in I$. To verify that w is regular set $j\in\operatorname{Argmax}_{i\in I}\|U_ix-x\|$ and $\beta=\min_{i\in I}w_i$ in Definition 9.23.

(b) Constant weights for violated constraints, i.e.,

$$w_i(x):=\begin{cases}\dfrac{w_i}{\sum_{j\in I(x)}w_j}, & \text{for } i\in I(x),\\ 0, & \text{for} i\notin I(x),\end{cases} \tag{9.57}$$

where $w=(w_1,w_2,\ldots,w_m)\in\operatorname{ri}\Delta_m$. A specific example is

$$w_i(x):=\begin{cases}1/m(x), & \text{for } i\in I(x),\\ 0, & \text{for} i\notin I(x).\end{cases} \tag{9.58}$$

To verify that w is regular set $j\in\operatorname{Argmax}_{i\in I}\|U_ix-x\|$ and $\beta=\min_{i\in I}w_i$ in Definition 9.23.

(c) Weights proportional to $\|U_i x - x\|$, i.e.,

$$w_i(x) = \begin{cases} \dfrac{\|U_i x - x\|}{\sum\limits_{j\in I}\|U_j x - x\|}, & \text{for } x \notin \bigcap_{i\in I} \operatorname{Fix} U_i, \\ 0, & \text{for } x \in \bigcap_{i\in I} \operatorname{Fix} U_i. \end{cases} \tag{9.59}$$

To verify, set $j \in \operatorname{Argmax}_{i\in I} \|U_i x - x\|$ and $\beta = 1/m$ in Definition 9.23.

(d) Weight functions $w : \mathscr{H} \to \Delta_m$ satisfying the condition

$$w_i(x) \geq \delta \text{ for } i \in I(x) \tag{9.60}$$

for some constant $\delta > 0$. To verify, choose $j(x) \in \operatorname{Argmax}_{i\in I} \|U_i x - x\|$ and set $\beta = \delta$ in Definition 9.23. These weight functions were applied by Combettes in [19, Sect. III] and in [20, Sect. 1]. Observe that the weight functions defined by (9.56) and by (9.57) satisfy (9.60).

(e) Weight functions $w : \mathscr{H} \to \Delta_m$ for which $w_i(x) = 0$ for all $x \in \mathscr{H}$ and for all $i \notin J_\gamma(x)$, where

$$J_\gamma(x) = \{j \in I \mid \|U_j x - x\| \geq \gamma \max_{i\in I} \|U_i x - x\|\}, \tag{9.61}$$

for some $\gamma \in (0,1]$. To verify, set $j = j(x) \in J_\gamma(x)$ with $\omega_j(x) \geq 1/m$ and $\beta = \gamma^2/m$ in Definition 9.23. The existence of such j follows from the fact that $w_i(x) \geq 0$ for all $i \in J_\gamma(x)$ and $\sum_{i\in J_\gamma(x)} w_i(x) = 1$. Specific examples are obtained as follows:

(i) When $U_i = P_{C_i}$ for a closed convex subset $C_i \subset \mathscr{H}$, $i \in I$, and

$$w_i(x) = \begin{cases} 1, & \text{if } i = \operatorname{argmax}_{j\in I} \|U_j x - x\| \\ 0, & \text{otherwise.} \end{cases} \tag{9.62}$$

In this case, w defines a *remotest set control* (for the definition, see [27, (3')] or [17, Sect. 5.1]).

(ii) When $U_i = P_{C_i}$ for a closed convex subset $C_i \subset \mathscr{H}$, $i \in I$, and

$$w_i(x) = \begin{cases} 1, & \text{if } i = j(x) \\ 0, & \text{otherwise,} \end{cases} \tag{9.63}$$

where $j(x) \in J_\gamma(x)$ for some $\gamma \in (0,1]$. In this case w is an *approximately remotest set control* (for the definition, see [27, (3)] or [17, Sect. 5.1]).

Definition 9.25. Let $\mathscr{V}^k = \{V_i^k\}_{i\in J_k}$ be a sequence of cutters $V_i^k : \mathscr{H} \to \mathscr{H}$, $i \in J_k = \{1, 2, \ldots, m_k\}$, $k \geq 0$, and let the sequence $\{x^k\}_{k=0}^\infty$ be generated by the recurrence (9.44). We say that a sequence of appropriate weight functions $w^k : \mathscr{H} \to \Delta_{m_k}$ (applied to the sequence of families $\mathscr{V}^k$) is

- *Regular* (with respect to the family $\mathscr{U} = \{U_i\}_{i\in I}$) if there is a constant $\beta \in (0,1]$ such that w^k are β-regular for all $k \geq 0$,
- *Approximately regular* (with respect to the family $\mathscr{U} = \{U_i\}_{i\in I}$) if there exists a sequence $i_k \in J_k$ such that the following implication holds

$$\lim_{k\to\infty} w_{i_k}^k(x^k)\|V_{i_k}^k x^k - x^k\|^2 = 0 \Longrightarrow \lim_{k\to\infty} \|U_i x^k - x^k\| = 0 \text{ for all } i \in I, \qquad (9.64)$$

- *Approximately semi-regular* (with respect to the family $\mathscr{U} = \{U_i\}_{i\in I}$) if there exists a sequence $i_k \in J_k$ such that the following implication holds

$$\lim_{k\to\infty} w_{i_k}^k(x^k)\|V_{i_k}^k x^k - x^k\|^2 = 0 \Longrightarrow \liminf_{k\to\infty} \|U_i x^k - x^k\| = 0 \text{ for all } i \in I. \qquad (9.65)$$

Example 9.26. Here are examples of weight functions which are approximately regular or approximately semi-regular with respect to the family $\mathscr{U} = \{U_i\}_{i\in I}$.

(a) A regular sequence of weight functions is approximately regular.
(b) A sequence containing a regular subsequence of weight functions is approximately semi-regular.
(c) Let $\{x^k\}_{k=0}^\infty$ be a sequence generated by the recurrence (9.46), where $\mathscr{V}^k = \mathscr{U}$ and $w^k = \delta_{i_k}$. We call the sequence $\{i_k\}_{k=0}^\infty$ a *control sequence* (see [13, Definition 3.2]). Recurrence (9.46) can be written as follows

$$x^{k+1} = x^k + \lambda_k \left(U_{i_k} x^k - x^k\right). \qquad (9.66)$$

Implication (9.64) takes the form

$$\lim_{k\to\infty} \|U_{i_k} x^k - x^k\| = 0 \Longrightarrow \lim_{k\to\infty} \|U_i x^k - x^k\| = 0 \text{ for all } i \in I. \qquad (9.67)$$

If (9.67) is satisfied we say that the control sequence $\{i_k\}_{k=0}^\infty$ is *approximately regular*. If we set $U_i = P_{C_i}$ for a closed convex subset $C_i \subset \mathscr{H}$, $i \in I$, then implication (9.67) can be written in the form

$$\lim_{k\to\infty} \|P_{C_{i_k}} x^k - x^k\| = 0 \Longrightarrow \lim_{k\to\infty} \max_{i\in I} \|P_{C_i} x^k - x^k\| = 0. \qquad (9.68)$$

A sequence $\{i_k\}_{k=0}^\infty$ satisfying (9.68) is called *approximately remotest set control* (see [27, Sect. 1]).
(d) (Combettes [19, Sect. II D]). Let I_k be a nonempty subset of I, $k \geq 0$. Suppose that there is a constant $s \geq 1$ such that

$$I = I_k \cup I_{k+1} \cup \ldots \cup I_{k+s-1} \text{ for all } k \geq 0. \qquad (9.69)$$

Let $U_i = P_{C_i}$, where $C_i \subset \mathscr{H}$ is closed and convex. Let $\{x^k\}_{k=0}^\infty$ be a sequence generated by the recurrence (9.46), where $\mathscr{V}^k = \mathscr{U} = \{U_i\}_{i\in I}$, $\lambda_k \in [\varepsilon, 2-\varepsilon]$ for some $\varepsilon \in (0,1)$, and $w^k \in \Delta_m$ is a weight vector such that $\sum_{i\in I_k} w_i^k = 1$ and

$w_i^k \geq \delta > 0$ for all $i \in I_k \cap I(x^k)$, $k \geq 0$, and $I(x) = \{i \in I \mid x \notin C_i\}$. Bauschke and Borwein called a sequence of weights satisfying (9.69) with $I_k = \{i \in I \mid w_i^k > 0\}$ an *intermittent control* (see [4, Definition 3.18]). The recurrence (9.46) can be written in the form $x^{k+1} = T_k x^k$, where $T_k = \mathrm{Id} + \lambda_k(V_k - \mathrm{Id})$ and $V_k = \sum_{i \in I_k} w_i^k P_{C_i}$, or, equivalently, in the form

$$x^{k+1} = x^k + \lambda_k \left(\sum_{i \in I_k} w_i^k P_{C_i} x^k - x^k \right). \tag{9.70}$$

One can show that V_k is a cutter. We show that $\{w^k\}_{k=0}^{\infty}$ is approximately regular. Let $i \in I$ be arbitrary and let $r_k \in \{0, 1, \ldots, s-1\}$ be such that $i \in I_{k+r_k}$, $k \geq 0$. By the triangle inequality, we have

$$\begin{aligned} \|x^{k+r_k} - x^k\| &\leq \sum_{i=0}^{r_k - 1} \|x^{k+i+1} - x^{k+i}\| = \sum_{i=0}^{r_k-1} \|T_{k+i} x^{k+i} - x^{k+i}\| \\ &\leq \sum_{i=0}^{s-1} \|T_{k+i} x^{k+i} - x^{k+i}\|, \end{aligned} \tag{9.71}$$

for $k \geq 0$. Since T_k are λ_k-relaxed cutters and $\lambda_k \in [\varepsilon, 2 - \varepsilon]$, Lemma 9.22 (iii) yields $\lim_{k\to\infty} \|T_{k+i} x^{k+i} - x^{k+i}\| = 0$, $i = 1, 2, \ldots, s-1$, consequently, $\|x^{k+r_k} - x^k\| \to 0$. Further, by the definition of the metric projection and by the triangle inequality, we have

$$\|P_{C_i} x^k - x^k\| \leq \|P_{C_i} x^{k+r_k} - x^k\| \leq \|P_{C_i} x^{k+r_k} - x^{k+r_k}\| + \|x^{k+r_k} - x^k\|. \tag{9.72}$$

Let $j_k \in I_k$ be such that $\|P_{C_{j_k}} x^k - x^k\| = \max_{j \in I_k} \|P_{C_j} x^k - x^k\|$, $k \geq 0$. Let $\lim_{k\to\infty} w_{j_k}^k \|P_{C_{j_k}} x^k - x^k\|^2 = 0$. Since $w_{j_k}^k \geq \delta$ for $j_k \in I(x^k)$ we have

$$\lim_{k\to\infty} \|P_{C_{j_k}} x^k - x^k\| = 0. \tag{9.73}$$

Since $i \in I_{k+r_k}$ we have

$$\|P_{C_i} x^{k+r_k} - x^{k+r_k}\| \leq \|P_{C_{j_{k+r_k}}} x^{k+r_k} - x^{k+r_k}\|.$$

consequently, $\lim_{k\to\infty} \|P_{C_i} x^{k+r_k} - x^{k+r_k}\| = 0$. The inequalities (9.71) and (9.72) yield now $\lim_{k\to\infty} \|P_{C_i} x^k - x^k\| = 0$, i.e., $\{w^k\}_{k=0}^{\infty}$ is approximately regular.

(e) Let $\mathscr{H} = \mathbb{R}^n$, let $U_i : \mathscr{H} \to \mathscr{H}$, $i \in I$, be cutters having a common fixed point and let $\liminf_{k\to\infty} \lambda_k(2 - \lambda_k) > 0$. Consider a sequence generated by the recurrence (9.66) with a *repetitive control* $\{i_k\}_{k=0}^{\infty} \subset I$, i.e., a control for which the subset $K_i = \{k \geq 0 \mid i_k = i\}$ is infinite for any $i \in I$ (see., e.g., [1, Sect. 3]). It is clear that $\mathbb{N}_0 = \{0, 1, 2, 3, \ldots\} = K_1 \cup K_2 \cup \ldots \cup K_m$ and that $K_i \cap J_j = \emptyset$ for all $i, j \in I, i \neq j$.

The control $\{i_k\}_{k=0}^{\infty}$ is approximately semi-regular. This follows from inequality (9.26) which guarantees that

$$\sum_{k=0}^{\infty} \lambda_k(2-\lambda_k)\|U_{i_k}x^k - x^k\|^2 < \infty. \tag{9.74}$$

Note that the series above is absolutely convergent, thus,

$$\sum_{i=1}^{m}\sum_{k\in K_i} \lambda_k(2-\lambda_k)\|U_i x^k - x^k\|^2 = \sum_{k=0}^{\infty} \lambda_k(2-\lambda_k)\|U_{i_k}x^k - x^k\|^2 < \infty. \tag{9.75}$$

Therefore,

$$\sum_{k\in K_i} \lambda_k(2-\lambda_k)\|U_i x^k - x^k\|^2 < \infty \text{ for all } i \in I \tag{9.76}$$

and

$$\lim_{k\to\infty, k\in K_i} \lambda_k(2-\lambda_k)\|U_i x^k - x^k\|^2 = 0 \text{ for all } i \in I. \tag{9.77}$$

Since $\liminf_{k\to\infty} \lambda_k(2-\lambda_k) > 0$, we have $\lim_{k\to\infty, k\in K_i} \|U_i x^k - x^k\| = 0$ for all $i \in I$, consequently,

$$\liminf_{k\to\infty} \|U_i x^k - x^k\| = 0 \tag{9.78}$$

for all $i \in I$, and $\{i_k\}_{k=0}^{\infty}$ is approximately semi-regular. One can prove that the approximate semi-regularity also holds for sequences generated by (9.46), where $J_k = I$ and $\mathscr{V} = \mathscr{U}$ and the sequence of weight functions $\{w^k\}_{k=0}^{\infty}$ has the property $\sum_{i\in I_k} w_i^k = 1$ for $I_k \subset I$, $k \geq 0$ and $w_i^k > \delta > 0$ for $i \in I_k$, and $i \in I_k$ for infinitely many k, $i \in I$. Note that a repetitive control is a special case of a sequence $\{w^k\}_{k=0}^{\infty}$ having the above property.

Theorem 9.27. *Suppose that:*

- $U_i : \mathscr{H} \to \mathscr{H}$, $i \in I$, *are cutters having a common fixed point,*
- $U_i - \mathrm{Id}$ *are demiclosed at* 0, $i \in I$,
- $\mathscr{V}^k = \{V_i^k\}_{i\in J_k}$ *are families of cutters* $V_i^k : \mathscr{H} \to \mathscr{H}$, $i \in J_k$, *with the property* $\bigcap_{i\in J_k} \operatorname{Fix} V_i^k \supset \bigcap_{i\in I} \operatorname{Fix} U_i$, $k \geq 0$,
- $\{w^k\}_{k=0}^{\infty} : \mathscr{H} \to \Delta_{|J_k|}$ *is a sequence of appropriate weight functions,*
- $\liminf_{k\to\infty} \lambda_k(2-\lambda_k) > 0$,
- $\{x^k\}_{k=0}^{\infty}$ *is generated by the recurrence (9.46).*

If the sequence of weight functions $\{w^k\}_{k=0}^{\infty}$ *applied to the sequence of families* $\mathscr{V}^k$:

(i) Is approximately regular with respect to the family $\mathscr{U} = \{U_i\}_{i\in I}$ *then* $\{x^k\}_{k=0}^{\infty}$ *converges weakly to a common fixed point of* U_i, $i \in I$;
(ii) Is approximately semi-regular with respect to the family $\mathscr{U} = \{U_i\}_{i\in I}$ *and* $\mathscr{H}$ *is finite-dimensional, then* $\{x^k\}_{k=0}^{\infty}$ *converges to a common fixed point of* U_i, $i \in I$.

Proof. Let $V^k : \mathcal{H} \to \mathcal{H}$ be defined by

$$V^k x = \sum_{i \in J_k} w_i^k(x) V_i^k x \tag{9.79}$$

and let T_k be the λ_k-relaxation of the operator V^k, i.e.,

$$T_k x = V_{\lambda_k}^k x = x + \lambda_k (V^k x - x). \tag{9.80}$$

The operators V^k are cutters,

$$\operatorname{Fix} T_k = \operatorname{Fix} V^k = \bigcap_{i \in J_k} \operatorname{Fix} V_i^k \supset \bigcap_{i \in I} \operatorname{Fix} U_i$$

and T_k are strongly quasi-nonexpansive, $k \geq 0$, (see Lemma 9.22), consequently, $\bigcap_{k=0}^{\infty} \operatorname{Fix} T_k \supset \bigcap_{i \in I} \operatorname{Fix} U_i$. Let $\varepsilon > 0$ be such that

$$\liminf_{k \to \infty} \lambda_k \geq \varepsilon \quad \text{and} \quad \liminf_{k \to \infty} (2 - \lambda_k) \geq \varepsilon \tag{9.81}$$

and let $z \in \bigcap_{i \in I} \operatorname{Fix} U_i$. For sufficiently large k we have $2 - \lambda_k \geq \varepsilon/2$ and $\frac{2-\lambda_k}{\lambda_k} \geq \varepsilon/4$. Now, it follows from Lemma 9.22 that, for sufficiently large k,

$$\begin{aligned}
\|x^{k+1} - z\|^2 &= \|T_k x^k - z\|^2 \\
&\leq \|x^k - z\|^2 - \lambda_k (2 - \lambda_k) \sum_{i \in J_k} w_i^k(x^k) \|V_i^k x^k - x^k\|^2 \\
&\leq \|x^k - z\|^2 - \lambda_k (2 - \lambda_k) \|V^k x^k - x^k\|^2 \\
&= \|x^k - z\|^2 - \frac{2 - \lambda_k}{\lambda_k} \|T_k x^k - x^k\|^2 \\
&\leq \|x^k - z\|^2 - \frac{\varepsilon}{4} \|T_k x^k - x^k\|^2.
\end{aligned} \tag{9.82}$$

Therefore, $\{\|x^k - z\|\}_{k=0}^{\infty}$ decreases and $\sum_{i \in J_k} w_i^k(x^k) \|V_i^k x^k - x^k\|^2 \to 0$. Consequently,

$$w_{i_k}^k(x^k) \|V_{i_k}^k x^k - x^k\|^2 \to 0 \tag{9.83}$$

for arbitrary $i_k \in J_k$.

(i) Suppose that $\{w^k\}_{k=0}^{\infty}$ is approximately regular with respect to the family $\mathcal{U} = \{U_i\}_{i \in I}$. Let $i_k \in J_k$, $k \geq 0$, be such that the implication (9.64) holds. Then (9.83) yields $\lim_{k \to \infty} \|U_i x^k - x^k\| = 0$ for all $i \in I$. Let x^* be a weak cluster point of $\{x^k\}_{k=0}^{\infty}$ and $\{x^{n_k}\}_{k=0}^{\infty}$ be a subsequence of $\{x^k\}_{k=0}^{\infty}$ such that $x^{n_k} \rightharpoonup x^*$ as $k \to \infty$. The demiclosedness of $U_i - \mathrm{Id}$ at 0, $i \in I$, yields that $x^* \in \bigcap_{i \in I} \operatorname{Fix} U_i$. Since x^* is an arbitrary weak cluster point of $\{x^k\}_{k=0}^{\infty}$ and $\{x^k\}_{k=0}^{\infty}$ is Fejér-monotone with respect to $\bigcap_{i\in I}\operatorname{Fix}U_i$, the weak convergence of the whole sequence $\{x^k\}_{k=0}^{\infty}$ to x^* follows from [7, Lemma 6] (see also [4, Theorem 2.16 (ii)]).

(ii) Suppose that $\mathscr{H}$ is finite-dimensional and $\{w^k\}_{k=0}^{\infty}$ is approximately semi-regular with respect to the family $\mathscr{U}=\{U_i\}_{i\in I}$. Let $i_k\in J_k$, $k\geq 0$, be such that the implication (9.65) holds. Let $i\in I$. Then (9.83) yields $\liminf_{k\to\infty}\|U_ix^k-x^k\|=0$. Consequently, $\lim_{k\to\infty}\|U_ix^{n_k}-x^{n_k}\|=0$ for a subsequence $\{x^{n_k}\}_{k=0}^{\infty}\subset\{x^k\}_{k=0}^{\infty}$. Since $\{x^{n_k}\}_{k=0}^{\infty}$ is bounded, a subsequence $\{x^{m_{n_k}}\}_{k=0}^{\infty}\subset\{x^{n_k}\}_{k=0}^{\infty}$ exists which converges to a point $x^*\in X$. Since $U_i-\mathrm{Id}$ is closed at 0, we have $x^*\in\operatorname{Fix}U_i$. The convergence of the whole sequence $\{x^k\}_{k=0}^{\infty}$ to x^* follows now from [4, Theorem 2.16 (v)]. The number $i\in I$ is arbitrary, so $x^*\in\bigcap_{i\in I}\operatorname{Fix}U_i$. ■

Remark 9.28. Bauschke and Borwein [4, Sect. 3, page 378] consider algorithms which are similar to (9.46), where $J_k=I$, $\lambda_k=1$, V_i^k is replaced by a firmly nonexpansive operator U_i^k with $\operatorname{Fix}U_i^k\supset\operatorname{Fix}U_i$, $i\in I$, $k\geq 0$, and $\bigcap_{i\in I}\operatorname{Fix}U_i\neq\emptyset$. They assumed that these algorithms are focusing, strongly focusing or linearly focusing (see [4, Definitions 3.7 and 4.8]). These assumptions differ from the assumptions on the regularity, approximate regularity or approximate semi-regularity, but they play a similar role in the proof of convergence of sequences generated by the considered algorithms. The recurrence considered by Bauschke and Borwein has the form

$$x^{k+1}=\sum_{i\in I}v_i^k\left(x^k+\mu_i^k\left(U_i^kx^k-x^k\right)\right), \tag{9.84}$$

where $\{\mu_i^k\}_{k=0}^{\infty}\subset[0,2]$ are sequences of relaxation parameters, $i\in I$, and $\{v^k\}_{k=0}^{\infty}\subset\Delta_m$ is a sequence of weight vectors (see [4, page 378]). Note that (9.84) can be written in the form

$$x^{k+1}=x^k+\lambda_k\left(\sum_{i\in I}w_i^kU_i^kx^k-x^k\right), \tag{9.85}$$

where $\lambda_k=\sum_{i\in I}\mu_i^kv_i^k$ and $w_i^k=\mu_i^kv_i^k/\lambda_k$. This transformation maintains the assumption $\liminf_{k\to\infty}\mu_i^k(2-\mu_i^k)>0$, $i\in I$, i.e., if the sequences $\{\mu_i^k\}_{k=0}^{\infty}$, $i\in I$, satisfies this assumption then $\liminf_{k\to\infty}\lambda_k(2-\lambda_k)>0$. Furthermore, if the sequence of weight vectors $\{v^k\}_{k=0}^{\infty}$ applied to the recurrence (9.84) is regular (approximately regular, approximately semi-regular) and $\liminf_{k\to\infty}\mu_i^k(2-\mu_i^k)>0$, $i\in I$, then the sequence of weight vectors $\{w^k\}_{k=0}^{\infty}$ applied to the recurrence (9.85) is regular (approximately regular, approximately semi-regular). Bauschke and Borwein proved the weak convergence of sequences $\{x^k\}_{k=0}^{\infty}$ generated by (9.84) to a point $x\in\bigcap_{i\in I}\operatorname{Fix}U_i$ under the assumptions that (i) the algorithm is focusing and intermittent and (ii) that $\liminf_{k\to\infty,v_i^k>0}v_i^k>0$ for all $i\in I$ (see [4, Theorem 3.20]). Assumption (ii) applied to sequences generated by (9.84) is equivalent to the following assumption (ii)' $\liminf_{k\to\infty,w_i^k>0}w_i^k>0$ applied to sequences generated by (9.85). Note, however, that assumptions (i) as well as (ii)' do not appear in Theorem 9.27. Assumptions similar to those in [4, Theorem 3.20] can be also found in [19, equalities (15)–(17)].

In the following examples we suppose that $C_i \subset \mathcal{H}$, $i \in I$, are closed and convex and that $C = \bigcap_{i \in I} \neq \emptyset$.

Example 9.29. Consider the recurrence (9.46), where $J_k = I$ for all $k \geq 0$, $V_i^k = P_{C_i}$, $i \in I$, $\lambda_k = 1$, $k \geq 0$, the sequence of weight functions $\{w^k\}_{k=0}^\infty$ is constant, $w^k = w$, $k \geq 0$, and $w : \mathbb{R}^n \to \Delta_m$ has the form

$$w_i(x) = \begin{cases} \dfrac{v_i}{\sum_{j \in I(x)} v_j}, & \text{for } i \in I(x), \\ 0, & \text{for } i \notin I(x), \end{cases} \tag{9.86}$$

where $v = (v_1, v_2, \ldots, v_m) \in \operatorname{ri} \Delta_m$ and $I(x) = \{i \in I \mid x \notin C_i\}$. Since w is regular (see Example 9.24 (b)), it is approximately regular and it follows from Theorem 9.27(i) that $x^k \rightharpoonup x^* \in C$. This convergence was proved by Iusem and De Pierro [28, Corollary 4] for $\mathcal{H} = \mathbb{R}^n$. Note, however, that in finite-dimensional case the convergence holds for any sequence $\{w^k\}_{k=0}^\infty$ containing a subsequence of β-regular weight functions, where $\beta > 0$, e.g., if $w^k = w$ for infinitely many $k \geq 0$.

Example 9.30. Aharoni and Censor [2, Theorem 1] consider the recurrence (9.46), where $\mathcal{H} = \mathbb{R}^n$, $J_k = I$ for all $k \geq 0$, $V_i^k = P_{C_i}$, $i \in I$, $\lambda_k \in [\varepsilon, 2-\varepsilon]$, where $\varepsilon \in (0,1)$, $w^k \in \Delta_m$ with $\sum_{k=0}^\infty w_i^k = +\infty$, $i \in I$. By Lemma 9.22, for any $z \in C$ we have

$$\|x^{k+1} - z\|^2 \leq \|x^0 - z\|^2 - \sum_{l=0}^{k} \lambda_l (2 - \lambda_l) \sum_{i=1}^{m} w_i^l \|P_{C_i} x^l - x^l\|^2. \tag{9.87}$$

Consequently,

$$\sum_{i=1}^{m} \sum_{k=0}^{\infty} \lambda_k (2 - \lambda_k) w_i^k(x^k) \|P_{C_i} x^k - x^k\|^2 = \sum_{k=0}^{\infty} \lambda_k (2 - \lambda_k) \sum_{i=1}^{m} w_i^k \|P_{C_i} x^k - x^k\|^2 < +\infty \tag{9.88}$$

and

$$\sum_{k=0}^{\infty} \lambda_k (2 - \lambda_k) w_i^k \|P_{C_i} x^k - x^k\|^2 < +\infty \tag{9.89}$$

for any $i \in I$. The assumption $\liminf_{k \to \infty} \lambda_k (2 - \lambda_k) > 0$ yields

$$\sum_{k=0}^{\infty} w_i^k \|P_{C_i} x^k - x^k\|^2 < +\infty, \tag{9.90}$$

$i \in I$. Since $\sum_{k=0}^\infty w_i^k = +\infty$, we have $\liminf_{k \to \infty} \|P_{C_i} x^k - x^k\| = 0$, $i \in I$, i.e., w^k is approximately semi-regular. Theorem 9.27(ii) yields now the convergence $x^k \to x^* \in C$.

Example 9.31. Butnariu and Censor [8, Theorem 4.4] consider the recurrence (9.46), where $\mathcal{H} = \mathbb{R}^n$, $J_k = I$, $V_i = P_{C_i}$, $i \in I$, $\liminf_{k\to\infty} \lambda_k > 0$, $\limsup_{k\to\infty} \lambda_k < 2$, $w^k \in \Delta_m$ has a subsequence converging to a point $w^* \in \operatorname{ri}\Delta_m$. Let $\varepsilon > 0$ be such that $w_i^* > \varepsilon$ for all $i \in I$. Then there exists a subsequence $\{w^{n_k}\}_{k=0}^{\infty} \subset \{w^k\}_{k=0}^{\infty}$ such that $w_i^{n_k} > \varepsilon/2$ for all $i \in I$ and $k \in \mathbb{N}$, consequently, $\{w^{n_k}\}_{k=0}^{\infty}$ is $\frac{\varepsilon}{2}$-regular. Therefore, $\{w^k\}_{k=0}^{\infty}$ is approximately semi-regular. Theorem 9.27(ii) yields now $\lim_{k\to\infty} x^k = x^* \in C$. If we suppose that all cluster points of $\{w^k\}_{k=0}^{\infty}$ belong to $\operatorname{ri}\Delta_m$ then $\{w^k\}_{k=0}^{\infty}$ is approximately regular, consequently the weak convergence $x^k \rightharpoonup x^*$ holds in general Hilbert spaces.

Example 9.32. Consider the recurrence (9.46), where $J_k = I$ for all $k \geq 0$,

$$\liminf_{k\to\infty} \lambda_k(2-\lambda_k) > 0, \tag{9.91}$$

$U_i = P_{C_i}$ for closed and convex subsets $C_i \subset \mathcal{H}$, $i \in I$, with $C = \bigcap_{i\in I} C_i \neq \emptyset$ and V_i^k are cutters satisfying the inequality

$$\|V_i^k x^k - x^k\| \geq \alpha \|P_{C_i} x^k - x^k\|, \tag{9.92}$$

$i \in I$, for some $\alpha > 0$ and such that $C \subset \bigcap_{i\in I} \operatorname{Fix} V_i^k$, $k \geq 0$. Furthermore, suppose that the sequence of weight vectors w^k satisfies the following conditions:

(i) $\limsup_{k\to\infty} w_i^k > 0$, $i \in I$,
(ii) $w_i^k \|P_{C_i} x^k - x^k\| \neq 0$ implies $w_i^k > \delta > 0$.

Inequality (9.92) and (i) and (ii) guarantee that the sequence of weights $\{w^k\}_{k=0}^{\infty}$ is regular and thus all assumptions of Theorem 9.27(i) are satisfied. Therefore, $x^k \rightharpoonup x^* \in C$. This convergence was proved by Flåm and Zowe [25, Theorem 1] in case $\mathcal{H} = \mathbb{R}^n$. Actually, they have considered a recurrence which can be reduced to (9.46). We omit the details.

Results similar to Theorem 9.27 also hold for sequences generated by extrapolated simultaneous cutters. Before formulating our next theorem, we prove some auxiliary results. The following lemma is an extension of Lemma 9.22. A part of this lemma can be found in [21, Proposition 2.4], where w is a constant weight function with positive coordinates.

Lemma 9.33. *Let $V_i : \mathcal{H} \to \mathcal{H}$ be cutters having a common fixed point, $i \in J = \{1,2,\ldots,l\}$, let $w : \mathcal{H} \to \Delta_l$ be an appropriate weight function and let $\sigma : \mathcal{H} \to (0,+\infty)$ be a step-size function defined by*

$$\sigma(x) = \begin{cases} \dfrac{\sum_{i=1}^{l} w_i(x)\|V_i x - x\|^2}{\|\sum_{i=1}^{l} w_i(x) V_i x - x\|^2}, & \text{if } x \notin \bigcap_{i\in J} \operatorname{Fix} V_i, \\ 1, & \text{otherwise,} \end{cases} \tag{9.93}$$

and let $V_\sigma := \mathrm{Id} + \sigma(\sum_{i=1}^{l} w_i V_i - \mathrm{Id})$ *be a generalized relaxation of the simultaneous cutter* $V = \sum_{i=1}^{l} w_i V_i$*. Then* $\operatorname{Fix} V_\sigma = \bigcap_{i \in J} \operatorname{Fix} V_i$*, the operator* V_σ *is a cutter and* V_σ *is an extrapolation of* V*. Consequently, for all* $\lambda \in (0,2)$*, the operator* $V_{\sigma,\lambda}$ *is* $\frac{2-\lambda}{\lambda}$*-strongly quasi-nonexpansive and*

$$\|V_{\sigma,\lambda}x - z\|^2 \leq \|x-z\|^2 - \lambda(2-\lambda)\sigma^2(x)\|Vx-x\|^2 \tag{9.94}$$

for all $\lambda \in [0,2]$*,* $x \in \mathcal{H}$ *and* $z \in \operatorname{Fix} V$*.*

Proof. Lemma 9.22 (i) and the positivity of the step-size function σ yield $\operatorname{Fix} V_\sigma = \operatorname{Fix} V = \bigcap_{i \in J} \operatorname{Fix} V_i$. Let $x \in \mathcal{H}$ and $z \in \operatorname{Fix} V_\sigma$. We prove that

$$\langle z - x, V_\sigma x - x\rangle \geq \|V_\sigma x - x\|^2, \tag{9.95}$$

which is equivalent to V_σ being a cutter; see Lemma 9.4(i). The inequality is clear for $x \in \operatorname{Fix} V_\sigma$. For $x \notin \operatorname{Fix} V_\sigma$ we have

$$\begin{aligned}\langle z-x, Vx-x\rangle &= \left\langle z-x, \sum_{i\in J} w_i(x)(V_i x - x)\right\rangle \\ &= \sum_{i \in J} w_i(x)\langle z-x, V_i x - x\rangle \\ &\geq \sum_{i\in J} w_i(x)\|V_i x - x\|^2 \\ &= \sigma(x)\|Vx-x\|^2,\end{aligned} \tag{9.96}$$

thus,

$$\langle z-x, Vx-x\rangle \geq \sigma(x)\|Vx-x\|^2, \tag{9.97}$$

which is equivalent to (9.95). By the convexity of the function $\|\cdot\|^2$ we have $\sigma(x) \geq 1$, i.e., V_σ is an extrapolation of V. Lemma 9.4 (ii) and the fact $V_{\sigma,\lambda} = (V_\sigma)_\lambda$ yield now the $\frac{2-\lambda}{\lambda}$-strong quasi-nonexpansivity of $V_{\sigma,\lambda}$. Inequality (9.94) follows from the equality $V_{\sigma,\lambda}x - x = \lambda\sigma(x)(Vx-x)$. ■

For a family of cutters $\mathcal{V} = \{V_i\}_{i\in J}$ and for an appropriate weight function $w : \mathcal{H} \to \Delta_{|J|}$ denote

$$\sigma_w(x) = \frac{\sum_{i\in J} w_i(x)\|V_i x - x\|^2}{\|\sum_{i \in J} w_i(x) V_i x - x\|^2}, \tag{9.98}$$

where $x \notin \bigcap_{i\in J} \operatorname{Fix} V_i$. By Lemma 9.22, $\bigcap_{i \in J} \operatorname{Fix} V_i = \operatorname{Fix} V$, where $V = \sum_{i\in J} w_i V_i$, and $\sigma_w(x)$ is well-defined.

Definition 9.34. Let $V_i : \mathcal{H} \to \mathcal{H}$, $i \in J$, be cutters with a common fixed point and let $w : \mathcal{H} \to \Delta_{|J|}$ be a weight function which is appropriate with respect to the family

$\mathscr{V} = \{V_i\}_{i\in J}$. We say that the step-size function $\sigma : \mathscr{H} \to (0, +\infty)$ is *α-admissible* with respect to the family $\mathscr{V}$, where $\alpha \in (0,1]$, or, shortly, *admissible*, if

$$\alpha\sigma_w(x) \leq \sigma(x) \leq \sigma_w(x) \tag{9.99}$$

for all $x \notin \bigcap_{i\in J} \operatorname{Fix} V_i$.

Theorem 9.35. *Suppose that:*

- $U_i : \mathscr{H} \to \mathscr{H}$, *$i \in I$, are cutters having a common fixed point,*
- $U_i - \mathrm{Id}$, *$i \in I$, are demiclosed at* 0,
- $\mathscr{V}^k = \{V_i^k\}_{i\in J_k}$ *are families of cutters $V_i^k : \mathscr{H} \to \mathscr{H}$, $i \in J_k$, with the properties* $\bigcap_{i\in J_k} \operatorname{Fix} V_i^k \supset \bigcap_{i\in I} \operatorname{Fix} U_i$, *and* $\max_{i\in J_k} \|V_i^k x - x\| \leq \gamma \max_{i\in I} \|U_i x - x\|$ *for all* $x \in \mathscr{H}$, $k \geq 0$, *and for some constant* $\gamma > 0$,
- $\{w^k\}_{k=0}^\infty : \mathscr{H} \to \Delta_{|J_k|}$ *is a sequence of appropriate weight functions,*
- *The step-size* $\sigma_k : \mathscr{H} \to (0, +\infty)$ *is α-admissible with respect to $\mathscr{V}^k$, $k \geq 0$, for some* $\alpha \in (0,1]$,
- $\liminf_{k\to\infty} \lambda_k(2 - \lambda_k) > 0$,
- $\{x^k\}_{k=0}^\infty$ *is generated by the recurrence (9.44).*

If the sequence of weight functions $\{w^k\}_{k=0}^\infty$ applied to the sequence of families $\mathscr{V}^k$:

(i) Is regular with respect to the family $\mathscr{U} = \{U_i\}_{i\in I}$ then $\{x^k\}_{k=0}^\infty$ converges weakly to a common fixed point of U_i, $i \in I$;

(ii) Contains a subsequence which is regular with respect to the family $\mathscr{U} = \{U_i\}_{i\in I}$ and $\mathscr{H}$ is finite-dimensional, then $\{x^k\}_{k=0}^\infty$ converges to a common fixed point of U_i, $i \in I$.

Proof. Let $V^k : \mathscr{H} \to \mathscr{H}$ be defined by

$$V^k x = \sum_{i\in J_k} w_i^k(x) V_i^k x \tag{9.100}$$

and let T_k be a generalized relaxation of the operator V^k, i.e.,

$$T_k x = V_{\sigma_k,\lambda_k}^k x = x + \lambda_k \sigma_k(x)(V^k x - x). \tag{9.101}$$

The operators V^k are cutters and $\operatorname{Fix} T_k = \operatorname{Fix} V^k = \bigcap_{i\in J_k} \operatorname{Fix} V_i^k$ (see Lemma 9.22). Consequently, $\bigcap_{k=0}^\infty \operatorname{Fix} T_k \supset \bigcap_{i\in I} \operatorname{Fix} U_i$. Let $\varepsilon > 0$ and $k_0 \in \mathbb{N}$ be such that $\lambda_k \in [\varepsilon, 2-\varepsilon]$ for $k \geq k_0$. By Lemma 9.33 the operator $V_{\sigma_{w_k}}^k$ is a cutter. Now, the second inequality in (9.99) and (9.5) which remains true also for $\lambda : \mathscr{H} \to [0,1]$ yield that $V_{\sigma_k}^k$ is a cutter, consequently T_k is a λ_k-relaxed cutter, $k \geq 0$. Lemma 9.33 also implies that

$$\|x^{k+1} - z\|^2 \leq \|x^k - z\|^2 - \frac{2-\lambda_k}{\lambda_k}\|T_k x^k - x^k\|^2 \tag{9.102}$$

for all $z \in \bigcap_{i=1}^m \operatorname{Fix} U_i$. Therefore, $\{x^k\}_{k=0}^\infty$ is bounded, $\{\|x^k - z\|\}_{k=0}^\infty$ is monotone and $\lim_{k\to\infty} \|T_k x^k - x^k\| = 0$.

(i) Let $\beta \in (0,1]$, $k_1 \geq k_0$ and $j_k \in J_k$ be such that

$$w_{j_k}(x)\|V_{j_k}^k x - x\|^2 \geq \beta \max_{i\in I} \|U_i x - x\|^2 \tag{9.103}$$

for any $x \in \mathscr{H}$ and for $k \geq k_1$. Since σ_k is α-admissible, the norm is a convex function and $\|V_j^k x^k - x^k\| \leq \gamma \max_i \|U_i x^k - x^k\|$ for all $j \in J_k$, we have

$$\begin{aligned}
\|T_k x^k - x^k\| &= \lambda_k \sigma_k(x^k)\|V^k x^k - x^k\| \\
&\geq \lambda_k \alpha \frac{\sum_{i\in J_k} w_i^k(x^k)\|V_i^k x^k - x^k\|^2}{\|\sum_{i\in J_k} w_i^k(x^k) V_i^k x^k - x^k\|} \\
&\geq \lambda_k \alpha \frac{w_{j_k}^k(x^k)\|V_{j_k}^k x^k - x^k\|^2}{\sum_{i\in J_k} w_i^k(x^k)\|V_i^k x^k - x^k\|} \\
&\geq \frac{\lambda_k \alpha}{\gamma} \frac{\beta \max_{i\in I}\|U_i x^k - x^k\|^2}{\left(\sum_{i\in J_k} w_i^k(x^k)\right) \max_{i\in I}\|U_i x^k - x^k\|} \\
&= \frac{\varepsilon\alpha\beta}{\gamma} \max_{i\in I}\|U_i x^k - x^k\|,
\end{aligned} \tag{9.104}$$

and $\lim_{k\to\infty} \|U_i x^k - x^k\| = 0$ for all $i \in I$. Therefore, condition (9.16) is satisfied for $U_k = T_k$ and $S = U_i$, $i \in I$. We have proved that all assumptions of Theorem 9.9(i) are satisfied for $S = U_i$, $i \in I$. Therefore, $\{x^k\}_{k=0}^\infty$ converges weakly to a common fixed point of U_i, $i \in I$.

(ii) Suppose that $\mathscr{H}$ is finite-dimensional and $\{w^k\}_{k=0}^\infty$ contains an approximately β-regular subsequence $\{w^{n_k}\}_{k=0}^\infty$. Let $\beta \in (0,1]$, $k_1 \geq k_0$ and $j_{n_k} \in I$ be such that

$$v_{j_{n_k}}(x)\|V_{j_{n_k}}^{n_k} x - x\|^2 \geq \beta \max_{i\in I} \|U_i x - x\|^2. \tag{9.105}$$

Similarly to (i), one can prove that

$$\|T_{n_k} x^{n_k} - x^{n_k}\| \geq \frac{\varepsilon\alpha\beta}{\gamma} \max_{i\in I} \|U_i x^{n_k} - x^{n_k}\|. \tag{9.106}$$

Therefore, $\liminf_{k\to\infty} \|U_i x^k - x^k\| = 0$ for all $i \in I$. If we set $U_k = T_k$ and $S = U_i$, $i \in I$, in Theorem 9.9(ii), we obtain the convergence of $\{x^k\}_{k=0}^\infty$ to a fixed point of U_i for all $i \in I$. ■

Remark 9.36. Combettes considers an algorithm which is similar to (9.44) with $J_k = I$, $w^k = w \in \operatorname{ri} \Delta_m$, $V_i^k = P_{C_i^k}$, where $C_i^k \supset C_i$ are closed and convex, $i \in I$, $k \geq 0$, and with a constant sequence of step-size functions $\sigma_k = \sigma_w$ given by

$$\sigma_w(x) = \frac{\sum_{i\in I} w_i \|P_{C_i} x - x\|^2}{\|\sum_{i\in I} w_i (P_{C_i} x - x)\|^2} \tag{9.107}$$

for $x \notin C = \bigcap_{i \in I} C_i$ (see [19, (33)–(36)]). He proves there weak convergence of sequences generated by this algorithm to a point $x \in C$ under the assumption that the algorithm is focusing (see [19, Theorem 2]). However, the assumption $w \in \operatorname{ri} \Delta_m$ is a special case of a regular sequence of weight functions and the step-size function σ_w, given by (9.107) is a special case of a sequence of α-admissible step-sizes which are considered in Theorem 9.35.

Remark 9.37. Results closely related to Theorems 9.27(ii) and 9.35(ii) appear in Kiwiel [29, Theorem 5.1], for the case $\mathcal{H} = \mathbb{R}^n$. Kiwiel applies some assumptions on weights and on the operators [29, Assumption 3.10] which differ from the assumptions in Theorems 9.27(ii) and 9.35 on the approximate semi-regularity. Our Theorems 9.27 and 9.35 show the importance of the regularity, approximate regularity and the approximate semi-regularity in both the finite- and the infinite-dimensional cases.

Example 9.38. Dos Santos' [24, Sect. 5] work is related to ours as follows. Let $c_i : \mathcal{H} \to \mathbb{R}$ be continuous and convex, let $C_i = \{x \in \mathcal{H} \mid c_i(x) \le 0\}$, $i \in I$ and let $C = \bigcap_{i=1}^m C_i \neq \emptyset$. Define $U_i : \mathcal{H} \to \mathcal{H}$ by

$$U_i x = \begin{cases} x - \dfrac{(c_i(x))_+}{\|g_i(x)\|^2} g_i(x), & \text{if } g_i(x) \neq 0, \\ x, & \text{if } g_i(x) = 0, \end{cases} \tag{9.108}$$

where a_+ denotes a nonnegative part of a real number a, i.e., $a_+ = \max\{0, a\}$, $g_i(x) \in \partial c_i(x) := \{g \in \mathcal{H} \mid \langle g, y - x\rangle \le c_i(y) - c_i(x), \text{ for all } y \in \mathcal{H}\}$ is a subgradient of the function c_i at the point x, $i \in I$. This operator U_i is called the *subgradient projection* onto C_i, $i \in I$. It follows from the definition of the subgradient that U_i is a cutter. Note that $\operatorname{Fix} U_i = C_i$, and thus $\bigcap_{i=1}^m \operatorname{Fix} U_i \neq \emptyset$. Suppose that the subgradients g_i are bounded on bounded subsets, $i \in I$ (this holds if, e.g., $\mathcal{H} = \mathbb{R}^n$). Then the operator $U_i - \mathrm{Id}$ is demiclosed at 0, $i \in I$. Indeed, let $x^k \rightharpoonup x^*$ and $\lim_{k\to\infty} \|U_i x^k - x^k\| = 0$. Then we have

$$\lim_{k\to\infty} \|U_i x^k - x^k\| = \lim_{k\to\infty} \frac{(c_i(x^k))_+}{\|g_i(x^k)\|} = 0. \tag{9.109}$$

The sequence $\{x^k\}_{k=0}^\infty$ is bounded due to its weak convergence. Condition (9.109) and the boundedness of $g_i(x^k)$ imply the convergence $\lim_{k\to\infty} c_i(x^k)_+ = 0$. Since c_i is weakly lower semi-continuous, we have $c_i(x^*) = 0$, i.e., $U_i - \mathrm{Id}$ is demiclosed at 0. Consider an extrapolated simultaneous subgradient projection method, i.e., a method which generates sequences $\{x^k\}_{k=0}^\infty$ defined by the recurrence (9.44) where $V_i^k = U_i$, w^k is a sequence of appropriate weight functions, $\liminf_{k\to\infty} \lambda_k(2-\lambda_k) > 0$ and $\sigma_k : \mathcal{H} \to (0, +\infty)$ is a sequence of step-size functions defined by

$$\sigma_k(x) = \frac{\sum_{i=1}^m w_i^k(x) \left(\dfrac{(c_i(x))_+}{\|g_i(x)\|} \right)^2}{\left\| \sum_{i=1}^m w_i^k(x) \dfrac{(c_i(x))_+}{\|g_i(x)\|^2} g_i(x) \right\|^2}. \tag{9.110}$$

Note that

$$U_ix - x = -\frac{(c_i(x))_+}{\|g_i(x)\|^2} g_i(x), \tag{9.111}$$

and so,

$$\sigma_k(x) = \sigma_w(x) = \frac{\sum_{i \in J} w_i^k(x) \|U_i x - x\|^2}{\|\sum_{i \in J} w_i^k(x) U_i x - x\|^2}, \tag{9.112}$$

and σ_k are 1-admissible. If we suppose that the sequence of weight functions $\{w^k\}_{k=0}^{\infty}$ is regular then, by Theorem 9.35(i) the sequence $\{x^k\}_{k=0}^{\infty}$ converges weakly to a point $x^* \in C$. Dos Santos [24] considers positive constant weights $w \in \operatorname{ri} \Delta_m$ and proves the convergence in the finite-dimensional case.

Acknowledgements We thank two anonymous referees for their constructive comments. This work was partially supported by Award Number R01HL070472 from the National Heart, Lung and Blood Institute and by United States-Israel Binational Science Foundation (BSF) grant No. 2009012.

References

1. Aharoni, R., Berman, A., Censor, Y.: An interior point algorithm for the convex feasibility problem. Advances in Applied Mathematics **4**, 479–489 (1983)
2. Aharoni, R., Censor, Y.: Block-iterative projection methods for parallel computation of solutions to convex feasibility problems. Linear Algebra and Its Applications **120**, 165–175 (1989)
3. Appleby, G., Smolarski, D.C.: A linear acceleration row action method for projecting onto subspaces. Electronic Transactions on Numerical Analysis **20**, 253–275 (2005)
4. Bauschke, H.H., Borwein, J.M.: On projection algorithms for solving convex feasibility problems. SIAM Review **38**, 367–426 (1996)
5. Bauschke, H.H., Combettes, P.L.: A weak to strong convergence principle for Fejér-monotone methods in Hilbert spaces. Mathematics of Operation Research **26**, 248–264 (2001)
6. Berinde, V.: Iterative Approximation of Fixed Points. Springer-Verlag, Berlin (2007)
7. Browder, F.E.: Convergence Theorems for Sequences of Nonlinear Operators in Banach Spaces. Math. Zeitschr. **100**, 201–225 (1967)
8. Butnariu, D., Censor, Y.: On the behavior of a block-iterative projection method for solving convex feasibility problems. Intern. J. Computer Math. **34**, 79–94 (1990)
9. Byrne, C.: A unified treatment of some iterative algorithms in signal processing and image reconstruction. Inverse Problems **20**, 103–120 (2004)
10. Cegielski, A.: Relaxation Methods in Convex Optimization Problems. Monographs, Vol. **67**, Institute of Mathematics, Higher College of Engineering, Zielona Góra (1993) (in Polish)
11. Cegielski, A.: A generalization of the Opial's Theorem. Control and Cybernetics **36**, 601–610 (2007)
12. Cegielski, A.: Generalized relaxations of nonexpansive operators and convex feasibility problems. Contemporary Mathematics **513**, 111–123 (2010)
13. Censor, Y.: Row-action methods for huge and sparse systems and their applications. SIAM Review **23**, 444–466 (1981)
14. Censor, Y., Segal, A.: On the string averaging method for sparse common fixed point problems. International Transactions in Operational Research **16**, 481–494 (2009)
15. Censor, Y., Segal, A.: The split common fixed point problem for directed operators. Journal of Convex Analysis **16**, 587–600 (2009)

16. Censor, Y., Segal, A.: On string-averaging for sparse problems and on the split common fixed point problem. Contemporary Mathematics **513**, 125–142 (2010)
17. Censor, Y., Zenios, S.A.: Parallel Optimization: Theory, Algorithms and Applications. Oxford University Press, New York, NY, USA (1997)
18. Cimmino, G.: Calcolo approssimato per le soluzioni dei sistemi di equazioni lineari. La Ricerca Scientifica XVI Series II, Anno IX **1**, 326–333 (1938)
19. Combettes, P.L.: Convex set theoretic image recovery by extrapolated iterations of parallel subgradient projections. IEEE Transactions on Image Processing **6**, 493–506 (1997)
20. Combettes, P.L.: Hilbertian convex feasibility problems: Convergence of projection methods. Appl. Math. Optim. **35**, 311–330 (1997)
21. Combettes, P.L.: Quasi-Fejérian analysis of some optimization algorithms. In: D. Butnariu, Y. Censor and S. Reich (eds.) *Inherently Parallel Algorithms in Feasibility and Optimization and Their Applications*, Elsevier Science Publishers, Amsterdam, The Netherlands, 115–152 (2001)
22. Crombez, G.: A geometrical look at iterative methods for operators with fixed points. Numerical Functional Analysis and Optimization **26**, 157–175 (2005)
23. De Pierro, A.R.: Metodos de projeção para a resolução de sistemas gerais de equações algébricas lienaers. Tese de Doutoramento, Instituto de Matemática, Universidade Federal do Rio de Janeiro (IM-UFRJ) (1981)
24. Dos Santos, L.T.: A parallel subgradient projections method for the convex feasibility problem. J. Comp. and Applied Math **18**, 307–320 (1987)
25. Flåm, S.D., Zowe, J.: Relaxed outer projections, weighted averages and convex feasibility. BIT **30**, 289–300 (1990)
26. Goebel, K., Reich, S.: Uniform Convexity, Hyperbolic Geometry, and Nonexpansive Mappings. Marcel Dekker, New York and Basel (1984)
27. Gurin, L.G., Polyak, B.T., Raik, E.V.: The method of projection for finding the common point in convex sets. Zh. Vychisl. Mat. i Mat. Fiz. **7** 1211–1228 (1967) (in Russian). English translation in: USSR Comput. Math. Phys. **7**, 1–24 (1967)
28. Iusem, A.N., De Pierro, A.R.: Convergence results for an accelerated nonlinear Cimmino algorithm. Numerische Mathematik **49**, 367–378 (1986)
29. Kiwiel, K.C.: Block-iterative surrogate projection methods for convex feasibility problems. Linear Algebra and Applications **215**, 225–259 (1995)
30. Opial, Z.: Weak convergence of the sequence of successive approximations for nonexpansive mappings. Bull. Amer. Math. Soc. **73**, 591–597 (1967)
31. Schott, D.: A general iterative scheme with applications to convex optimization and related fields. Optimization **22**, 885–902 (1991)
32. Segal, A.: Directed Operators for Common Fixed Point Problems and Convex Programming Problems. Ph.D. Thesis, University of Haifa, Haifa, Israel (2008)
33. Zaknoon, M.: Algorithmic Developments for the Convex Feasibility Problem. Ph.D. Thesis, University of Haifa, Haifa, Israel (2003)
34. Zhao, J., Yang, Q.: Several solution methods for the split feasibility problem. Inverse Problems **21**, 1791–1799 (2005)

Chapter 10
Proximal Splitting Methods in Signal Processing

Patrick L. Combettes and Jean-Christophe Pesquet

Abstract The proximity operator of a convex function is a natural extension of the notion of a projection operator onto a convex set. This tool, which plays a central role in the analysis and the numerical solution of convex optimization problems, has recently been introduced in the arena of inverse problems and, especially, in signal processing, where it has become increasingly important. In this paper, we review the basic properties of proximity operators which are relevant to signal processing and present optimization methods based on these operators. These proximal splitting methods are shown to capture and extend several well-known algorithms in a unifying framework. Applications of proximal methods in signal recovery and synthesis are discussed.

Keywords Alternating-direction method of multipliers · Backward–backward algorithm · Convex optimization · Denoising · Douglas–Rachford algorithm · Forward–backward algorithm · Frame · Landweber method · Iterative thresholding · Parallel computing · Peaceman–Rachford algorithm · Proximal algorithm · Restoration and reconstruction · Sparsity · Splitting

AMS 2010 Subject Classification: 90C25, 65K05, 90C90, 94A08

10.1 Introduction

Early signal processing methods were essentially linear, as they were based on classical functional analysis and linear algebra. With the development of nonlinear analysis in mathematics in the late 1950s and early 1960s (see the bibliographies of [6, 142]) and the availability of faster computers, nonlinear techniques have slowly

P.L. Combettes (✉)
UPMC Université Paris 06, Laboratoire Jacques-Louis Lions – UMR CNRS 7598,
75005 Paris, France
e-mail: plc@math.jussieu.fr

H.H. Bauschke et al. (eds.), *Fixed-Point Algorithms for Inverse Problems in Science and Engineering*, Springer Optimization and Its Applications 49,
DOI 10.1007/978-1-4419-9569-8_10,

become prevalent. In particular, convex optimization has been shown to provide efficient algorithms for computing reliable solutions in a broadening spectrum of applications.

Many signal processing problems can *in fine* be formulated as convex optimization problems of the form

$$\underset{x\in\mathbb{R}^N}{\text{minimize}}\ f_1(x)+\cdots+f_m(x), \tag{10.1}$$

where $f_1,\ldots,f_m$ are convex functions from $\mathbb{R}^N$ to $]-\infty,+\infty]$. A major difficulty that arises in solving this problem stems from the fact that, typically, some of the functions are not differentiable, which rules out conventional smooth optimization techniques. In this paper, we describe a class of efficient convex optimization algorithms to solve (10.1). These methods proceed by *splitting* in that the functions $f_1, \ldots, f_m$ are used individually so as to yield an easily implementable algorithm. They are called *proximal* because each nonsmooth function in (10.1) is involved via its proximity operator. Although proximal methods, which can be traced back to the work of Martinet [98], have been introduced in signal processing only recently [46, 55], their use is spreading rapidly.

Our main objective is to familiarize the reader with proximity operators, their main properties, and a variety of proximal algorithms for solving signal and image processing problems. The power and flexibility of proximal methods will be emphasized. In particular, it will be shown that a number of apparently unrelated, well-known algorithms (e.g., iterative thresholding, projected Landweber, projected gradient, alternating projections, alternating-direction method of multipliers, alternating split Bregman) are special instances of proximal algorithms. In this respect, the proximal formalism provides a unifying framework for analyzing and developing a broad class of convex optimization algorithms. Although many of the subsequent results are extendible to infinite-dimensional spaces, we restrict ourselves to a finite-dimensional setting to avoid technical digressions.

The paper is organized as follows. Proximity operators are introduced in Sect. 10.2, where we also discuss their main properties and provide examples. In Sects. 10.3 and 10.4, we describe the main proximal splitting algorithms, namely the forward-backward algorithm and the Douglas–Rachford algorithm. In Sect. 10.5, we present a proximal extension of Dykstra's projection method which is tailored to problems featuring strongly convex objectives. Composite problems involving linear transformations of the variables are addressed in Sect. 10.6. The algorithms discussed so far are designed for $m = 2$ functions. In Sect. 10.7, we discuss parallel variants of these algorithms for problems involving $m \geq 2$ functions. Concluding remarks are given in Sect. 10.8.

10.1.1 Notation

We denote by $\mathbb{R}^N$ the usual N-dimensional Euclidean space, by $\|\cdot\|$ its norm, and by I the identity matrix. Standard definitions and notation from convex analysis will be used [13, 87, 114]. The domain of a function $f\colon \mathbb{R}^N \to]-\infty,+\infty]$ is $\operatorname{dom} f = \{x \in \mathbb{R}^N \mid f(x) < +\infty\}$. $\Gamma_0(\mathbb{R}^N)$ is the class of lower semicontinuous convex functions from $\mathbb{R}^N$ to $]-\infty,+\infty]$ such that $\operatorname{dom} f \neq \varnothing$. Let $f \in \Gamma_0(\mathbb{R}^N)$. The conjugate of f is the function $f^* \in \Gamma_0(\mathbb{R}^N)$ defined by

$$f^*\colon \mathbb{R}^N \to]-\infty,+\infty] : u \mapsto \sup_{x\in\mathbb{R}^N} x^\top u - f(x), \tag{10.2}$$

and the subdifferential of f is the set-valued operator

$$\partial f\colon \mathbb{R}^N \to 2^{\mathbb{R}^N} : x \mapsto \big\{u \in \mathbb{R}^N \mid (\forall y \in \mathbb{R}^N)\ (y-x)^\top u + f(x) \le f(y)\big\}. \tag{10.3}$$

Let C be a nonempty subset of $\mathbb{R}^N$. The indicator function of C is

$$\iota_C\colon x \mapsto \begin{cases} 0, & \text{if } x \in C;\\ +\infty, & \text{if } x \notin C, \end{cases} \tag{10.4}$$

the support function of C is

$$\sigma_C = \iota_C^*\colon \mathbb{R}^N \to]-\infty,+\infty] : u \mapsto \sup_{x\in C} u^\top x, \tag{10.5}$$

the distance from $x \in \mathbb{R}^N$ to C is $d_C(x) = \inf_{y\in C}\|x-y\|$, and the relative interior of C (i.e., interior of C relative to its affine hull) is the nonempty set denoted by $\operatorname{ri} C$. If C is closed and convex, the projection of $x \in \mathbb{R}^N$ onto C is the unique point $P_C x \in C$ such that $d_C(x) = \|x - P_C x\|$.

10.2 From Projection to Proximity Operators

One of the first widely used convex optimization splitting algorithms in signal processing is projection onto convex sets (POCS) [31, 42, 141]. This algorithm is employed to recover/synthesize a signal satisfying simultaneously several convex constraints. Such a problem can be formalized within the framework of (10.1) by letting each function f_i be the indicator function of a nonempty closed convex set C_i modeling a constraint. This reduces (10.1) to the classical *convex feasibility problem* [31, 42, 44, 86, 93, 121, 122, 128, 141]

$$\text{find } x \in \bigcap_{i=1}^m C_i. \tag{10.6}$$

The POCS algorithm [25, 141] activates each set C_i individually by means of its projection operator P_{C_i}. It is governed by the updating rule

$$x_{n+1} = P_{C_1} \cdots P_{C_m} x_n. \tag{10.7}$$

When $\bigcap_{i=1}^m C_i \neq \varnothing$ the sequence $(x_n)_{n\in\mathbb{N}}$ thus produced converges to a solution to (10.6) [25]. Projection algorithms have been enriched with many extensions of this basic iteration to solve (10.6) [10, 43, 45, 90]. Variants have also been proposed to solve more general problems, e.g., that of finding the projection of a signal onto an intersection of convex sets [22,47,137]. Beyond such problems, however, projection methods are not appropriate and more general operators are required to tackle (10.1). Among the various generalizations of the notion of a convex projection operator that exist [10, 11, 44, 90], proximity operators are best suited for our purposes.

The projection $P_C x$ of $x \in \mathbb{R}^N$ onto the nonempty closed convex set $C \subset \mathbb{R}^N$ is the solution to the problem

$$\underset{y\in\mathbb{R}^N}{\text{minimize}} \;\; \iota_C(y) + \frac{1}{2}\|x-y\|^2. \tag{10.8}$$

Under the above hypotheses, the function ι_C belongs to $\Gamma_0(\mathbb{R}^N)$. In 1962, Moreau [101] proposed the following extension of the notion of a projection operator, whereby the function ι_C in (10.8) is replaced by an arbitrary function $f \in \Gamma_0(\mathbb{R}^N)$.

Definition 10.1 (Proximity operator). Let $f \in \Gamma_0(\mathbb{R}^N)$. For every $x \in \mathbb{R}^N$, the minimization problem

$$\underset{y\in\mathbb{R}^N}{\text{minimize}} \;\; f(y) + \frac{1}{2}\|x-y\|^2 \tag{10.9}$$

admits a unique solution, which is denoted by $\mathrm{prox}_f x$. The operator $\mathrm{prox}_f \colon \mathbb{R}^N \to \mathbb{R}^N$ thus defined is the *proximity operator* of f.

Let $f \in \Gamma_0(\mathbb{R}^N)$. The proximity operator of f is characterized by the inclusion

$$(\forall (x,p) \in \mathbb{R}^N \times \mathbb{R}^N) \quad p = \mathrm{prox}_f x \quad \Leftrightarrow \quad x - p \in \partial f(p), \tag{10.10}$$

which reduces to

$$(\forall (x,p) \in \mathbb{R}^N \times \mathbb{R}^N) \quad p = \mathrm{prox}_f x \quad \Leftrightarrow \quad x - p = \nabla f(p) \tag{10.11}$$

if f is differentiable. Proximity operators have very attractive properties that make them particularly well suited for iterative minimization algorithms. For instance, prox_f is firmly nonexpansive, i.e.,

$$\begin{aligned}(\forall x \in \mathbb{R}^N)(\forall y \in \mathbb{R}^N) \quad & \|\mathrm{prox}_f x - \mathrm{prox}_f y\|^2 + \|(x - \mathrm{prox}_f x) - (y - \mathrm{prox}_f y)\|^2 \\ & \leq \|x-y\|^2,\end{aligned} \tag{10.12}$$

and its fixed point set is precisely the set of minimizers of f. Such properties allow us to envision the possibility of developing algorithms based on the proximity operators $(\mathrm{prox}_{f_i})_{1\leq i\leq m}$ to solve (10.1), mimicking to some extent the way convex feasibility algorithms employ the projection operators $(P_{C_i})_{1\leq i\leq m}$ to solve (10.6). As shown in Table 10.1, proximity operators enjoy many additional properties. One will find in

Table 10.1 Properties of proximity operators [27, 37, 53–55, 102]: $\varphi \in \Gamma_0(\mathbb{R}^N)$; $C \subset \mathbb{R}^N$ is nonempty, closed, and convex; $x \in \mathbb{R}^N$

	Property	$f(x)$	$\operatorname{prox}_f x$
i	Translation	$\varphi(x-z)$, $z \in \mathbb{R}^N$	$z+\operatorname{prox}_\varphi(x-z)$
ii	Scaling	$\varphi(x/\rho)$, $\rho \in \mathbb{R} \smallsetminus \{0\}$	$\rho \operatorname{prox}_{\varphi/\rho^2}(x/\rho)$
iii	Reflection	$\varphi(-x)$	$-\operatorname{prox}_\varphi(-x)$
iv	Quadratic Perturbation	$\varphi(x)+\alpha\|x\|^2/2+u^\top x+\gamma$ $u \in \mathbb{R}^N$, $\alpha \geq 0$, $\gamma \in \mathbb{R}$	$\operatorname{prox}_{\varphi/(\alpha+1)}\big((x-u)/(\alpha+1)\big)$
v	Conjugation	$\varphi^*(x)$	$x-\operatorname{prox}_\varphi x$
vi	Squared distance	$\frac{1}{2}d_C^2(x)$	$\frac{1}{2}(x+P_C x)$
vii	Moreau envelope	$\widetilde{\varphi}(x)=\inf\limits_{y\in\mathbb{R}^N}\varphi(y)+\frac{1}{2}\|x-y\|^2$	$\frac{1}{2}(x+\operatorname{prox}_{2\varphi}x)$
viii	Moreau complement	$\frac{1}{2}\|\cdot\|^2-\widetilde{\varphi}(x)$	$x-\operatorname{prox}_{\varphi/2}(x/2)$
ix	Decomposition in an orthonormal basis $(b_k)_{1\leq k\leq N}$	$\sum_{k=1}^N \phi_k(x^\top b_k)$ $\phi_k \in \Gamma_0(\mathbb{R})$	$\sum_{k=1}^N \operatorname{prox}_{\phi_k}(x^\top b_k)b_k$
x	Semi-orthogonal linear transform	$\varphi(Lx)$ $L \in \mathbb{R}^{M\times N}$, $LL^\top=\nu I$, $\nu>0$	$x+\nu^{-1}L^\top(\operatorname{prox}_{\nu\varphi}(Lx)-Lx)$
xi	Quadratic function	$\gamma\|Lx-y\|^2/2$ $L\in\mathbb{R}^{M\times N}$, $\gamma>0$, $y\in\mathbb{R}^M$	$(I+\gamma L^\top L)^{-1}(x+\gamma L^\top y)$
xii	Indicator function	$\iota_C(x)=\begin{cases}0 & \text{if } x\in C\\ +\infty & \text{otherwise}\end{cases}$	$P_C x$
xiii	Distance function	$\gamma d_C(x)$, $\gamma>0$	$\begin{cases}x+\gamma(P_C x-x)/d_C(x) & \text{if } d_C(x)>\gamma\\ P_C x & \text{otherwise}\end{cases}$
xiv	Function of distance	$\phi(d_C(x))$ $\phi\in\Gamma_0(\mathbb{R})$ even, differentiable at 0 with $\phi'(0)=0$	$\begin{cases}x+\left(1-\dfrac{\operatorname{prox}_\phi d_C(x)}{d_C(x)}\right)(P_C x-x) & \text{if } x\notin C\\ x & \text{otherwise}\end{cases}$
xv	Support function	$\sigma_C(x)$	$x-P_C x$
xvi	Thresholding	$\sigma_C(x)+\phi(\|x\|)$ $\phi\in\Gamma_0(\mathbb{R})$ even and not constant	$\begin{cases}\dfrac{\operatorname{prox}_\phi d_C(x)}{d_C(x)}(x-P_C x) & \text{if } d_C(x)>\max\operatorname{Argmin}\phi\\ x-P_C x & \text{otherwise}\end{cases}$

Table 10.2 closed-form expressions of the proximity operators of various functions in $\Gamma_0(\mathbb{R})$ (in the case of functions such as $|\cdot|^p$, proximity operators implicitly appear in several places, e.g., [3, 4, 35]).

Table 10.2 Proximity operator of $\phi \in \Gamma_0(\mathbb{R})$; $\alpha \in \mathbb{R}$, $\kappa > 0$, $\underline{\kappa} > 0$, $\overline{\kappa} > 0$, $\omega > 0$, $\underline{\omega} < \overline{\omega}$, $q > 1$, $\tau \geq 0$ [37, 53, 55]

	$\phi(x)$	$\operatorname{prox}_\phi x$
i	$\iota_{[\underline{\omega},\overline{\omega}]}(x)$	$P_{[\underline{\omega},\overline{\omega}]}\, x$
ii	$\sigma_{[\underline{\omega},\overline{\omega}]}(x) = \begin{cases} \underline{\omega}x & \text{if } x < 0 \\ 0 & \text{if } x = 0 \\ \overline{\omega}x & \text{otherwise} \end{cases}$	$\operatorname{soft}_{[\underline{\omega},\overline{\omega}]}(x) = \begin{cases} x - \underline{\omega} & \text{if } x < \underline{\omega} \\ 0 & \text{if } x \in [\underline{\omega},\overline{\omega}] \\ x - \overline{\omega} & \text{if } x > \overline{\omega} \end{cases}$
iii	$\psi(x) + \sigma_{[\underline{\omega},\overline{\omega}]}(x)$ $\psi \in \Gamma_0(\mathbb{R})$ differentiable at 0 $\psi'(0) = 0$	$\operatorname{prox}_\psi\big(\operatorname{soft}_{[\underline{\omega},\overline{\omega}]}(x)\big)$
iv	$\max\{\lvert x\rvert - \omega, 0\}$	$\begin{cases} x & \text{if } \lvert x\rvert < \omega \\ \operatorname{sign}(x)\omega & \text{if } \omega \leq \lvert x\rvert \leq 2\omega \\ \operatorname{sign}(x)(\lvert x\rvert - \omega) & \text{if } \lvert x\rvert > 2\omega \end{cases}$
v	$\kappa\lvert x\rvert^q$	$\operatorname{sign}(x)p$, where $p \geq 0$ and $p + q\kappa p^{q-1} = \lvert x\rvert$
vi	$\begin{cases} \kappa x^2 & \text{if } \lvert x\rvert \leq \omega/\sqrt{2\kappa} \\ \omega\sqrt{2\kappa}\lvert x\rvert - \omega^2/2 & \text{otherwise} \end{cases}$	$\begin{cases} x/(2\kappa+1) & \text{if } \lvert x\rvert \leq \omega(2\kappa+1)/\sqrt{2\kappa} \\ x - \omega\sqrt{2\kappa}\operatorname{sign}(x) & \text{otherwise} \end{cases}$
vii	$\omega\lvert x\rvert + \tau\lvert x\rvert^2 + \kappa\lvert x\rvert^q$	$\operatorname{sign}(x)\operatorname{prox}_{\kappa\lvert\cdot\rvert^q/(2\tau+1)} \dfrac{\max\{\lvert x\rvert - \omega, 0\}}{2\tau+1}$
viii	$\omega\lvert x\rvert - \ln(1 + \omega\lvert x\rvert)$	$(2\omega)^{-1}\operatorname{sign}(x)\Big(\omega\lvert x\rvert - \omega^2 - 1 + \sqrt{\big\lvert\omega\lvert x\rvert - \omega^2 - 1\big\rvert^2 + 4\omega\lvert x\rvert}\Big)$
ix	$\begin{cases} \omega x & \text{if } x \geq 0 \\ +\infty & \text{otherwise} \end{cases}$	$\begin{cases} x - \omega & \text{if } x \geq \omega \\ 0 & \text{otherwise} \end{cases}$
x	$\begin{cases} -\omega x^{1/q} & \text{if } x \geq 0 \\ +\infty & \text{otherwise} \end{cases}$	$p^{1/q}$, where $p > 0$ and $p^{2q-1} - xp^{q-1} = q^{-1}\omega$
xi	$\begin{cases} \omega x^{-q} & \text{if } x > 0 \\ +\infty & \text{otherwise} \end{cases}$	$p > 0$ such that $p^{q+2} - xp^{q+1} = \omega q$
xii	$\begin{cases} x\ln(x) & \text{if } x > 0 \\ 0 & \text{if } x = 0 \\ +\infty & \text{otherwise} \end{cases}$	$W(e^{x-1})$, where W is the Lambert W-function
xiii	$\begin{cases} -\ln(x - \underline{\omega}) + \ln(-\underline{\omega}) & \text{if } x \in \,]\underline{\omega}, 0] \\ -\ln(\overline{\omega} - x) + \ln(\overline{\omega}) & \text{if } x \in \,]0, \overline{\omega}[\\ +\infty & \text{otherwise} \end{cases}$ $\underline{\omega} < 0 < \overline{\omega}$	$\begin{cases} \frac{1}{2}\Big(x + \underline{\omega} + \sqrt{\lvert x - \underline{\omega}\rvert^2 + 4}\Big) & \text{if } x < 1/\underline{\omega} \\ \frac{1}{2}\Big(x + \overline{\omega} - \sqrt{\lvert x - \overline{\omega}\rvert^2 + 4}\Big) & \text{if } x > 1/\overline{\omega} \\ 0 & \text{otherwise} \end{cases}$ (see Fig. 10.2)

(continued)

Table 10.2 (continued)

	$\phi(x)$	$\operatorname{prox}_\phi x$
xiv	$\begin{cases} -\kappa\ln(x)+\tau x^2/2+\alpha x & \text{if } x>0 \\ +\infty & \text{otherwise} \end{cases}$	$\dfrac{1}{2(1+\tau)}\Big(x-\alpha+\sqrt{\lvert x-\alpha\rvert^2+4\kappa(1+\tau)}\Big)$
xv	$\begin{cases} -\kappa\ln(x)+\alpha x+\omega x^{-1} & \text{if } x>0 \\ +\infty & \text{otherwise} \end{cases}$	$p>0$ such that $p^3+(\alpha-x)p^2-\kappa p=\omega$
xvi	$\begin{cases} -\kappa\ln(x)+\omega x^q & \text{if } x>0 \\ +\infty & \text{otherwise} \end{cases}$	$p>0$ such that $q\omega p^q+p^2-xp=\kappa$
xvii	$\begin{cases} -\underline{\kappa}\ln(x-\underline{\omega})-\overline{\kappa}\ln(\overline{\omega}-x) & \text{if } x\in\,]\underline{\omega},\overline{\omega}[\\ +\infty & \text{otherwise} \end{cases}$	$p\in\,]\underline{\omega},\overline{\omega}[$ such that $p^3-(\underline{\omega}+\overline{\omega}+x)p^2+(\underline{\omega}\overline{\omega}-\underline{\kappa}-\overline{\kappa}+(\underline{\omega}+\overline{\omega})x)\,p=\underline{\omega}\overline{\omega}x-\underline{\omega}\overline{\kappa}-\overline{\omega}\underline{\kappa}$

From a signal processing perspective, proximity operators have a very natural interpretation in terms of denoising. Let us consider the standard denoising problem of recovering a signal $\overline{x}\in\mathbb{R}^N$ from an observation

$$y=\overline{x}+w, \tag{10.13}$$

where $w\in\mathbb{R}^N$ models noise. This problem can be formulated as (10.9), where $\|\cdot-y\|^2/2$ plays the role of a data fidelity term and where f models a priori knowledge about $\overline{x}$. Such a formulation derives in particular from a Bayesian approach to denoising [21, 124, 126] in the presence of Gaussian noise and of a prior with a log-concave density $\exp(-f)$.

10.3 Forward–Backward Splitting

In this section, we consider the case of $m=2$ functions in (10.1), one of which is smooth.

Problem 10.2. Let $f_1\in\Gamma_0(\mathbb{R}^N)$, let $f_2\colon\mathbb{R}^N\to\mathbb{R}$ be convex and differentiable with a β-Lipschitz continuous gradient ∇f_2, i.e.,

$$(\forall(x,y)\in\mathbb{R}^N\times\mathbb{R}^N)\quad \|\nabla f_2(x)-\nabla f_2(y)\|\le\beta\|x-y\|, \tag{10.14}$$

where $\beta\in\,]0,+\infty[$. Suppose that $f_1(x)+f_2(x)\to+\infty$ as $\|x\|\to+\infty$. The problem is to

$$\underset{x\in\mathbb{R}^N}{\text{minimize}}\; f_1(x)+f_2(x). \tag{10.15}$$

It can be shown [55] that Problem 10.2 admits at least one solution and that, for any $\gamma\in\,]0,+\infty[$, its solutions are characterized by the fixed point equation

$$x=\operatorname{prox}_{\gamma f_1}\big(x-\gamma\nabla f_2(x)\big). \tag{10.16}$$

This equation suggests the possibility of iterating

$$x_{n+1} = \underbrace{\operatorname{prox}_{\gamma_n f_1}}_{\text{backward step}} \underbrace{\left(x_n - \gamma_n \nabla f_2(x_n)\right)}_{\text{forward step}} \tag{10.17}$$

for values of the step-size parameter γ_n in a suitable bounded interval. This type of scheme is known as a *forward–backward* splitting algorithm for, using the terminology used in discretization schemes in numerical analysis [132], it can be broken up into a forward (explicit) gradient step using the function f_2, and a backward (implicit) step using the function f_1. The forward–backward algorithm finds its roots in the projected gradient method [94] and in decomposition methods for solving variational inequalities [99, 119]. More recent forms of the algorithm and refinements can be found in [23,40,48,85,130]. Let us note that, on the one hand, when $f_1 = 0$, (10.17) reduces to the *gradient method*

$$x_{n+1} = x_n - \gamma_n \nabla f_2(x_n) \tag{10.18}$$

for minimizing a function with a Lipschitz continuous gradient [19,61]. On the other hand, when $f_2 = 0$, (10.17) reduces to the *proximal point algorithm*

$$x_{n+1} = \operatorname{prox}_{\gamma_n f_1} x_n \tag{10.19}$$

for minimizing a nondifferentiable function [26, 48, 91, 98, 115]. The forward–backward algorithm can therefore be considered as a combination of these two basic schemes. The following version incorporates relaxation parameters $(\lambda_n)_{n\in\mathbb{N}}$.

Algorithm 10.3 (Forward–backward algorithm).
Fix $\varepsilon \in]0, \min\{1, 1/\beta\}[$, $x_0 \in \mathbb{R}^N$
For $n = 0, 1, \ldots$

$$\left\lfloor \begin{array}{l} \gamma_n \in [\varepsilon, 2/\beta - \varepsilon] \\ y_n = x_n - \gamma_n \nabla f_2(x_n) \\ \lambda_n \in [\varepsilon, 1] \\ x_{n+1} = x_n + \lambda_n(\operatorname{prox}_{\gamma_n f_1} y_n - x_n). \end{array} \right. \tag{10.20}$$

Proposition 10.4. [55] *Every sequence $(x_n)_{n\in\mathbb{N}}$ generated by Algorithm 10.3 converges to a solution to Problem 10.2.*

The above forward–backward algorithm features varying step-sizes $(\gamma_n)_{n\in\mathbb{N}}$ but its relaxation parameters $(\lambda_n)_{n\in\mathbb{N}}$ cannot exceed 1. The following variant uses constant step-sizes and larger relaxation parameters.

Algorithm 10.5 (Constant-step forward–backward algorithm).
Fix $\varepsilon \in]0, 3/4[$ and $x_0 \in \mathbb{R}^N$
For $n = 0, 1, \ldots$

$$\left\lfloor \begin{array}{l} y_n = x_n - \beta^{-1}\nabla f_2(x_n) \\ \lambda_n \in [\varepsilon, 3/2-\varepsilon] \\ x_{n+1} = x_n + \lambda_n(\mathrm{prox}_{\beta^{-1}f_1} y_n - x_n). \end{array} \right. \tag{10.21}$$

Proposition 10.6. [13] *Every sequence $(x_n)_{n\in\mathbb{N}}$ generated by Algorithm 10.5 converges to a solution to Problem 10.2.*

Although they may have limited impact on actual numerical performance, it may be of interest to know whether linear convergence rates are available for the forward-backward algorithm. In general, the answer is negative: even in the simple setting of Example 10.12 below, linear convergence of the iterates $(x_n)_{n\in\mathbb{N}}$ generated by Algorithm 10.3 fails [9, 139]. Nonetheless, it can be achieved at the expense of additional assumptions on the problem [10,24,40,61,92,99,100,115,119,144].

Another type of convergence rate is that pertaining to the objective values $(f_1(x_n)+f_2(x_n))_{n\in\mathbb{N}}$. This rate has been investigated in several places [15,24,83] and variants of Algorithm 10.3 have been developed to improve it [15, 16, 84, 104, 105,131,136] in the spirit of classical work by Nesterov [106]. It is important to note that the convergence of the sequence of iterates $(x_n)_{n\in\mathbb{N}}$, which is often crucial in practice, is no longer guaranteed in general in such variants. The proximal gradient method proposed in [15,16] assumes the following form.

Algorithm 10.7 (Beck–Teboulle proximal gradient algorithm).
Fix $x_0 \in \mathbb{R}^N$, set $z_0 = x_0$ and $t_0 = 1$
For $n = 0, 1, \ldots$

$$\left\lfloor \begin{array}{l} y_n = z_n - \beta^{-1}\nabla f_2(z_n) \\ x_{n+1} = \mathrm{prox}_{\beta^{-1}f_1} y_n \\ t_{n+1} = \dfrac{1+\sqrt{4t_n^2+1}}{2} \\ \lambda_n = 1 + \dfrac{t_n - 1}{t_{n+1}} \\ z_{n+1} = x_n + \lambda_n(x_{n+1} - x_n). \end{array} \right. \tag{10.22}$$

While little is known about the actual convergence of sequences produced by Algorithm 10.7, the $O(1/n^2)$ rate of convergence of the objective function they achieve is optimal [103], although the practical impact of such property is not always manifest in concrete problems (see Fig. 10.1 for a comparison with the Forward–Backward algorithm).

Proposition 10.8. [15] *Assume that, for every $y \in \mathrm{dom}\, f_1$, $\partial f_1(y) \neq \varnothing$, and let x be a solution to Problem 10.2. Then every sequence $(x_n)_{n\in\mathbb{N}}$ generated by Algorithm 10.7 satisfies*

$$(\forall n \in \mathbb{N}\smallsetminus\{0\}) \quad f_1(x_n) + f_2(x_n) \leq f_1(x) + f_2(x) + \frac{2\beta\|x_0 - x\|^2}{(n+1)^2}. \tag{10.23}$$

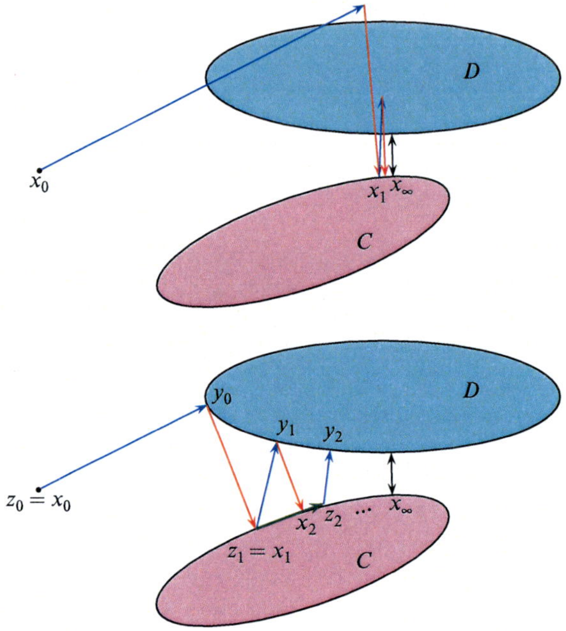

Fig. 10.1 Forward–backward vs. Beck–Teboulle: As in Example 10.12, let C and D be two closed convex sets and consider the problem (10.30) of finding a point x_∞ in C at minimum distance from D. Let us set $f_1 = \iota_C$ and $f_2 = d_D^2/2$. *Top*: The forward–backward algorithm with $\gamma_n \equiv 1.9$ and $\lambda_n \equiv 1$. As seen in Example 10.12, it reduces to the alternating projection method (10.31). *Bottom*: The Beck–Teboulle algorithm

Other variations of the forward–backward algorithm have also been reported to yield improved convergence profiles [20, 70, 97, 134, 135].

Problem 10.2 and Proposition 10.4 cover a wide variety of signal processing problems and solution methods [55]. For the sake of illustration, let us provide a few examples. For notational convenience, we set $\lambda_n \equiv 1$ in Algorithm 10.3, which reduces the updating rule to (10.17).

Example 10.9 (Projected gradient). In Problem 10.2, suppose that $f_1 = \iota_C$, where C is a closed convex subset of $\mathbb{R}^N$ such that $\{x \in C \mid f_2(x) \leq \eta\}$ is nonempty and bounded for some $\eta \in \mathbb{R}$. Then we obtain the constrained minimization problem

$$\underset{x \in C}{\text{minimize}} \; f_2(x). \tag{10.24}$$

Since $\text{prox}_{\gamma f_1} = P_C$ (see Table 10.1xii), the forward–backward iteration reduces to the *projected gradient method*

$$x_{n+1} = P_C\big(x_n - \gamma_n \nabla f_2(x_n)\big), \quad \varepsilon \leq \gamma_n \leq 2/\beta - \varepsilon. \tag{10.25}$$

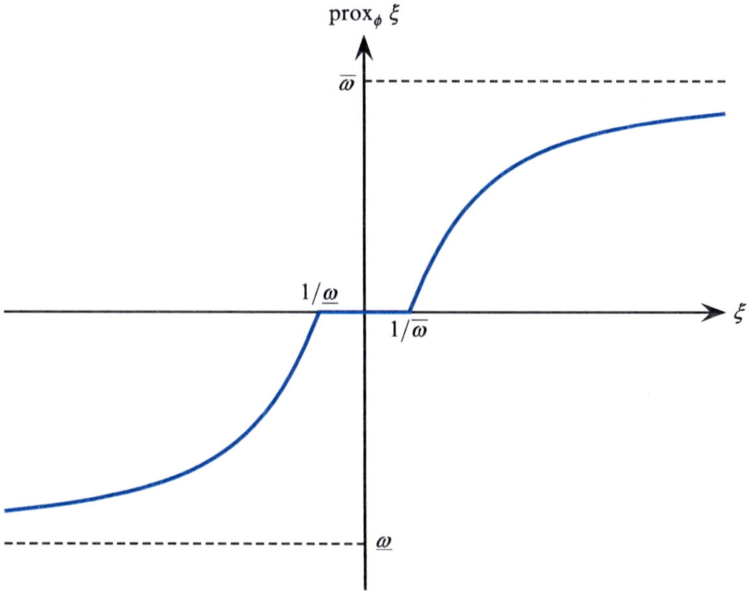

Fig. 10.2 Proximity operator of the function

$$\phi\colon \mathbb{R} \to \,]-\infty,+\infty]: \xi \mapsto \begin{cases} -\ln(\xi-\underline{\omega})+\ln(-\underline{\omega}) & \text{if } \xi \in \,]\underline{\omega},0] \\ -\ln(\overline{\omega}-\xi)+\ln(\overline{\omega}) & \text{if } \xi \in \,]0,\overline{\omega}[\\ +\infty & \text{otherwise.} \end{cases}$$

The proximity operator thresholds over the interval $[1/\underline{\omega}, 1/\overline{\omega}]$, and saturates at $-\infty$ and $+\infty$ with asymptotes at $\underline{\omega}$ and $\overline{\omega}$, respectively (see Table 10.2xiii and [53])

This algorithm has been used in numerous signal processing problems, in particular in total variation denoising [34], in image deblurring [18], in pulse shape design [50], and in compressed sensing [73].

Example 10.10 (Projected Landweber). In Example 10.9, setting $f_2\colon x \mapsto \|Lx - y\|^2/2$, where $L \in \mathbb{R}^{M\times N} \smallsetminus \{0\}$ and $y \in \mathbb{R}^M$, yields the constrained least-squares problem

$$\underset{x\in C}{\text{minimize}}\ \frac{1}{2}\|Lx-y\|^2. \tag{10.26}$$

Since $\nabla f_2\colon x \mapsto L^\top(Lx-y)$ has Lipschitz constant $\beta = \|L\|^2$, (10.25) yields the *projected Landweber method* [68]

$$x_{n+1} = P_C\big(x_n + \gamma_n L^\top(y - Lx_n)\big), \ \ \varepsilon \le \gamma_n \le 2/\|L\|^2 - \varepsilon. \tag{10.27}$$

This method has been used in particular in computer vision [89] and in signal restoration [129].

Example 10.11 (Backward–backward algorithm). Consider $f, g \in \Gamma_0(\mathbb{R}^N)$ and the problem

$$\underset{x \in \mathbb{R}^N}{\text{minimize}} \; f(x) + \widetilde{g}(x), \tag{10.28}$$

where $\widetilde{g}$ is the Moreau envelope of g (see Table 10.1vii), and suppose that $f(x) + \widetilde{g}(x) \to +\infty$ as $\|x\| \to +\infty$. This is a special case of Problem 10.2 with $f_1 = f$ and $f_2 = \widetilde{g}$. Since $\nabla f_2 \colon x \mapsto x - \operatorname{prox}_g x$ has Lipschitz constant $\beta = 1$ [55, 102], Proposition 10.4 with $\gamma_n \equiv 1$ asserts that the sequence $(x_n)_{n \in \mathbb{N}}$ generated by the *backward–backward algorithm*

$$x_{n+1} = \operatorname{prox}_f(\operatorname{prox}_g x_n) \tag{10.29}$$

converges to a solution to (10.28). Detailed analyses of this scheme can be found in [1, 14, 48, 108].

Example 10.12 (Alternating projections). In Example 10.11, let f and g be respectively the indicator functions of nonempty closed convex sets C and D, one of which is bounded. Then (10.28) amounts to finding a signal x in C at closest distance from D, i.e.,

$$\underset{x \in C}{\text{minimize}} \; \frac{1}{2} d_D^2(x). \tag{10.30}$$

Moreover, since $\operatorname{prox}_f = P_C$ and $\operatorname{prox}_g = P_D$, (10.29) yields the *alternating projection method*

$$x_{n+1} = P_C(P_D x_n), \tag{10.31}$$

which was first analyzed in this context in [41]. Signal processing applications can be found in the areas of spectral estimation [80], pulse shape design [107], wavelet construction [109], and signal synthesis [140].

Example 10.13 (Iterative thresholding). Let $(b_k)_{1 \le k \le N}$ be an orthonormal basis of $\mathbb{R}^N$, let $(\omega_k)_{1 \le k \le N}$ be strictly positive real numbers, let $L \in \mathbb{R}^{M \times N} \smallsetminus \{0\}$, and let $y \in \mathbb{R}^M$. Consider the ℓ^1–ℓ^2 problem

$$\underset{x \in \mathbb{R}^N}{\text{minimize}} \; \sum_{k=1}^{N} \omega_k |x^\top b_k| + \frac{1}{2} \|Lx - y\|^2. \tag{10.32}$$

This type of formulation arises in signal recovery problems in which y is the observed signal and the original signal is known to have a sparse representation in the basis $(b_k)_{1 \le k \le N}$, e.g., [17, 20, 56, 58, 72, 73, 125, 127]. We observe that (10.32) is a special case of (10.15) with

$$\begin{cases} f_1 \colon x \mapsto \sum_{1 \le k \le N} \omega_k |x^\top b_k| \\ f_2 \colon x \mapsto \|Lx - y\|^2/2. \end{cases} \tag{10.33}$$

Since $\operatorname{prox}_{\gamma f_1} : x \mapsto \sum_{1 \le k \le N} \operatorname{soft}_{[-\gamma\omega_k, \gamma\omega_k]}(x^\top b_k)\, b_k$ (see Tables 10.1viii and 10.2ii), it follows from Proposition 10.4 that the sequence $(x_n)_{n\in\mathbb{N}}$ generated by the *iterative thresholding algorithm*

$$x_{n+1} = \sum_{k=1}^{N} \xi_{k,n} b_k, \quad \text{where}$$

$$\begin{cases} \xi_{k,n} = \operatorname{soft}_{[-\gamma_n\omega_k, \gamma_n\omega_k]}\big(x_n + \gamma_n L^\top(y - Lx_n)\big)^\top b_k \\ \varepsilon \le \gamma_n \le 2/\|L\|^2 - \varepsilon, \end{cases} \tag{10.34}$$

converges to a solution to (10.32).

Additional applications of the forward–backward algorithm in signal and image processing can be found in [28–30, 32, 36, 37, 53, 55, 57, 74].

10.4 Douglas–Rachford Splitting

The forward-backward algorithm of Sect. 10.3 requires that one of the functions be differentiable, with a Lipschitz continuous gradient. In this section, we relax this assumption.

Problem 10.14. Let f_1 and f_2 be functions in $\Gamma_0(\mathbb{R}^N)$ such that

$$(\operatorname{ri}\operatorname{dom} f_1) \cap (\operatorname{ri}\operatorname{dom} f_2) \neq \varnothing \tag{10.35}$$

and $f_1(x) + f_2(x) \to +\infty$ as $\|x\| \to +\infty$. The problem is to

$$\underset{x\in\mathbb{R}^N}{\text{minimize}}\ f_1(x) + f_2(x). \tag{10.36}$$

What is nowadays referred to as the *Douglas–Rachford algorithm* goes back to a method originally proposed in [60] for solving matrix equations of the form $u = Ax + Bx$, where A and B are positive-definite matrices (see also [132]). The method was transformed in [95] to handle nonlinear problems and further improved in [96] to address monotone inclusion problems. For further developments, see [48, 49, 66].

Problem 10.14 admits at least one solution and, for any $\gamma \in \,]0, +\infty[$, its solutions are characterized by the two-level condition [52]

$$\begin{cases} x = \operatorname{prox}_{\gamma f_2} y \\ \operatorname{prox}_{\gamma f_2} y = \operatorname{prox}_{\gamma f_1}(2\operatorname{prox}_{\gamma f_2} y - y), \end{cases} \tag{10.37}$$

which motivates the following scheme.

Algorithm 10.15 (Douglas–Rachford algorithm).
Fix $\varepsilon \in]0,1[$, $\gamma > 0$, $y_0 \in \mathbb{R}^N$
For $n = 0,1,\ldots$

$$
\left\lfloor
\begin{array}{l}
x_n = \operatorname{prox}_{\gamma f_2} y_n \\
\lambda_n \in [\varepsilon, 2-\varepsilon] \\
y_{n+1} = y_n + \lambda_n\big(\operatorname{prox}_{\gamma f_1}(2x_n - y_n) - x_n\big).
\end{array}
\right.
\tag{10.38}
$$

Proposition 10.16. [52] *Every sequence* $(x_n)_{n\in\mathbb{N}}$ *generated by Algorithm 10.15 converges to a solution to Problem 10.14.*

Just like the forward–backward algorithm, the Douglas–Rachford algorithm operates by splitting since it employs the functions f_1 and f_2 separately. It can be viewed as more general in scope than the forward–backward algorithm in that it does not require that any of the functions have a Lipschitz continuous gradient. However, this observation must be weighed against the fact that it may be more demanding numerically as it requires the implementation of two proximal steps at each iteration, whereas only one is needed in the forward–backward algorithm. In some problems, both may be easily implementable (see Fig. 10.3 for an example) and it is not clear a priori which algorithm may be more efficient.

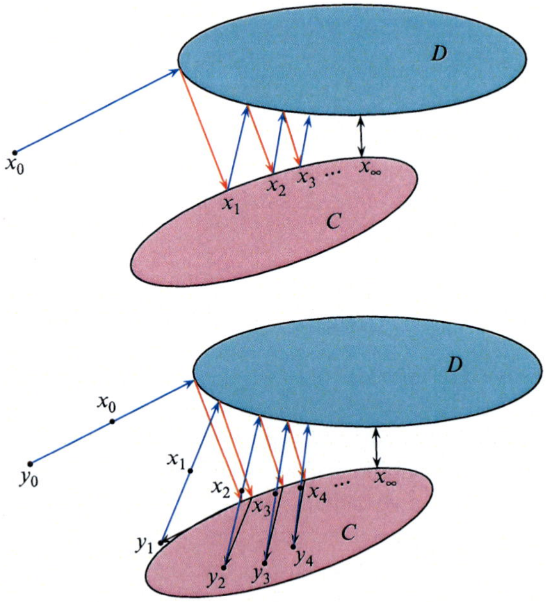

Fig. 10.3 Forward–backward vs. Douglas–Rachford: As in Example 10.12, let C and D be two closed convex sets and consider the problem (10.30) of finding a point x_∞ in C at minimum distance from D. Let us set $f_1 = \iota_C$ and $f_2 = d_D^2/2$. *Top*: The forward–backward algorithm with $\gamma_n \equiv 1$ and $\lambda_n \equiv 1$. As seen in Example 10.12, it assumes the form of the alternating projection method (10.31). *Bottom*: The Douglas–Rachford algorithm with $\gamma = 1$ and $\lambda_n \equiv 1$. Table 10.1xii yields $\operatorname{prox}_{f_1} = P_C$ and Table 10.1vi yields $\operatorname{prox}_{f_2}\colon x \mapsto (x + P_D x)/2$. Therefore, the updating rule in Algorithm 10.15 reduces to $x_n = (y_n + P_D y_n)/2$ and $y_{n+1} = P_C(2x_n - y_n) + y_n - x_n = P_C(P_D y_n) + y_n - x_n$

Applications of the Douglas–Rachford algorithm to signal and image processing can be found in [38, 52, 62, 63, 117, 118, 123].

The limiting case of the Douglas–Rachford algorithm in which $\lambda_n \equiv 2$ is the *Peaceman–Rachford algorithm* [48, 66, 96]. Its convergence requires additional assumptions (for instance, that f_2 be strictly convex and real-valued) [49].

10.5 Dykstra-Like Splitting

In this section we consider problems involving a quadratic term penalizing the deviation from a reference signal r.

Problem 10.17. Let f and g be functions in $\Gamma_0(\mathbb{R}^N)$ such that $\operatorname{dom} f \cap \operatorname{dom} g \neq \varnothing$, and let $r \in \mathbb{R}^N$. The problem is to

$$\underset{x \in \mathbb{R}^N}{\text{minimize}} \; f(x) + g(x) + \frac{1}{2}\|x - r\|^2. \tag{10.39}$$

It follows at once from (10.9) that Problem 10.17 admits a unique solution, namely $x = \operatorname{prox}_{f+g} r$. Unfortunately, the proximity operator of the sum of two functions is usually intractable. To compute it iteratively, we can observe that (10.39) can be viewed as an instance of (10.36) in Problem 10.14 with $f_1 = f$ and $f_2 = g + \|\cdot - r\|^2/2$. However, in this Douglas–Rachford framework, the additional qualification condition (10.35) needs to be imposed. In the present setting we require only the minimal feasibility condition $\operatorname{dom} f \cap \operatorname{dom} g \neq \varnothing$.

Algorithm 10.18 (Dykstra-like proximal algorithm).
Set $x_0 = r$, $p_0 = 0$, $q_0 = 0$
For $n = 0, 1, \ldots$

$$\left\lfloor \begin{array}{l} y_n = \operatorname{prox}_g(x_n + p_n) \\ p_{n+1} = x_n + p_n - y_n \\ x_{n+1} = \operatorname{prox}_f(y_n + q_n) \\ q_{n+1} = y_n + q_n - x_{n+1}. \end{array} \right. \tag{10.40}$$

Proposition 10.19. [12] *Every sequence $(x_n)_{n\in\mathbb{N}}$ generated by Algorithm 10.18 converges to the solution to Problem 10.17.*

Example 10.20 (Best approximation). Let f and g be the indicator functions of closed convex sets C and D, respectively, in Problem 10.17. Then the problem is to find the best approximation to r from $C \cap D$, i.e., the projection of r onto $C \cap D$. In this case, since $\operatorname{prox}_f = P_C$ and $\operatorname{prox}_g = P_D$, the above algorithm reduces to Dykstra's projection method [22, 64].

Example 10.21 (Denoising). Consider the problem of recovering a signal $\overline{x}$ from a noisy observation $r = \overline{x} + w$, where w models noise. If f and g are functions in $\Gamma_0(\mathbb{R}^N)$ promoting certain properties of $\overline{x}$, adopting a least-squares data fitting objective leads to the variational denoising problem (10.39).

10.6 Composite Problems

We focus on variational problems with $m = 2$ functions involving explicitly a linear transformation.

Problem 10.22. Let $f \in \Gamma_0(\mathbb{R}^N)$, let $g \in \Gamma_0(\mathbb{R}^M)$, and let $L \in \mathbb{R}^{M \times N} \smallsetminus \{0\}$ be such that $\operatorname{dom} g \cap L(\operatorname{dom} f) \neq \varnothing$ and $f(x) + g(Lx) \to +\infty$ as $\|x\| \to +\infty$. The problem is to

$$\underset{x \in \mathbb{R}^N}{\text{minimize}} \; f(x) + g(Lx). \tag{10.41}$$

Our assumptions guarantee that Problem 10.22 possesses at least one solution. To find such a solution, several scenarios can be contemplated.

10.6.1 *Forward–Backward Splitting*

Suppose that in Problem 10.22 g is differentiable with a τ-Lipschitz continuous gradient (see (10.14)). Now set $f_1 = f$ and $f_2 = g \circ L$. Then f_2 is differentiable and its gradient

$$\nabla f_2 = L^\top \circ \nabla g \circ L \tag{10.42}$$

is β-Lipschitz continuous, with $\beta = \tau \|L\|^2$. Hence, we can apply the forward–backward splitting method, as implemented in Algorithm 10.3. As seen in (10.20), it operates with the updating rule

$$\left\lfloor \begin{array}{l} \gamma_n \in [\varepsilon, 2/(\tau\|L\|^2) - \varepsilon] \\ y_n = x_n - \gamma_n L^\top \nabla g(Lx_n) \\ \lambda_n \in [\varepsilon, 1] \\ x_{n+1} = x_n + \lambda_n(\operatorname{prox}_{\gamma_n f} y_n - x_n). \end{array} \right. \tag{10.43}$$

Convergence is guaranteed by Proposition 10.4.

10.6.2 *Douglas–Rachford Splitting*

Suppose that in Problem 10.22 the matrix L satisfies

$$LL^\top = \nu I, \quad \text{where} \quad \nu \in \left]0, +\infty\right[\tag{10.44}$$

and $(\operatorname{ri} \operatorname{dom} g) \cap \operatorname{ri} L(\operatorname{dom} f) \neq \varnothing$. Let us set $f_1 = f$ and $f_2 = g \circ L$. As seen in Table 10.1x, $\operatorname{prox}_{f_2}$ has a closed-form expression in terms of prox_g and we can therefore apply the Douglas–Rachford splitting method (Algorithm 10.15). In this scenario, the updating rule reads

$$\left\lfloor \begin{array}{l} x_n = y_n + \nu^{-1} L^\top \big(\mathrm{prox}_{\gamma\nu g}(Ly_n) - Ly_n\big) \\ \lambda_n \in [\varepsilon, 2-\varepsilon] \\ y_{n+1} = y_n + \lambda_n \big(\mathrm{prox}_{\gamma f}(2x_n - y_n) - x_n\big). \end{array} \right. \tag{10.45}$$

Convergence is guaranteed by Proposition 10.16.

10.6.3 Dual Forward–Backward Splitting

Suppose that in Problem 10.22 $f = h + \|\cdot - r\|^2/2$, where $h \in \Gamma_0(\mathbb{R}^N)$ and $r \in \mathbb{R}^N$. Then (10.41) becomes

$$\underset{x\in\mathbb{R}^N}{\mathrm{minimize}}\;\; h(x) + g(Lx) + \frac{1}{2}\|x - r\|^2, \tag{10.46}$$

which models various signal recovery problems, e.g., [33, 34, 51, 59, 112, 138]. If (10.44) holds, $\mathrm{prox}_{g\circ L}$ is decomposable, and (10.46) can be solved with the Dykstra-like method of Sect. 10.5, where $f_1 = h + \|\cdot - r\|^2/2$ (see Table 10.1iv) and $f_2 = g\circ L$ (see Table 10.1x). Otherwise, we can exploit the nice properties of the Fenchel–Moreau–Rockafellar dual of (10.46), solve this dual problem by forward–backward splitting, and recover the unique solution to (10.46) [51].

Algorithm 10.23 (Dual forward–backward algorithm).
Fix $\varepsilon \in \left]0, \min\{1, 1/\|L\|^2\}\right[$, $u_0 \in \mathbb{R}^M$
For $n = 0, 1, \ldots$

$$\left\lfloor \begin{array}{l} x_n = \mathrm{prox}_h(r - L^\top u_n) \\ \gamma_n \in \left[\varepsilon, 2/\|L\|^2 - \varepsilon\right] \\ \lambda_n \in [\varepsilon, 1] \\ u_{n+1} = u_n + \lambda_n \big(\mathrm{prox}_{\gamma_n g^*}(u_n + \gamma_n L x_n) - u_n\big). \end{array} \right. \tag{10.47}$$

Proposition 10.24. [51] *Assume that* $(\mathrm{ri}\,\mathrm{dom}\,g) \cap \mathrm{ri}\,L(\mathrm{dom}\,h) \neq \varnothing$. *Then every sequence* $(x_n)_{n\in\mathbb{N}}$ *generated by the dual forward–backward Algorithm 10.23 converges to the solution to* (10.46).

10.6.4 Alternating-Direction Method of Multipliers

Augmented Lagrangian techniques are classical approaches for solving Problem 10.22 [77, 79] (see also [75, 78]). First, observe that (10.41) is equivalent to

$$\underset{\substack{x\in\mathbb{R}^N,\, y\in\mathbb{R}^M \\ Lx=y}}{\mathrm{minimize}}\;\; f(x) + g(y). \tag{10.48}$$

The *augmented Lagrangian* of index $\gamma \in \left]0,+\infty\right[$ associated with (10.48) is the saddle function

$$\mathscr{L}_\gamma\colon \mathbb{R}^N \times \mathbb{R}^M \times \mathbb{R}^M \to \left]-\infty,+\infty\right]$$
$$(x,y,z) \mapsto f(x) + g(y) + \frac{1}{\gamma} z^\top (Lx - y) + \frac{1}{2\gamma}\|Lx-y\|^2. \tag{10.49}$$

The alternating-direction method of multipliers consists in minimizing $\mathscr{L}_\gamma$ over x, then over y, and then applying a proximal maximization step with respect to the Lagrange multiplier z. Now suppose that

$$L^\top L \text{ is invertible and } (\operatorname{ri}\operatorname{dom} g) \cap \operatorname{ri} L(\operatorname{dom} f) \neq \varnothing. \tag{10.50}$$

By analogy with (10.9), if we denote by prox_f^L the operator which maps a point $y \in \mathbb{R}^M$ to the unique minimizer of $x \mapsto f(x) + \|Lx-y\|^2/2$, we obtain the following implementation.

Algorithm 10.25 (Alternating-direction method of multipliers (ADMM)).
Fix $\gamma > 0$, $y_0 \in \mathbb{R}^M$, $z_0 \in \mathbb{R}^M$
For $n = 0,1,\ldots$

$$\left\lfloor \begin{array}{l} x_n = \operatorname{prox}_{\gamma f}^L (y_n - z_n) \\ s_n = Lx_n \\ y_{n+1} = \operatorname{prox}_{\gamma g}(s_n + z_n) \\ z_{n+1} = z_n + s_n - y_{n+1}. \end{array}\right. \tag{10.51}$$

The convergence of the sequence $(x_n)_{n\in\mathbb{N}}$ thus produced under assumption (10.50) has been investigated in several places, e.g., [75, 77, 78]. It was first observed in [76] that the ADMM algorithm can be derived from an application of the Douglas-Rachford algorithm to the dual of (10.41). This analysis was pursued in [66], where the convergence of $(x_n)_{n\in\mathbb{N}}$ to a solution to (10.41) is shown. Variants of the method relaxing the requirements on L in (10.50) have been proposed [5, 39].

In image processing, ADMM was applied in [81] to an ℓ_1 regularization problem under the name "alternating split Bregman algorithm." Further applications and connections are found in [2, 69, 117, 143].

10.7 Problems with $m \geq 2$ Functions

We return to the general minimization problem (10.1).

Problem 10.26. Let $f_1,\ldots,f_m$ be functions in $\Gamma_0(\mathbb{R}^N)$ such that

$$(\operatorname{ri}\operatorname{dom} f_1) \cap \cdots \cap (\operatorname{ri}\operatorname{dom} f_m) \neq \varnothing \tag{10.52}$$

and $f_1(x)+\cdots+f_m(x) \to +\infty$ as $\|x\| \to +\infty$. The problem is to

$$\underset{x\in\mathbb{R}^N}{\text{minimize}}\; f_1(x)+\cdots+f_m(x). \tag{10.53}$$

Since the methods described so far are designed for $m=2$ functions, we can attempt to reformulate (10.53) as a 2-function problem in the m-fold product space

$$\mathcal{H} = \mathbb{R}^N \times\cdots\times \mathbb{R}^N \tag{10.54}$$

(such techniques were introduced in [110, 111] and have been used in the context of convex feasibility problems in [10, 43, 45]). To this end, observe that (10.53) can be rewritten in $\mathcal{H}$ as

$$\underset{\substack{(x_1,\dots,x_m)\in\mathcal{H}\\ x_1=\cdots=x_m}}{\text{minimize}}\; f_1(x_1)+\cdots+f_m(x_m). \tag{10.55}$$

If we denote by $x=(x_1,\dots,x_m)$ a generic element in $\mathcal{H}$, (10.55) is equivalent to

$$\underset{x\in\mathcal{H}}{\text{minimize}}\; \iota_D(x)+f(x), \tag{10.56}$$

where

$$\begin{cases} D = \big\{(x,\dots,x)\in\mathcal{H} \mid x\in\mathbb{R}^N\big\}\\ f\colon x\mapsto f_1(x_1)+\cdots+f_m(x_m). \end{cases} \tag{10.57}$$

We are thus back to a problem involving two functions in the larger space $\mathcal{H}$. In some cases, this observation makes it possible to obtain convergent methods from the algorithms discussed in the preceding sections. For instance, the following parallel algorithm was derived from the Douglas–Rachford algorithm in [54] (see also [49] for further analysis and connections with Spingarn's splitting method [120]).

Algorithm 10.27 (Parallel proximal algorithm (PPXA)).

Fix $\varepsilon\in\,]0,1[$, $\gamma>0$, $(\omega_i)_{1\le i\le m}\in\,]0,1]^m$ such that
$\sum_{i=1}^m \omega_i = 1$, $y_{1,0}\in\mathbb{R}^N,\dots,y_{m,0}\in\mathbb{R}^N$
Set $x_0=\sum_{i=1}^m \omega_i y_{i,0}$
For $n=0,1,\dots$

$$\left\lfloor\begin{array}{l} \text{For } i=1,\dots,m\\ \quad\big\lfloor\; p_{i,n} = \operatorname{prox}_{\gamma f_i/\omega_i} y_{i,n}\\ p_n = \displaystyle\sum_{i=1}^m \omega_i p_{i,n}\\ \varepsilon\le\lambda_n\le 2-\varepsilon\\ \text{For } i=1,\dots,m\\ \quad\big\lfloor\; y_{i,n+1} = y_{i,n}+\lambda_n\big(2p_n-x_n-p_{i,n}\big)\\ x_{n+1} = x_n+\lambda_n(p_n-x_n). \end{array}\right.$$

Proposition 10.28. [54] *Every sequence* $(x_n)_{n\in\mathbb{N}}$ *generated by Algorithm 10.27 converges to a solution to Problem 10.26.*

Example 10.29 (Image recovery). In many imaging problems, we record an observation $y \in \mathbb{R}^M$ of an image $\overline{z} \in \mathbb{R}^K$ degraded by a matrix $L \in \mathbb{R}^{M\times K}$ and corrupted by noise. In the spirit of a number of recent investigations (see [37] and the references therein), a tight frame representation of the images under consideration can be used. This representation is defined through a synthesis matrix $F^\top \in \mathbb{R}^{K\times N}$ (with $K \leq N$) such that $F^\top F = \nu I$, for some $\nu \in \,]0,+\infty[$. Thus, the original image can be written as $\overline{z} = F^\top \overline{x}$, where $\overline{x} \in \mathbb{R}^N$ is a vector of frame coefficients to be estimated. For this purpose, we consider the problem

$$\underset{x\in C}{\text{minimize}} \;\; \frac{1}{2}\|LF^\top x - y\|^2 + \Phi(x) + \text{tv}(F^\top x), \tag{10.58}$$

where C is a closed convex set modeling a constraint on $\overline{z}$, the quadratic term is the standard least-squares data fidelity term, Φ is a real-valued convex function on $\mathbb{R}^N$ (e.g., a weighted ℓ^1 norm) introducing a regularization on the frame coefficients, and tv is a discrete total variation function aiming at preserving piecewise smooth areas and sharp edges [116]. Using appropriate gradient filters in the computation of tv, it is possible to decompose it as a sum of convex functions $(\text{tv}_i)_{1\leq i\leq q}$, the proximity operators of which can be expressed in closed form [54, 113]. Thus, (10.58) appears as a special case of (10.53) with $m = q+3$, $f_1 = \iota_C$, $f_2 = \|LF^\top \cdot - y\|^2/2$, $f_3 = \Phi$, and $f_{3+i} = \text{tv}_i(F^\top \cdot)$ for $i \in \{1,\ldots,q\}$. Since a tight frame is employed, the proximity operators of f_2 and $(f_{3+i})_{1\leq i\leq q}$ can be deduced from Table 10.1x. Thus, the PPXA algorithm is well suited for solving this problem numerically.

A product space strategy can also be adopted to address the following extension of Problem 10.17.

Problem 10.30. Let $f_1, \ldots, f_m$ be functions in $\Gamma_0(\mathbb{R}^N)$ such that $\text{dom} f_1 \cap \cdots \cap \text{dom} f_m \neq \varnothing$, let $(\omega_i)_{1\leq i\leq m} \in \,]0,1]^m$ be such that $\sum_{i=1}^m \omega_i = 1$, and let $r \in \mathbb{R}^N$. The problem is to

$$\underset{x\in\mathbb{R}^N}{\text{minimize}} \;\; \sum_{i=1}^m \omega_i f_i(x) + \frac{1}{2}\|x - r\|^2. \tag{10.59}$$

Algorithm 10.31 (Parallel Dykstra-like proximal algorithm).
Set $x_0 = r$, $z_{1,0} = x_0$, $\ldots$, $z_{m,0} = x_0$
For $n = 0, 1, \ldots$

$$\left\lfloor \begin{array}{l} \text{For } i = 1,\ldots,m \\ \left\lfloor p_{i,n} = \text{prox}_{f_i} z_{i,n} \right. \\ x_{n+1} = \sum_{i=1}^m \omega_i p_{i,n} \\ \text{For } i = 1,\ldots,m \\ \left\lfloor z_{i,n+1} = x_{n+1} + z_{i,n} - p_{i,n}. \right. \end{array} \right. \tag{10.60}$$

Proposition 10.32. [49] *Every sequence* $(x_n)_{n\in\mathbb{N}}$ *generated by Algorithm 10.31 converges to the solution to Problem 10.30.*

Next, we consider a composite problem.

Problem 10.33. For every $i \in \{1,\dots,m\}$, let $g_i \in \Gamma_0(\mathbb{R}^{M_i})$ and let $L_i \in \mathbb{R}^{M_i\times N}$. Assume that

$$(\exists q \in \mathbb{R}^N) \quad L_1 q \in \operatorname{ri}\operatorname{dom} g_1, \dots, L_m q \in \operatorname{ri}\operatorname{dom} g_m, \tag{10.61}$$

that $g_1(L_1x)+\cdots+g_m(L_mx) \to +\infty$ as $\|x\| \to +\infty$, and that $Q = \sum_{1\le i\le m} L_i^\top L_i$ is invertible. The problem is to

$$\underset{x\in\mathbb{R}^N}{\text{minimize}} \; g_1(L_1x)+\cdots+g_m(L_mx). \tag{10.62}$$

Proceeding as in (10.55) and (10.56), (10.62) can be recast as

$$\underset{\substack{x\in\mathcal{H},\, y\in\mathcal{G}\\ y=Lx}}{\text{minimize}} \; \iota_D(x)+g(y), \tag{10.63}$$

where

$$\begin{cases} \mathcal{H} = \mathbb{R}^N \times\cdots\times \mathbb{R}^N, \; \mathcal{G} = \mathbb{R}^{M_1}\times\cdots\times\mathbb{R}^{M_m}\\ L\colon \mathcal{H}\to\mathcal{G}\colon x\mapsto (L_1x_1,\dots,L_mx_m)\\ g\colon \mathcal{G}\to\,]-\infty,+\infty]\colon y\mapsto g_1(y_1)+\cdots+g_m(y_m). \end{cases} \tag{10.64}$$

In turn, a solution to (10.62) can be obtained as the limit of the sequence $(x_n)_{n\in\mathbb{N}}$ constructed by the following algorithm, which can be derived from the alternating-direction method of multipliers of Sect. 10.6.4 (alternative parallel offsprings of ADMM exist, see for instance [65]).

Algorithm 10.34 (Simultaneous-direction method of multipliers (SDMM)). Fix $\gamma > 0$, $y_{1,0}\in\mathbb{R}^{M_1},\dots,y_{m,0}\in\mathbb{R}^{M_m}$, $z_{1,0}\in\mathbb{R}^{M_1},\dots,z_{m,0}\in\mathbb{R}^{M_m}$

$$\begin{array}{l} \text{For } n = 0,1,\ldots\\ \left\lfloor \begin{array}{l} x_n = Q^{-1}\sum_{i=1}^m L_i^\top(y_{i,n}-z_{i,n})\\ \text{For } i = 1,\ldots,m\\ \left\lfloor \begin{array}{l} s_{i,n} = L_i x_n\\ y_{i,n+1} = \operatorname{prox}_{\gamma g_i}(s_{i,n}+z_{i,n})\\ z_{i,n+1} = z_{i,n}+s_{i,n}-y_{i,n+1} \end{array}\right. \end{array}\right. \end{array} \tag{10.65}$$

This algorithm was derived from a slightly different viewpoint in [118] with a connection with the work of [71]. In these papers, SDMM is applied to deblurring in the presence of Poisson noise. The computation of x_n in (10.65) requires the solution of a positive-definite symmetric system of linear equations. Efficient methods for solving such systems can be found in [82]. In certain situations, fast Fourier diagonalization is also an option [2, 71].

In the above algorithms, the proximal vectors, as well as the auxiliary vectors, can be computed simultaneously at each iteration. This parallel structure is useful when the algorithms are implemented on multicore architectures. A parallel proximal algorithm is also available to solve multicomponent signal processing problems [27]. This framework captures in particular problem formulations found in [7, 8, 80, 88, 133]. Let us add that an alternative splitting framework applicable to (10.53) was recently proposed in [67].

10.8 Conclusion

We have presented a panel of convex optimization algorithms sharing two main features. First, they employ proximity operators, a powerful generalization of the notion of a projection operator. Second, they operate by splitting the objective to be minimized into simpler functions that are dealt with individually. These methods are applicable to a wide class of signal and image processing problems ranging from restoration and reconstruction to synthesis and design. One of the main advantages of these algorithms is that they can be used to minimize nondifferentiable objectives, such as those commonly encountered in sparse approximation and compressed sensing, or in hard-constrained problems. Finally, let us note that the variational problems described in (10.39), (10.46), and (10.59), consist of computing a proximity operator. Therefore, the associated algorithms can be used as a subroutine to compute approximately proximity operators within a proximal splitting algorithm, provided the latter is error tolerant (see [48, 49, 51, 66, 115] for convergence properties under approximate proximal computations). An application of this principle can be found in [38].

Acknowledgements This work was supported by the Agence Nationale de la Recherche under grants ANR-08-BLAN-0294-02 and ANR-09-EMER-004-03.

References

1. Acker, F., Prestel, M.A.: Convergence d'un schéma de minimisation alternée. Ann. Fac. Sci. Toulouse V. Sér. Math. **2**, 1–9 (1980)
2. Afonso, M.V., Bioucas-Dias, J.M., Figueiredo, M.A.T.: Fast image recovery using variable splitting and constrained optimization. IEEE Trans. Image Process. **19** 2345–2356 (2010)
3. Antoniadis, A., Fan, J.: Regularization of wavelet approximations. J. Amer. Statist. Assoc. **96**, 939–967 (2001)
4. Antoniadis, A., Leporini, D., Pesquet, J.C.: Wavelet thresholding for some classes of non-Gaussian noise. Statist. Neerlandica **56**, 434–453 (2002)
5. Attouch, H., Soueycatt, M.: Augmented Lagrangian and proximal alternating direction methods of multipliers in Hilbert spaces – applications to games, PDE's and control. Pacific J. Optim. **5**, 17–37 (2009)
6. Aubin, J.P.: Optima and Equilibria – An Introduction to Nonlinear Analysis, 2nd edn. Springer, New York (1998)

7. Aujol, J.F., Chambolle, A.: Dual norms and image decomposition models. Int. J. Computer Vision **63**, 85–104 (2005)
8. Aujol, J.F., Aubert, G., Blanc-Féraud, L., Chambolle, A.: Image decomposition into a bounded variation component and an oscillating component. J. Math. Imaging Vision **22**, 71–88 (2005)
9. Bauschke, H.H., Borwein, J.M.: Dykstra's alternating projection algorithm for two sets. J. Approx. Theory **79**, 418–443 (1994)
10. Bauschke, H.H., Borwein, J.M.: On projection algorithms for solving convex feasibility problems. SIAM Rev. **38**, 367–426 (1996)
11. Bauschke, H.H., Combettes, P.L.: A weak-to-strong convergence principle for Fejér-monotone methods in Hilbert spaces. Math. Oper. Res. **26**, 248–264 (2001)
12. Bauschke, H.H., Combettes, P.L.: A Dykstra-like algorithm for two monotone operators. Pacific J. Optim. **4**, 383–391 (2008)
13. Bauschke, H.H., Combettes, P.L.: Convex Analysis and Monotone Operator Theory in Hilbert Spaces. Springer (2011)
14. Bauschke, H.H., Combettes, P.L., Reich, S.: The asymptotic behavior of the composition of two resolvents. Nonlinear Anal. **60**, 283–301 (2005)
15. Beck, A., Teboulle, M.: A fast iterative shrinkage-thresholding algorithm for linear inverse problems. SIAM J. Imaging Sci. **2**, 183–202 (2009)
16. Beck, A., Teboulle, M.: Fast gradient-based algorithms for constrained total variation image denoising and deblurring problems. IEEE Trans. Image Process. **18**, 2419–2434 (2009)
17. Bect, J., Blanc-Féraud, L., Aubert, G., Chambolle, A.: A ℓ^1 unified variational framework for image restoration. Lecture Notes in Comput. Sci. **3024**, 1–13 (2004)
18. Benvenuto, F., Zanella, R., Zanni, L., Bertero, M.: Nonnegative least-squares image deblurring: improved gradient projection approaches. Inverse Problems **26**, 18 (2010). Art. 025004
19. Bertsekas, D.P., Tsitsiklis, J.N.: Parallel and Distributed Computation: Numerical Methods. Athena Scientific, Belmont, MA (1997)
20. Bioucas-Dias, J.M., Figueiredo, M.A.T.: A new TwIST: Two-step iterative shrinkage/thresholding algorithms for image restoration. IEEE Trans. Image Process. **16**, 2992–3004 (2007)
21. Bouman, C., Sauer, K.: A generalized Gaussian image model for edge-preserving MAP estimation. IEEE Trans. Image Process. **2**, 296–310 (1993)
22. Boyle, J.P., Dykstra, R.L.: A method for finding projections onto the intersection of convex sets in Hilbert spaces. Lecture Notes in Statist. **37**, 28–47 (1986)
23. Bredies, K.: A forward-backward splitting algorithm for the minimization of non-smooth convex functionals in Banach space. Inverse Problems **25**, 20 (2009). Art. 015005
24. Bredies, K., Lorenz, D.A.: Linear convergence of iterative soft-thresholding. J. Fourier Anal. Appl. **14**, 813–837 (2008)
25. Brègman, L.M.: The method of successive projection for finding a common point of convex sets. Soviet Math. Dokl. **6**, 688–692 (1965)
26. Brézis, H., Lions, P.L.: Produits infinis de résolvantes. Israel J. Math. **29**, 329–345 (1978)
27. Briceño-Arias, L.M., Combettes, P.L.: Convex variational formulation with smooth coupling for multicomponent signal decomposition and recovery. Numer. Math. Theory Methods Appl. **2**, 485–508 (2009)
28. Cai, J.F., Chan, R.H., Shen, Z.: A framelet-based image inpainting algorithm. Appl. Comput. Harm. Anal. **24**, 131–149 (2008)
29. Cai, J.F., Chan, R.H., Shen, L., Shen, Z.: Convergence analysis of tight framelet approach for missing data recovery. Adv. Comput. Math. **31**, 87–113 (2009)
30. Cai, J.F., Chan, R.H., Shen, L., Shen, Z.: Simultaneously inpainting in image and transformed domains. Numer. Math. **112**, 509–533 (2009)
31. Censor, Y., Zenios, S.A.: Parallel Optimization: Theory, Algorithms and Applications. Oxford University Press, New York (1997)
32. Chaâri, L., Pesquet, J.C., Ciuciu, P., Benazza-Benyahia, A.: An iterative method for parallel MRI SENSE-based reconstruction in the wavelet domain. Med. Image Anal. **15**, 185–201 (2011)

33. Chambolle, A.: An algorithm for total variation minimization and applications. J. Math. Imaging Vision **20**, 89–97 (2004)
34. Chambolle, A.: Total variation minimization and a class of binary MRF model. Lecture Notes in Comput. Sci. **3757**, 136–152 (2005)
35. Chambolle, A., DeVore, R.A., Lee, N.Y., Lucier, B.J.: Nonlinear wavelet image processing: Variational problems, compression, and noise removal through wavelet shrinkage. IEEE Trans. Image Process. **7**, 319–335 (1998)
36. Chan, R.H., Setzer, S., Steidl, G.: Inpainting by flexible Haar-wavelet shrinkage. SIAM J. Imaging Sci. **1**, 273–293 (2008)
37. Chaux, C., Combettes, P.L., Pesquet, J.C., Wajs, V.R.: A variational formulation for frame-based inverse problems. Inverse Problems **23**, 1495–1518 (2007)
38. Chaux, C., Pesquet, J.C., Pustelnik, N.: Nested iterative algorithms for convex constrained image recovery problems. SIAM J. Imaging Sci. **2**, 730–762 (2009)
39. Chen, G., Teboulle, M.: A proximal-based decomposition method for convex minimization problems. Math. Programming **64**, 81–101 (1994)
40. Chen, G.H.G., Rockafellar, R.T.: Convergence rates in forward-backward splitting. SIAM J. Optim. **7**, 421–444 (1997)
41. Cheney, W., Goldstein, A.A.: Proximity maps for convex sets. Proc. Amer. Math. Soc. **10**, 448–450 (1959)
42. Combettes, P.L.: The foundations of set theoretic estimation. Proc. IEEE **81**, 182–208 (1993)
43. Combettes, P.L.: Inconsistent signal feasibility problems: Least-squares solutions in a product space. IEEE Trans. Signal Process. **42**, 2955–2966 (1994)
44. Combettes, P.L.: The convex feasibility problem in image recovery. In: P. Hawkes (ed.) Advances in Imaging and Electron Physics, vol. 95, pp. 155–270. Academic, New York (1996)
45. Combettes, P.L.: Convex set theoretic image recovery by extrapolated iterations of parallel subgradient projections. IEEE Trans. Image Process. **6**, 493–506 (1997)
46. Combettes, P.L.: Convexité et signal. In: Proc. Congrès de Mathématiques Appliquées et Industrielles SMAI'01, pp. 6–16. Pompadour, France (2001)
47. Combettes, P.L.: A block-iterative surrogate constraint splitting method for quadratic signal recovery. IEEE Trans. Signal Process. **51**, 1771–1782 (2003)
48. Combettes, P.L.: Solving monotone inclusions via compositions of nonexpansive averaged operators. Optimization **53**, 475–504 (2004)
49. Combettes, P.L.: Iterative construction of the resolvent of a sum of maximal monotone operators. J. Convex Anal. **16**, 727–748 (2009)
50. Combettes, P.L., Bondon, P.: Hard-constrained inconsistent signal feasibility problems. IEEE Trans. Signal Process. **47**, 2460–2468 (1999)
51. Combettes, P.L., Dinh Dũng, Vũ, B.C.: Dualization of signal recovery problems. Set-Valued Var. Anal. **18**, 373–404 (2010)
52. Combettes, P.L., Pesquet, J.C.: A Douglas-Rachford splitting approach to nonsmooth convex variational signal recovery. IEEE J. Selected Topics Signal Process. **1**, 564–574 (2007)
53. Combettes, P.L., Pesquet, J.C.: Proximal thresholding algorithm for minimization over orthonormal bases. SIAM J. Optim. **18**, 1351–1376 (2007)
54. Combettes, P.L., Pesquet, J.C.: A proximal decomposition method for solving convex variational inverse problems. Inverse Problems **24**, 27 (2008). Art. 065014
55. Combettes, P.L., Wajs, V.R.: Signal recovery by proximal forward-backward splitting. Multiscale Model. Simul. **4**, 1168–1200 (2005)
56. Daubechies, I., Defrise, M., De Mol, C.: An iterative thresholding algorithm for linear inverse problems with a sparsity constraint. Comm. Pure Appl. Math. **57**, 1413–1457 (2004)
57. Daubechies, I., Teschke, G., Vese, L.: Iteratively solving linear inverse problems under general convex constraints. Inverse Probl. Imaging **1**, 29–46 (2007)
58. De Mol, C., Defrise, M.: A note on wavelet-based inversion algorithms. Contemp. Math. **313**, 85–96 (2002)
59. Didas, S., Setzer, S., Steidl, G.: Combined ℓ_2 data and gradient fitting in conjunction with ℓ_1 regularization. Adv. Comput. Math. **30**, 79–99 (2009)

60. Douglas, J., Rachford, H.H.: On the numerical solution of heat conduction problems in two or three space variables. Trans. Amer. Math. Soc. **82**, 421–439 (1956)
61. Dunn, J.C.: Convexity, monotonicity, and gradient processes in Hilbert space. J. Math. Anal. Appl. **53**, 145–158 (1976)
62. Dupé, F.X., Fadili, M.J., Starck, J.L.: A proximal iteration for deconvolving Poisson noisy images using sparse representations. IEEE Trans. Image Process. **18**, 310–321 (2009)
63. Durand, S., Fadili, J., Nikolova, M.: Multiplicative noise removal using L1 fidelity on frame coefficients. J. Math. Imaging Vision **36**, 201–226 (2010)
64. Dykstra, R.L.: An algorithm for restricted least squares regression. J. Amer. Stat. Assoc. **78**, 837–842 (1983)
65. Eckstein, J.: Parallel alternating direction multiplier decomposition of convex programs. J. Optim. Theory Appl. **80**, 39–62 (1994)
66. Eckstein, J., Bertsekas, D.P.: On the Douglas-Rachford splitting method and the proximal point algorithm for maximal monotone operators. Math. Programming **55**, 293–318 (1992)
67. Eckstein, J., Svaiter, B.F.: General projective splitting methods for sums of maximal monotone operators. SIAM J. Control Optim. **48**, 787–811 (2009)
68. Eicke, B.: Iteration methods for convexly constrained ill-posed problems in Hilbert space. Numer. Funct. Anal. Optim. **13**, 413–429 (1992)
69. Esser, E.: Applications of Lagrangian-based alternating direction methods and connections to split Bregman (2009). ftp://ftp.math.ucla.edu/pub/camreport/cam09-31.pdf
70. Fadili, J., Peyré, G.: Total variation projection with first order schemes. IEEE Trans. Image Process. **20**, 657–669 (2011)
71. Figueiredo, M.A.T., Bioucas-Dias, J.M.: Deconvolution of Poissonian images using variable splitting and augmented Lagrangian optimization. In: Proc. IEEE Workshop Statist. Signal Process. Cardiff, UK (2009). http://arxiv.org/abs/0904.4872
72. Figueiredo, M.A.T., Nowak, R.D.: An EM algorithm for wavelet-based image restoration. IEEE Trans. Image Process. **12**, 906–916 (2003)
73. Figueiredo, M.A.T., Nowak, R.D., Wright, S.J.: Gradient projection for sparse reconstruction: Application to compressed sensing and other inverse problems. IEEE J. Selected Topics Signal Process. **1**, 586–597 (2007)
74. Fornasier, M.: Domain decomposition methods for linear inverse problems with sparsity constraints. Inverse Problems **23**, 2505–2526 (2007)
75. Fortin, M., Glowinski, R. (eds.): Augmented Lagrangian Methods: Applications to the Numerical Solution of Boundary-Value Problems. North-Holland, Amsterdam (1983)
76. Gabay, D.: Applications of the method of multipliers to variational inequalities. In: M. Fortin, R. Glowinski (eds.) Augmented Lagrangian Methods: Applications to the Numerical Solution of Boundary-Value Problems, pp. 299–331. North-Holland, Amsterdam (1983)
77. Gabay, D., Mercier, B.: A dual algorithm for the solution of nonlinear variational problems via finite elements approximations. Comput. Math. Appl. **2**, 17–40 (1976)
78. Glowinski, R., Le Tallec, P. (eds.): Augmented Lagrangian and Operator-Splitting Methods in Nonlinear Mechanics. SIAM, Philadelphia (1989)
79. Glowinski, R., Marrocco, A.: Sur l'approximation, par éléments finis d'ordre un, et la résolution, par pénalisation-dualité, d'une classe de problèmes de Dirichlet non linéaires. RAIRO Anal. Numer. **2**, 41–76 (1975)
80. Goldburg, M., Marks II, R.J.: Signal synthesis in the presence of an inconsistent set of constraints. IEEE Trans. Circuits Syst. **32**, 647–663 (1985)
81. Goldstein, T., Osher, S.: The split Bregman method for $L1$-regularized problems. SIAM J. Imaging Sci. **2**, 323–343 (2009)
82. Golub, G.H., Van Loan, C.F.: Matrix Computations, 3rd edn. Johns Hopkins University Press, Baltimore, MD (1996)
83. Güler, O.: On the convergence of the proximal point algorithm for convex minimization. SIAM J. Control Optim. **20**, 403–419 (1991)
84. Güler, O.: New proximal point algorithms for convex minimization. SIAM J. Optim. **2**, 649–664 (1992)

85. Hale, E.T., Yin, W., Zhang, Y.: Fixed-point continuation for l_1-minimization: methodology and convergence. SIAM J. Optim. **19**, 1107–1130 (2008)
86. Herman, G.T.: Fundamentals of Computerized Tomography – Image Reconstruction from Projections, 2nd edn. Springer, London (2009)
87. Hiriart-Urruty, J.B., Lemaréchal, C.: Fundamentals of Convex Analysis. Springer, New York (2001)
88. Huang, Y., Ng, M.K., Wen, Y.W.: A fast total variation minimization method for image restoration. Multiscale Model. Simul. **7**, 774–795 (2008)
89. Johansson, B., Elfving, T., Kozlovc, V., Censor, Y., Forssén, P.E., Granlund, G.: The application of an oblique-projected Landweber method to a model of supervised learning. Math. Comput. Modelling **43**, 892–909 (2006)
90. Kiwiel, K.C., Łopuch, B.: Surrogate projection methods for finding fixed points of firmly nonexpansive mappings. SIAM J. Optim. **7**, 1084–1102 (1997)
91. Lemaire, B.: The proximal algorithm. In: J.P. Penot (ed.) New Methods in Optimization and Their Industrial Uses, International Series of Numerical Mathematics, vol. 87, pp. 73–87. Birkhäuser, Boston, MA (1989)
92. Lemaire, B.: Itération et approximation. In: Équations aux Dérivées Partielles et Applications, pp. 641–653. Gauthiers-Villars, Paris (1998)
93. Lent, A., Tuy, H.: An iterative method for the extrapolation of band-limited functions. J. Math. Anal. Appl. **83**, 554–565 (1981)
94. Levitin, E.S., Polyak, B.T.: Constrained minimization methods. U.S.S.R. Comput. Math. Math. Phys. **6**, 1–50 (1966)
95. Lieutaud, J.: Approximation d'Opérateurs par des Méthodes de Décomposition. Thèse, Université de Paris (1969)
96. Lions, P.L., Mercier, B.: Splitting algorithms for the sum of two nonlinear operators. SIAM J. Numer. Anal. **16**, 964–979 (1979)
97. Loris, I., Bertero, M., De Mol, C., Zanella, R., Zanni, L.: Accelerating gradient projection methods for ℓ^1-constrained signal recovery by steplength selection rules. Appl. Comput. Harm. Anal. **27**, 247–254 (2009)
98. Martinet, B.: Régularisation d'inéquations variationnelles par approximations successives. Rev. Française Informat. Rech. Opér. **4**, 154–158 (1970)
99. Mercier, B.: Topics in Finite Element Solution of Elliptic Problems. No. 63 in Lectures on Mathematics. Tata Institute of Fundamental Research, Bombay (1979)
100. Mercier, B.: Inéquations Variationnelles de la Mécanique. No. 80.01 in Publications Mathématiques d'Orsay. Université de Paris-XI, Orsay, France (1980)
101. Moreau, J.J.: Fonctions convexes duales et points proximaux dans un espace hilbertien. C. R. Acad. Sci. Paris Sér. A Math. **255**, 2897–2899 (1962)
102. Moreau, J.J.: Proximité et dualité dans un espace hilbertien. Bull. Soc. Math. France **93**, 273–299 (1965)
103. Nemirovsky, A.S., Yudin, D.B.: Problem Complexity and Method Efficiency in Optimization. Wiley, New York (1983)
104. Nesterov, Yu.: Smooth minimization of non-smooth functions. Math. Program. **103**, 127–152 (2005)
105. Nesterov, Yu.: Gradient methods for minimizing composite objective function. CORE discussion paper 2007076, Université Catholique de Louvain, Center for Operations Research and Econometrics (2007)
106. Nesterov, Yu.E.: A method of solving a convex programming problem with convergence rate $o(1/k^2)$. Soviet Math. Dokl. **27**, 372–376 (1983)
107. Nobakht, R.A., Civanlar, M.R.: Optimal pulse shape design for digital communication systems by projections onto convex sets. IEEE Trans. Communications **43**, 2874–2877 (1995)
108. Passty, G.B.: Ergodic convergence to a zero of the sum of monotone operators in Hilbert space. J. Math. Anal. Appl. **72**, 383–390 (1979)
109. Pesquet, J.C., Combettes, P.L.: Wavelet synthesis by alternating projections. IEEE Trans. Signal Process. **44**, 728–732 (1996)

110. Pierra, G.: Éclatement de contraintes en parallèle pour la minimisation d'une forme quadratique. Lecture Notes in Comput. Sci. **41**, 200–218 (1976)
111. Pierra, G.: Decomposition through formalization in a product space. Math. Programming **28**, 96–115 (1984)
112. Potter, L.C., Arun, K.S.: A dual approach to linear inverse problems with convex constraints. SIAM J. Control Optim. **31**, 1080–1092 (1993)
113. Pustelnik, N., Chaux, C., Pesquet, J.C.: Parallel proximal algorithm for image restoration using hybrid regularization. IEEE Trans. Image Process., to appear. http://arxiv.org/abs/0911.1536
114. Rockafellar, R.T.: Convex Analysis. Princeton University Press, Princeton, NJ (1970)
115. Rockafellar, R.T.: Monotone operators and the proximal point algorithm. SIAM J. Control Optim. **14**, 877–898 (1976)
116. Rudin, L.I., Osher, S., Fatemi, E.: Nonlinear total variation based noise removal algorithms. Physica D **60**, 259–268 (1992)
117. Setzer, S.: Split Bregman algorithm, Douglas-Rachford splitting and frame shrinkage. Lecture Notes in Comput. Sci. **5567**, 464–476 (2009)
118. Setzer, S., Steidl, G., Teuber, T.: Deblurring Poissonian images by split Bregman techniques. J. Vis. Commun. Image Represent. **21**, 193–199 (2010)
119. Sibony, M.: Méthodes itératives pour les équations et inéquations aux dérivées partielles non linéaires de type monotone. Calcolo **7**, 65–183 (1970)
120. Spingarn, J.E.: Partial inverse of a monotone operator. Appl. Math. Optim. **10**, 247–265 (1983)
121. Stark, H. (ed.): Image Recovery: Theory and Application. Academic, San Diego, CA (1987)
122. Stark, H., Yang, Y.: Vector Space Projections : A Numerical Approach to Signal and Image Processing, Neural Nets, and Optics. Wiley, New York (1998)
123. Steidl, G., Teuber, T.: Removing multiplicative noise by Douglas-Rachford splitting methods. J. Math. Imaging Vision **36**, 168–184 (2010)
124. Thompson, A.M., Kay, J.: On some Bayesian choices of regularization parameter in image restoration. Inverse Problems **9**, 749–761 (1993)
125. Tibshirani, R.: Regression shrinkage and selection via the lasso. J. Royal. Statist. Soc. B **58**, 267–288 (1996)
126. Titterington, D.M.: General structure of regularization procedures in image reconstruction. Astronom. and Astrophys. **144**, 381–387 (1985)
127. Tropp, J.A.: Just relax: Convex programming methods for identifying sparse signals in noise. IEEE Trans. Inform. Theory **52**, 1030–1051 (2006)
128. Trussell, H.J., Civanlar, M.R.: The feasible solution in signal restoration. IEEE Trans. Acoust., Speech, Signal Process. **32**, 201–212 (1984)
129. Trussell, H.J., Civanlar, M.R.: The Landweber iteration and projection onto convex sets. IEEE Trans. Acoust., Speech, Signal Process. **33**, 1632–1634 (1985)
130. Tseng, P.: Applications of a splitting algorithm to decomposition in convex programming and variational inequalities. SIAM J. Control Optim. **29**, 119–138 (1991)
131. Tseng, P.: On accelerated proximal gradient methods for convex-concave optimization (2008). http://www.math.washington.edu/~tseng/papers/apgm.pdf
132. Varga, R.S.: Matrix Iterative Analysis, 2nd edn. Springer, New York (2000)
133. Vese, L.A., Osher, S.J.: Image denoising and decomposition with total variation minimization and oscillatory functions. J. Math. Imaging Vision **20**, 7–18 (2004)
134. Vonesh, C., Unser, M.: A fast thresholded Landweber algorithm for wavelet-regularized multidimensional deconvolution. IEEE Trans. Image Process. **17**, 539–549 (2008)
135. Vonesh, C., Unser, M.: A fast multilevel algorithm for wavelet-regularized image restoration. IEEE Trans. Image Process. **18**, 509–523 (2009)
136. Weiss, P., Aubert, G., Blanc-Féraud, L.: Efficient schemes for total variation minimization under constraints in image processing. SIAM J. Sci. Comput. **31**, 2047–2080 (2009)
137. Yamada, I., Ogura, N., Yamashita, Y., Sakaniwa, K.: Quadratic optimization of fixed points of nonexpansive mappings in Hilbert space. Numer. Funct. Anal. Optim. **19**, 165–190 (1998)

138. Youla, D.C.: Generalized image restoration by the method of alternating orthogonal projections. IEEE Trans. Circuits Syst. **25**, 694–702 (1978)
139. Youla, D.C.: Mathematical theory of image restoration by the method of convex projections. In: H. Stark (ed.) Image Recovery: Theory and Application, pp. 29–77. Academic, San Diego, CA (1987)
140. Youla, D.C., Velasco, V.: Extensions of a result on the synthesis of signals in the presence of inconsistent constraints. IEEE Trans. Circuits Syst. **33**, 465–468 (1986)
141. Youla, D.C., Webb, H.: Image restoration by the method of convex projections: Part 1 – theory. IEEE Trans. Medical Imaging **1**, 81–94 (1982)
142. Zeidler, E.: Nonlinear Functional Analysis and Its Applications, vol. I–V. Springer, New York (1985–1990)
143. Zhang, X., Burger, M., Bresson, X., Osher, S.: Bregmanized nonlocal regularization for deconvolution and sparse reconstruction (2009). SIAM J. Imaging Sci. **3**, 253–276 (2010)
144. Zhu, C.Y.: Asymptotic convergence analysis of the forward-backward splitting algorithm. Math. Oper. Res. **20**, 449–464 (1995)

Chapter 11
Arbitrarily Slow Convergence of Sequences of Linear Operators: A Survey

Frank Deutsch and Hein Hundal

Abstract This is a survey (without proofs except for verifying a few new facts) of the slowest possible *rate of convergence* of a sequence of linear operators that converges *pointwise* to a linear operator. A sequence of linear operators (L_n) is said to converge to a linear operator L *arbitrarily slowly* (resp., *almost arbitrarily slowly*) provided that (L_n) converges to L pointwise, and for each sequence of real numbers $(\phi(n))$ converging to 0, there exists a point $x = x_\phi$ such that $\|L_n(x) - L(x)\| \geq \phi(n)$ for all n (resp., for infinitely many n). Two main "lethargy" theorems are prominent in this study, and they have numerous applications. The first lethargy theorem (Theorem 11.16) characterizes almost arbitrarily slow convergence. Applications of this lethargy theorem include the fact that a large class of polynomial operators (e.g., Bernstein, Hermite–Fejer, Landau, Fejer, and Jackson operators) all converge almost arbitrarily slowly to the identity operator. Also all the classical quadrature rules (e.g., the composite Trapezoidal Rule, composite Simpson's Rule, and Gaussian quadrature) converge almost arbitrarily slowly to the integration functional. The second lethargy theorem (Theorem 11.21) gives useful sufficient conditions that guarantee arbitrarily slow convergence. In the particular case when the sequence of linear operators is generated by the powers of a single linear operator, there is a "dichotomy" theorem (Theorem 11.27) which states that either there is linear (fast) convergence or arbitrarily slow convergence; no other type of convergence is possible. Some applications of the dichotomy theorem include generalizations and sharpening of (1) the von Neumann–Halperin cyclic projections theorem, (2) the rate of convergence for intermittently (i.e., "almost" randomly) ordered projections, and (3) a theorem of Xu and Zikatanov.

Keywords Arbitrarily slow convergence · Higher powers of linear operators · Cyclic projections · Alternating projections · Randomly ordered projections · Intermittently ordered projections · Subspace corrections · Finite elements · Domain decomposition · Multigrid method · Rate of convergence · Bernstein

F. Deutsch (✉)
Department of Mathematics, Penn State University, University Park, PA 16802, USA
e-mail: deutsch@math.psu.edu

H.H. Bauschke et al. (eds.), *Fixed-Point Algorithms for Inverse Problems in Science and Engineering*, Springer Optimization and Its Applications 49,
DOI 10.1007/978-1-4419-9569-8_11,

polynomial operators · Hermite–Fejer operators · Landau operators · Fejer operators · Jackson operators · The Trapezoidal rule · Simpson's rule · Gaussian quadrature

AMS 2010 Subject Classification: 40A05, 41A25, 41A36, 41A65, 47N10.

11.1 Introduction

There are some important algorithms in analysis that are all special cases of the following type. Let (L_n) be sequence of bounded linear operators from one normed linear space X to another Y, and suppose that the sequence converges *pointwise* to a bounded linear operator L, that is,

$$L(x) := \lim_{n\to\infty} L_n(x) \qquad \text{for each } x \in X.$$

A natural and practical question that arises then is: what can be said about the *rate* of this convergence? This is an interesting and important question that first seems to have been studied in a systematic way in two recent papers of the authors [20] and [21], and independently in a paper of Badea, Grivaux, and Müller [3] who focused their study on the case of *powers* of a single linear operator (i.e., $L_n = T^n$ for some linear operator $T : X \to X$). However, the motivation for these papers came in turn from the following two papers: Bauschke et al. [7] and Bauschke et al. [9]. This survey will highlight what is known about this question.

In Sect. 11.2, relationships between the various kinds of convergence for a sequence of linear operators are exhibited. The phrase "arbitrarily slow convergence" has appeared in several papers. But in many of these, no precise definition was given. But even the precise definitions differed in a significant way. However, Schock [50] did give such a definition for a special class of methods for obtaining approximate solutions to a particular linear operator equation, and his definition (extended to the general setting) is seen to be equivalent to "almost arbitrarily slow" convergence (Lemma 11.12).

In Sect. 11.3, the "first lethargy theorem" (Theorem 11.16) *characterizes* almost arbitrarily slow convergence. Briefly, the sequence converges almost arbitrarily slowly if and only if it converges pointwise, but not in norm. In Sect. 11.7, as applications of Theorem 11.16, it is seen that the Bernstein, Hermite–Fejer, Landau, Fejer, and Jackson operators all converge almost arbitrarily slowly to the identity operator. In fact, the Bernstein and Hermite–Fejer operators even converge arbitrarily slowly to the identity operator. Similarly, in Sect. 11.8, it is seen that all the classical numerical quadrature rules (e.g., the composite Trapezoidal Rule, the composite Simpson's Rule, and Gaussian quadrature) converge almost arbitrarily slowly to the definite integral functional.

Section 11.4 highlights the "second lethargy theorem" (Theorem 11.21). It provides essential sufficient conditions guaranteeing that the sequence (L_n) converges to L arbitrarily slowly. Furthermore, Theorem 11.21 is the basis for all the main results and applications in Sects. 11.5, 11.9–11.11.

In Sect. 11.5, the important special case when the sequence (L_n) is generated by the powers of a *single* linear operator T, i.e., $L_n = T^n$ for each n is considered. The main result here is a "dichotomy theorem" (Theorem 11.27), which shows that, in the case of powers, there are exactly two different kinds of convergence possible: either linear (possibly finite) or arbitrarily slow. There are no intermediate types of pointwise convergence possible. The arbitrarily slow variants developed by Badea et al. [3] are presented in this section along with the main results of [3].

In Sect. 11.6, the second lethargy theorem (Theorem 11.21) is compared with a classical "lethargy theorem" of Bernstein.

In Sects. 11.9–11.11, applications of the dichotomy theorem are given that sharpen and improve (1) the von Neumann–Halperin theorem, (2) a result on *intermittently ordered* projections, and (3) one of the two main results of Xu and Zikatanov [54].

Recall some common notation. If H is a Hilbert space and M is a closed (linear) subspace, the *orthogonal projection* onto M is denoted by P_M. It is well-known that P_M is linear, has norm one (unless $M = \{0\}$), and $P_M(x)$ is the unique point in M closest to x:

$$\|x - P_M(x)\| = d(x,M) := \inf_{y \in M} \|x - y\|.$$

The *orthogonal complement* of M is the set

$$M^\perp := \{x \in H \mid \langle x, m \rangle = 0 \text{ for all } m \in M\}.$$

Further, if T is any bounded linear mapping from one normed linear space X into another Y, then the *kernel* or *null space* of T is the set

$$\ker T := \mathscr{N}(T) := \{x \in X \mid T(x) = 0\}.$$

All other undefined notation and terminology is standard and can be found, e.g., in [12].

11.2 Types of Convergence

In this section, it is assumed that $X(\neq \{0\})$ and Y are normed linear spaces over the same scalar field ($\mathbb{R}$ or $\mathbb{C}$) and $\mathscr{B}(X,Y)$ denotes the normed linear space of all bounded linear operators L from X to Y with its usual norm

$$\|L\| := \sup_{x \neq 0} \frac{\|L(x)\|}{\|x\|},$$

(where the same notation is used for the norm in X, Y, and $\mathscr{B}(X,Y)$).

Let the sequence (L_n) and L be in $\mathscr{B}(X,Y)$.

Definition 11.1. The sequence (L_n) is said to converge to L in *norm* (resp., *pointwise*) provided that $\lim_n \|L_n - L\| = 0$ (resp., $\lim_n \|L_n(x) - L(x)\| = 0$ for each $x \in X$).

Let $\mathscr{O}$ denote the collection of all real-valued functions on the positive integers $\mathbb{N} = \{1,2,3,\dots\}$ that converge to 0. That is,

$$\mathscr{O} := \{\phi \mid \phi : \mathbb{N} \to \mathbb{R},\ \lim_n \phi(n) = 0\}. \tag{11.1}$$

Definition 11.2. Let $\phi \in \mathscr{O}$. The sequence (L_n) converges to L *pointwise with order* ϕ provided that for each $x \in X$ there exists a constant $c_x > 0$ such that $\|L_n(x) - L(x)\| \leq c_x \phi(n)$ for each $n \in \mathbb{N}$.

Using the "big O" notation (see, e.g., [38, p. 16]), this can be rephrased by saying that (L_n) converges to L pointwise with order ϕ provided that $\|L_n(x) - L(x)\| = O(\phi(n))$ for each $x \in X$.

Definition 11.3. The sequence (L_n) is said to converge to L *linearly* if there exist constants $\alpha \in [0,1)$ and $c \in \mathbb{R}$ such that $\|L_n - L\| \leq c\alpha^n$ for each n.

In "big O" notation, this can be rephrased as (L_n) converges to L linearly provided $\|L_n - L\| = O(\alpha^n)$ for some $\alpha \in [0,1)$. (Some authors call this "geometric" convergence, and in [3] it is called "quick uniform convergence.")

The relationship between these types of convergence is easily described.

Lemma 11.4. *Consider the following statements.*

(1) (L_n) converges to L linearly.
(2) (L_n) converges to L in norm.
(3) (L_n) converges to L pointwise with order ϕ for some $\phi \in \mathscr{O}$.
(4) (L_n) converges to L pointwise.

Then $(1) \Rightarrow (2) \Rightarrow (3) \Rightarrow (4)$. *In general, none of these implications is reversible.*

Example 11.5 (Convergence in norm does not imply linear convergence). Let $\mathbb{R}$ denote the real line with the absolute value norm. Define $L_n : \mathbb{R} \to \mathbb{R}$ by $L_n(x) = (1/n)x$ for each n. Then $\|L_n\| = 1/n$ for each n so (L_n) converges to 0 in norm. If (L_n) converged to 0 linearly, there would exist constants c and $\alpha \in [0,1)$ such that $\|L_n\| \leq c\alpha^n$ for each n. It follows that $1 \leq cn\alpha^n$ for each n. But the right side of this inequality converges to 0 by L'Hospital's rule, and this is absurd.

Despite the last statement of Lemma 11.4, *when X is complete, statements* (2) *and* (3) *of Lemma 11.4 are indeed equivalent.* This is the content of the next result.

Theorem 11.6 ([20, Theorem 2.6]). *Let X be a Banach space, Y a normed linear space, and let (L_n) and L be in $\mathscr{B}(X,Y)$. Then (L_n) converges to L in norm if and only if (L_n) converges to L pointwise with order ϕ for some $\phi \in \mathscr{O}$.*

Example 11.18 below shows that completeness of X cannot be omitted in Theorem 11.6.

Two types of very slow pointwise convergence were defined in [20] and [21].

Definition 11.7. The sequence (L_n) converges to L *arbitrarily slowly* (resp., *almost arbitrarily slowly*) if the following two conditions are satisfied:

(1) $L_n(x) \to L(x)$ for each $x \in X$,
(2) For each $\phi \in \mathscr{C}$, there exists $x = x_\phi \in X$ such that

$$\|L_n(x) - L(x)\| \geq \phi(n) \text{ for each } n \in \mathbb{N} \text{ (resp., for infinitely many } n \in \mathbb{N}).$$

By Theorem 11.9 below, the definition of arbitrarily slow convergence is equivalent to one that was first given by Bauschke et al. [7] (see also [9]). Note that arbitrarily slow convergence of (L_n) to L is just pointwise convergence that can be made slowest possible. Clearly, if (L_n) converges arbitrarily slowly to L, then it must also converge almost arbitrarily slowly. (Example 11.20 below shows that the converse is false.)

Remark 11.8. An *equivalent* definition of arbitrarily slow (resp., almost arbitrarily slow) convergence is obtained by replacing the fundamental set $\mathscr{C}$ defined in (11.1) by the more restrictive set

$$\widetilde{\mathscr{C}} := \{\phi \mid \phi : \mathbb{N} \to (0,\infty),\ \phi(n+1) \leq \phi(n) \text{ for each } n,\ \lim_n \phi(n) = 0\}. \tag{11.2}$$

That is, unlike $\mathscr{C}$, the functions in $\widetilde{\mathscr{C}}$ are also strictly positive and decreasing.

More precisely, the following theorem holds.

Theorem 11.9 ([20, Theorem 2.9]). *A sequence of linear operators (L_n) converges to L arbitrarily slowly (resp., almost arbitrarily slowly) if and only if*

(1) $L_n(x) \to L(x)$ for each $x \in X$, and
(2) For each $\psi \in \widetilde{\mathscr{C}}$, there exists $x = x_\psi \in X$ such that

$$\|L_n(x) - L(x)\| \geq \psi(n) \textit{ for each } n \in \mathbb{N} \textit{ (resp., for infinitely many } n \in \mathbb{N}).$$

Remark 11.10. The proof of this theorem is an easy consequence of the fact that if $\phi \in \mathscr{C}$, then the function ψ defined by $\psi(n) := \max\{1/n, \sup_{i\geq n} \phi(i)\}$ is in $\widetilde{\mathscr{C}}$ and $\psi(n) \geq \phi(n)$ for all n.

Schock [50] defined arbitrarily slow convergence for certain methods which yield approximate solutions to a particular linear operator equation. In the present more general setting, his definition can be rephrased as follows.

Definition 11.11. The sequence (L_n) converges to L *Schock slowly* if (L_n) converges to L pointwise and, for each $\phi \in \widetilde{\mathscr{O}}$, there exists $x = x_\phi \in X$ such that

$$\limsup_n \left(\frac{\|L_n(x) - L(x)\|}{\phi(n)} \right) = \infty. \tag{11.3}$$

The next lemma shows in particular that Schock slow convergence is equivalent to almost arbitrarily slow convergence.

Lemma 11.12 ([20, Lemmas 2.11 and 2.12]). *The following statements are equivalent:*

(1) (L_n) *converges to L almost arbitrarily slowly;*
(2) (L_n) *converges to L pointwise, but not pointwise with order* ϕ *for any* $\phi \in \widetilde{\mathscr{O}}$*;*
(3) (L_n) *converges to L pointwise, and for each* $\phi \in \widetilde{\mathscr{O}}$ *there exists* $x = x_\phi \in X$ *such that*

$$\limsup_n \left(\frac{\|L_n(x) - L(x)\|}{\phi(n)} \right) > 0; \tag{11.4}$$

(4) (L_n) *converges to L Schock slowly.*

Badea et al. [3] have defined three variants of arbitrarily slow convergence. Although these were originally given in [3] only for powers of a single linear operator $T : X \to X$, for comparison purposes we state them below in our more general setting.

Definition 11.13. [3] Let (L_n) converge to L pointwise. Then (L_n) is said to converge to L (ASCi), where $i = 1, 2, 3$, according to the following conditions:

(ASC1) For each $\varepsilon > 0$ and every $\phi \in \mathscr{O}$, there exists $x = x_\phi \in X$ such that $\|x\| < \max_n \phi(n) + \varepsilon$ and $\|L_n(x) - L(x)\| \geq \phi(n)$ for all $n \in \mathbb{N}$.
(ASC2) For each $\phi \in \mathscr{O}$, there exists a dense subset of points $x \in X$ such that $\|L_n(x) - L(x)\| \geq \phi(n)$ "eventually" (i.e., for n sufficiently large).
(ASC3) For each $\phi \in \mathscr{O}$, there exist $x = x_\phi \in X$ and $y^* = y^*_\phi \in Y^*$ (the dual of Y) such that $\Re e\, y^*(L_n(x) - L(x)) \geq \phi(n)$ for all $n \in \mathbb{N}$.

Just as in Remark 11.10, these definitions are equivalent to those in which the set $\mathscr{O}$ is replaced by the (smaller) set $\widetilde{\mathscr{O}}$.

Lemma 11.14. *If* (L_n) *converges to L either* (ASC1) *or* (ASC3)*, then* (L_n) *converges to L arbitrarily slowly. The reverse implication is false in general. Finally, contrary to the first statement,* (ASC2) *convergence does* not *imply arbitrarily slow convergence.*

Proof. If the convergence is (ASC1), the result is obvious. If the convergence is (ASC3), then for each $\phi \in \mathscr{O}$, there exist $x \in X$ and $y^* \in Y^*$ such that

$\Re e\, y^*(L_n(x) - L(x)) \geq \phi(n)$ for all $n \in \mathbb{N}$. Then $\phi(n) \leq |y^*(L_n(x) - L(x))| \leq \|y^*\|\|L_n(x) - L(x)\| = \|L_n(\|y^*\|x) - L(\|y^*\|x)\|$. Thus the element $\|y^*\|x$ works in the definition of arbitrarily slow convergence.

The next example (Example 11.15) shows that the reverse implication is false in general.

Finally, let (L_n) be any sequence that converges to L (ASC2). Defining a new sequence (T_n) by $T_1 = L$ and $T_{n+1} = L_n$ for all $n \geq 1$, it is clear that (T_n) also converges to L (ASC2). Thus for any $\phi \in \mathscr{O}$ with $\phi(1) > 0$, we have that $\|T_1(x) - L(x)\| = 0 < \phi(1)$ for all x and so (T_n) does not converge to L arbitrarily slowly. ■

Example 11.15. Let $X = \ell_2$ with $\{e_1, e_2, \ldots\}$ denoting the canonical orthonormal basis. For each $n \in \mathbb{N}$, define $L_n : \ell_2 \to \ell_2$ by

$$L_n(x) := \frac{1}{2}\sum_{i=n}^{\infty} \langle x, e_i \rangle e_i.$$

Then (L_n) converges to 0 arbitrarily slowly, but does not converge (ASC1).

Proof. Since each $x \in X$ has the representation $x = \sum_1^{\infty} \langle x, e_i \rangle e_i$, it is easy to verify that (L_n) converges to 0 pointwise and $\|L_n\| = 1/2$ for each n. By Theorem 11.16 below, (L_n) converges to 0 almost arbitrarily slowly. Since, for each $x \in X$,

$$\|L_{n+1}(x)\|^2 = \frac{1}{4}\sum_{i=n+1}^{\infty} |\langle x, e_i \rangle|^2 \leq \frac{1}{4}\sum_{i=n}^{\infty} |\langle x, e_i \rangle|^2 = \|L_n(x)\|^2,$$

it follows by Theorem 11.21 below that (L_n) converges to 0 arbitrarily slowly.

By way of contradiction, suppose (L_n) converged to 0 (ASC1). Let the function $\phi \in \mathscr{O}$ be defined by $\phi(n) = 1/n$ for each n and let $\varepsilon = 1/2$. Then there exists $x \in X$ such that $\|x\| < \sup_n \phi(n) + \varepsilon \; (= \phi(1) + 1/2 = 3/2)$ and $\|L_n(x)\| \geq \phi(n)$ for all n. It follows in particular that

$$1 = \phi(1) \leq \|L_1(x)\| = \frac{1}{2}\|x\| < \frac{1}{2}\left(\frac{3}{2}\right) = \frac{3}{4},$$

which is absurd. ■

11.3 A Characterization of Almost Arbitrarily Slow Convergence

The *first lethargy theorem* (Theorem 11.16) characterizes almost arbitrarily slow convergence. It is the basis for Theorem 11.42 and hence for virtually all the main results of Sects. 11.7 and 11.8.

Theorem 11.16 (First Lethargy Theorem) ([20, Theorem 3.1]). *Let X be a Banach space, Y a normed linear space, and let (L_n) and L be in $\mathscr{B}(X,Y)$. Then the following statements are equivalent:*

(1) (L_n) *converges to L almost arbitrarily slowly;*
(2) (L_n) *converges to L pointwise but not in norm, that is,* $\limsup_n \|L_n - L\| > 0$;
(3) (L_n) *converges to L pointwise, but not pointwise with order ϕ for any $\phi \in \widetilde{\mathscr{O}}$.*

When the domain space X is finite-dimensional, pointwise convergence and norm convergence are equivalent. In particular, almost arbitrarily slow convergence, arbitrarily slow convergence, (ASC1) convergence, and (ASC3) convergence are phenomena that can happen only in *infinite*-dimensional spaces.

Theorem 11.17 ([20, Theorem 2.12]). *Suppose X is finite-dimensional and L_n, L are linear operators from X to Y. Then (L_n) converges to L pointwise if and only if it converges in norm.*

In particular, it is never possible to have arbitrarily slow convergence, almost arbitrarily slow convergence, (ASC1) *convergence, or* (ASC3) *convergence when X is finite-dimensonal.*

The following example shows that completeness of X cannot be omitted from the hypothesis of the First Lethargy Theorem or Theorem 11.6.

Example 11.18 (The completeness of X is essential for almost arbitrarily slow convergence) [20, Example 3.2]). Let X denote the dense subspace of ℓ_2 consisting of those $x \in \ell_2$ with finite support, i.e., $\langle x, e_n \rangle = 0$ for all n sufficiently large, where (e_n) is the canonical orthonormal basis in ℓ_2. Define $L_n : X \to X$ by $L_n(x) = \langle x, e_n \rangle e_n$. Then:

(1) (L_n) converges to 0 pointwise and $\|L_n\| = 1$ for each n.
(2) (L_n) does not converge to 0 in norm.
(3) (L_n) does not converge to 0 almost arbitrarily slowly.
(4) (L_n) converges to 0 pointwise with order ϕ for some $\phi \in \mathscr{O}$.
(5) (L_n) converges to 0 pointwise with order ϕ for every $\phi \in \widetilde{\mathscr{O}}$.

Consequently, the hypothesis that X be complete *cannot* be omitted in the Lethargy theorem or in Theorem 11.6.

The next consequence of the First Lethargy Theorem will be the basis for all the applications in Sects. 11.7 and 11.8.

Lemma 11.19 ([20, Theorem 4.1]). *Let X be a Banach space, Y a normed linear space, and let (L_n) and L be in $\mathscr{B}(X,Y)$. Suppose that there exists $\rho > 0$ such that*

$$\ker L_n \cap \{x \in X \mid \|L(x)\| \geq \rho \|x\|\} \neq \{0\} \text{ for infinitely many } n, \tag{11.5}$$

where $\ker L_n := \{x \in X \mid L_n(x) = 0\}$. *If (L_n) converges to L pointwise, then (L_n) converges to L almost arbitrarily slowly.*

The following example shows that, in general, arbitrarily slow convergence is not the same as almost arbitrarily slow convergence.

Example 11.20 (Almost arbitrarily slow convergence does not imply arbitrarily slow convergence) ([20, Example 3.4]). For each $n \in \mathbb{N}$, let $L_n : \ell_2 \to \ell_2$ be defined by $L_n(x) := \langle x, e_n \rangle e_n$. Here (e_n) denotes the canonical orthonormal basis for ℓ_2, i.e., e_n is 1 in the nth coordinate, and 0 elsewhere. Then $\|L_n\| = 1$ for each n, $L_n(x) \to 0$ for each x, and (L_n) converges to 0 almost arbitrarily slowly, but not arbitrarily slowly.

11.4 Arbitrarily Slow Convergence: A Useful Sufficient Condition

The *second lethargy theorem* (Theorem 11.21) states that, under hypotheses that are essential, (L_n) converges to L arbitrarily slowly. It will be the basis for all the main results in the Sects. 11.5, 11.6, 11.9–11.11.

Theorem 11.21 (Second Lethargy Theorem) ([21, Theorem 3,3]). *Let X be a Banach space, Y a normed linear space, and let $L, L_1, L_2, \ldots$ be bounded linear operators in $\mathscr{B}(X,Y)$. Suppose that (L_n) converges to L almost arbitrarily slowly and satisfies the following* monotonicity *condition:*

$$\|L_{n+1}(x) - L(x)\| \le \|L_n(x) - L(x)\| \text{ for each } n \in \mathbb{N} \text{ and } x \in X. \tag{11.6}$$

Then (L_n) converges to L arbitrarily slowly.

Remark 11.22. The Lethargy Theorem is *best possible* in the sense that *none* of the hypotheses is superfluous. More precisely, the theorem is false in general if either of the following hypotheses is omitted: the almost arbitrarily slow convergence of (L_n) to L or the monotonicity condition (11.6).

To see that the "almost arbitrarily slow convergence" hypothesis cannot be dropped, see Example 11.15. To see that the monotonicity condition (11.6) cannot be dropped, see Example 11.20.

Remark 11.23. (1) It is worth mentioning that the monotonicity condition (11.6) is related to the *Fejér monotonicity condition* which has been shown to useful in convexity and optimization (see, e.g., [6] and [13]). (Recall that if C is a closed convex set in X, then a sequence (x_n) in X is said to be *Fejér monotone with respect to C* if $\|x_{n+1} - c\| \le \|x_n - c\|$ for each $c \in C$.) Indeed, using this terminology, the condition (11.6) may be restated as: for each $x \in X$, the sequence $(L_n(x))$ is Fejér monotone with respect to $L(x)$.

(2) Note that condition (11.6) implies the monotonicity

$$\|L_{n+1} - L\| \le \|L_n - L\| \text{ for all } n \in \mathbb{N}. \tag{11.7}$$

However, Example 11.20 shows that relation (11.7) does *not* imply the relation (11.6).

11.5 Trichotomy for Powers of an Operator

The simplest application of the Second Lethargy theorem (Theorem 11.21), and the most useful for some of the later applications, occurs when the sequence (L_n) is generated by the *powers of a single nonexpansive operator.* This will follow as a consequence of the Second Lethargy Theorem and the fact that for powers of operators, the monotonicity condition (11.6) automatically holds.

This case may be stated as a *trichotomy* theorem for powers of a linear operator.

Theorem 11.24 (Trichotomy) ([21, Theorem 4.2]). *Let X be a Banach space and $T : X \to X$ be a linear operator with $\|T\| \leq 1$. Then exactly one of the following three statements holds:*

(1) $\|T^{n_1}\| < 1$ for some n_1; in this case, (T^n) converges to 0 linearly.
(2) $\|T^n\| = 1$ for all n and $T^n(x) \to 0$ for each $x \in X$; in this case, (T^n) converges to 0 arbitrarily slowly.
(3) $\|T^n\| = 1$ for all n and $T^n(x) \not\to 0$ for some $x \in X$.

In contrast to Example 11.20, it turns out that for powers of a nonexpansive operator, almost arbitrarily slow convergence and arbitrarily slow convergence are the same since the monotonicity condition (11.6) is automatic in this case.

Corollary 11.25 ([20, Corollary 4.3]). *Let X be complete and $T \in \mathscr{B}(X,X)$ with $\|T\| \leq 1$. Then (T^n) converges to 0 arbitrarily slowly if and only if (T^n) converges to 0 almost arbitrarily slowly.*

If one drops the hypothesis that $\|T\| \leq 1$ in the Trichotomy Theorem but adds some other conditions, the main result of Müller [40] applies.

Theorem 11.26 (Müller [40]). *Let X be a Banach space which does not contain c_0, and let $T \in \mathscr{B}(X,X)$ be such that 1 is in the spectrum of T and (T^n) converges pointwise to 0. Then for each $\phi \in \mathscr{C}$, there exist $x = x_\phi \in X$ and $x^* \in X^*$ such that*

$$\Re e x^*(T^n(x)) \geq \phi(n) \text{ for all } n \in \mathbb{N}.$$

In particular, (T^n) converges to 0 arbitrarily slowly.

In all the applications of the trichotomy theorem that are made below, the condition $T^n(x) \to 0$ for all $x \in X$ is known (or can be shown) to hold. In this particular case, the trichotomy theorem reduces to the following *dichotomy theorem.*

Theorem 11.27 (Dichotomy) ([20, Theorem 4.4]). *Let X be a Banach space and $T : X \to X$ be a linear operator with $\|T\| \leq 1$ and $T^n(x) \to 0$ for each $x \in X$. Then exactly one of the following two statements holds:*

(1) There exists $n_1 \in \mathbb{N}$ such that $\|T^{n_1}\| < 1$, and (T^n) converges to 0 linearly.
(2) $\|T^n\| = 1$ for each $n \in \mathbb{N}$, and (T^n) converges to 0 arbitrarily slowly.

If one drops the hypothesis that $\|T\| \leq 1$ in the dichotomy theorem, then the following result of Badea et al. governs this situation.

Theorem 11.28 (Badea, Grivaux, Müller Dichotomy) ([3, Theorem 2.1]). *Let X be a Banach space, and let $T \in \mathscr{B}(X,X)$ be such that (T^n) converges pointwise to $T_0 \in \mathscr{B}(X,X)$. Then exactly one of the following two statements holds:*

(1) (T^n) converges linearly to T_0.
(2) (T^n) converges to T_0 (ASC1).

Further, in statement (2), (ASC1) *may be replaced by* (ASC2).

Moreover, (T^n) converges linearly to T_0 if and only if for each λ in the scalar field with $|\lambda| = 1$, the range of $(\lambda I - T)$ is closed.

Example 11.29 ([21, Example 4.5]). Let $L : \ell_2 \to \ell_2$ denote the *left-shift* operator. That is, for each $x = \sum_{i=1}^{\infty} \langle x, e_i \rangle e_i \in \ell_2$,

$$L(x) = \sum_{i=2}^{\infty} \langle x, e_i \rangle e_{i-1},$$

where $\{e_i \mid i = 1, 2, \ldots\}$ is the canonical orthonormal basis in ℓ_2: $e_i(j) = \delta_{ij}$, Kronecker's delta. Then (L^n) converges to 0 arbitrarily slowly.

Corollary 11.30 ([21, Corollary 4.6]). *Let X be a Banach space and $T : X \to X$ a linear operator with $\|T\| \leq 1$. Then (T^n) converges to 0 linearly if and only if $\|T^{n_1}\| < 1$ for some $n_1 \in \mathbb{N}$.*

This follows immediately from the Trichotomy Theorem 11.24.

Corollary 11.31. *Let X be a Banach space and $T : X \to X$ a linear operator with $\|T\| \leq 1$. Then the following statements are equivalent:*

(1) (T^n) converges to 0 arbitrarily slowly;
(2) (T^n) converges to 0 almost arbitrarily slowly;
(3) (T^n) converges to 0 pointwise and $\|T^n\| = 1$ for each n;
(4) (T^n) converges to 0 pointwise, but (T^n) does not converge to 0 pointwise with order ϕ for any $\phi \in \mathscr{O}$;
(5) (T^n) converges to 0 (ASC1)*;*
(6) (T^n) converges to 0 (ASC2).

Proof. The equivalence of the first four statements is from [21, Corollary 4.7].

Now suppose that (1) holds. Then, by Theorem 11.27, $\|T^n\| = 1$ for all n and (T^n) does not converge linearly to 0. By Theorem 11.28, (T^n) converges to 0 both (ASC1) and (ASC2). Thus (1) implies both (5) and (6).

If either (5) or (6) holds, then by Theorem 11.28, (T^n) does not converge linearly to 0. By Theorem 11.27, (T^n) converges to 0 arbitrarily slowly. Thus (1) holds. ∎

When X is finite-dimensional, pointwise convergence and norm convergence coincide by Theorem 11.17. Hence, by appealing to the Dichotomy Theorem 11.27, we immediately obtain the following result.

Corollary 11.32 ([21, Corollary 4.8]). *Let X be finite-dimensional and let $T \in \mathscr{B}(X,X)$ with $\|T\| \leq 1$. Then the following statements are equivalent:*

(1) (T^n) converges to 0 pointwise;
(2) There exists an integer n_1 such that $\|T^{n_1}\| < 1$;
(3) (T^n) converges to 0 linearly.

(Alternately, Corollary 11.32 can be derived from the Jordan Decomposition Theorem.)

With stronger conditions on the operator T, stronger dichotomy results are available. (See [41] for the basic spectral theory and terminology used here.) There is one other type of arbitrarily slow convergence that was defined in [3] specifically for a Hilbert space.

Definition 11.33. ([3]) If H is a Hilbert space and $T, T_0 \in \mathscr{B}(H,H)$, then (T^n) *converges to* T_0 (ASCH) provided that for each $\varepsilon > 0$ and for each $\phi \in \mathscr{O}$, there exists $x = x_{\phi,\varepsilon} \in H$ such that $\|x\| < \max_n \phi(n) + \varepsilon$ and $\Re e \langle T^n(x) - T_0(x), x \rangle \geq \phi(n)$ for all $n \in \mathbb{N}$.

It is easy to see that for powers of a single linear operator on a Hilbert space, (ASCH) implies (ASC3). Whether the converse implication holds in this situation is unclear.

Theorem 11.34 ([3, Theorem 2.3]). *Let X be a Banach space and let $T \in \mathscr{B}(X,X)$ be a power bounded, mean ergodic operator with spectrum $\sigma(T)$ contained in $\{\lambda \in \mathbb{C} \mid |\lambda| < 1$ or $\lambda = 1\}$. Then (T^n) converges pointwise to an operator $T_0 \in \mathscr{B}(X,X)$. Moreover, the following statements hold.*

(1) Either (T^n) converges linearly to T_0 or (T^n) converges to T_0 (ASC1).
(2) Either (T^n) converges linearly to T_0 or (T^n) converges to T_0 (ASC2).
(3) If X contains no isomorphic copy of c_0, then either (T^n) converges linearly to T_0 or (T^n) converges to T_0 (ASC3).
(4) If $X = H$ is a Hilbert space, then either (T^n) converges linearly to T_0 or (T^n) converges to T_0 (ASCH).

Moreover, in all of these statements, (T^n) converges linearly to T_0 if and only if the range of $(I - T)$ is closed.

The next result was established in a *complex* Banach space. It would be of some interest to know whether this restriction can be dropped.

Theorem 11.35 ([3, Theorem 2.4]). *Let X be a complex Banach space and let $P_1, P_2, \ldots, P_r$ be $r \geq 2$ projections on X (i.e., $P_i^2 = P_i$). Let T be in the convex multiplicative semigroup generated by $P_1, \ldots, P_r$. That is, T is a convex combination of terms each of which is a product with factors in $P_1, \ldots, P_r$. Suppose one of the following three conditions holds.*

(1) The space X is uniformly convex and each P_j $(j = 1,2,\dots,r)$ is a norm one projection.
(2) The space X^* is uniformly convex and each P_j $(j = 1,2,\dots,r)$ is a norm one projection.
(3) The space X is reflexive and for each j there exists $r_j \in (0,1)$ such that $\|P_j - r_jI\| \leq 1 - r_j$. In particular, this holds if each P_j is hermitian, $1 \leq j \leq r$.

Then, the sequence (T^n) converges pointwise to some $T_0 \in \mathscr{B}(X,X)$ and the following dichotomies hold:

(4) Either (T^n) converges linearly to T_0 or (T^n) converges to T_0 (ASC1).
(5) Either (T^n) converges linearly to T_0 or (T^n) converges to T_0 (ASC2).
(6) Either (T^n) converges linearly to T_0 or (T^n) converges to T_0 (ASC3).

Moreover, if $X = H$ is a Hilbert space, then either (T^n) converges linearly to T_0 or (T^n) converges to T_0 (ASCH).

11.6 The Bernstein Lethargy Theorem

In this section, the Second Lethargy Theorem 11.21 is compared with the classical lethargy theorem of Bernstein.

Let $\{x_1, x_2, x_3, \dots\}$ be a set of linearly independent elements in a normed linear space X with the property that each $x \in X$ can be approximated arbitrarily well by elements in the linear space spanned by the x_n's. That is, for each $x \in X$ and each $\varepsilon > 0$, there exist scalars α_i for $i = 1,2,\dots,n$ such that $y = \sum_1^n \alpha_i x_i$ satisfies $\|x - y\| < \varepsilon$. Note that such a space must be *separable*, that is, it must contain a countable dense set (viz., all linear combinations with rational coefficients). We define the associated sequence of linear subspaces M_n by

$$M_n := \operatorname{span}\{x_1, x_2, \dots, x_n\} \text{ for } n = 1,2,\dots.x$$

In particular, for each n, $\dim M_n = n$, $M_n \subset M_{n+1}$, and $X = \overline{\cup_1^\infty M_n}$.

The distance from any $x \in X$ to M_n is denoted by

$$d(x, M_n) := \inf_{y \in M_n} \|x - y\|.$$

Then the Bernstein lethargy theorem may be stated as follows.

Theorem 11.36 (Bernstein Lethargy Theorem). *Let X be a Banach space and assume (M_n) is an increasing sequence of subspaces with $\dim M_n = n$ and $X = \overline{\cup_1^\infty M_n}$. For each $\phi \in \widetilde{\mathscr{O}}$, there exists $x = x_\phi \in X$ such that*

$$d(x, M_n) = \phi(n) \text{ for each } n \in \mathbb{N}.$$

Bernstein [10] actually proved this result in the case when $X = C[a,b]$ in 1938, but Timan [52, pp. 41–43] observed that it holds in the more general case as stated above. (See also Davis [15, p. 322].)

How does the Bernstein Lethargy Theorem compare with the Second Lethargy Theorem 11.21? In general, a direct comparison is not possible since the latter is phrased in terms of linear operators, while the former is phrased in terms of distances to finite-dimensional subspaces. There is one case, however, where a reasonable comparison is possible. This is when X is a (separable) Hilbert space since then $d(x,M_n) = \|x - P_{M_n}(x)\|$. In this case, Bernstein's Lethargy Theorem can be stated in the following form.

Theorem 11.37 (Bernstein Lethargy Theorem: Hilbert Space Case). *Let H be a Hilbert space and let (M_n) be an increasing sequence of subspaces such that $\dim M_n = n$ and $H = \overline{\cup_1^\infty M_n}$. Then for each $\phi \in \widetilde{\mathscr{O}}$ there exists $x = x_\phi \in H$ such that*

$$\|x - P_{M_n}(x)\| = \phi(n) \text{ for all } n \in \mathbb{N}. \tag{11.8}$$

It should be mentioned that in Theorem 11.37, it is also not hard to show that (P_{M_n}) converges pointwise to the identity operator I and, in particular, that (P_{M_n}) converges to I arbitrarily slowly. An even stronger version of Theorem 11.37 was recently established.

Theorem 11.38 ([21, Theorem 5.3]). *Let H be a Hilbert space and let (M_n) be a sequence of closed (not necessarily finite-dimensional) subspaces in H having the property that $\{0\} \neq M_n \subset M_{n+1}$, $M_n \neq M_{n+1}$, and let $M := \overline{\cup_1^\infty M_n}$. Then (P_{M_n}) converges pointwise to P_M, and for each $\phi \in \widetilde{\mathscr{O}}$ there exists $x = x_\phi \in H$ such that*

$$\|P_{M_n}(x) - P_M(x)\| = \phi(n) \text{ for each } n \in \mathbb{N}. \tag{11.9}$$

In particular, (P_{M_n}) converges arbitrarily slowly to P_M.

Remark 11.39. (1) Comparing Theorems 11.37 and 11.38, it is seen that in Theorem 11.38, the closed subspaces are not necessarily increasing by one dimension at each step as in Theorem 11.37, and they can even be infinite-dimensional. Second, the closure of the union of the subspaces in Theorem 11.38 does not have to be the whole space as in Theorem 11.37.

(2) It is worth noting that Theorem 11.38 is no longer valid if the hypothesis that $M_n \neq M_{n+1}$ for all n is dropped. For if $M_m = M_{m+1}$ for some m, then *every* $x \in X$ must satisfy

$$\|P_{M_m}(x) - P_M(x)\| = \|P_{M_{m+1}}(x) - P_M(x)\|. \tag{11.10}$$

Hence if $\phi \in \widetilde{\mathscr{O}}$ is chosen so that $\phi(m) > \phi(m+1)$, then because of (11.10), it follows that (11.9) is impossible to hold simultaneously for both m and $m+1$ no matter which x is chosen.

However, if one is only interested in concluding arbitrarily slow convergence, then the hypothesis of Theorem 11.38 can be further weakened.

Theorem 11.40 ([21, Theorem 5.5]). *Let H be a Hilbert space and let (M_n) be any nondecreasing sequence of closed subspaces (not necessarily finite-dimensional) such that the closed subspace $M := \overline{\cup_1^\infty M_n}$ is infinite-dimensional and $M \neq M_n$ for every n. Then (P_{M_n}) converges to P_M arbitrarily slowly.*

Remark 11.41. (1) Note that the main difference in the hypotheses of Theorems 11.38 and 11.40 is that in the latter, it is *not* assumed that $M_n \neq M_{n+1}$ for each n.

(2) Theorem 11.40 is best possible in the sense that if either of the two hypotheses (M is infinite-dimensional, or $M \neq M_n$ for all n) is dropped, then the conclusion fails. For if M were finite-dimensional, then by Theorem 11.17 (taking $X = M$), the sequence of projections could not converge arbitrarily slowly. While if $M = M_n$ for some n, then $M = M_n$ for all n sufficiently large. Hence, it follows that $P_{M_n} = P_M$ for all n sufficiently large, and so for any $\phi \in \widetilde{\mathscr{O}}$ and any $x \in H$, we have $\|P_{M_n}x - P_M x\| = 0 < \phi(n)$ for all n large, so arbitrarily slow convergence is not possible.

11.7 Application to Positive Linear Operators

In this section, it is seen that all the standard linear approximating methods for uniformly approximating continuous functions on an interval suffer from the same type of slow convergence; indeed, they are all almost arbitrarily slowly converging. These include the Bernstein polynomial operators, the Hermite–Fejer polynomial operators, Landau operators, Fejer operators, and Jackson operators, among others. An excellent source of information about approximating continuous functions by positive linear operators are the lecture notes by DeVore [24]. All of the applications will follow by appealing to the following easy consequences of the First Lethargy Theorem 11.16.

Theorem 11.42 ([20, Theorem 4.1]). *Let X be a Banach space and let $(L_n) \subset \mathscr{B}(X,X)$. Suppose that* $\ker L_n \neq \{0\}$ *for each n. If (L_n) converges pointwise to the identity operator I, then (L_n) converges to I almost arbitrarily slowly.*

Remark 11.43. In general, the hypothesis that the kernels of the L_n be nontrivial *cannot* be omitted in Theorem 11.42. In fact, in case $\ker L_n = \{0\}$ for each n and (L_n) converges pointwise to I, then there is *nothing* that can be concluded about the rate of convergence. This is a consequence of the following example, where it is seen that virtually any type of convergence is possible.

Example 11.44 ([20, Example 4.3]). Let $X = \ell_2$, let $\{e_n\}$ be the canonical orthonormal basis in X, and for each $n \in \mathbb{N}$, let $E_n = \text{span}\{e_1, e_2, \ldots, e_n\}$. Fix any $\phi \in \widetilde{\mathscr{O}}$ and define the following linear operators on X for each $n \in \mathbb{N}$:

(i) $L_n := \frac{1}{2}(I + P_{E_n})$;

(ii) $T_n := \frac{1}{2}(I + P_n)$, where $P_n := P_{(\text{span}\, e_n)^\perp}$;

(iii) $U_n := (\phi(n) + 1)I$.

Then

$$\ker L_n = \ker T_n = \ker U_n = \{0\} \text{ for each } n \in \mathbb{N},$$

and all three sequences of operators (L_n), (T_n), and (U_n) converge pointwise to the identity I. Further,

(1) (L_n) converges to I arbitrarily slowly;
(2) (T_n) converges to I almost arbitrarily slowly, but not arbitrarily slowly;
(3) (U_n) converges to I pointwise with order ϕ.

From Theorem 11.42, the following result is obtained.

Theorem 11.45. *Let X be an infinite-dimensional Banach space, $(L_n) \subset \mathscr{B}(X,X)$, and suppose that the range of each L_n is finite-dimensional. If (L_n) converges to I pointwise, then (L_n) converges to I almost arbitrarily slowly.*

For the remainder of this section and the next, $C[a,b]$ will denote the Banach space of all real-valued continuous functions f on the interval $[a,b]$ with the maximum norm: $\|f\| = \max_{t\in[a,b]} |f(t)|$.

Example 11.46 (Bernstein Operators). Define the Bernstein operators $B_n : C[0,1] \to C[0,1]$ by

$$(B_n f)(t) := \sum_{k=0}^{n} f\left(\frac{k}{n}\right)\binom{n}{k} t^k(1-t)^{n-k} \text{ for all } t \in [0,1].$$

Clearly, the range of B_n lies in the space $\mathscr{P}_n$ of all polynomials of degree at most n, hence is finite-dimensional. Moreover, it is well known (see, e.g., [15, p. 108 ff] or [24, p. 24 ff]) that $(B_n f)$ converges uniformly to f for each $f \in C[0,1]$. In other words, (B_n) converges pointwise to the identity operator. From Theorem 11.45, it follows that (B_n) *converges to the identity operator almost arbitrarily slowly.*

It is noteworthy that in fact it can be shown that *the Bernstein operators (B_n) converge* arbitrarily slowly, *not just almost arbitrarily slowly, to the identity operator.* (The proof of this fact does not seem to follow from any of the general results stated thus far. However, in [20] it is claimed that a direct elementary but rather lengthy proof of an even more general fact is available.)

Example 11.47 (Hermite–Fejer Operators). Fix any $n \in \mathbb{N}$. For $t \in [-1,1]$, let $T_n(t) = \cos(n \arccos t)$ be the Chebshev polynomial of degree n. The zeros of T_n are

$$t_{n,k} := \cos\left(\frac{2k-1}{2n}\pi\right) \quad (k = 1,2,\ldots,n),$$

and all lie in the open interval $(-1,1)$. Define the Hermite–Fejer Operators H_n on $C[-1,1]$ as follows: $H_n f$ is the polynomial of degree at most $2n-1$ that interpolates to f at the n points $t_{n,k}$ and whose derivative at each of these points is 0. More explicitly,

$$(H_n f)(t) = \sum_{k=1}^{n} f(t_{n,k})(1 - t_{n,k}t) \left(\frac{T_n(t)}{n(t - t_{n,k})} \right)^2 \text{ for all } t \in [-1,1].$$

The range of H_n is contained in the subspace of all polynomials of degree at most $2n-1$, hence is finite-dimensional. It is well-known (see, e.g., [15, pp. 118–121] or [24, pp. 42–44]) that $H_n f \to f$ for each $f \in C[-1,1]$. That is, (H_n) converges pointwise to the identity operator I. From Theorem 11.45, it follows that (H_n) *converges to the identity operator almost arbitrarily slowly.*

Just as for the Bernstein operators, it was claimed in [20] that a direct elementary (but lengthy) proof can be given showing that (H_n) *converges to the identity operator* arbitrarily slowly, *not just almost arbitrarily slowly.*

Example 11.48 (Landau Operators). For each $n \in \mathbb{N}$, define the Landau Operator L_n on $C[-1/2,1/2]$ by

$$(L_n f)(t) := c_n \int_{-1/2}^{1/2} f(s)[1-(s-t)^2]^n \mathrm{d}s \text{ for all } t \in [-1/2,1/2], \tag{11.11}$$

where

$$c_n = \left(\int_{-1}^{1} (1-s^2)^n \mathrm{d}s \right)^{-1}.$$

The range of L_n is contained in the subspace of polynomials of degree at most $2n$, and so the range is finite-dimensional. It is well-known (see [24, p. 26 ff]) that $L_n f \to f$ for each $f \in C[-1/2,1/2]$. That is, (L_n) converges pointwise to the identity operator. It follows from Theorem 11.45 that (L_n) *converges to the identity operator almost arbitrarily slowly.* We do not know if the Landau operators converge arbitrarily slowly.

Example 11.49 (Fejer Operators). Let $C_{2\pi}$ denote the Banach space of all real-valued continuous 2π-periodic functions on $\mathbb{R}$ with the maximum norm. For each $f \in C_{2\pi}$, let $S_n(f)$ denote the nth partial sum of the Fourier series for f. For each $n \in \mathbb{N}$, define the Fejer Operator F_n on $C_{2\pi}$ by

$$F_n(f) := \frac{1}{n+1}[S_0(f) + S_1(f) + \cdots + S_n(f)]. \tag{11.12}$$

Clearly, the range of F_n is contained in the subspace of trigonometric polynomials of degree at most n, so is finite-dimensional. It is well-known (see [24, p. 22 ff]) that $F_n(f) \to f$ for each $f \in C_{2\pi}$. In other words, (F_n) converges pointwise to the identity operator. It follows from Theorem 11.45 that (F_n) *converges to the identity operator almost arbitrarily slowly.* Note also that (S_n) and (F_n) converge pointwise to the identity operator in the space $L_2[-\pi,\pi]$ (see, e.g., [34, Theorems 16.31 and 18.28]), and thus both (S_n) and (F_n) converge to the identity operator almost arbitrarily slowly in this space.

Conjecture 11.50. The Fejer operators converge arbitrarily slowly (not just almost arbitrarily slowly) to the identity operator.

Example 11.51 (Jackson Operators). For each $n \in \mathbb{N}$, define the Jackson Operator J_n on $C_{2\pi}$ by the convolution

$$(J_n f)(t) = (f \star K_n)(t) := \int_{-\pi}^{\pi} f(s) K_n(t-s) \mathrm{d}s \text{ for all } t \in \mathbb{R}, \tag{11.13}$$

where

$$K_n(t) = a_n \left[\frac{\sin((n+1)(t/2))}{\sin(t/2)} \right]^4$$

and a_n is chosen so that $\frac{1}{\pi} \int_{-\pi}^{\pi} K_n(t) \mathrm{d}t = 1$.

The range of J_n is contained in the subspace of trigonometric polynomials of degree at most $2n$, hence is finite-dimensional, and $J_n(f) \to f$ for each $f \in C_{2\pi}$ (see [24, p. 23 ff]). That is, (J_n) converges pointwise to the identity operator. It follows from Theorem 11.45 that (J_n) *converges to the identity operator almost arbitrarily slowly.* We do not know if the Jackson operators converge arbitrarily slowly.

All of the above examples have the following common properties. They are examples of "positive" linear operators that converge pointwise. Recall that a linear operator L from $C[a,b]$ into itself is called *positive* if $L(f) \geq 0$ whenever $f \geq 0$. Bohman and Korovkin have independently established the following result (see, e.g., [24, p. 27ff]).

Theorem 11.52 (Bohman–Korovkin). *Let (L_n) be positive linear operators from $C[a,b]$ into $C[a,b]$. Then $\|L_n(f) - f\| \to 0$ for each $f \in C[a,b]$ if and only if $\|L_n(e_i) - e_i\| \to 0$ for $i = 0,1,2$, where $e_i(t) := t^i$.*

In other words, (L_n) converges to the identity operator pointwise if it converges pointwise for just the three functions e_i. By using Theorem 11.45, the Bohman-Korovkin Theorem may be quantified in the case that the range of each L_n is finite-dimensional.

Theorem 11.53 ([20, Theorem 4.12]). *Let (L_n) be positive linear operators from $C[a,b]$ into itself such that the range of each L_n is finite-dimensional. Then (L_n) converges to the identity operator almost arbitrarily slowly if and only if $\|L_n(e_i) - e_i\| \to 0$ for $i = 0,1,2$.*

Theorem 11.53 cannot be strengthened to *arbitrarily slow convergence* because of the following fact: *If there exists k such that $L_k = 0$ in the sequence (L_n), then (L_n) does not converge arbitrarily slowly.*

Remark 11.54. For other examples of this type, the reader is advised to consult the book of Korovkin [39] or the notes of DeVore [24]. Included there are also many rate of convergence results for these operators.

11.8 Application to Quadrature Rules

In this section, it is observed that the First Lethargy Theorem implies that the standard quadrature rules like the Trapezoidal Rule, Simpson's Rule, and Gaussian quadrature all share the same type of slow convergence: namely, almost arbitrarily slow convergence. Just as in the last section, $C[a,b]$ will denote the Banach space of all continuous functions on the interval $[a,b]$ with the maximum norm.

A general class of quadrature rules in the space $C[a,b]$ can be described as follows. Suppose that for each $n \in \mathbb{N}$, there is some $N_n \in \mathbb{N}$, weights $w_{n,k}$ $(k = 1,2,\ldots,N_n)$, and points $a \le t_{n,1} < t_{n,2} < \cdots < t_{n,N_n} \le b$. Define a quadrature rule Q_n for any $f \in C[a,b]$ by setting

$$Q_n(f) = \sum_{k=1}^{N_n} w_{n,k} f(t_{n,k}), \tag{11.14}$$

and define the integral operator Q on $C[a,b]$ by

$$Q(f) = \int_a^b f(t)\mathrm{d}t. \tag{11.15}$$

Lemma 11.55 ([20, Lemma 5.1]). *If Q_n and Q are defined as in (11.14) and (11.15), then*

$$\|Q_n - Q\| \ge \frac{1}{2}(b-a) \text{ for each } n. \tag{11.16}$$

Theorem 11.56 ([20, Theorem 5.2]). *Let Q_n and Q be defined as in (11.14) and (11.15). Then (Q_n) converges to Q pointwise if and only if (Q_n) converges to Q almost arbitrarily slowly.*

It is not known whether or not almost arbitrarily slow convergence in Theorem 11.56 can be replaced by arbitrarily slow convergence. Since most of the classical quadrature rules have positive weights and are exact for constants, it would be of interest to know the answer to this question in that particular case (i.e., when $w_{n,k} \ge 0$ and $\sum_{k=1}^{N_n} w_{n,k} = b-a$).

All the classical numerical quadrature rules are of the type (11.14). A few popular examples are listed below. (For a more detailed description of these and other quadrature rules, and the motivation behind the derivation of these rules, see, e.g., [15] and [38].)

(The Composite Trapezoidal Rule). In the formula (11.14), let $t_{n,k} = a + [(b-a)/n]k$ for $k = 0,1,\ldots,n$ and $w_{n,k} = (b-a)/n$ for $k \ne 0,n$, and $w_{n,k} = (b-a)/(2n)$ otherwise. The resulting quadrature formula Q_n is called the composite Trapezoidal Rule. It is exact for polynomials of degree ≤ 1. It is well-known that (see, e.g., [15, Chap. 14]), $Q_n(f) \to Q(f)$ for each $f \in C[a,b]$. By Theorem 11.56, *the composite Trapezoidal Rule Q_n converges to the integral Q almost arbitrarily slowly.*

(The composite Simpson's Rule). In the formula (11.14), let $t_{n,i} = a + ih$, where $h = (b-a)/(2n)$ for $0 \le i \le 2n$ and $w_{n,o} = h/3 = w_{n,2n}$, $w_{n,2i-2} = 2h/3$ for $i = 2,\dots,n$ and $w_{n,2i-1} = 4h/3$ for $1 \le i \le n$. The resulting quadrature rule is called the composite Simpson's Rule. It is exact for polynomials of degree ≤ 3. Again (see, e.g., [15, Chap. 14]) it is known that $Q_n(f) \to Q(f)$ for each $f \in C[a,b]$. By Theorem 11.56, *the composite Simpson's Rule* (Q_n) *converges to the integral* Q *almost arbitrarily slowly.*

(Gaussian Quadrature). Here define Q on $C[a,b]$ by

$$Q(f) = \int_a^b w(t)f(t)\mathrm{d}t,$$

where w is a positive *weight function*. By this it is meant that w is continuous on the open interval (a,b) and positive there, and w is (Lebesgue) integrable on $[a,b]$. It can be shown (see, e.g., [15, pp. 342 ff] or [38, pp. 528 ff]) that there exist n weights $w_{n,k}$ and n points $t_{n,k}$ in $[a,b]$ for $1 \le k \le n$ (that may be explicitly computed) such that the Gaussian Quadrature formula Q_n defined on $C[a,b]$ by

$$Q_n(f) = \sum_1^n w_{n,k} f(t_{n,k})$$

is exact for all polynomials p of degree $\le 2n-1$. That is, $Q_n(p) = Q(p)$ for any polynomial p of degree $\le 2n-1$. By an application of the Weierstrass polynomial approximation theorem, it follows that $Q_n(f) \to Q(f)$ for each $f \in C[a,b]$. Moreover, the same proof as used in Lemma 11.55 shows that in this more general situation Lemma 11.55 still holds (using $\int_a^b w(t)\mathrm{d}t > 0$ instead of $b-a$.) Thus, by Theorem 11.56 *the Gaussian Quadrature Rules* (Q_n) *converge to the integral* Q *almost arbitrarily slowly.*

11.9 Application to Cyclic Projections

In this section, an application of the Dichotomy Theorem 11.27 is made to cyclic projections in Hilbert space or, more precisely, to the von Neumann–Halperin theorem. The von Neumann–Halperin theorem has had many far-reaching applications in at least a dozen different areas of mathematics including solving linear equations, linear prediction theory, image restoration, and computed tomography (see the survey [17], or the book [19, Chap. 9] for more details and references).

Theorem 11.57 (von Neumann–Halperin). *Let* $M_1, M_2, \dots, M_r$ *be closed subspaces of the Hilbert space* H *and* $M = \cap_1^r M_i$. *Then, for each* $x \in H$,

$$\lim_{n\to\infty} \|(P_{M_r}P_{M_{r-1}}\cdots P_{M_1})^n(x) - P_M(x)\| = 0.$$

In the two-subpace case ($r = 2$), this result was first proved by von Neumann in 1933 (but wasn't published until 1949 and 1950 [44, 45]). Halperin [32] extended the von-Neumann theorem to any number $r \geq 2$ of subspaces. The von Neumann theorem (i.e., the $r = 2$ case) was discovered independently by several authors including Aronszajn [2], Nakano [42], Wiener [53], Powell [47], Gordon et al. [31], and Hounsfield [35] – the Nobel Prize winning inventor of the EMI scanner.

Note that since

$$(P_{M_r}P_{M_{r-1}}\cdots P_{M_1})^n - P_M = (P_{M_r\cap M^\perp}\cdots P_{M_1\cap M^\perp})^n \tag{11.17}$$

(see, e.g., [19, Lemma 9.30]), we have the following error estimate in the von Neumann–Halperin theorem.

Lemma 11.58. *Let $M_1, M_2, \ldots, M_r$ be closed subspaces in the Hilbert space H. Then, for each $x \in H$,*

$$\|(P_{M_r}P_{M_{r-1}}\cdots P_{M_1})^n(x) - P_M(x)\| \leq c^n\|x\|, \tag{11.18}$$

where $c := \|P_{M_r\cap M^\perp}\cdots P_{M_1\cap M^\perp}\|$.

Given any $x \in H$, let $x_n := (P_{M_r}P_{M_{r-1}}\cdots P_{M_1})^n(x)$ for each $n \in \mathbb{N}$. The von Neumann–Halperin theorem shows that the sequence (x_n) always converges to $P_M(x)$. However, the theorem says nothing about the *rate* of convergence. To say something about this, the following fact is needed.

Lemma 11.59. *Let $M_1, M_2, \ldots, M_r$ be closed subspaces of the Hilbert space H and $M := \cap_1^r M_i$. Then the following statements are equivalent:*

(1) $\sum_1^r M_i^\perp$ is closed;
(2) $\|P_{M_r\cap M^\perp}P_{M_{r-1}\cap M^\perp}\cdots P_{M_1\cap M^\perp}\| < 1$;
(3) There exists $\alpha \in [0,1)$ such that

$$\|(P_{M_r}P_{M_{r-1}}\cdots P_{M_1})^n - P_M\| = \|(P_{M_r\cap M^\perp}P_{M_{r-1}\cap M^\perp}\cdots P_{M_1\cap M^\perp})^n\| \leq \alpha^n$$

for each $n \in \mathbb{N}$.

Bauschke et al. [7, Theorem 3.7.4] proved the equivalence of (1) and (2) in this lemma. The remainder was observed in [21, Fact 6.2].

Also, it clear that the α that works in (3) is any scalar satisfying

$$\|P_{M_r\cap M^\perp}P_{M_{r-1}\cap M^\perp}\cdots P_{M_1\cap M^\perp}\| \leq \alpha < 1.$$

In particular, $\alpha = \|P_{M_r\cap M^\perp}P_{M_{r-1}\cap M^\perp}\cdots P_{M_1\cap M^\perp}\|$ works when the sum in (1) is closed. In [3, Theorem 4.4], an upper bound was given for the expression

$$\|(P_{M_r}P_{M_{r-1}}\cdots P_{M_1})^n - P_M\|,$$

but it is not as sharp as the expression $\|P_{M_r\cap M^\perp}P_{M_{r-1}\cap M^\perp}\cdots P_{M_1\cap M^\perp}\|^n$ as given in Lemma 11.59.

Remark 11.60. It is worth mentioning that the statement (1) of Lemma 11.59 has an equivalent formulation because: $\sum_1^r M_i^\perp$ is closed if and only if the collection of subspaces $\{M_1, M_2, \ldots, M_r\}$ has the "strong CHIP" property (This is an immediate consequence of the proof of Example 10.5 in [19]). The strong CHIP was shown to be a fundamental property that arose, for example, in constrained interpolation [22, 23], convex optimization [18], and various kinds of "regularity" and Jameson's property (G) [8]. For more detail and references, see the historical notes on page 283–285 of [19].

The *cyclic projections algorithm* for the subspaces $\{M_1, M_2, \ldots, M_r\}$ is the algorithm that generates, starting with any $x \in H$, the sequence

$$x_0 := x, \quad \text{and} \quad x_n := P_{M_{[n]}}(x_{n-1}) \quad \text{for each } n \in \mathbb{N},$$

where $[n]$ is the function "mod r" with values in $\{1, 2, \ldots, r\}$. That is,

$$[n] = \{1, 2, \ldots, r\} \cap \{n - kr \mid k = 1, 2, \ldots\}.$$

In particular, $x_{nr} = (P_{M_r} P_{M_{r-1}} \cdots P_{M_1})^n(x)$. In this terminology, the von Neumann–Halperin theorem shows that, for each $x \in H$, the cyclic projections algorithm generates a sequence that converges to $P_M(x)$.

One important corollary of the Dichotomy Theorem 11.27 is what is called here the *von Neumann–Halperin dichotomy*.

Theorem 11.61 (von Neumann-Halperin Dichotomy) ([21, Theorem 6.4]). *Let $M_1, M_2, \ldots, M_r$ be closed subspaces of the Hilbert space H and $M := \cap_1^r M_i$. Then exactly one of the following two statements holds.*

(1) $\sum_1^r M_i^\perp$ is closed. Then $((P_{M_r} P_{M_{r-1}} \cdots P_{M_1})^n)$ converges to P_M linearly.
(2) $\sum_1^r M_i^\perp$ is not closed. Then $((P_{M_r} P_{M_{r-1}} \cdots P_{M_1})^n)$ converges to P_M arbitrarily slowly.

Remark 11.62. (1) In the special case of two subspaces ($r = 2$), this result was stated by Bauschke et al. [7]. Bauschke et al. [9] found an error in the proof of [7] that invalidated the proof of the theorem, but they showed that the theorem was nevertheless true by providing an alternate proof. Briefly, the case of $r = 2$ in Theorem 11.61 was due to [7] and [9] by a substantially different and more involved proof than is given in [21].
(2) We suspect that when $\sum_1^r M_i^\perp$ is not closed and $\phi \in \widetilde{\mathscr{O}}$, then the $x = x_\phi$ that satisfies $\|(P_{M_r} \cdots P_{M_1})^n(x) - P_M(x)\| \geq \phi(n)$ for all n must in general be chosen from $M^\perp \setminus \sum_1^r M_i^\perp$.

In a complex Hilbert space, Badea, Grivaux, and Müller have independently characterized when the sum of the orthogonal complements is not closed in addition to other equivalences.

Theorem 11.63 ([3, Theorem 4.1]). *Let $M_1, M_2, \dots, M_r$ be closed subspaces of the complex Hilbert space H, let $M := \cap_1^r M_i$, and $T = P_{M_r} P_{M_{r-1}} \cdots P_{M_1}$. Then the following statements are equivalent:*

(1) The range of $T - I$ is not closed.
(2) (T^n) converges to P_M (ASC1).
(3) $\|T - P_M\| = 1$.
(4) $\sum_1^r M_i^\perp$ is not closed.

Since the equivalence of (3) and (4) holds in any (real or complex) Hilbert space by Lemma 11.59 (using (11.17)), we believe that Theorem 11.63 is valid in any Hilbert space.

By a simple translation argument, Theorem 11.61 can be easily generalized to the case of affine sets (i.e., translates of subspaces).

Theorem 11.64 (Affine Sets Dichotomy) ([21, Theorem 6.6]). *Let $V_1, V_2, \dots, V_r$ be closed affine sets in the Hilbert space H with $V := \cap_1^r V_i \neq \emptyset$. Then exactly one of the following two statements holds.*

(1) $\sum_1^r (V_i - V_i)^\perp$ is closed. Then $((P_{V_r} P_{V_{r-1}} \cdots P_{V_1})^n)$ converges to P_V linearly.
(2) $\sum_1^r (V_i - V_i)^\perp$ is not closed. Then $((P_{V_r} P_{V_{r-1}} \cdots P_{V_1})^n)$ converges to P_V arbitrarily slowly.

Proof. The proof of this theorem given in [21, Theorem 6.6] is incomplete. This is because we have only defined linear convergence and arbitrarily slow convergence for *linear* mappings, and the mappings P_{V_i} and products of these are *nonlinear* in general. The definition of arbitrarily slow convergence (Definition 11.7) is exactly the same for nonlinear maps. Recall that in [21, Theorem 6.6], that if $v \in V$, we observed that for each i, there exist unique subspaces M_i (in fact, $M_i = V_i - V_i$) such that $V_i = M_i + v$. Further, we observed there that

$$(P_{V_r} P_{V_{r-1}} \cdots P_{V_1})^n (x) - P_V(x) = (P_{M_r} P_{M_{r-1}} \cdots P_{M_1})^n (x - v) - P_M(x - v).$$

Since $\sum_1^r (V_i - V_i)^\perp$ is closed if and only if $\sum_1^r M_i^\perp$ is closed, it follows from Theorem 11.61 that if $\sum_1^r (V_i - V_i)^\perp$ is closed, then there exist constants $c > 0$ and $\alpha \in [0, 1)$ such that $\|(P_{M_r} P_{M_{r-1}} \cdots P_{M_1})^n - P_M\| \leq c\alpha^n$. Hence, for all $x \in H$ with $\|x\| \leq 1$, we have

$$\|(P_{V_r} P_{V_{r-1}} \cdots P_{V_1})^n (x) - P_V(x)\| \leq c\alpha^n \|x - v\| \leq \tilde{c}\alpha^n, \tag{11.19}$$

where $\tilde{c} = c(1 + \|v\|)$. Thus if we extend the definition of a linear operator to a nonlinear one Q on H by *defining* the norm of Q by $\|Q\| := \sup\{\|Q(x)\| \mid \|x\| \leq 1\}$, then we see from (11.19) that $(P_{V_r} P_{V_{r-1}} \cdots P_{V_1})^n$ converges to P_V linearly. The rest of the proof is now clear. ∎

Since in a finite-dimensional space, every subspace is closed, it follows that *in $\mathbb{R}^n$, alternating projections always converge linearly.*

The following two corollaries will be useful for comparison to the results of Xu and Zikatanov in Sect. 11.11. If in the von Neumann–Halperin dichotomy theorem one replaces each subspace M_i by its orthogonal complement $M_i^{\perp}$ and recalls the well-known facts that $P_{M_i^{\perp}} = I - P_{M_i}$, $M_i^{\perp\perp} = M_i$, and $\cap_1^r M_i^{\perp} = (\sum_1^r M_i)^{\perp}$ (see, e.g., [19]), then

Corollary 11.65 ([21, Corollary 9.7]). *Let $M_1, M_2, \ldots, M_r$ be closed subspaces in the Hilbert space H and let $M := \overline{\sum_1^r M_i}$. Then exactly one of the following two statements holds.*

(1) *$\sum_1^r M_i$ is closed. Then $([(I - P_{M_r})(I - P_{M_{r-1}}) \cdots (I - P_{M_1})]^n)$ converges to $I - P_M$ linearly.*
(2) *$\sum_1^r M_i$ is not closed. Then $([(I - P_{M_r})(I - P_{M_{r-1}}) \cdots (I - P_{M_1})]^n)$ converges to $I - P_M$ arbitrarily slowly.*

By a proof analogous to that of Theorem 11.64, the following consequence of Corollary 11.65 that is actually more general than Corollary 11.65 is obtained.

Theorem 11.66 ([21, Theorem 6.8]). *Let $V_1, V_2, \ldots, V_r$ be closed affine sets in H with $V := \cap_1^r V_i \neq \emptyset$. Then exactly one of the following two statements holds.*

(1) *$\sum_1^r (V_i - V_i)$ is closed. Then $[(I - P_{V_r})(I - P_{V_{r-1}}) \cdots (I - P_{V_1})]^n$ converges to $I - P_V$ linearly.*
(2) *$\sum_1^r (V_i - V_i)$ is not closed. Then $[(I - P_{V_r})(I - P_{V_{r-1}}) \cdots (I - P_{V_1})]^n$ converges to $I - P_V$ arbitrarily slowly.*

11.10 Application to Intermittent Projections

The Dichotomy Theorem 11.27 can be applied to intermittent or "almost" randomly ordered projections. Throughout this section H will always denote a Hilbert space and $M_1, \ldots, M_r$ will be a collection of r closed subspaces in H with $M := \cap_1^r M_i$.

Definition 11.67. A function $\sigma : \mathbb{N} \to \{1, 2, \ldots, r\}$ is called a *random selection* for $\{1, 2, \ldots, r\}$ if for each $n \in \mathbb{N}$, there exists $N(n) \in \mathbb{N}$ such that

$$\{\sigma(n), \sigma(n+1), \ldots, \sigma(n + N(n))\} = \{1, 2, \ldots, r\}. \tag{11.20}$$

The following is easy to verify.

Lemma 11.68. *Let $\sigma : \mathbb{N} \to \{1, 2, \ldots, r\}$. Then the following statements are equivalent:*

(1) *σ is a random selection for $\{1, 2, \ldots, r\}$;*
(2) *The range of σ is $\{1, 2, \ldots, r\}$ and σ assumes each value in its range infinitely often.*

A *random product* of the projections $P_{M_1}, P_{M_2}, \ldots, P_{M_r}$ is the sequence (S_n), where

$$S_n := P_{M_{\sigma(n)}} P_{M_{\sigma(n-1)}} \cdots P_{M_{\sigma(1)}} \quad (n = 1, 2, \ldots), \tag{11.21}$$

and where σ is a random selection for $\{1, 2, \ldots, r\}$.

Recall that a sequence (x_n) in H is said to converge *weakly* to $x \in H$ provided that

$$\lim_{n \to \infty} \langle x_n, z \rangle = \langle x, z \rangle \text{ for each } z \in H.$$

Theorem 11.69 (Amemiya and Ando [1]). *If (S_n) is the random product of projections (11.21), then for each $x \in H$ the sequence $(S_n(x))$ converges weakly to $P_M(x)$.*

For some far-reaching generalizations of Theorem 11.69, see Dye et al. [28].

Apparently, it is still unknown whether or not the convergence in Theorem 11.69 must be in norm. However, when certain additional conditions are imposed on either the subspaces M_i or the function σ, then norm convergence in Theorem 11.69 is indeed guaranteed.

One result along these lines is the following.

Proposition 11.70 (Bauschke [5, Example 3.8]). *Let (S_n) be the random product of projections (11.21). If $\sum_{i \in J} M_i^{\perp}$ is closed for each nonempty subset J of $\{1, 2, \ldots, r\}$, then*

$$\lim_n \|S_n(x) - P_M(x)\| = 0 \quad \textit{for each } x \in H.$$

We do not know whether this result is valid under the weaker condition that only $\sum_1^r M_i^{\perp}$ be closed. However, with an additional condition on the function σ, then the answer is affirmative.

Definition 11.71 ([6, Definition 3.18]). A random selection function $\sigma : \mathbb{N} \to \{1, 2, \ldots, r\}$ is called an *intermittent selection* for $\{1, 2, \ldots, r\}$ if there exists $N_1 \in \mathbb{N}$ such that, for each $n \in \mathbb{N}$,

$$\{\sigma(n), \sigma(n+1), \ldots, \sigma(n+N_1)\} = \{1, 2, \ldots, r\}. \tag{11.22}$$

Note that an intermittent selection is a random selection with the property that for each $n \in \mathbb{N}$, the $N(n)$ that works in the definition of random selection does *not* depend on n, but is some fixed N_1 that works for all n.

An *intermittent product* of the projections $P_{M_1}, P_{M_2}, \ldots, P_{M_n}$ is the sequence (S_n), where

$$S_n := P_{M_{\sigma(n)}} P_{M_{\sigma(n-1)}} \cdots P_{M_{\sigma(1)}} \quad (n = 1, 2, \ldots), \tag{11.23}$$

and where $\sigma : \mathbb{N} \to \{1, 2, \ldots, r\}$ is an intermittent selection for $\{1, 2, \ldots, r\}$.

Note that if we define $\sigma(n) := [n]$, where $[\cdot]$ is the function "mod r," i.e.,

$$[n] := \{n - kr \mid k = 0, 1, 2, \ldots\} \cap \{1, 2, \ldots, r\},$$

then the function σ satisfies the above hypothesis (with $N_1 = r-1$) and

$$S_{rn} = (P_{M_r} P_{M_{r-1}} \cdots P_{M_1})^n \tag{11.24}$$

is just the sequence of "cyclically" ordered projections that appeared in the von Neumann–Halperin theorem of the preceding section.

The following fact, which generalizes the von Neumann–Halperin cyclic projections theorem to intermittently ordered projections, is needed.

Fact 11.72 (Hundal and Deutsch [36, Subspace case of Theorem 3.1]). Let $\sigma : \mathbb{N} \to \{1, 2, \ldots, r\}$ be an intermittent selection function, and let S_n be the intermittent product (11.23). Then

$$\lim_n \|S_n(x) - P_M(x)\| = 0 \quad \text{for each } x \in H.$$

Theorem 11.73. *Assume the hypothesis of Fact 11.72 with σ an intermittent selection. Then exactly one of the following two statements holds:*

(1) $\sum_1^r M_i^\perp$ is closed; then (S_n) converges to P_M linearly.
(2) $\sum_1^r M_i^\perp$ is not closed; then (S_n) converges to P_M arbitrarily slowly.

Remark 11.74. (1) Statement (1) of Theorem 11.73 is a consequence of a result of Bauschke and Borwein [6, Theorem 5.7]. Statement (2) is from [21, Theorem 7.7].
(2) Note that Theorem 11.73 is a generalization of the Von Neumann-Halperin Dichotomy (Theorem 11.61).

11.11 Application to a Xu–Zikatanov Theorem

In their fundamental paper [54], Xu and Zikatanov showed a beautiful connection between the method of alternating projections and the method of subspace corrections in a Hilbert space. The method of subspace corrections is applied in the area of finite element analysis and is also referred to as domain decomposition or the multigrid method. In this section one of their two main results (viz., [54, Theorem 4.7]) can be improved by using the Dichotomy Theorem 11.27.

Let H be a Hilbert space, let $V_1, V_2, \ldots, V_r$ be closed subspaces of H, and let $T_i : H \to V_i$ be bounded linear mappings satisfying the following two assumptions for each $i = 1, 2, \ldots, r$:

(A1) The range of T_i is V_i and $T_i|_{V_i} : V_i \to V_i$ is an isomorphism.
(A2) $\|T_i(x)\|^2 \leq \omega \langle T_i(x), x\rangle$ for each $x \in H$ and some constant $\omega \in (0, 2)$.

In particular, assumption (A2) guarantees that $I - T_i$ is nonexpansive: $\|I - T_i\| \leq 1$. (In the applications that are made in [54], the T_i may be regarded as *approximations* to the projections P_{V_i}. Typically, the T_i correspond to damped Jacobi, Gauss-Seidel, or successive overrelaxation methods applied at different mesh resolutions.)

Let

$$E := (I - T_r)(I - T_{r-1}) \cdots (I - T_1), \tag{11.25}$$

$$\mathrm{Fix}(E) := \{x \in H \mid E(x) = x\}, \text{ and} \tag{11.26}$$

$$\mathcal{N}(T_i) := \{x \in H \mid T_i(x) = 0\}. \tag{11.27}$$

Lemma 11.75 (Xu–Zikatanov [54, Lemma 4.4]).

$$M := \mathrm{Fix}(E) = \bigcap_1^r \mathcal{N}(T_i) = \bigcap_1^r V_i^{\perp}, \text{and} \tag{11.28}$$

$$V := M^{\perp} = \overline{\sum_1^r V_i}. \tag{11.29}$$

Theorem 11.76 (Xu–Zikatanov [54, Theorem 4.6]). *The following two statements are equivalent:*

(1) $\sum_1^r V_i$ *is closed;*
(2) $\|EP_V\| < 1$.

Next note the identities

$$\begin{aligned} E^n - (I - P_V) &= E^n - P_M = (E - P_M)^n = [E(I - P_M)]^n \\ &= (EP_{M^{\perp}})^n = (EP_V)^n. \end{aligned} \tag{11.30}$$

The second equality $E^n - P_M = (E - P_M)^n$ follows from the fact that $P_M E = P_M = EP_M$, which in turn is a consequence of the (not obvious) fact that $T_i = T_i P_{V_i} = P_{V_i} T_i$ (see [54, (2.10)]), and hence that $P_M T_i = P_M P_{V_i} T_i = 0$ and $T_i P_M = P_i P_{V_i} P_M = 0$ for all i since $P_{V_i} P_M = 0 = P_M P_{V_i}$.

Theorem 11.77 (Xu–Zikatanov [54, Theorem 4.7]).

$$\lim_{n \to \infty} \|E^n(x) - (I - P_V)(x)\| = 0 \text{ for each } x \in H, \tag{11.31}$$

or equivalently,

$$\lim_{n \to \infty} \|(EP_V)^n(x)\| = 0 \text{ for each } x \in H. \tag{11.32}$$

Since $\|EP_V\| \leq 1$, it follows from the Dichotomy Theorem 11.27 and Theorems 11.76 and 11.77 that the following dichotomy pertaining to the Xu–Zikatanov theory is obtained.

Theorem 11.78 (Xu-Zikatanov Dichotomy) ([21, Theorem 8.4]). *Exactly one of the following two statements holds.*

(1) $\sum_1^r V_i$ *is closed. Then* (E^n) *converges to* $(I - P_V)$ *linearly.*
(2) $\sum_1^r V_i$ *is not closed. Then* (E^n) *converges to* $(I - P_V)$ *arbitrarily slowly.*

It should be noted that this result is somewhat more general than Theorem 11.66 since here the T_i need not be equal to P_{V_i}, but need only be a certain kind of approximation to P_{V_i}.

11.12 Further Applications

A Google search for the term "arbitrarily slow convergence" brings up hits in the areas of probability and statistics (density estimation [25, 26, 29], the central limit theorem [49], and Gibbs sampling [30]), machine learning/classifiers [14, 16, 27], numerical methods for inverse problems [33, 43, 51], control theory [46], optimization [48], finite element analysis [11], random matrices [4], and analysis [37]. The Second Lethargy Theorem and arguments similar to the proof of the Second Lethargy Theorem can be used to reproduce many of the arbitrarily slow or almost arbitrarily slow convergence results for these applications.

Acknowledgements We are grateful to the two referees for raising some points that helped us to make the paper more complete and readable. We are also grateful to Heinz Bauschke who originally pointed out the paper [3] to us that we were unaware of at the time.

References

1. Amemiya, I., Ando, T.: Convergence of random products of contractions in Hilbert space. Acta Sci. Math. (Szeged) **26**, 239–244 (1965)
2. Aronszajn, N.: Theory of reproducing kernels. Trans. Amer. Math. Soc., **68**, 337–403 (1950)
3. Badea, C., Grivaux, S., Müller, V.: The rate of convergence in the method of alternating projections. St. Petersburg Math. J. **22**, (2010). Announced in C. R. Math. Acad. Sci. Paris **348**, 53–56 (2010)
4. Bai, Z.D., Yin, Y.Q.: Necessary and sufficient conditions for almost sure convergence of the largest eigenvalue of a Wigner matrix. Ann. Probability **16**, 1729–1741 (1988)
5. Bauschke, H.H.: A norm convergence result on random products of relaxed projections in Hilbert space. Trans. Amer. Math. Soc. **347**, 1365–1373 (1995)
6. Bauschke, H.H., Borwein, J.M.: On projection algorithms for solving convex feasibility problems. SIAM Review **38**, 367–426 (1996)
7. Bauschke, H.H., Borwein, J.M., Lewis, A.S.: The method of cyclic projections for closed convex sets in Hilbert space. Contemporary Mathematics **204**, 1–38 (1997)
8. Bauschke, H.H., Borwein, J.M., Li, W.: The strong conical hull intersection property, bounded linear regularity, Jameson's property(G), and error bounds in convex optimization. Math. Programming (Series A) **86**, 135–160 (1999)
9. Bauschke, H.H., Deutsch, F., Hundal, H.: Characterizing arbitrarily slow convergence in the method of alternating projections. Intl. Trans. in Op. Res. **16**, 413–425 (2009)

10. Bernstein, S.N.: On the inverse problem of the theory of the best approximation of continuous functions. Sochineniya **II**, 292–294 (1938)
11. Boland, J.M., Nicolaides, R.A.: Stable and semistable low order finite elements for viscous flows. SIAM J. Numer. Anal. **22**, 474–492 (1985)
12. Cheney, W.: Analysis for Applied Mathematics. Graduate Texts in Mathematics #208, Springer, New York (2001)
13. Combettes, P.L.: Fejér-monotonicity in convex optimization. In: C.A. Floudas and P.M. Pardalos (eds.) Encyclopedia of Optimization, Kluwer Acad. Pub. (2000)
14. Cover, T.M.: Rates of convergence for nearest neighbor procedures. Proc. Hawaii Intl. Conf. Systems Sciences, 413–415 (1968)
15. Davis, P.J.: Interpolation and Approximation. Blaisdell, New York (1963)
16. Deroïan, F.: Formation of social networks and diffusion of innovations. Research Policy **31**, 835–846 (2002)
17. Deutsch, F.: The method of alternating orthogonal projections. In: S.P. Singh (ed.) Approximation Theory, Spline Functions and Applications. Kluwer Academic Publishers, The Netherlands, 105–121 (1992)
18. Deutsch, F.: The role of the strong conical hull intersection property in convex optimization and approximation. In: C.K. Chui and L.L. Schumaker (eds.) Approximation Theory IX, Vanderbilt University Press, Nashville, TN, 143–150 (1998)
19. Deutsch, F.: Best Approximation in Inner Product Spaces. Springer, New York (2001)
20. Deutsch, F., Hundal, H.: Slow convergence of sequences of linear operators I: Almost arbitrarily slow convergence. J. Approx. Theory **162**, 1701–1716 (2010)
21. Deutsch, F., Hundal, H.: Slow convergence of sequences of linear operators II: Arbitrarily slow convergence. J. Approx. Theory **162**, 1717–1738 (2010)
22. Deutsch, F., Ubhaya, V.A., Ward, J.D., Xu, Y.: Constrained best approximation in Hilbert space III. Applications to n-convex functions. Constr. Approx. **12**, 361–384 (1996)
23. Deutsch, F., Li, W., Ward, J.D.: A dual approach to constrained interpolation from a convex subset of Hilbert space. J. Approx. Theory **80**, 381–405 (1997)
24. DeVore, R.: The Approximation of Continuous Functions by Positive Linear Operators. Lecture Notes in Mathematics # 293, Springer, New York (1972)
25. Devroye, L.: On arbitrarily slow rates of global convergence in density estimation. Probability Theory and Related Fields **62**, 475–483 (1983)
26. Devroye, L.: Another proof of a slow convergence result of Birgé. Statistics and Probability Letters **23**, 63–67 (1995)
27. Devroye, L., Györfi, L., Logosi, G.: A Probabilistic Theory of Pattern Recognition. Springer, New York (1996)
28. Dye, J., Khamsi, M.A., Reich, S.: Random products of contractions in Banach spaces. Trans. Amer. Math. Soc. **325**, 87–99 (1991)
29. Ghosal, S.: Convergence rates for density estimation with Bernstein polynomials. Ann. Statistics **29**, 1264–1280 (2001)
30. Golightly, A., Wilkinson, D.J.: Bayesian inference for nonlinear multivariate diffusion models observed with error. Computational Statistics and Data Analysis **52**, 1674–1693 (2008)
31. Gordon, R., Bender, R., Herman, G.T.: Algebraic reconstruction techniques (ART) for three-dimensional electron microscopy and X-ray photography. J. Theoretical Biol. **29**, 471–481 (1970)
32. Halperin, I.: The product of projection operators. Acta Sci. Math. (Szeged) **23**, 96–99 (1962)
33. Hanke, M., Neubauer, A., Scherzer, O.: A convergence analysis of the Landweber iteration for nonlinear ill-posed problems. Numerische Math. **72**, 21–37 (1995)
34. Hewitt, E., Stromberg, K.: Real and Abstract Analysis. Springer, New York, (1965)
35. Hounsfield, G.N.: Computerized transverse axial scanning (tomography); Part I Description of system. British J. Radiol. **46**, 1016–1022 (1973)
36. Hundal, H., Deutsch, F.: Two generalizations of Dykstra's cyclic projections algorithm. Math. Programming **77**, 335–355 (1997)
37. Jahnke, H.N. (ed.): A History of Analysis. History of Mathematics **24**. Amer. Math. Soc., Providence, RI, London Math. Soc., London (2003)

38. Kincaid, D., Cheney, W.: Numerical Analysis, 2nd edn. Brooks/Cole, New York (1996)
39. Korovkin, P.P.: Linear Operators and Approximation Theory. Hindustan Publ. Corp. (India), Delhi (1960)
40. Müller, V.: Power bounded operators and supercyclic vectors II. Proc.Amer. Math. Soc. **133**, 2997–3004 (2005)
41. Müller, V.: Spectral Theory of Linear Operators and Spectral Systems in Banach Algebras, 2nd edn. Operator Theory: Advances and Applications **139**, Birkhauser, Basel (2007)
42. Nakano, H.: Spectal Theory in the Hilbert Space. Japan Soc. Promotion Sc., Tokyo (1953)
43. Neubauer, A.: On converse and saturation results for Tikhhonov regularization of linear ill-posed problems. SIAM J. Numer. Anal. **34**, 517–527 (1997)
44. von Neumann, J.: On rings of operators. Reduction theory. Ann. of Math. **50**, 401–485 (1949)
45. von Neumann, J.: Functional Operators-Vol. II. The Geometry of Orthogonal Spaces. Annals of Math. Studies #22, Princeton University Press, Princeton, NJ (1950) [This is a reprint of mimeographed lecture notes first distributed in 1933.]
46. Olshevsky, A., Tsitsiklis, J.N.: Convergence rates in distributed consensus and averaging. Proc. IEEE Conf. Decision Control, San Diego, CA, 3387–3392 (2006)
47. Powell, M.J.D.: A new algorithm for unconstrained optimization. In: J.B. Rosen, O.L. Mangasarian, and K. Ritter (eds.) Nonlinear Programming, Academic, New York (1970)
48. Ratschek, H., Rokne, J.G.: Efficiency of a global optimization algorithm. SIAM J. Numer. Anal. **24**, 1191–1201 (1987)
49. Rhee, W., Talagrand, M.: Bad rates of convergence for the central limit theorem in Hilbert space. Ann. Prob. **12**, 843–850 (1984)
50. Schock, E.: Arbitrarily slow convergence, uniform convergence and superconvergence of Galerkin-like methods. IMA Jour. Numerical Anal. **5**, 153–160 (1985)
51. Schock, E.: Semi-iterative methods for the approximated solutions of ill-posed problems. Numer. Math. **50**, 263–271 (1987)
52. Timan, A.F.: Theory of Approximation of Functions of a Real Variable. MacMillan, New York (1963)
53. Wiener, N.: On the factorization of matrices. Comment. Math. Helv. **29**, 97–111 (1955)
54. Xu, J., Zikatanov, L.: The method of alternating projections and the method of subspace corrections in Hilbert space. J. Amer. Math. Soc. **15**, 573–597 (2002)

Chapter 12
Graph-Matrix Calculus for Computational Convex Analysis

Bryan Gardiner and Yves Lucet

Abstract We introduce a new family of algorithms for computing fundamental operators arising from convex analysis. The new algorithms rely on the fact that the graph of the subdifferential of most convex operators depends linearly on the graph of the subdifferential of the function. By storing the subdifferential information, the computation of the conjugate is reduced to a matrix multiplication. We explain how other operators can be computed similarly, and present numerical experiments that compare graph-matrix calculus algorithms with piecewise-linear quadratic algorithms from computational convex analysis (CCA), and with a bundle method using warmstarting. Our results show that the new algorithms are an order of magnitude faster. They also add subdifferential calculus to our numerical library, and are very simple to implement.

Keywords Computer-Aided convex analysis · Computational convex analysis · Convex function · Fenchel conjugate · Legendre–Fenchel transform · Proximal average · Proximal mapping · Subdifferential operator.

AMS 2010 Subject Classification: 90C25, 26A51, 26B25, 47H05, 52A41

12.1 Introduction

The birth of computational convex analysis (CCA) algorithms can be traced back to the introduction of fast algorithms to compute the (Legendre–Fenchel) conjugate [7, 9, 17, 25, 27], although key ideas were introduced much earlier [24, Paragraph 5c]. While other transforms such as the Moreau–Yosida approximate can be deduced from these algorithms, the study of monotone operators

---

Y. Lucet (✉)
Computer Science, I. K. Barber School, University of British Columbia Okanagan, Kelowna, B.C. V1V 1V7, Canada
e-mail: yves.lucet@ubc.ca

H.H. Bauschke et al. (eds.), *Fixed-Point Algorithms for Inverse Problems in Science and Engineering*, Springer Optimization and Its Applications 49,
DOI 10.1007/978-1-4419-9569-8_12,

using the Rockafellar and Fitzpatrick functions require specialized algorithms [11]. The CCA numerical library provides a numerical implementation of most of these algorithms and is available from [8].

The performance of algorithms in CCA [17–21] was challenged with the introduction of the proximal average. The origin of this operator can be traced back to Moreau [24] when he proved that the set of proximal mappings is convex. The proximal average was explicitly defined in 2004 [2]. It has been studied in [3–6, 16, 23], and extended in [1, 13]. Its numerical computation is challenging as it requires computing the composition of several operators. It is currently performed with a family of algorithms based on the class of piecewise linear-quadratic (PLQ) functions [22].

The linear relationship between the graph of the subdifferential of the most common convex analysis operators such as the (Legendre–Fenchel) conjugate was noticed in [12] while basing computational algorithms on such properties was suggested in [21]. By storing the subdifferential data, the computation of the conjugate is reduced to a multiplication by a 2×2 matrix. We show that other operators can be computed similarly, and compare numerical algorithms based on graph-matrix calculus (named GPH algorithms in the following, GPH standing for graph) with PLQ algorithms (which we call PLQ algorithms) introduced in [22], and with the linear-time Legendre transform (LLT) algorithm (named LLT algorithm in the following) from [18]. We also compare them with warmstarting using a bundle method (named OPT algorithms in the paper). Graph-matrix calculus algorithms are particularly well suited for implementation in mathematical software like Matlab and Scilab, which are optimized for matrix operations but penalize heavily the usage of loops. They naturally extend known numerical algorithms by providing easy manipulation of the subdifferential.

The paper is organized as follows. Section 12.1 introduces the paper; Section 12.2 fixes the notations and recalls the definitions of the main operators in convex analysis including two self-dual smoothing operators. Section 12.3 recalls Goebel's graph-matrix calculus [12, p. 181] and provides the matrices associated with most unary and binary operators. Section 12.4 proposes a GPH data structure, which is used in Sect. 12.5 to present linear-time algorithms for the main convex operators. In Sect. 12.6, we compare the running time of PLQ, GPH, OPT, and LLT algorithms and emphasize the advantages of the GPH algorithms. Finally, Sect. 12.7 summarizes our results and proposes future directions.

12.2 Preliminaries

First, we fix our notations. We will adopt Matlab and Scilab matrix notation to write the matrix

$$G = \begin{bmatrix} x \\ s \\ y \end{bmatrix}, \tag{12.1}$$

where x, s, and y are row vectors as $G = [x; s; y]$.

We consider functions $f : \mathbb{R}^d \to \mathbb{R} \cup \{+\infty\}$. The set of points where the function takes finite values is called the effective domain and denoted $\operatorname{dom} f$. A function f is proper if $\operatorname{dom} f$ is nonempty. Unless otherwise stated all functions considered are proper lower semicontinuous (lsc) convex. We denote by $\mathbb{R}^+$ the set of positive real numbers, $\operatorname{ri} S$ the relative interior of a set S, $\mathrm{Id} : \mathbb{R}^d \to \mathbb{R}^d$ the identity mapping, $\|\cdot\|$ the Euclidean norm, $\langle \cdot, \cdot \rangle$ the standard dot product, and $\Box$ the inf-convolution operator defined by

$$f \Box g(x) = \inf_{y \in \mathbb{R}^d} [f(y) + g(x-y)],$$

while

$$\partial f(x) = \{s \in \mathbb{R}^d \ : \ \forall y \in \mathbb{R}^d, \ f(y) \geq f(x) + \langle s, y-x \rangle\}$$

denotes the convex subdifferential, and

$$\operatorname{gph} \partial f = \{(x,s) \in \mathbb{R}^d \times \mathbb{R}^d \ : \ s \in \partial f(x)\}$$

the graph of the subdifferential. More generally, we denote $\operatorname{gph} P$ the graph of an operator $P : \mathbb{R}^d \rightrightarrows \mathbb{R}^d$, i.e.,

$$\operatorname{gph} P = \{(x,s) \in \mathbb{R}^d \times \mathbb{R}^d \ : \ s \in P(x)\}.$$

We write the (Legendre–Fenchel) conjugate as

$$f^*(s) = \sup_{x \in \mathbb{R}^d} [\langle s, x \rangle - f(x)],$$

and the Moreau(-Yosida) envelope of a function f with parameter $\lambda > 0$ as

$$e_\lambda f(x) = (f \Box \lambda^{-1} q)(x) = \min_{y \in \mathbb{R}^d} \left[f(y) + \frac{\|x-y\|^2}{2\lambda} \right],$$

where q is the quadratic kernel $q(x) = \|x\|^2/2$. The proximal mapping is the application that assigns to a point x the minimizers in the definition of the Moreau envelope

$$\operatorname{Prox}_\lambda f(x) = \operatorname*{Argmin}_{y \in \mathbb{R}^d} \left[f(y) + \frac{\|x-y\|^2}{2\lambda} \right].$$

The epi-multiplication operator is defined by

$$\alpha \star f = \begin{cases} \alpha f(\cdot/\alpha) & \text{if } \alpha > 0, \\ \iota_{\{0\}} & \text{if } \alpha = 0, \end{cases}$$

where $\iota_{\{0\}}$ is the indicator function: $\iota_{\{0\}}(x) = 0$ if $x = 0$ otherwise it is $+\infty$.

The *proximal average* is the function $\mathscr{P}(f_0, f_1; \lambda_0, \lambda_1) : \mathbb{R}^d \to]-\infty, +\infty]$ defined by

$$\mathscr{P}(f_0, f_1; \lambda_0, \lambda_1) = \left(\lambda_0 \left(f_0 + \frac{1}{\mu} q\right)^* + \lambda_1 \left(f_1 + \frac{1}{\mu} q\right)^*\right)^* - \frac{1}{\mu} q,$$

where the functions $f_0, f_1 : \mathbb{R}^d \to]-\infty, +\infty]$ are proper lsc convex functions, $\mu > 0$, and $\lambda_0, \lambda_1 > 0$ with $\lambda_0 + \lambda_1 = 1$. When there is no ambiguity on $f_0, f_1, \lambda_0, \lambda_1$, we will simplify the notation to $\mathscr{P}$. Note that the proximal average can be written as the constrained minimization problem [5, Formula (20)]

$$\mathscr{P}(f_0, f_1; \lambda_0, \lambda_1)(\xi) = \inf_{\lambda_0 y_0 + \lambda_1 y_1 = \xi} \left[\lambda_0 f_0(y_0) + \lambda_1 f_1(y_1) + \frac{\lambda_0 \lambda_1}{2\mu} \|y_0 - y_1\|^2\right],$$

which is the basis to generalizing the proximal average as the kernel average [1]. When the infimum in the definition of the proximal average is reached at y_0, y_1, we say the proximal average is *exact* at $\xi = \lambda_0 y_0 + \lambda_1 y_1$. We will use the following formula later [5, Proposition 4.3]

$$\mathscr{P}(f_0, f_1; \lambda_0, \lambda_1)(\xi) = \inf_{\lambda_0 y_0 + \lambda_1 y_1 = \xi} \left[\sum_{i=0}^{1} (\lambda_i (f_i(y_i) + q(y_i))) - q(\xi)\right]. \quad (12.2)$$

The proximal average is a very useful convex operator as it is a convex function [5, Corollary 5.2] that satisfies $\mathscr{P}(f_0, f_1; \lambda_0, \lambda_1) = -e_\mu(-\lambda_0 e_\mu f_0 - \lambda_1 e_\mu f_1)$ [13, Formula (4)], [5, Theorem 8.3], and $\operatorname{dom} \mathscr{P}(f_0, f_1; \lambda_0, \lambda_1) = \lambda_0 \operatorname{dom} f_0 + \lambda_1 \operatorname{dom} f_1$ [5, Theorem 4.6]. Moreover, the proximal mapping of the proximal average is the convex combination of the proximal mappings: $\operatorname{Prox} \mathscr{P}(f_0, f_1; \lambda_0, \lambda_1) = \lambda_0 \operatorname{Prox} f_0 + \lambda_1 \operatorname{Prox} f_1$ [5, Theorem 6.7], and the conjugate of the proximal average is the proximal average of the conjugate [5, Theorem 5.1]

$$[\mathscr{P}(f_0, f_1; \lambda_0, \lambda_1)]^* = \mathscr{P}(f_0^*, f_1^*; \lambda_0, \lambda_1).$$

Finally, we recall the self-dual smoothing operator

$$s_\lambda f = (1 - \lambda^2) e_\lambda f + \lambda q,$$

introduced in [12], and another self-dual operator

$$T_\lambda f = \mathscr{P}(f, q; 1 - \lambda, \lambda)$$

introduced in [6, 23].

We will work with PLQ functions, which were described in [26] and implemented in [22]. For univariate functions, they are extended-valued piecewise quadratic functions defined on a finite number of intervals which are continuous on the interior of their effective domain. In other words, they are defined for $x \in [x_i, x_{i+1}]$ by $p(x) = a_i x^2 + b_i x + c_i$, where $a_i, b_i \in \mathbb{R}$, $c_i \in \mathbb{R} \cup \{+\infty\}$, $x_i \in \mathbb{R} \cup \{-\infty, +\infty\}$ with $-\infty = x_0 < x_1 < \cdots < x_n < x_{n+1} = +\infty$.

12.3 Goebel's Graph-Matrix Calculus

We now collect rules, most of them introduced in [12, p. 181], on how the graph of the subdifferential mapping ∂f is transformed under basic operations on f.

Assume $f : \mathbb{R}^d \to \mathbb{R} \cup \{+\infty\}$ is a proper lsc convex function. The following calculus rules, which we call Goebel's Graph-matrix calculus for unary operators, on the graph of the subdifferential hold for any $\alpha > 0$, $0 < \lambda < 1$, and $\beta \geq 0$.

$$\operatorname{gph}\partial(f^*) = \begin{bmatrix} 0 & \mathrm{Id} \\ \mathrm{Id} & 0 \end{bmatrix} \operatorname{gph}\partial f \qquad \operatorname{gph}\partial(\alpha f) = \begin{bmatrix} \mathrm{Id} & 0 \\ 0 & \alpha\,\mathrm{Id} \end{bmatrix} \operatorname{gph}\partial f$$

$$\operatorname{gph}\partial(\alpha \star f) = \begin{bmatrix} \alpha\,\mathrm{Id} & 0 \\ 0 & \mathrm{Id} \end{bmatrix} \operatorname{gph}\partial f \quad \operatorname{gph}\partial(f(\alpha\cdot)) = \begin{bmatrix} \alpha^{-1}\,\mathrm{Id} & 0 \\ 0 & \alpha\,\mathrm{Id} \end{bmatrix} \operatorname{gph}\partial f \tag{12.3}$$

$$\operatorname{gph}\partial(f+\beta q) = \begin{bmatrix} \mathrm{Id} & 0 \\ \beta\,\mathrm{Id} & \mathrm{Id} \end{bmatrix} \operatorname{gph}\partial f \qquad \operatorname{gph}\partial e_\alpha(f) = \begin{bmatrix} \mathrm{Id} & \alpha\,\mathrm{Id} \\ 0 & \mathrm{Id} \end{bmatrix} \operatorname{gph}\partial f$$

$$\operatorname{gph}\operatorname{Prox}_\alpha(f) = \begin{bmatrix} \mathrm{Id} & \alpha\,\mathrm{Id} \\ \mathrm{Id} & 0 \end{bmatrix} \operatorname{gph}\partial f \qquad \operatorname{gph}\partial s_\lambda f = \begin{bmatrix} \mathrm{Id} & \lambda\,\mathrm{Id} \\ \lambda\,\mathrm{Id} & \mathrm{Id} \end{bmatrix} \operatorname{gph}\partial f.$$

All these formulas follow directly from well-known convex analysis formulas except the last one, which was proven in [12, p. 181] based on the definition of $s_\lambda f$. The operator T_λ admits the following formula.

Lemma 12.1. *Under the above assumptions we have*

$$\operatorname{gph}\partial T_\lambda f = \begin{bmatrix} (1-\frac{\lambda}{2})\,\mathrm{Id} & \frac{\lambda}{2}\,\mathrm{Id} \\ \frac{\lambda}{2}\,\mathrm{Id} & (1-\frac{\lambda}{2})\,\mathrm{Id} \end{bmatrix} \operatorname{gph}\partial f. \tag{12.4}$$

Proof. We use [23, Theorem 4.3.2(ii)]

$$T_\lambda f = (1-\lambda)\left[f \square \left(\frac{2-\lambda}{\lambda}\right) q\right]\left(\frac{2}{2-\lambda}\cdot\right) + \frac{\lambda}{2-\lambda} q$$

to deduce

$$\begin{aligned}
\operatorname{gph}\partial T_\lambda f &= \begin{bmatrix} \mathrm{Id} & 0 \\ \frac{\lambda}{2-\lambda}\mathrm{Id} & \mathrm{Id} \end{bmatrix} \operatorname{gph}\partial \left[(1-\lambda)\left(f\Box\left(\frac{2-\lambda}{\lambda}\right)q\right)\left(\frac{2}{2-\lambda}\cdot\right)\right] \\
&= \begin{bmatrix} \mathrm{Id} & 0 \\ \frac{\lambda}{2-\lambda}\mathrm{Id} & \mathrm{Id} \end{bmatrix} \begin{bmatrix} \frac{2-\lambda}{2}\mathrm{Id} & 0 \\ 0 & \frac{2}{2-\lambda}\mathrm{Id} \end{bmatrix} \operatorname{gph}\partial \left[(1-\lambda)\left(f\Box\left(\frac{2-\lambda}{\lambda}\right)q\right)\right] \\
&= \begin{bmatrix} \mathrm{Id} & 0 \\ \frac{\lambda}{2-\lambda}\mathrm{Id} & \mathrm{Id} \end{bmatrix} \begin{bmatrix} \frac{2-\lambda}{2}\mathrm{Id} & 0 \\ 0 & \frac{2}{2-\lambda}\mathrm{Id} \end{bmatrix} \begin{bmatrix} \mathrm{Id} & 0 \\ 0 & (1-\lambda)\mathrm{Id} \end{bmatrix} \operatorname{gph}\partial \left[f\Box\left(\frac{2-\lambda}{\lambda}\right)q\right] \\
&= \begin{bmatrix} \mathrm{Id} & 0 \\ \frac{\lambda}{2-\lambda}\mathrm{Id} & \mathrm{Id} \end{bmatrix} \begin{bmatrix} \frac{2-\lambda}{2}\mathrm{Id} & 0 \\ 0 & \frac{2}{2-\lambda}\mathrm{Id} \end{bmatrix} \begin{bmatrix} \mathrm{Id} & 0 \\ 0 & (1-\lambda)\mathrm{Id} \end{bmatrix} \begin{bmatrix} \mathrm{Id} & \frac{\lambda}{2-\lambda}\mathrm{Id} \\ 0 & \mathrm{Id} \end{bmatrix} \operatorname{gph}\partial f.
\end{aligned}$$

Multiplying the matrices gives the result. ■

Similar formulas hold for binary operators.

Lemma 12.2 (Graph-matrix calculus for binary operators). *Assume f_1 (resp. f_2) is a proper lsc convex function with $s_1 \in \partial f_1(x_1)$ (resp. $s_2 \in \partial f_2(x_2)$). The index i is in $\{1,2\}$. Consider points $x \in \mathbb{R}^d$ (primal space) and $s \in \mathbb{R}^d$ (dual space).*

(i) If $\operatorname{ri}\operatorname{dom} f_1 \cap \operatorname{ri}\operatorname{dom} f_2 \neq \emptyset$, then

$$(x,s) \in \operatorname{gph}\partial(f_1+f_2) \Leftrightarrow \exists (x_i,s_i) \in \operatorname{gph}\partial f_i \text{ such that } \begin{cases} x = x_1 = x_2, \\ s = s_1 + s_2. \end{cases}$$

(ii) If $\partial f_1(x_1) \cap \partial f_2(x_2) \neq \emptyset$, then

$$(x_1+x_2, s) \in \operatorname{gph}\partial(f_1 \Box f_2) \Leftrightarrow (x_i,s) \in \operatorname{gph}\partial f_i.$$

(iii) Take $\lambda_1 + \lambda_2 = 1$ with $\lambda_1, \lambda_2 > 0$. Assume $\mathscr{P}_\mu(f_1,f_2;\lambda_1,\lambda_2)$ is exact (i.e. the infimum is attained) at $x = \lambda_1 x_1 + \lambda_2 x_2$ with $x_i \in \operatorname{dom} f_i$. Then

$$(x,s) \in \operatorname{gph}\partial\mathscr{P}(f_1,f_2;\lambda_1,\lambda_2) \Leftrightarrow \begin{cases} x = \lambda_1 x_1 + \lambda_2 x_2, \\ s = x_1 + s_1 - x = x_2 + s_2 - x. \end{cases}$$

Proof. (i) From [14, Corollary XI.3.1.2], $\partial f_1(x) + \partial f_2(x) = \partial(f_1+f_2)(x)$, and the result follows.

(ii) From [14, Corollary XI.3.4.2], $\partial f_1(x_1) \cap \partial f_2(x_2) = \partial(f_1 \Box f_2)(x_1+x_2)$, and the result follows.

(iii) From [5, Theorem 7.1], $\partial\mathscr{P}(f_1,f_2;\lambda_1,\lambda_2)(x) = -x + \cap_{i:\lambda_i>0}(\partial f_i(x_i) + x_i)$, and the result follows. ■

Remark 12.3. An alternate proof of Lemma 12.1 can be obtained from Lemma 12.2(iii) using $f_1 = f$, $f_2 = q$, $\lambda_1 = 1-\lambda$, and $\lambda_2 = \lambda$ to obtain $(x,s) \in \operatorname{gph} T_\lambda$

if, and only if,

$$\begin{aligned} x &= (1-\lambda)x_1 + \lambda x_2 \\ s &= x_1 + s_1 - x \\ s &= x_2 + s_2 - x. \end{aligned}$$

Then using $s_2 = x_2$ we can solve the system to obtain

$$\begin{aligned} x &= \left(1 - \frac{\lambda}{2}\right) x_1 + \frac{\lambda}{2} s_1 \\ s &= \frac{\lambda}{2} x_1 + \left(1 - \frac{\lambda}{2}\right) s_1, \end{aligned}$$

which is Formula (12.4).

12.4 Graph-Matrix Representation of PLQ Functions

The key idea is to use a data structure that allows the computation of the subdifferential using a single matrix multiplication. We store the subdifferential data of a PLQ function, and evaluate the function by integration.

The following data structure will be used. A lsc proper convex PLQ function f of one variable is defined by $f(x) = a_i x^2 + b_i x + c_i$ for $x \in [x_{i-1}, x_i]$ and $i \in \{1, \ldots, n+1\}$, where $a_i, b_i \in \mathbb{R}$, $c_i \in \mathbb{R} \cup \{+\infty\}$, $x_i \in \mathbb{R} \cup \{-\infty, +\infty\}$ with $-\infty = x_0 < x_1 < \cdots < x_n < x_{n+1} = +\infty$. While in previous work the function f was stored as the matrix

$$\begin{bmatrix} x_1 & a_1 & b_1 & c_1 \\ \vdots & \vdots & \vdots & \vdots \\ x_n & a_n & b_n & c_n \\ +\infty & a_{n+1} & b_{n+1} & c_{n+1} \end{bmatrix},$$

(which was called the PLQ matrix), we will use a different data structure.

To take advantage of the results in the previous section, the function f will be stored as a matrix

$$\begin{bmatrix} \bar{x}_0 & x_1 & \cdots & x_n & \bar{x}_{n+1} \\ \bar{s}_0 & s_1 & \cdots & s_n & \bar{s}_{n+1} \\ \bar{y}_0 & y_1 & \cdots & y_n & \bar{y}_{n+1} \end{bmatrix}.$$

The above matrix will be called a GPH matrix. We relax the requirement that $x_i < x_{i+1}$ in the PLQ matrix by only requiring $-\infty < \bar{x}_0 \le x_1 \le \ldots \le x_n \le \bar{x}_{n+1} < +\infty$. The points x_i form a partition of the real line, the values s_i store the subdifferential data at the points x_i, and the values y_i store the value of f at x_i. Since the function f is piecewise quadratic, the graph of ∂f is piecewise affine. We store it as the piecewise affine graph going through (x_i, s_i) for $i = 1, \ldots, n$ (with $s_i \in \partial f(x_i)$). (Since the graph may have vertical parts, it may not be a function). We capture the information outside $[x_1, x_n]$ by adding two points $(\bar{x}_0, \bar{s}_0)$ and $(\bar{x}_{n+1}, \bar{s}_{n+1})$ defined as follows.

Case 1: $\operatorname{dom} f$ is unbounded from the left i.e., c_0 is finite. In that case, f is a quadratic function on $(-\infty, x_1]$. Then we define $\bar{x}_0 = x_1 - 1$, $\bar{s}_0 = 2a_0\bar{x}_0 + b_0$, and $\bar{y}_0 = f(\bar{x}_0) = a_0\bar{x}_0^2 + b_0\bar{x}_0 + c_0$.

Case 2: $\operatorname{dom} f$ is bounded from the left i.e., $c_0 = +\infty$. In that case, $f(x) = +\infty$ for $x < x_1$. So we set $\bar{x}_0 = x_1$, $\bar{s}_0 = s_1 - 1$ and $\bar{y}_0 = +\infty$.

The definition of $(\bar{x}_{n+1}, \bar{s}_{n+1})$ is done similarly. These values correspond to a linear extension outside $[x_1, x_n]$ i.e., any subdifferential value on $(-\infty, x_1]$ (resp. $[x_n, +\infty)$) is on the half line going through $(\bar{x}_0, \bar{s}_0)$ and ending at (x_1, s_1) (resp. $(\bar{x}_{n+1}, \bar{s}_{n+1})$ and (x_n, s_n)). The point $\bar{x}_0$ (resp. $\bar{x}_{n+1}$) is added specifically to store the subdifferential on $(-\infty, x_1]$ (resp. $[x_n, +\infty)$).

Remark 12.4. The value 1 used in the definition of $(\bar{x}_0, \bar{s}_0)$ is arbitrary as any positive value can be used. To reduce floating point errors, a better value may be $x_2 - x_1$ in case 1 and $s_2 - s_1$ in case 2. The value 1 is used for simplicity.

To recover values of the function f, we can integrate the subdifferential but still need one constant of integration. For ease of computation, we store all the values $y_i = f(x_i)$ for $i = 1, \ldots, n$. The additional values $\bar{y}_0$ and $\bar{y}_{n+1}$ are computed as above. Note that all values x_i and s_i are always finite for $i = 1, \ldots n$ as are $\bar{x}_0, \bar{s}_0, \bar{x}_{n+1}, \bar{s}_{n+1}$.

Example 12.5. The absolute value function may be stored as the matrix

$$\begin{bmatrix} -1 & 0 & 0 & 1 \\ -1 & -1 & 1 & 1 \\ 1 & 0 & 0 & 1 \end{bmatrix}.$$

The first and second row mean the graph of the subdifferential goes through the points $(-1,-1)$, $(0,-1)$, $(0,1)$, and $(1,1)$. The half-line going through $(-1,-1)$, and ending at $(0,-1)$ is extended to $(-\infty,-1]$, so the subdifferential is constant with value -1 on that interval. Similarly, it is constant with value 1 on $[1,+\infty)$. Finally, the first and last row give function values i.e., the function goes through the points $(-1,1)$, $(0,0)$, $(0,0)$, and $(1,1)$. Note that we store the point $(0,0)$ twice for convenience.

Example 12.6. The function $f(x) = \max(0, |x| - 1)$ may be stored as the matrix

$$\begin{bmatrix} -2 & -1 & -1 & 1 & 1 & 2 \\ -1 & -1 & 0 & 0 & 1 & 1 \\ 1 & 0 & 0 & 0 & 0 & 1 \end{bmatrix}.$$

Example 12.7. The function $f(x) = \max(0, x^2 - 1)$ may be stored as the matrix

$$\begin{bmatrix} -2 & -1 & -1 & 1 & 1 & 2 \\ -4 & -2 & 0 & 0 & 2 & 4 \\ 3 & 0 & 0 & 0 & 0 & 3 \end{bmatrix}.$$

There is no uniqueness of our representation. Values at x_i can be stored multiple times to capture multi-valuedness of the subdifferential and lower semicontinuity of the function, different points may be taken for $(\bar{x}_0, \bar{s}_0)$ and $(\bar{x}_{n+1}, \bar{s}_{n+1})$, and additional (redundant) points with abscissa not equal to x_i could be used in the GPH matrix. So different GPH matrices can represent the same function. While we could define a unique representation by imposing more constraints on the format, we would then have to normalize after each computation. By not normalizing we save computation time when calculating the composition of several operators. We also avoid coding a shape-preserving spline approximation algorithm. The downside is comparing two functions requires more work than a simple matrix comparison.

The following are "special" cases of our data structure. The function

$$\begin{bmatrix} -1 & 1 \\ 0 & 0 \\ 0 & 0 \end{bmatrix}$$

is the function identically 0 (the values -1 and 1 can be replaced by any nonequal values). The indicator function $\iota_{[-1,1]}$ of the interval $[-1, 1]$ may be written

$$\begin{bmatrix} -1 & -1 & 1 & 1 \\ -1 & 0 & 0 & 1 \\ \infty & 0 & 0 & \infty \end{bmatrix}.$$

A quadratic function $f(x) = ax^2 + bx + c$ defined everywhere is stored as

$$\begin{bmatrix} \bar{x}_0 & \bar{x}_1 \\ 2a\bar{x}_0 + b & 2a\bar{x}_1 + b \\ a\bar{x}_0^2 + b\bar{x}_0 + c & a\bar{x}_1^2 + b\bar{x}_1 + c \end{bmatrix}$$

with $\bar{x}_0 < \bar{x}_1$. Conversely, the GPH matrix

$$\begin{bmatrix} \bar{x}_0 & \bar{x}_1 \\ \bar{s}_0 & \bar{s}_1 \\ \bar{y}_0 & \bar{y}_1 \end{bmatrix}$$

with $\bar{x}_0 < \bar{x}_1$ represents a quadratic function $f(x) = ax^2 + bx + c$ with $a = (\bar{s}_1 - \bar{s}_0)/(2(\bar{x}_1 - \bar{x}_0))$, $b = \bar{s}_0 - 2a\bar{x}_0$ and $c = \bar{y}_0 - a\bar{x}_0^2 - b\bar{x}_0 = \bar{y}_1 - a\bar{x}_1^2 - b\bar{x}_1$. Finally, the indicator function $\iota_{\{\bar{x}\}} + \bar{y}$ of a point has GPH matrix

$$\begin{bmatrix} \bar{x} & \bar{x} & \bar{x} & \bar{x} \\ \alpha-1 & \alpha & \beta & \beta+1 \\ \infty & \bar{y} & \bar{y} & \infty \end{bmatrix}$$

provided $\alpha < \beta$ (the value 1 can be replaced with any positive number).

12.5 Computational Convex Analysis through Graph Calculus

Using Formulas (12.3) with the GPH matrix gives simple algorithms for the most common convex operators. For example, the conjugate of the function represented as the GPH matrix $G = [x; s; y]$ where x, s, and y are row vectors, admits the GPH matrix

$$\begin{bmatrix} \begin{bmatrix} 0 & 1 \\ 1 & 0 \end{bmatrix} * \begin{bmatrix} x \\ s \end{bmatrix} \\ s.*x-y \end{bmatrix} = \begin{bmatrix} s \\ x \\ s.*x-y \end{bmatrix},$$

where $.*$ represents the element-wise multiplication and $*$ the standard matrix multiplication. (Note that the operation $s.*x-y$ is always well defined as s_i and x_i are always finite even when $y_i = \infty$.)

Similarly, the scalar multiplication by $\alpha > 0$ of the same function represented by G has GPH matrix

$$\begin{bmatrix} \begin{bmatrix} 1 & 0 \\ 0 & \alpha \end{bmatrix} * \begin{bmatrix} x \\ s \end{bmatrix} \\ \alpha y \end{bmatrix} = \begin{bmatrix} x \\ \alpha s \\ \alpha y \end{bmatrix}.$$

The Moreau envelope can be computed as follows.

Proposition 12.8. *Given a convex PLQ function f in GPH matrix G given by Formula (12.1), its Moreau envelope $e_\lambda f$ admits the GPH matrix*

$$\begin{bmatrix} \begin{bmatrix} 1 & \lambda \\ 0 & 1 \end{bmatrix} * \begin{bmatrix} x \\ s \end{bmatrix} \\ f+\frac{\lambda}{2}s.\hat{}2 \end{bmatrix} = \begin{bmatrix} x+\lambda s \\ s \\ f+\frac{\lambda}{2}s.\hat{}2 \end{bmatrix},$$

where $.\hat{}$ is the elementwise square function.

Proof. Using the fact $e_\lambda f = (f^* + \lambda q)^*$ and Formulas (12.3) we deduce that f^* admits the GPH matrix $[s; x; s.*x-f]$ so the function $f^* + \lambda q$ can be represented as $[s; x+\lambda s; s.*x-f+\frac{\lambda}{2}s.\hat{}2]$. Taking the conjugate gives the GPH matrix of the Moreau envelope and finishes the proof. ■

We next compute a GPH matrix of the proximal average.

Proposition 12.9. *Given two convex PLQ functions f_1 (resp. f_2) in GPH matrix $G_1 = [x_1; s_1; y_1]$ (resp. $G_2 = [x_2; s_2; y_2]$) and $\lambda_1 = 1-\lambda$, $\lambda_2 = \lambda$ with $\lambda \in [0,1]$;*

the λ proximal average of f_1 with f_2 admits the GPH matrix $G = [x;s;y]$, where $x = \lambda_1 x_1 + \lambda_2 P x_1$, $s = x_1 + s_1 - x$,

$$y = \lambda_1 \left(y_1 + \frac{1}{2} x_1 .\hat{}2 \right) + \lambda_2 \left(y_{Px_1} + \frac{1}{2} P x_1 .\hat{}2 \right) - \frac{1}{2} x .\hat{}2$$

the operator P is defined by

$$P = (\mathrm{Id} + \partial f_2)^{-1} (\mathrm{Id} + \partial f_1),$$

and $y_{Px_1} = f_2(Px_1)$.

Proof. We parametrize the graph of the proximal average using x_1, so we need to find the corresponding x_2 for each value of x_1. From Lemma 12.2 we have $x_1 - x + s_1 = x_2 - x + s_2$ so we deduce $(\mathrm{Id} + \partial f_1)(x_1) = (\mathrm{Id} + \partial f_2)(x_2)$, thus $x_2 = P x_1$ is the corresponding point that makes the prox average exact. The value of y comes from Formula (12.2). ■

We note that Proposition 12.9 allows us to speed up computation by precomputing Px_1, $(y_1 + \frac{1}{2} x_1 .\hat{}2)$ and $(y_{Px_1} + \frac{1}{2} P x_1 .\hat{}2)$, which do not depend on λ. Then we can deduce the λ-dependent values x, s, and y. Applying such precomputation scheme, we plot the proximal average of $f_0(x) = x^2/2$ with $f_1(x) = \iota_{\{0\}}(x) + 1$ (this example was challenging to plot during the writing of [4] and motivated further research to speed up such computation [16, 22]) in Fig. 12.1. The PLQ algorithm took 117 s to generate the picture under Scilab 5.1.1, while with precomputation, using the GPH algorithm only took 33 s.

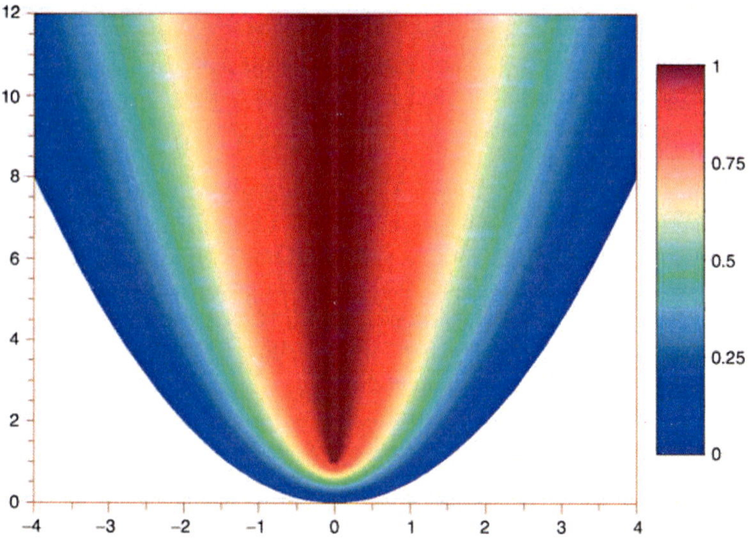

Fig. 12.1 The proximal average of $f_0(x) = x^2/2$ with $f_1(x) = \iota_{\{0\}}(x) + 1$. Each color corresponds to a specific value of $\lambda \in [0, 1]$

12.6 Comparing GPH Algorithms with PLQ Algorithms

The previous section contains extremely simple algorithms to compute the conjugate, the scalar multiplication, the Moreau envelope, and the proximal average. In particular, the computation of the conjugate is much simpler than previous algorithms, namely the (obsolete log-linear time) fast Fenchel transform [7, 9, 17, 25], and the (optimal) linear time algorithms: the LLT [18], the PLQ algorithm [22], the parametric Legendre transform (PLT) [15], and the parabolic envelope algorithm (PE) [10].

All numerical experiments were performed on an IBM Thinkpad Lenovo T60 running Intel Core Duo at 2Ghz with 2GB memory using Scilab 5.1.1 under Windows XP pro SP3.

In [19], we argued for the superiority of the PLQ algorithm over the LLT algorithm based on the ability of PLQ algorithms to model unbounded effective domains, and the closedness of the class of PLQ functions under common convex operators which makes the algorithms exact for these functions. The price to pay for these advantages is the more expensive computation time required by the PLQ algorithms over the LLT algorithm. The GPH algorithms bridge that gap by keeping all the properties of the PLQ algorithm with the speed of the LLT algorithm. Figure 12.2a shows the CPU time vs. n (the number of pieces) when computing the conjugate with the PLQ, GPH, and LLT algorithms: the PLQ algorithm is slower, while the GPH and LLT algorithms run in almost the same time. We included in Fig. 12.2b the computation time using a nonsmooth bundle method (n1fc1 as available through the optim function in Scilab) at each point warmstarted by the value at the previous point. The figure clearly shows that specialized algorithms PLQ, GPH, and LLT outperform a generic optimization solver.

The numerical computation of the Moreau envelope is compared in Fig. 12.3. The algorithms LLT and GPH share similar performances and are faster than the PLQ algorithm, which is itself faster than the OPT algorithm.

The numerical computation of the proximal average envelope is compared in Fig. 12.4. The graph clearly shows the GPH algorithm is faster (without precomputation as it does not apply here) and that for $n = 700$ the GPH algorithm runs 15 times faster than the PLQ algorithm and 95 times faster than the OPT algorithm.

We conclude this section with a couple of examples illustrating the capabilities of the numerical package. Figure 12.5 shows the subdifferential of a piecewise affine function. Adding the GPH algorithms expanded the existing numerical library to manipulate subdifferentials. Finally Fig. 12.6 illustrates the proximal point algorithm. It shows the proximal mapping associated with the quadratic function $f(x) = x^2$ computed using graph-matrix calculus, the line $y = x$, and the iterations of the proximal point algorithm. We clearly see convergence to the minimum of the function which is the only fixed point.

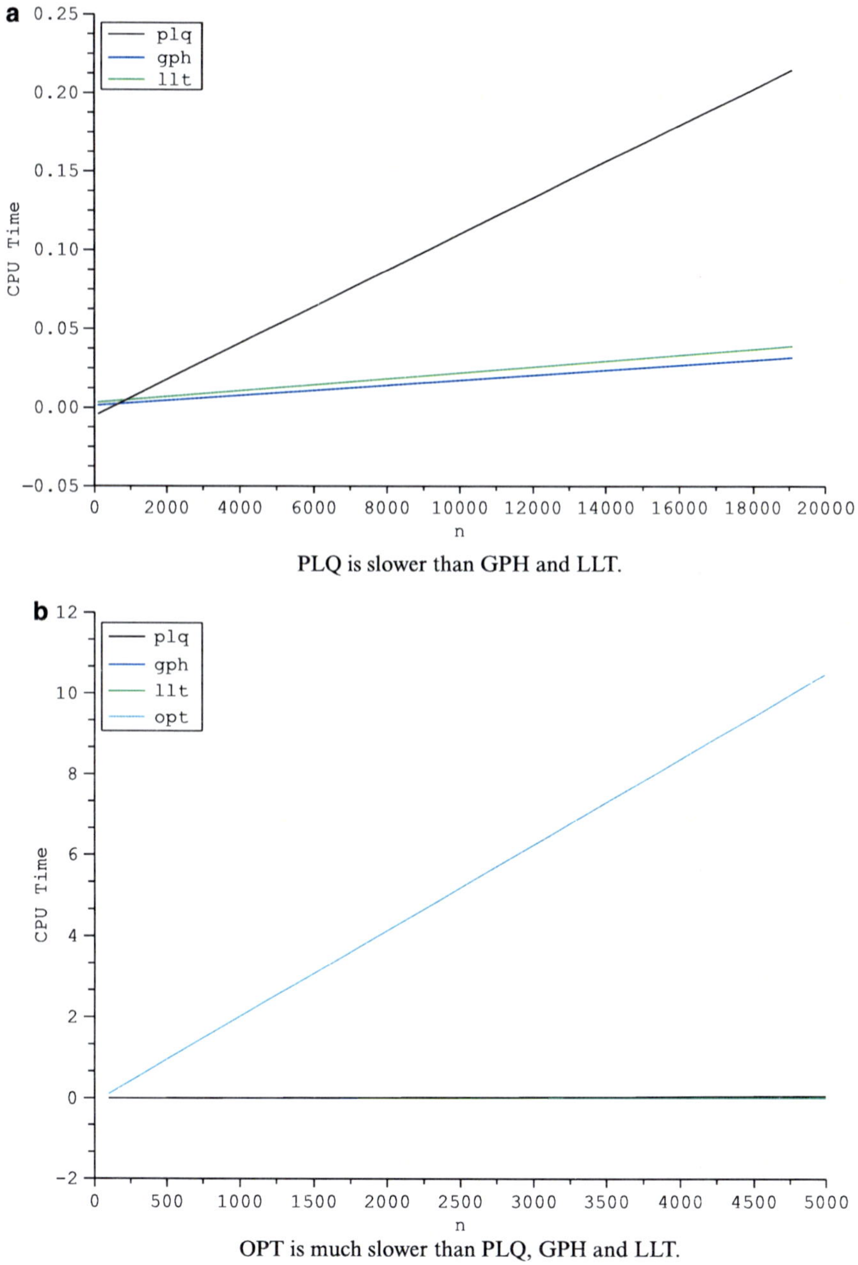

Fig. 12.2 Comparison of PLQ, GPH, and LLT algorithms for computing the conjugate of the function $f(x) = x^4$ approximated by n piecewise linear functions on $[-10, 10]$ (Least-square regression lines are shown). (**a**) GPH and LLT are on par while PLQ is slower. (**b**) OPT is much slower than the other three algorithms

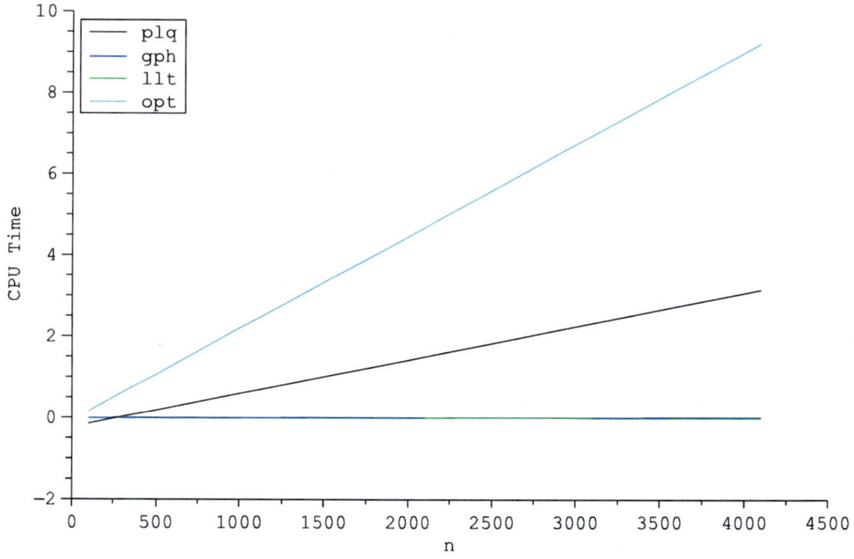

Fig. 12.3 Comparison of PLQ, GPH, LLT, and OPT algorithms for computing the Moreau envelope of the function $f(x) = x^4$ approximated by n piecewise linear functions on $[-10, 10]$ when $\lambda = 0.5$ (Least-square regression lines are shown)

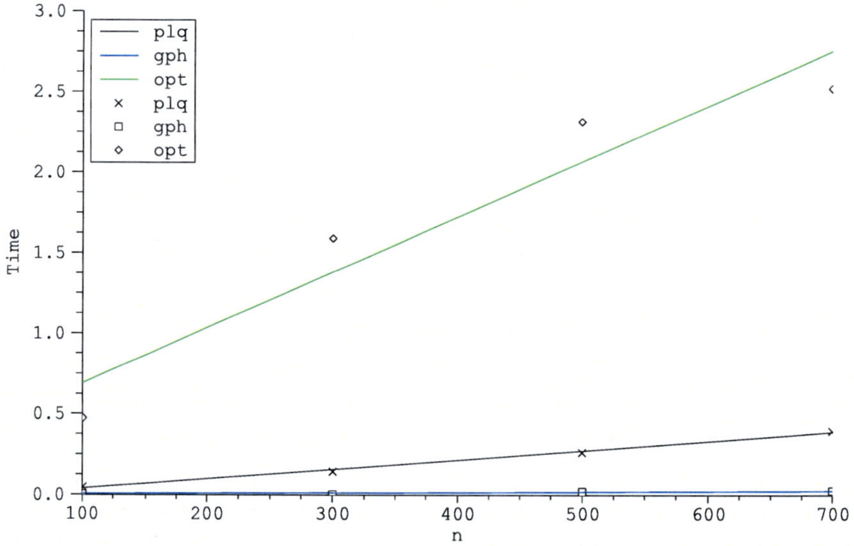

Fig. 12.4 Comparison of PLQ, GPH, and OPT algorithms for computing the proximal average of the function $f_1(x) = x^4$ with $f_2(x) = e^x$ approximated by n piecewise affine functions on $[-10, 10]$ when $\lambda = 0.5$ (Least-square regression lines are shown)

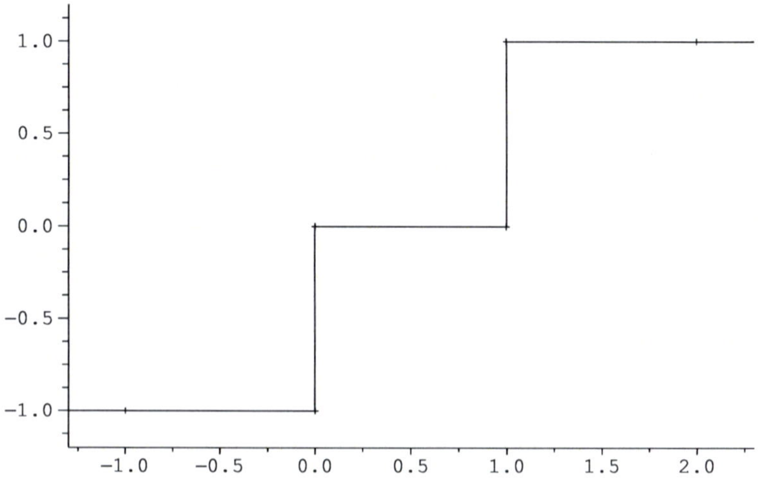

Fig. 12.5 Subdifferential of the function $f(x) = -x$ for $x \leq 0$, $f(x) = 0$ when $0 \leq x \leq 1$, and $f(x) = x - 1$ if $x \geq 1$. The points indicated belong to the GPH matrix of f

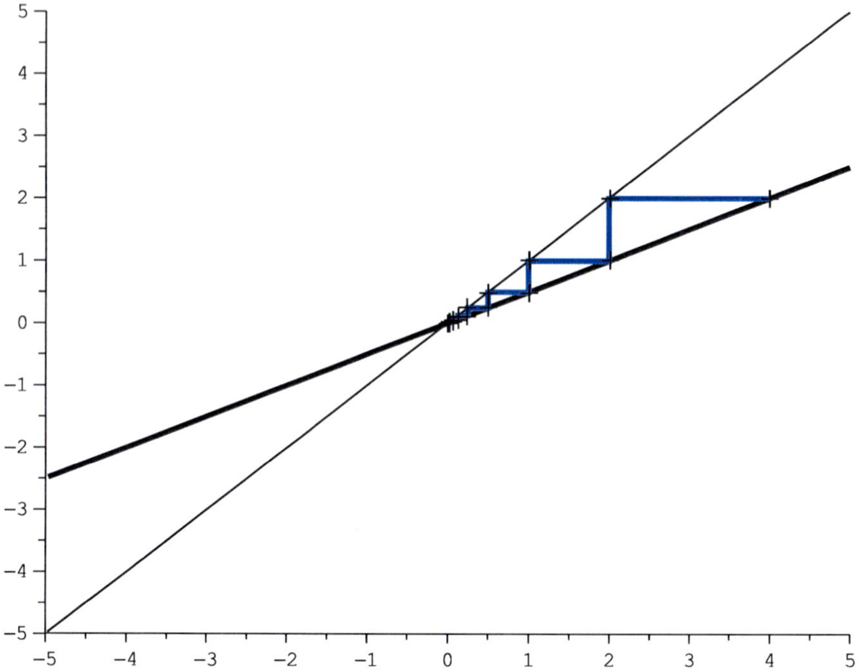

Fig. 12.6 Proximal mapping of the function $f(x) = x^2$ with the 16 iterations of the proximal point algorithm

12.7 Conclusion

We recalled Goebel's graph-matrix calculus formulas that relate the graph of the subdifferential of convex operators to the graph of the subdifferential of the function it is applied to. We proposed an intuitive data structure for storing PLQ functions in GPH format and deduced very simple linear-time algorithms to compute numerically the main convex operators. In addition, we provided numerical evidence that the algorithms are (an order of magnitude) faster than previously known PLQ algorithms while providing the same modeling advantages. We also showed that the PLQ, LLT, and GPH specialized algorithms are much faster than using a generic solver with warmstarting (named the OPT algorithms). The new GPH algorithms extend the current CCA numerical library by providing a natural manipulation of the subdifferential. They are especially efficient in matrix-based languages like Matlab and Scilab.

Future work will focus on extending the algorithms to handle nonconvex functions, and to provide a numerical library to manipulate bivariate functions.

Acknowledgements The authors thank the two referees for their careful reading of the manuscripts and their multiple comments, which resulted in correcting an error in Lemma 12.4.

Yves Lucet was partially supported by a Discovery grant from the Natural Sciences and Engineering Research Council of Canada.

References

1. Bauschke, H.H., Wang, X.: The kernel average of two convex functions and its application to the extension and representation of monotone operators. Trans. Amer. Math. Soc. **361**, 5947–5965 (2009)
2. Bauschke, H.H., Matoušková, E., Reich, S.: Projection and proximal point methods: Convergence results and counterexamples. Nonlinear Anal. **56**, 715–738 (2004)
3. Bauschke, H.H., Lucet, Y., Wang, X.: Primal-dual symmetric intrinsic methods for finding antiderivatives of cyclically monotone operators. SIAM J. Control Optim. **46**, 2031–2051 (2007)
4. Bauschke, H.H., Lucet, Y., Trienis, M.: How to transform one convex function continuously into another. SIAM Rev. **50**, 115–132 (2008)
5. Bauschke, H.H., Goebel, R., Lucet, Y., Wang, X.: The proximal average: Basic theory. SIAM J. Optim. **19**, 768–785 (2008)
6. Bauschke, H.H., Moffat, S.M., Wang, X.: Self-dual smooth approximations of convex functions via the proximal average. Tech. Rep. arXiv:1003.5866v1 [math.FA], UBC Okanagan (2010)
7. Brenier, Y.: Un algorithme rapide pour le calcul de transformées de Legendre–Fenchel discrètes. C. R. Acad. Sci. Paris Sér. I Math. **308**, 587–589 (1989)
8. Computational Convex Analysis library. https://people.ok.ubc.ca/ylucet/cca.html (1996–2009)
9. Corrias, L.: Fast Legendre–Fenchel transform and applications to Hamilton–Jacobi equations and conservation laws. SIAM J. Numer. Anal. **33**, 1534–1558 (1996)
10. Felzenszwalb, P.F., Huttenlocher, D.P.: Distance transforms of sampled functions. Tech. Rep. TR2004-1963, Cornell Computing and Information Science (2004)
11. Gardiner, B., Lucet, Y.: Numerical computation of Fitzpatrick functions. J. Convex Anal. **16**, 779–790 (2009)
12. Goebel, R.: Self-dual smoothing of convex and saddle functions. J. Convex Anal. **15**, 179–190 (2008)

13. Hare, W.: A proximal average for nonconvex functions: A proximal stability perspective. SIAM J. Optim. **20**, 650–666 (2009)
14. Hiriart-Urruty, J.B., Lemaréchal, C.: Convex Analysis and Minimization Algorithms, *Grundlehren der Mathematischen Wissenschaften [Fundamental Principles of Mathematical Sciences]*, vol. 305–306. Springer, Berlin (1993)
15. Hiriart-Urruty, J.B., Lucet, Y.: Parametric computation of the Legendre–Fenchel conjugate. J. Convex Anal. **14**, 657–666 (2007)
16. Koch, V., Johnstone, J., Lucet, Y.: Convexity of the proximal average. Tech. rep., University of British Columbia (2010). Accepted for publication in Journal of Optimization Theory and Applications
17. Lucet, Y.: A fast computational algorithm for the Legendre–Fenchel transform. Comput. Optim. Appl. **6**, 27–57 (1996)
18. Lucet, Y.: Faster than the Fast Legendre Transform, the Linear-time Legendre Transform. Numer. Algorithms **16**, 171–185 (1997)
19. Lucet, Y.: Fast Moreau envelope computation I: Numerical algorithms. Numer. Algorithms **43**, 235–249 (2006)
20. Lucet, Y.: New sequential exact Euclidean distance transform algorithms based on convex analysis. Image and Vision Computing **27**, 37–44 (2009)
21. Lucet, Y.: What shape is your conjugate? A survey of computational convex analysis and its applications. SIAM J. Optim. **20**, 216–250 (2009)
22. Lucet, Y., Bauschke, H.H., Trienis, M.: The piecewise linear-quadratic model for computational convex analysis. Comput. Optim. Appl. **43**, 95–118 (2009)
23. Moffat, S.M.: On the kernel average of n functions. Master's thesis, Department of Mathematics, University of British Columbia (2009)
24. Moreau, J.J.: Proximité et dualité dans un espace Hilbertien. Bull. Soc. Math. France **93**, 273–299 (1965)
25. Noullez, A., Vergassola, M.: A fast Legendre transform algorithm and applications to the adhesion model. J. Sci. Comput. **9**, 259–281 (1994)
26. Rockafellar, R.T., Wets, R.J.B.: Variational Analysis. Springer, Berlin (1998)
27. She, Z.S., Aurell, E., Frisch, U.: The inviscid Burgers equation with initial data of Brownian type. Comm. Math. Phys. **148**, 623–641 (1992)

Chapter 13
Identifying Active Manifolds in Regularization Problems

W.L. Hare

Abstract In 2009, Tseng and Yun [Math. Programming (Ser. B) 117, 387–423 (2009)] showed that the regularization problem of minimizing $f(x) + \|x\|_1$, where f is a $\mathscr{C}^2$ function and $\|x\|_1$ is the l_1 norm of x, can be approached by minimizing the sum of a quadratic approximation of f and the l_1 norm. We consider a generalization of this problem, in which the l_1 norm is replaced by a more general nonsmooth function that contains an underlying smooth substructure. In particular, we consider the problem

$$\min_x \{f(x) + P(x)\}, \tag{13.1}$$

where f is $\mathscr{C}^2$ and P is prox-regular and partly smooth with respect to an active manifold $\mathscr{M}$ (the l_1 norm satisfies these conditions.) We reexamine Tseng and Yun's algorithm in terms of active set identification, showing that their method will correctly identify the active manifold in a finite number of iterations. That is, after a finite number of iterations, all future iterates x^k will satisfy $x^k \in \mathscr{M}$. Furthermore, we confirm a conjecture of Tseng that, regardless of what technique is used to solve the original problem, the subproblem $p^k = \text{argmin}_p\{\langle \nabla f(x^k), p\rangle + \frac{r}{2}|x^k - p|^2 + P(p)\}$ will correctly identify the active manifold in a finite number of iterations.

Keywords Nonconvex optimization · Active constraint identification · Prox-regular · Partly smooth

AMS 2010 Subject Classification: Primary: 49K40, 65K05; Secondary: 52A30, 52A41, 90C53

W.L. Hare (✉)
Department of Mathematics and Statistics, UBC Okanagan Campus, Kelowna, B.C. V1V 1V7, Canada
e-mail: warren.hare@ubc.ca

H.H. Bauschke et al. (eds.), *Fixed-Point Algorithms for Inverse Problems in Science and Engineering*, Springer Optimization and Its Applications 49,
DOI 10.1007/978-1-4419-9569-8_13,

13.1 Introduction

In this work, we consider the problem of minimizing the sum of two functions over a finite dimensional Euclidean space,

$$\min_x \{f(x) + P(x)\}, \tag{13.2}$$

where f is $\mathscr{C}^2$ and P is nonsmooth, but contains some underlying smooth substructure. One common example arises in l_1 regularization problems,

$$\min_x \{f(x) + c\|x\|_1\}, \tag{13.3}$$

where $\|x\|_1$ is the l_1 norm of x. A brief survey of l_1 regularization problems and some applications can be found in the introduction to [7].

Recent work by Tseng and Yun [7] has suggested that one practical method to approach such a problem is to solve a sequence of quadratic approximation problems,

$$x^{k+1} = \min_x \{\langle \nabla f(x^k), x - x^k\rangle + (x - x^k)' H^k (x - x^k) + P(x)\}, \tag{13.4}$$

where H^k is an approximation to the Hessian of f. When P is well structured, such as the case when $P = c\|x\|_1$, this problem is easily solved; potentially having a closed form solution. Convergence theory and numerical testing can be found in [7].

In this work, we consider Tseng and Yun's method in terms of active manifold identification. In particular, we consider the case when P is *partly smooth* with respect to some manifold $\mathscr{M}$ containing $\bar{x} \in \operatorname{argmin}_x\{f(x) + P(x)\}$. We show that, under some conditions, all but a finite number of iterates will lie on the active manifold. In terms of l_1 regularization, this means that if $\bar{x} \in \operatorname{argmin}_x\{f(x) + c\|x\|_1\}$ then for k sufficiently large, $x_i^k = 0$ if and only if $\bar{x}_i = 0$.

Tseng and Yun's method separates the original problem (13.2) into two pieces, the smooth portion and the nonsmooth portion, that are treated distinctly different. Similar ideas often occur in constrained optimization, where objective functions and constrain sets are treated differently. Tseng and Yun's separation technique could be viewed as an analog to such techniques for constrained optimization by rephrasing the problem as

$$\min_x \{f(x) + r : 0 \geq P(x) - r\}. \tag{13.5}$$

Recently, Hare [1] showed that the active manifold of a constraint set could be identified by examining a proximal subproblem. Along similar lines, Tseng conjectured that the subproblem

$$p^k = \operatorname{argmin}_p \left\{ \langle \nabla f(x^k), p\rangle + \frac{r}{2}|x^k - p|^2 + P(p) \right\} \tag{13.6}$$

will correctly identify the active constraints of the problem (13.2) in a finite number of iterations, regardless of the method used to solve the problem (13.2). Theorem 13.9 provides an affirmative proof for this conjecture.

The remainder of this work is organized as follows. In Sect. 13.2, we provide the background required to understand this work. In particular, Sect. 13.2 includes the definitions of *prox-regular* and *partly smooth functions*. In Sect. 13.3, we examine the active manifold identification properties of iterates generated by (13.4) and (13.6). A brief conclusion, consisting primarily of an e-mail from Tseng which prompted this work, appears in Sect. 13.4.

13.2 Definitions and Notations

To keep this work brief, we provide only the definitions and background necessary to understand its two main results (Theorems 13.7 and 13.9). In general we follow the notation of [6], with one notable exception.

We shall define the *Moreau envelope*, and its corresponding (potentially empty) *proximal point mapping*, of a function f at a point x with respect to a parameter r as

$$e_r f(x) := \inf_y \left\{ f(y) + \frac{r}{2}|y-x|^2 \right\} \tag{13.7}$$

$$\operatorname{prox}_r f(x) := \operatorname{argmin}_y \left\{ f(y) + \frac{r}{2}|y-x|^2 \right\}, \tag{13.8}$$

where $|\cdot|$ is the usual Euclidean norm. (In [6, Definition 1.23] the parameter r is place in the denominator of the quadratic penalty term '$\frac{1}{2r}$'.) We say that a function is *prox-bounded* if there exists some $r > 0$ and point x such that $e_r f(x)$ is finite. If a function is prox-bounded, then (for r sufficiently large) $e_r f$ is finite-value everywhere [6, Example 1.24].

A useful lemma, from [3], regarding proximal points is reproduced next.

Lemma 13.1 (Tilting proximal points). *Let f be a proper lsc prox-bounded function and v be a vector. Then for r sufficiently large and any x*

$$\min_y \left\{ f(y) - \langle v,y\rangle + \frac{r}{2}|y-x|^2 \right\} + \langle x,v\rangle + \frac{1}{2r}|v|^2 = e_r f\left(x + \frac{1}{r}v\right), \tag{13.9}$$

$$\operatorname{argmin}_y \left\{ f(y) - \langle v,y\rangle + \frac{r}{2}|y-x|^2 \right\} = \operatorname{prox}_r f\left(x + \frac{1}{r}v\right). \tag{13.10}$$

Proof. Lemma 2.2 of [3] can be rephrased to this form. □

Following the notation of [6], we define the *regular normal cone* to a set S at a point $\bar{x} \in S$ as

$$\widehat{N}_S(\bar{x}) = \{v : \langle v, x-\bar{x}\rangle \le o(|x-\bar{x}|)\}, \tag{13.11}$$

and the *limiting normal cone* as

$$N_S(\bar{x}) = \limsup_{x\to\bar{x}} \widehat{N}_S(x). \tag{13.12}$$

A set is *regular* at $\bar{x}$ if these two cones coincide.

Corresponding to functions we define the *regular subdifferential* of a function f at a point $\bar{x}$ where $f(\bar{x})$ is finite as

$$\widehat{\partial} f(\bar{x}) := \{v \in \mathbf{R}^m : f(x) \geq f(\bar{x}) + \langle v, x-\bar{x}\rangle + o(|x-\bar{x}|)\} \tag{13.13}$$

and the *subdifferential*,

$$\partial f(\bar{x}) := \limsup_{x \to \bar{x},\ f(x) \to f(\bar{x})} \widehat{\partial} f(x). \tag{13.14}$$

A function is *regular* at $\bar{x}$ if its epi-graph is regular at $(\bar{x}, f(\bar{x}))$. If f and g are regular at x, then

$$\partial f(x) = \widehat{\partial} f(x) \text{ and } \partial(f(x)+g(x)) = \partial f(x) + \partial g(x) \tag{13.15}$$

(see [6, Corollary 8.11] and [6, Corollary 10.9] respectively).

Prox-regularity will provide us with a framework for working with nonconvex functions.

Definition 13.2 (prox-regularity). A function f is *prox-regular* at a point $\bar{x}$ for a subgradient $\bar{v} \in \partial f(\bar{x})$ if f is finite at $\bar{x}$, locally lower semi-continuous around $\bar{x}$, and there exists $\rho > 0$ such that

$$f(x') \geq f(x) + \langle v, x'-x\rangle - \frac{\rho}{2}|x'-x|^2, \tag{13.16}$$

whenever x and x' are near $\bar{x}$ with $f(x)$ near $f(\bar{x})$ and $v \in \partial f(x)$ is near $\bar{v}$. Further, f is *prox-regular* at $\bar{x}$ if it is prox-regular at $\bar{x}$ for every $v \in \partial f(\bar{x})$.

It is clear that all convex functions are prox-regular. Moreover, any function f such that $f + \frac{r}{2}|\cdot|^2$ is convex is prox-regular. Prox-regularity further includes the broad class of functions known as *strongly amenable* [5, Definition 2.4 and Proposition 2.5] and *lower-$\mathscr{C}^2$ functions* [5, Example 2.7]. In particular, function that is composed of the maximum of a finite number of smooth functions is prox-regular [5, Example 2.9].

Our framework for active manifolds will be partly smooth functions.

Definition 13.3 (Partly smooth). A function f is *partly smooth* at a point $\bar{x}$ relative to a set $\mathscr{M}$ containing $\bar{x}$ if $\mathscr{M}$ is a $\mathscr{C}^2$ manifold about $\bar{x}$ and:

i. **(Smoothness)** $f|_{\mathscr{M}}$ is a $\mathscr{C}^2$ function near $\bar{x}$;
ii. **(Regularity)** f is regular at all points $x \in \mathscr{M}$ near $\bar{x}$, with $\partial f(x) \neq \emptyset$;
iii. **(Sharpness)** the affine span of $\partial f(\bar{x})$ is a translate of $N_{\mathscr{M}}(\bar{x})$;
iv. **(Sub-continuity)** ∂f restricted to $\mathscr{M}$ is continuous at $\bar{x}$.

Further, a set S is *partly smooth* at a point $\bar{x} \in S$ relative to a manifold $\mathscr{M}$ if its indicator function maintains this property. For both cases we refer to $\mathscr{M}$ as the *active manifold*.

First developed in [4], the idea of partly smooth functions provides a unifying framework for optimization research into functions where the minimum lies upon an active manifold. Most notably, the idea of partly smooth functions captures functions that are composed of the maximum of a finite number of smooth functions, provided a standard constraint qualification holds.

Example 13.4 (Finite max functions). Let

$$g(x) := \max\{g_i(x) : i = 1, 2, \ldots, n\}, \tag{13.17}$$

where g_i are $\mathscr{C}^2$ functions around the point $\bar{x}$. Then g is prox-regular at $\bar{x}$ [5, Example 2.9].

Define the *active set* for g at a point x by

$$A_g(x) := \{i : g_i(x) = g(x)\}. \tag{13.18}$$

If that the set of all active gradients of g, $\{\nabla g_i(\bar{x}) : i \in A_g(\bar{x})\}$, is linearly independent, then [4, Corollary 4.8] shows that g is partly smooth at $\bar{x}$ relative to the manifold

$$\mathscr{M} := \{x : A_g(x) = A_g(\bar{x})\}. \tag{13.19}$$

Relevant to this work, it is easy to confirm that the l_1 norm is partly smooth and prox-regular.

Example 13.5 (l_1 norm). The l_1 norm is convex and therefore prox-regular at any point. Also, the l_1 norm is partly smooth with respect to the manifold

$$\mathscr{M} = \{x : A_1(x) = A_1(\bar{x})\}, \tag{13.20}$$

where A_1 is the active set of the l_1 norm: $A_1(x) := \{i : |x_i| = 0\}$ [4, p. 714].

One strength of partly smooth functions is the ability of algorithms to identify their active manifolds. That is, under some conditions, many algorithms have the property that after a finite number of iterations all future iterates will be contained in the active manifold. The next theorem, reproduced from [2, Theorem 5.3], captures this idea mathematically.

Theorem 13.6 (Identifying active manifold). *Let the function f be prox-regular at $\bar{x}$ and partly smooth there relative to the manifold $\mathscr{M}$ with $0 \in \operatorname{rint} \partial f(\bar{x})$. Suppose $x^k \to \bar{x}$ and $f(x^k) \to f(\bar{x})$. Then*

$$x^k \in \mathscr{M} \text{ for all large } k \tag{13.21}$$

if and only if

$$\operatorname{dist}(0, \partial f(x^k)) \to 0. \tag{13.22}$$

13.3 Active Manifold Identification

13.3.1 Algorithmic Manifold Identification

Recall, this work concerns itself with the optimization problem

$$\min_x \{f(x)+P(x)\}, \tag{13.23}$$

where $f \in \mathscr{C}^2$ and P is prox-regular and partly smooth. We first consider the algorithm proposed in [7], and show that, provided the algorithm converges and the approximate Hessians are bounded, it identifies the active manifold of P in a finite number of iterations. For detailed analysis on when the algorithm converges, see [7]. Note that Assumption 1 of [7] implies the approximate Hessians are bounded.

Theorem 13.7 (Identification via Tseng and Yun's algorithm). *Let $f \in \mathscr{C}^2$ and P be regular. Suppose $\bar{x} \in \operatorname{argmin}\{f(x)+P(x)\}$ and P is prox-regular at $\bar{x}$ and partly smooth there with respect to the manifold $\mathscr{M}$. Suppose iterates x^k are generated by solving the subproblem*

$$x^{k+1} \in \operatorname{argmin}_x\{\langle \nabla f(x^k), x-x^k\rangle + (x-x^k)' H^k (x-x^k) + P(x)\}, \tag{13.24}$$

where H^k is a sequence of positive definite matrices with $\|H^k\|$ bounded. Suppose iterates x^k converge to $\bar{x}$.

If $-\nabla f(\bar{x}) \in \operatorname{rint} \partial P(\bar{x})$, then $x^k \in \mathscr{M}$ for all k sufficiently large.

Proof. Notice that, as $f \in \mathscr{C}^2$, $F(x) = f(x)+P(x)$ is prox-regular at $\bar{x}$ and partly smooth there with respect to $\mathscr{M}$ [4, Corollary 4.6]. Since P is regular at $\bar{x}$ we have that $\partial F(x) = \nabla f(x) + \partial P(x)$, and $0 \in \operatorname{rint} \partial F(\bar{x})$. In order to apply Theorem 13.6 to F we must show $f(x^k)+P(x^k) \to f(\bar{x})+P(\bar{x})$ and $\operatorname{dist}(0, \partial(f(x^k)+P(x^k)) \to 0$ (notice $x^k \to \bar{x}$ by assumption).

To see that $f(x^k)+P(x^k) \to f(\bar{x})+P(\bar{x})$, first notice that for all k

$$f(\bar{x})+P(\bar{x}) = \min_x\{f(x)+P(x)\} \le f(x^k)+P(x^k), \tag{13.25}$$

so

$$f(\bar{x})+P(\bar{x}) \le \liminf_{k\to\infty} f(x^k)+P(x^k). \tag{13.26}$$

Next, notice that, as $x^{k+1} \in \operatorname{argmin}_x\{\langle \nabla f(x^k), x-x^k\rangle + (x-x^k)'H^k(x-x^k)+P(x)\}$, we have

$$\begin{aligned} &\langle \nabla f(x^k), x^{k+1}-x^k\rangle + (x^{k+1}-x^k)'H^k(x^{k+1}-x^k) + P(x^{k+1}) \\ &\le \langle \nabla f(x^k), \bar{x}-x^k\rangle + (\bar{x}-x^k)'H^k(\bar{x}-x^k) + P(\bar{x}). \end{aligned} \tag{13.27}$$

Passing to a limit in k, while noting $x^k \to \bar{x}$ and $\|H^k\|$ bounded, yields

$$\limsup_{k\to\infty} P(x^{k+1}) \le P(\bar{x}). \tag{13.28}$$

As $f \in \mathscr{C}^2$ this implies (with (13.26)) that

$$\limsup_{k\to\infty} f(x^k) + P(x^k) \le f(\bar{x}) + P(\bar{x}) \le \liminf_{k\to\infty} f(x^k) + P(x^k), \tag{13.29}$$

which proves $f(x^k) + P(x^k) \to f(\bar{x}) + P(\bar{x})$.

To see $\mathrm{dist}(0, \partial(f(x^k) + P(x^k)) \to 0$, notice that $x^{k+1} \in \mathrm{argmin}_x\{\langle \nabla f(x^k), x - x^k\rangle + (x - x^k)' H^k (x - x^k) + P(x)\}$ implies

$$\begin{aligned} 0 &\in \partial(\langle \nabla f(x^k), x - x^k\rangle + (x - x^k)' H^k (x - x^k) + P(x))(x^{k+1}) \\ 0 &\in \nabla f(x^k) + H^k(x^{k+1} - x^k) + \partial P(x^{k+1}) \\ -H^k(x^{k+1} - x^k) &\in \nabla f(x^k) + \partial P(x^{k+1}) \end{aligned} \tag{13.30}$$

Therefore,

$$\begin{aligned} \mathrm{dist}(0, \partial(f(x^k) + P(x^k))) &= \mathrm{dist}(0, \nabla f(x^k) + \partial P(x^k)) \\ &\le |-H^k(x^{k+1} - x^k)| \\ &\le \|H^k\| |x^{k+1} - x^k|. \end{aligned} \tag{13.31}$$

Since $\|H^k\|$ is bounded and $|x^{k+1} - x^k| \to 0$, we have $\mathrm{dist}(0, \partial(f(x^k) + P(x^k))) \to 0$.

The result now follows from Theorem 13.6. □

13.3.2 Manifold Identification via a Proximal Subproblem

In the paper [1], it was shown that the active manifold of a constraint set could be identified by inducing a proximal style subproblem. Tseng conjectured that a similar technique might work for problem (13.23). In particular, it was proposed that the subproblem

$$p^k = \mathrm{argmin}_p \left\{ \langle \nabla f(x^k), p\rangle + \frac{r}{2}|x^k - p|^2 + P(p) \right\} \tag{13.32}$$

would identify the active manifold of the function P in a finite number of iterations. We next confirm this conjecture, by proving that if x^k converges to a critical point of problem (13.23) and P is a prox-regular partly smooth function, then excluding a finite number of iterations all points p^k will lie on the active manifold of P.

The proof will hinge on the following lemma. Note that in Lemma 13.8 (as in Theorem 13.7 earlier) we assume that the sequence of points x^k converges to $\bar{x}$. This simply means that the results hold for any convergent algorithm.

Lemma 13.8. *Let $f \in \mathscr{C}^2$ and P be a regular prox-bounded function. Suppose $\bar{x} \in \operatorname{argmin}\{f(x)+P(x)\}$ and P is prox-regular at the point $\bar{x}-\frac{1}{r}\nabla f(\bar{x})$. Suppose the sequence of points x^k converges to $\bar{x}$, and consider a sequence of points*

$$p^k \in \operatorname{argmin}_p \left\{\langle \nabla f(x^k), p\rangle + \frac{r}{2}|p-x^k|^2 + P(p)\right\}. \tag{13.33}$$

If $r>0$ is sufficiently large, then the points p^k satisfy

(i) $p^k \to \bar{x}$
(ii) $P(p^k)+\langle \nabla f(\bar{x}), p^k\rangle \to P(\bar{x})+\langle \nabla f(\bar{x}),\bar{x}\rangle$
(iii) $\operatorname{dist}(0,\partial(P+\langle \nabla f(\bar{x}),\cdot\rangle)(p^k)\to 0$

Proof. To ease discussion, define

$$\tilde{P}(x) = P(x)+\langle \nabla f(\bar{x}), x\rangle. \tag{13.34}$$

Since $f \in \mathscr{C}^2$ and P is regular, $\partial\tilde{P}(\bar{x}) = \nabla f(\bar{x})+\partial P(\bar{x}) = \partial(f+P)(\bar{x})$. In particular, $0 \in \partial\tilde{P}(\bar{x})$, as $\bar{x} \in \operatorname{argmin}\{f(x)+P(x)\}$.

Part i. $p^k \to \bar{x}$

Applying Lemma 13.1 to (13.33) we see that

$$p^k = \operatorname{prox}_r P\left(x^k - \frac{1}{r}\nabla f(x^k)\right). \tag{13.35}$$

Since $x^k \to \bar{x}$ and $f \in \mathscr{C}^2$ we have $x^k - \frac{1}{r}\nabla f(x^k) \to \bar{x}-\frac{1}{r}\nabla f(\bar{x})$. Since P is prox-bounded and prox-regular at $\bar{x}-\frac{1}{r}\nabla f(\bar{x})$, for r sufficiently large the proximal point mapping is single-valued Lipschitz continuous in some neighbourhood of $\bar{x}-\frac{1}{r}\nabla f(\bar{x})$ [3, Theorem 2.4]. Therefore, we have p^k converges to some $\bar{p}$ with

$$\bar{p} = \operatorname{prox}_r P\left(\bar{x} - \frac{1}{r}\nabla f(\bar{x})\right). \tag{13.36}$$

By Lemma 13.1, we see that

$$\begin{aligned}
\operatorname{prox}_r P\left(\bar{x}-\frac{1}{r}\nabla f(\bar{x})\right) &= \operatorname{argmin}\left\{\langle \nabla f(\bar{x}), p\rangle + \frac{r}{2}|p-\bar{x}|^2 + P(p)\right\} \\
&= \operatorname{argmin}\left\{\tilde{P}(p)+\frac{r}{2}|p-\bar{x}|^2\right\} \\
&= \operatorname{prox}_r \tilde{P}(\bar{x}).
\end{aligned} \tag{13.37}$$

Since $0 \in \partial\tilde{P}(\bar{x})$ we have that $\bar{x} \in \operatorname{prox}_r\tilde{P}(\bar{x}) = \bar{p}$. Thus $\bar{p}=\bar{x}$, so $p^k \to \bar{x}$.

Part ii. $\tilde{P}(p^k) \to \tilde{P}(\bar{x})$

Since P is prox-regular at $\bar{x} - \frac{1}{r}\nabla f(\bar{x})$, for r sufficiently large the Moreau envelope is $\mathscr{C}^{1+}$ [3, Theorem 2.4]. In particular, this implies that

$$e_r P\left(x^k - \frac{1}{r}\nabla f(x^k)\right) \to e_r P\left(\bar{x} - \frac{1}{r}\nabla f(\bar{x})\right). \tag{13.38}$$

Applying Lemma 13.1 and noting that the minimum for these proximal envelopes is achieved at p^k and $\bar{x}$ respectively, we see that

$$P(p^k) - \langle \nabla f(x^k), p^k \rangle + \frac{r}{2}|p^k - x^k|^2 + \langle x^k, \nabla f(x^k) \rangle + \frac{1}{2r}|\nabla f(x^k)|^2 \tag{13.39}$$

converges to

$$P(\bar{x}) - \langle \nabla f(\bar{x}), \bar{x} \rangle + \frac{r}{2}|\bar{x} - \bar{x}|^2 + \langle \bar{x}, \nabla f(\bar{x}) \rangle + \frac{1}{2r}|\nabla f(\bar{x})|^2 \tag{13.40}$$

Since $p^k \to \bar{x}$, $x^k \to \bar{x}$, and $f \in \mathscr{C}^2$ this shows that

$$P(p^k) \to P(\bar{x}), \text{ and } \tilde{P}(p^k) \to \tilde{P}(\bar{x}) \tag{13.41}$$

Part iii. $\operatorname{dist}(0, \partial\tilde{P}(p^k)) \to 0$

Since $p^k \in \operatorname{argmin}_p\{\langle \nabla f(x^k), p \rangle + \frac{r}{2}|x^k - p|^2 + P(p)\}$ we have for each k

$$\begin{aligned} 0 &\in \nabla f(x^k) + r(x^k - p^k) + \partial P(p^k) \\ -r(x^k - p^k) + \nabla f(\bar{x}) - \nabla f(x^k) &\in \nabla f(\bar{x}) + \partial P(p^k) \\ -r(x^k - p^k) + \nabla f(\bar{x}) - \nabla f(x^k) &\in \partial\tilde{P}(p^k). \end{aligned} \tag{13.42}$$

Since $p^k \to \bar{x}$, $x^k \to \bar{x}$, and $f \in \mathscr{C}^2$ this yields

$$\operatorname{dist}(0, \partial\tilde{P}(p^k)) \leq r|x^k - p^k| + |\nabla f(\bar{x}) - \nabla f(x^k)| \to 0. \tag{13.43}$$

The proof now following easily. □

Theorem 13.9 (Identification of $\mathscr{M}$ via sub-problem (13.33)). *Let $f \in \mathscr{C}^2$ and P be regular and prox-bounded. Suppose $\bar{x} \in \operatorname{argmin}\{f(x) + P(x)\}$ and P is prox-regular at the point $\bar{x} - \frac{1}{r}\nabla f(\bar{x})$. Suppose f is partly smooth at $\bar{x}$ relative to a manifold $\mathscr{M}$. Suppose the sequence of points x^k converge to $\bar{x}$, and consider the sequence of points p^k generated by sub-problem (13.33).*

If $r > 0$ is sufficiently large and $-\nabla f(\bar{x}) \in \operatorname{rint}\partial P(\bar{x})$, then $p^k \in \mathscr{M}$ for all k sufficiently large.

Proof. Using $\tilde{P} = P(x) + \langle \nabla f(\bar{x}), x \rangle$ as before, we note that $\tilde{P}$ is partly smooth at $\bar{x}$ with respect to the same manifold $\mathscr{M}$ by [4, Corollary 4.6]. Furthermore, $0 \in \operatorname{rint}\partial\tilde{P}(\bar{x})$. Finally, $\tilde{P}$ is prox-regular at the point $\bar{x}$ by [6, Example 13.35]. Lemma 13.8 and Theorem 13.6 now combine to complete the proof. □

13.4 Conclusion

On May 26th, 2009, Paul Tseng sent the following e-mail:

```
Almost forgot..

Instead of projecting onto the feasible set, another
type of active set identification arises in
compressed sensing or, more generally,
    min  f(x) + P(x),
where P(x) = max_i g_i(x), say, and f is smooth.
One can consider an analog of projection, namely,
    min_p  < f'(x), p > + r*|p-x|^2/2  + P(p)
and do active identification accordingly.  When P is
separable, as in the case of l_1-norm, this
decomposes and often has closed form solution. This
approach was used in my paper with Sangwoon Yun on
CGD method in the context of l_1-regularized
optimization, so P(x)=|x|_1. It helped to accelerate
the method on ill-conditioned problems.  This
is much more efficient than rewriting the original
problem as a smooth constrained problem
    min  f(x) + z    s.t.   g_i(x) <= z, i=1,...,m,
and then projecting onto the feasible set.  (In
general, projecting on to the feasible set is
expensive, unless the set has simple structure
like a simplex or a box.)  Results on active set
identification for smooth constrained optimization
should be extendable to this setting.

Paul
```

In this work we show that the algorithm developed by Tseng and Yun in [7] also finitely identifies active manifolds, and furthermore provide an affirmative proof to Tseng's conjecture above.

References

1. Hare, W.L.: A proximal method for identifying active manifolds. Comput. Optim. Appl. **43**, 295–306 (2009)
2. Hare, W.L., Lewis, A.S.: Identifying active constraints via partial smoothness and prox-regularity. J. Convex Anal. **11**, 251–266 (2004)
3. Hare, W.L., Poliquin, R.A.: Prox-regularity and stability of the proximal mapping. J. Convex Anal **14**, 589–606 (2007)
4. Lewis, A.S.: Active sets, nonsmoothness, and sensitivity. SIAM J. Optim. **13**, 702–725 (2003)

5. Poliquin, R.A., Rockafellar, R.T.: Prox-regular functions in variational analysis. Trans. Amer. Math. Soc. **348**, 1805–1838 (1996)
6. Rockafellar, R.T., Wets, R.J-B: Variational Analysis. Springer, Berlin (1998)
7. Tseng, P., Yun, S.: A coordinate gradient descent method for nonsmooth separable minimization. Math. Programming (Ser. B) **117**, 387–423 (2009)

Chapter 14
Approximation Methods for Nonexpansive Type Mappings in Hadamard Manifolds

Genaro López and Victoria Martín-Márquez

Abstract Nonexpansive type mappings defined on Hadamard manifolds and iterative methods for approximating fixed points of these mappings are surveyed. The close relationship with monotone vector fields is pointed out and some numerical examples are included.

Keywords Hadamard manifold · Fixed point · Nonexpansive mapping · Iterative method · Monotone vector field · Resolvent

AMS 2010 Subject Classification: 47H09, 47H14, 65K05, 90C25

14.1 Introduction

Many problems arising in different areas of mathematics, such as optimization and differential equations, can be modeled by the equation

$$x = T(x),$$

where T is a nonlinear operator defined in a metric space. The solutions to this equation are called fixed points of T. If T is a contraction (i.e., there exist $\alpha \in (0,1)$ such that $d(T(x),T(y)) \leq \alpha d(x,y)$, $\forall x,y \in X$) defined on a complete metric space X, the Banach contraction principle establishes that T has a unique fixed point and, for any $x \in X$, the sequence of Picard iterates $\{T^n x\}$ converges to the fixed point of T. However, if the mapping T is a nonexpansive mapping, that is,

$$d(T(x),T(y)) \leq d(x,y),\ \forall x,y \in X,$$

V. Martín-Márquez (✉)
Department of Mathematical Analysis, University of Seville, 41012 Seville, Spain
e-mail: victoriam@us.es

H.H. Bauschke et al. (eds.), *Fixed-Point Algorithms for Inverse Problems in Science and Engineering*, Springer Optimization and Its Applications 49,
DOI 10.1007/978-1-4419-9569-8_14,

then we must assume additional conditions on T and/or the underlying space to ensure the existence of fixed points. Since the sixties, the study of the class of nonexpansive mappings is one of the major and most active research areas of nonlinear analysis. This is due to the connection with the geometry of Banach spaces along with the relevance of these mappings in the theory of monotone and accretive operators.

If we denote by X^* the dual space of a Banach space X, a set-valued operator $A : X \to 2^{X^*}$, with domain $\mathscr{D}(A)$, is said to be monotone if

$$\langle x^* - y^*, x - y \rangle \geq 0, \quad \forall x, y \in \mathscr{D}(A) \text{ and } x^* \in A(x),\ y^* \in A(y).$$

On the other hand, a set-valued operator $A : X \to 2^X$ is said to be *accretive* if

$$\langle x^* - y^*, j(x - y) \rangle \geq 0, \quad \forall\, x, y \in \mathscr{D}(A) \text{ and } x^* \in A(x),\ y^* \in A(y),$$

where $j(x - y) \in J(x - y)$ and J denotes the normalized duality mapping. One of the most relevant facts in the theory of monotone and accretive operators is that the two classes of operators coincide in the setting of Hilbert spaces. The concepts of monotonicity and accretivity have turned out to be very powerful in diverse fields such as operator theory, numerical analysis, convex optimization and partial differential equations; see [1, 2, 15, 56, 72]. For instance, the class of monotone operators is broad enough to cover subdifferentials of convex functions, which are operators of increasing importance in optimization theory. Recall that, given an extended real-valued function $f : X \to \mathbb{R} \cup \{+\infty\}$, the subdifferential of f is the set-valued operator $\partial f : X \to 2^{X^*}$ defined by

$$\partial f(x_0) = \{x^* \in X^* : f(x) \geq f(x_0) + \langle x - x_0, x^* \rangle,\ \forall x \in X\},$$

for any $x_0 \in X$. Suppose that the domain of f, $\mathscr{D}(f) := \{x \in X : f(x) < \infty\}$, has nonempty interior. If f is proper, lower semicontinuous and convex, then $\partial f(x) \neq \emptyset$ for all $x \in \operatorname{Int} \mathscr{D}(f)$ and the subdifferential ∂f is monotone; see [15]. This fact establishes an equivalence between convex minimization problems and the search for zeros of monotone operators. A zero of an operator A is a point $x \in X$ such that $0 \in A(x)$.

In the setting of a Hilbert space the relationship between the theory of monotone operators and the theory of nonexpansive mappings is basically determined by two facts: (1) if T is a nonexpansive mapping then the complementary operator $I - T$ is monotone and (2) the resolvent of a monotone operator A is nonexpansive. Moreover, in both cases the fixed point set of the nonexpansive mapping coincides with the set of zeros of the monotone operator.

The resolvent of a monotone operator in the setting of a Banach space was originally defined by Brezis et al. in [3] though was first implicitly studied by Browder [7]. They set up the fundamental properties of the resolvent, with special emphasis on the strong connection between its fixed points and the zeros of the monotone operator. From this starting point, the study of the asymptotic behavior of the resolvent operator has awakened the interest of many researchers. See, for instance, [12,33,61] and references therein. In the framework of a Hilbert space H, given a monotone operator $A : H \to 2^H$, the resolvent of A of order $\lambda > 0$ is the single-valued mapping $J_\lambda : \mathscr{D}(J_\lambda) \subseteq H \to H$ defined by

$$J_\lambda(x) = (I + \lambda A)^{-1}(x),$$

for any $x \in \mathscr{D}(J_\lambda)$, where $\mathscr{D}(J_\lambda) = \mathscr{R}(I + \lambda A)$ the range of $I + \lambda A$. It is straightforward to check that $A^{-1}(0) = \mathrm{Fix}\,(J_\lambda)$, where $\mathrm{Fix}\,(J_\lambda)$ denotes the fixed point set of J_λ. Moreover, the resolvent is not just nonexpansive but also firmly nonexpansive; that is,

$$\|J_\lambda(x) - J_\lambda(y)\|^2 \leq \langle x - y, J_\lambda(x) - J_\lambda(y)\rangle,$$

for all $x, y \in \mathscr{D}(J_\lambda)$; see, for instance, [12]. Additionally, the resolvent has full domain when A is assumed to be maximal monotone; in other words, when the graph of A is not properly contained in the graph of any other monotone operator. Thus, the problems of existence and approximation of zeros of maximal monotone operators can be formulated as the corresponding problems for fixed points of firmly nonexpansive mappings. It is this approach, applicable to other related problems as well, which renders firmly nonexpansive mappings an important tool in monotone operator and optimization theory.

In the interface between monotone operators and nonexpansive type mappings another class of nonlinear mappings appears, the so-called pseudo-contractive mappings. Recall that a mapping $T : H \to 2^H$ is said to be pseudo-contractive if, for any $r > 0$,

$$\|x - y\| \leq \|(1 + r)(x - y) - r(u - v)\|, \ \forall\, x, y \in H, \ u \in T(x), \ v \in T(y).$$

This concept was introduced independently by Browder and Petryshyn [8] and Kato [32]. They proved that a mapping T is pseudo-contractive if and only if the complementary operator $I - T$ is monotone. This means that the problem of solving an equation for monotone operators may be formulated as a fixed point problem of a pseudo-contractive mapping.

Concerning the fixed point approximation problem, we recall that the sequence of Picard iterates $\{T^n x\}$ converges for contractions on complete metric spaces. However, if T is nonexpansive, even when it has a fixed point, this sequence $\{T^n x\}$ does not converge in general. For this reason, in the last decades, the development of feasible iterative methods for approximating fixed points of a nonexpansive mapping T has been of particular importance. For instance, [11, 13] constitute nice surveys about the asymptotic behavior of nonexpansive mappings in Hilbert and Banach spaces.

This extensive theory dealing with nonexpansive mappings and monotone operators has mainly been developed in the framework of Banach spaces. Out of the setting of linear vector spaces, some concepts and techniques have been extended to other metric spaces. In particular, in the setting of Riemannian manifolds, relevant advances have been made in this direction. The study of optimization methods to solve minimization problems on Riemannian manifolds has been the subject of many works. It has opened a new way to solve nonconvex constrained minimization problems in Euclidean spaces by rewriting them as convex problems on Riemannian manifolds; see [18,19,21,22]. In the study of this problem and other related ones (see [53]) several classes of monotone vector fields have been introduced (see [18,49,51] for single-valued vector fields and [17, 39, 41, 45, 70] for set-valued vector fields) and convergence properties of iterative methods have been presented (see, for instance, [23]).

Riemannian manifolds constitute a broad and fruitful framework for the development of different fields. However, most of the extended methods previously mentioned require the Riemannian manifold to have nonpositive sectional curvature. This is an important property, which is enjoyed by a large class of Riemannian manifolds, and it is strong enough to imply tight topological restrictions and rigidity phenomena (cf. [59]). Particularly, Hadamard manifolds, which are complete simply connected and finite dimensional Riemannian manifolds of nonpositive sectional curvature, have become a suitable setting for diverse disciplines. A Hadamard manifold is an example of hyperbolic space and geodesic space, more precisely, a Busemann nonpositive curvature (NPC) space and a CAT(0) space; see [4, 30, 35].

The aim of this paper is to survey up to day the main results concerning the existence and approximation of fixed points of nonexpansive type mappings as well as the connection with monotone and accretive operators in the setting of Hadamard manifolds.

14.2 Theoretical Framework

The object of this section is to familiarize the reader with the classical language and some fundamental theorems in Hadamard manifolds, needed to understand the content of this paper. For this aim we introduce some concepts and results well-known on Riemannian geometry and, in particular, the objects and facts that characterize the Hadamard manifolds. A complete description of these concepts can be found in any textbook on Riemannian geometry, for instance [20, 66].

14.2.1 Riemannian Manifolds

The Riemannian geometry can be seen as a natural development of the differential geometry of surfaces in $\mathbb{R}^3$. Then, departing from a differentiable manifold M of dimension n, we can introduce a way of measuring the length of tangent vectors by means of an inner product, which leads us to have special curves behaving as if they were "the straight lines" of M.

Definition 14.1. Let $\gamma: (-\varepsilon,\varepsilon) \to M$ be a *smooth curve* in M; that is a function of class $\mathscr{C}^\infty$. Suppose that $\gamma(0) = x \in M$, and let D be the set of functions on M that are differentiable at x. The *tangent vector* to the curve γ at $t = 0$ is a function $\gamma'(0) : D \to \mathbb{R}$ given by

$$\gamma'(0)f = \left.\frac{\mathrm{d}(f\circ\gamma)}{\mathrm{d}t}\right|_{t=0}, \quad f \in D.$$

And we say that a tangent vector at x is a tangent vector at $t = 0$ of some curve $\gamma: (-\varepsilon,\varepsilon) \to M$ with $\gamma(0) = x$. The set of all tangent vectors to M at x, denoted by T_xM, forms a vector space of dimension n called tangent space of M at x. The set $TM = \bigcup_{x\in M} T_xM$ provided with a differentiable structure is a differentiable manifold and will be called the *tangent bundle* of M. A vector field A on M is a mapping of M into the tangent bundle TM, that is, it associates to each point $x \in M$ a vector $A(x) \in T_xM$.

Definition 14.2. A *Riemannian metric* on a differential manifold M is a correspondence which associates to each point $x \in M$ an inner product $\langle\,,\rangle$, (that is, a symmetric bilinear positive-definite form) on the tangent space T_xM, which varies differentiably in the following sense: for any vector fields X and Y, which are differentiable in a neighborhood V of M, the function $\langle X,Y\rangle$ is differentiable on V. A differentiable manifold with a Riemannian metric will be called *Riemannian manifold* and its corresponding norm will be denoted by $\|\cdot\|$.

Definition 14.3. Given a smooth curve $\gamma: [a,b] \to M$ joining x to y, that is, $\gamma(a) = x$ and $\gamma(b) = y$, we can define the *length* of γ by using the metric as

$$l(\gamma) = \int_a^b \|\gamma'(t)\|\mathrm{d}t.$$

Then the *Riemannian distance* $d(x,y)$, which induces the original topology on M, is defined minimizing this length over the set of all such curves joining x to y,

$$d(x,y) := \inf\{l(\gamma) : \gamma \text{ joining } x \text{ to } y\}.$$

Remark 14.4. From now on, M is assumed to be connected so that the set of curves joining x to y is always nonempty.

Definition 14.5. Let ∇ be the *Levi-Civita connection* associated to $(M,\langle\,,\rangle)$ and $\nabla_X Y$ the *covariant derivative* of Y by X (see [66] for more details). Given a smooth curve γ in M a vector field X is said to be *parallel* along γ if $\nabla_{\gamma'} X = 0$. If γ' itself is parallel along γ, we say that γ is a *geodesic*, and in this case $\|\gamma'\|$ is constant. When $\|\gamma'\| = 1$, γ is called *normalized*. A geodesic joining x to y in M is said to be *minimal* if its length equals $d(x,y)$.

Note that given a point $x \in M$ and $u \in T_xM$ there exists a neighborhood U of u in TM such that for any $v \in U$ we have a unique geodesic γ defined on an interval satisfying $\gamma(0) = x$ and $\gamma'(0) = v$. We denote this geodesic, starting at x with velocity v, by $\gamma_v(.,x)$.

Definition 14.6. The *parallel transport* on the tangent bundle TM along γ with respect to ∇ is defined by

$$P_{\gamma,\gamma(b),\gamma(a)}(v) := A(\gamma(b)),\ \forall a,b \in \mathbb{R} \text{ and } v \in T_{\gamma(a)}M,$$

where A is the unique vector field satisfying $\nabla_{\gamma'(t)}A = 0$ for all t and $A(\gamma(a)) = v$.

Remark 14.7. It can be proved that for any $a, b \in \mathbb{R}$, $P_{\gamma,\gamma(b),\gamma(a)}$ is an isometry from $T_{\gamma(a)}M$ to $T_{\gamma(b)}M$. Note that, for any a, b, b_1, $b_2 \in \mathbb{R}$,

$$P_{\gamma,\gamma(b_2),\gamma(b_1)} \circ P_{\gamma,\gamma(b_1),\gamma(a)} = P_{\gamma,\gamma(b_2),\gamma(a)} \quad \text{and} \quad P^{-1}_{\gamma,\gamma(b),\gamma(a)} = P_{\gamma,\gamma(a),\gamma(b)}.$$

For the sake of simplicity, we will write $P_{y,x}$ instead of $P_{\gamma,y,x}$ in the case when γ is a minimal geodesic joining x to y and no confusion arises.

Definition 14.8. A Riemannian manifold M is said to be complete, if for any point $x \in M$, all geodesics emanating from y are defined for all $t \in \mathbb{R}$.

By Hopf–Rinow Theorem we know that if M is a complete Riemannian manifold then any pair of points in M can be joined by a minimal geodesic. Moreover, a complete Riemannian manifold (M,d) is a complete metric space and bounded closed subsets are compact. The concept of completeness allows us to study the global behavior of a Riemannian manifold M by looking at how geodesics run on M.

Definition 14.9. Assuming that M is a complete Riemannian manifold, the *exponential map* at $x \in M$, $\exp_x : T_xM \to M$ is defined by

$$\exp_x v = \gamma_v(1,x),\ v \in T_xM,$$

where we recall that $\gamma_v(.,x)$ is the geodesic starting at x with velocity v. Then, for any value of t, $\exp_x tv = \gamma_v(t,x)$. Note that the map $\exp_x$ is differentiable on T_xM for any $x \in M$.

14.2.2 Hadamard Manifolds

The notion of sectional curvature in a Riemannian manifold plays an important role in the development of geometry. This concept measures in some sense the amount that a Riemannian manifold deviates from being Euclidean. It was introduced by Riemann as a natural generalization of the Gaussian curvature of surfaces. A few years later, an explicit formula was given by Christoffel by using the Levi-Civita connection. We do not include the technical definition of sectional curvature; see references [20, 66] for explicit definitions. However, we are concerned with Riemannian manifolds of nonpositive sectional curvature, whose basic geometrical characterization is gathered in Proposition 14.12.

Definition 14.10. A complete simply connected Riemannian manifold of nonpositive sectional curvature is called a *Hadamard manifold.*

Throughout the remainder of this paper, we will always assume that M is an m-dimensional Hadamard manifold. The following well-known result, essential on Riemannian geometry, can be found, for example, in [66, page 221, Theorem 4.1].

Proposition 14.11. *Let $x \in M$. Then, $\exp_x : T_xM \to M$ is a diffeomorphism, and for any two points $x,y \in M$ there exists a unique normalized geodesic joining x to y, which is a minimal geodesic.*

This proposition says that M is diffeomorphic to the Euclidean space $\mathbb{R}^m$. Thus, M has the same topology and differential structure as $\mathbb{R}^m$. Moreover, Hadamard manifolds and Euclidean spaces have some similar geometrical properties.

One of the most important characterizations of Hadamard manifolds is described in the following proposition, which can be taken from [66, page 223, Proposition 4.5]. Recall that a geodesic triangle $\Delta(x_1,x_2,x_3)$ of a Riemannian manifold is a set consisting of three points x_1, x_2, x_3, and three minimal geodesics joining these points.

Proposition 14.12. *Let $\Delta(x_1x_2x_3)$ be a geodesic triangle in M. Denote, for each $i = 1,2,3$ (mod 3), by $\gamma_i : [0,l_i] \to M$ the geodesic joining x_i to x_{i+1}, and set $l_i := l(\gamma_i)$, $\alpha_i := \angle(\gamma_i'(0), -\gamma_{i-1}'(l_{i-1}))$. Then*

$$\alpha_1 + \alpha_2 + \alpha_3 \le \pi, \tag{14.1}$$

$$l_i^2 + l_{i+1}^2 - 2l_i l_{i+1} \cos\alpha_{i+1} \le l_{i-1}^2. \tag{14.2}$$

In terms of the distance and the exponential map, the inequality (14.2) can be rewritten as

$$d^2(x_i,x_{i+1}) + d^2(x_{i+1},x_{i+2}) - 2\langle \exp_{x_{i+1}}^{-1} x_i, \exp_{x_{i+1}}^{-1} x_{i+2}\rangle \le d^2(x_{i-1},x_i), \tag{14.3}$$

since

$$\langle \exp_{x_{i+1}}^{-1} x_i, \exp_{x_{i+1}}^{-1} x_{i+2}\rangle = d(x_i,x_{i+1})d(x_{i+1},x_{i+2})\cos\alpha_{i+1}.$$

The following relation between geodesic triangles and triangles in $\mathbb{R}^2$ can be found in [4, page 24].

Lemma 14.13. *Let $\Delta(x,y,z)$ be a geodesic triangle in M Hadamard space. Then, there exists $x',y',z' \in \mathbb{R}^2$ such that*

$$d(x,y) = \|x'-y'\|,\ d(y,z) = \|y'-z'\|,\ d(z,x) = \|z'-x'\|.$$

The triangle $\Delta(x',y',z')$ is called the comparison triangle of the geodesic triangle $\Delta(x,y,z)$, which is unique up to isometry of M. The next result taken from [40] shows the relation between a geodesic triangle and its comparison triangle involving angles and distances between points.

Lemma 14.14. *Let $\Delta(x,y,z)$ be a geodesic triangle in a Hadamard space M and $\Delta(x',y',z')$ be its comparison triangle.*

(1) Let α,β,γ (resp. α',β',γ') be the angles of $\Delta(x,y,z)$ (resp. $\Delta(x',y',z')$) at the vertices x,y,z (resp. x',y',z'). Then, the following inequalities hold:

$$\alpha' \geq \alpha,\ \beta' \geq \beta,\ \gamma' \geq \gamma. \tag{14.4}$$

(2) Let r be a point in the geodesic joining x to y and r' its comparison point in the interval $[x',y']$, that is, $d(r,x) = \|r'-x'\|$ and $d(r,y) = \|r'-y'\|$. Then

$$d(z,r) \leq \|z'-r'\|. \tag{14.5}$$

The following lemma is a consequence of the inequality (14.5) and the parallelogram identity in a Euclidean space $\mathbb{R}^n$:

$$\|x-y\|^2 + \|x+y\|^2 = 2(\|x\|^2 + \|y\|^2), \tag{14.6}$$

for all $x,y \in \mathbb{R}^n$.

Lemma 14.15. *For all $x,y,z \in M$ and $m \in M$ with $d(x,m) = d(y,m) = d(x,y)/2$, one has*

$$d^2(z,m) \leq \frac{1}{2}\, d^2(z,x) + \frac{1}{2}\, d^2(z,y) - \frac{1}{4}\, d^2(x,y). \tag{14.7}$$

From the well-known "law of cosines" in $\mathbb{R}^2$ and inequality (14.5), we deduce the following inequality, which is a general characteristic of the spaces with nonpositive curvature (see [4]).

Proposition 14.16. *For any $x,y,z \in M$ the following inequality holds,*

$$\langle \exp_x^{-1} y, \exp_x^{-1} z\rangle + \langle \exp_y^{-1} x, \exp_y^{-1} z\rangle \geq d^2(x,y).$$

Let us introduce now some fundamental notions and results of convex analysis in Hadamard manifolds, as well as other metric properties. Some references on this topic are [66, 68, 69].

Definition 14.17. A subset $C \subseteq M$ is said to be *convex* if for any two points x and y in C, the geodesic joining x to y is contained in C, that is, if $\gamma\colon [a,b] \to M$ is a geodesic such that $x = \gamma(a)$ and $y = \gamma(b)$, then $\gamma((1-t)a+tb) \in C$ for all $t \in [0,1]$.

As in linear metric spaces, we can define a projection map onto closed convex sets.

Definition 14.18. The *projection* onto a set C is the set-valued mapping defined by

$$P_C(x) = \{x^* \in C : d(x,x^*) \leq d(x,y) \text{ for all } y \in C\},\ \forall x \in M.$$

Proposition 14.19. [38,69] *For any point $x \in M$, given a closed convex set $C \subseteq M$, $P_C(x)$ is a singleton and, $z = P_C(x)$ if and only if, for all $y \in C$,*

$$\langle \exp_z^{-1} x,\ \exp_z^{-1} y\rangle \leq 0.$$

From now on, C will denote a nonempty closed convex set in M, unless explicitly stated otherwise. We denote the extended real line by $\overline{\mathbb{R}} := \mathbb{R} \cup \{+\infty\}$.

Definition 14.20. Let $f : M \to \overline{\mathbb{R}}$ be a proper extended real-valued function with domain $\mathscr{D}(f) := \{x \in M :\ f(x) \neq +\infty\}$. The function f is said to be convex if for any geodesic γ in M, the composition function $f \circ \gamma : \mathbb{R} \to \overline{\mathbb{R}}$ is convex, that is,

$$(f \circ \gamma)(ta + (1-t)b) \leq t(f \circ \gamma)(a) + (1-t)(f \circ \gamma)(b)$$

for any $a, b \in \mathbb{R}$ and $0 \leq t \leq 1$.

Definition 14.21. The *subdifferential* of a function $f : M \to \overline{\mathbb{R}}$ at $x \in M$ is the set-valued mapping $\partial f : M \to 2^{TM}$ defined by

$$\partial f(x) = \{u \in T_x M : f(y) \geq f(x) - \langle u, \exp_x^{-1} y\rangle,\ \forall y \in M\},$$

and its elements are called *subgradients*.

The subdifferential $\partial f(x)$ at a point $x \in M$ is a closed convex (possibly empty) set. The existence of subgradients for convex functions is guaranteed by the following proposition taken from [22].

Proposition 14.22. *Let M be a Hadamard manifold and $f : M \to \mathbb{R}$ be convex. Then, for any $x \in M$, the subdifferential $\partial f(x)$ of f at x is nonempty. That is, the domain of the subdifferential $\mathscr{D}(\partial f) = M$.*

The following proposition describes the convexity property of the distance function (cf. [66, page 222, Proposition 4.3]).

Proposition 14.23. *Let $d : M \times M \to \mathbb{R}$ be the distance function. Then $d(\cdot,\cdot)$ is a convex function with respect to the product Riemannian metric, that is, given any pair of geodesics $\gamma_1 : [0,1] \to M$ and $\gamma_2 : [0,1] \to M$, for all $t \in [0,1]$,*

$$d(\gamma_1(t), \gamma_2(t)) \leq (1-t)d(\gamma_1(0), \gamma_2(0)) + td(\gamma_1(1), \gamma_2(1)).$$

In particular, for each $x \in M$, the function $d(\cdot, x) : M \to \mathbb{R}$ is a convex function on M.

Examples of nonconvex problems which can be transformed into convex problems by choosing an appropriate Riemannian metric were given in [19].

Example 14.24. Consider the nonconvex Rosenbrock's banana function $f : \mathbb{R}^2 \to \mathbb{R}$ defined by

$$f(x_1,x_2) = (1-x_1)^2 + 100(x_2 - x_1^2)^2.$$

Endowing $\mathbb{R}^2$ with the Riemannian metric $G : \mathbb{R}^2 \to S^n_{++}$, defined by

$$\begin{pmatrix} 1+4x_1^2 & -2x_1 \\ -2x_1 & 1 \end{pmatrix}$$

we obtain the Riemannian manifold M_G which is complete and of constant curvature 0. Then the function f can be proved to be convex in M_G.

14.2.3 Monotone and Accretive Vector Fields on Hadamard Manifolds

Let $\mathscr{X}(M)$ denote the set of all set-valued vector fields $A : M \to 2^{TM}$ with domain

$$\mathscr{D}(A) = \{x \in M : A(x) \neq \emptyset\}.$$

The concept of monotonicity for single-valued vector fields on Riemannian manifolds was introduced by Németh in [51]. In [23], the gradients of convex functions were proved to be an example of monotone vector fields. Likewise, the complementary vector field of a mapping T was introduced and proved to be monotone when T is nonexpansive in [52]. For more examples and relations between different kinds of generalized monotone vector fields in Riemannian manifolds see [18, 49, 50].

Monotone set-valued vector fields were first studied in [17] where it was shown that the subdifferential operator of a Riemannian convex function is a monotone set-valued vector field. The notion of maximal monotonicity for set-valued vector fields was given in [39]. This and the previous concepts in the setting of Hadamard manifolds are gathered in the following definition, though they could be written in the general framework of Riemannian manifolds in terms of geodesics.

Definition 14.25. A vector field $A \in \mathscr{X}(M)$ is said to be

- *Monotone* if for any $x, y \in \mathscr{D}(A)$,

$$\langle u, \exp_x^{-1} y\rangle \leq \langle v, -\exp_y^{-1} x\rangle, \ \forall u \in A(x) \text{ and } \forall v \in A(y); \qquad (14.8)$$

- *Strictly monotone* if for any $x, y \in \mathscr{D}(A)$ with $x \neq y$, the strict inequality in (14.8) holds;
- *Strongly monotone* if there exists $\rho > 0$ such that, for any $x, y \in \mathscr{D}(A)$,

$$\langle u, \exp_x^{-1} y\rangle - \langle v, -\exp_y^{-1} x\rangle \leq -\rho d^2(x,y), \ \forall u \in A(x) \text{ and } \forall v \in A(y); \qquad (14.9)$$

- *Maximal monotone* if it is monotone and for any $x \in M$ and $u \in T_xM$, this implication holds:

$$\langle u, \exp_x^{-1} y \rangle \leq \langle v, -\exp_y^{-1} x \rangle, \ \forall y \in \mathscr{D}(A) \text{ and } v \in A(y) \Rightarrow u \in A(x). \quad (14.10)$$

To characterize the maximal monotone vector fields, the notion of upper semicontinuity as well as local boundedness for operators in Banach spaces (cf. [67, page 55]) have been extended to the setting of Hadamard manifolds (cf. [39]).

Definition 14.26. Given $A \in \mathscr{X}(M)$ and $x_0 \in \mathscr{D}(A)$, the vector field A is said to be

- *Upper semicontinuous* at x_0 if for any open set V satisfying $A(x_0) \subseteq V \subseteq T_{x_0}M$, there exists an open neighborhood $U(x_0)$ of x_0 such that $P_{x_0,x}A(x) \subseteq V$ for any $x \in U(x_0)$;
- *Locally bounded* at x_0 if there exists an open neighborhood $U(x_0)$ of x_0 such that the set $\cup_{x \in U(x_0)} A(x)$ is bounded;
- Upper semicontinuous (resp. locally bounded) on M if it is upper semicontinuous (resp. locally bounded) at each $x_0 \in \mathscr{D}(A)$.

Recall that the maximal monotonicity and the upper semicontinuity are equivalent for a set-valued operator with closed and convex values in a Hilbert space (cf. [56]). This result was extended, in [39], to set-valued vector fields with full domain on a Hadamard manifold. The key of this fact is that any maximal monotone vector field with full domain can be proved to be locally bounded.

Theorem 14.27. *Suppose that $A \in \mathscr{X}(M)$ is a monotone vector field with $\mathscr{D}(A) = M$. Then the following statements are equivalent.*

(i) A is maximal monotone.
(ii) A is upper semicontinuous on M and $A(x)$ is closed and convex for each $x \in M$.

The classical notion of accretivity on Banach spaces was extended to vector fields on Hadamard manifolds in [70].

Definition 14.28. Given $\alpha > 0$, a vector field $A \in \mathscr{X}(M)$ is said to be

- *Accretive* if for any $x, y \in \mathscr{D}(A)$ and each $r \geq 0$ we have that

$$d(x,y) \leq d(\exp_x(ru), \exp_y(rv)), \text{ for each } u \in A(x) \text{ and } v \in A(y); \quad (14.11)$$

- *α-strongly accretive* if for any $x, y \in \mathscr{D}(A)$ and each $r \geq 0$ we have that

$$(1+\alpha r)d(x,y) \leq d(\exp_x(ru), \exp_y(rv)), \text{ for each } u \in A(x) \text{ and } v \in A(y); \quad (14.12)$$

- *m-accretive* if it is accretive and

$$\bigcup_{x \in \mathscr{D}(A)} \left(\bigcup_{u \in A(x)} \exp_x u \right) = M. \quad (14.13)$$

Note that these definitions make also sense in the setting of Riemannian manifolds (cf. [70]). However, it is in the particular case of a Hadamard manifold where the notions of accretivity and monotonicity can be proved to be equivalent. On the other hand, in [29], Iwamiya and Okochi introduced an alternative definition of monotonicity in terms of the derivative of the distance function between geodesics in a more general Riemannian manifold; this one was proved to coincide with the definition of accretive vector fields on a Hadamard manifold; see [70].

Theorem 14.29. *Let $A \in \mathscr{X}(M)$ and $\alpha > 0$. Then the following assertions hold.*

(i) A is accretive if and only if A is monotone.
(ii) A is α-strongly accretive if and only if A is α-strongly monotone.
(iii) If A is m-accretive, then A is maximal monotone.
(iv) Conversely, if A is maximal monotone and $\mathscr{D}(A) = M$, then A is m-accretive.

Nmeth, in [52], introduced the notion of complementary vector field of a single-valued mapping to provide a relationship between nonexpansive mappings and monotone vector fields. The same concept can be defined in the set-valued case.

Definition 14.30. Let $T : C \subseteq M \to 2^M$. The vector field $A \in \mathscr{X}(M)$ defined by

$$A(x) = -\exp_x^{-1} T(x), \tag{14.14}$$

for any $x \in C$, is said to be the complementary vector field of T.

Theorem 14.31. *[52] Given a nonexpansive mapping $T : C \subseteq M \to M$, its complementary vector field A is monotone.*

The existence of singularities of vector fields is a relevant problem which numerous applications in other areas. In particular, in the setting of Hadamard manifolds, it is crucial for the resolvent operator of a vector field to have good properties.

Definition 14.32. Given $A \in \mathscr{X}(M)$, we say that $x \in \mathscr{D}(A)$ is a *singularity* of A if $0 \in A(x)$. The set of all singularities of A is denoted by $A^{-1}(0)$.

Concerning the existence of singularities for monotone vector fields, as a direct consequence of Definition 14.25, it is first deduced that any strictly monotone vector field A has at most one singularity. In [18, 23], it was proved that differentiable strongly monotone single-valued vector fields on Hadamard manifolds with $\mathscr{D}(A) = M$ have at least one singularity; thus, since the strong monotonicity implies the strictly monotonicity, existence and uniqueness are ensured. This result was improved and extended to the set-valued case for maximal strongly monotone vector fields, by using the equivalence established in Theorem 14.27; see [39].

Theorem 14.33. *Let $A \in \mathscr{X}(M)$ be a maximal strongly monotone vector field with $\mathscr{D}(A) = M$. Then there exists a unique singularity of A.*

The notion of resolvent in the setting of a Hadamard manifold was introduced in [41].

Definition 14.34. Given $\lambda > 0$, the *resolvent* of $A \in \mathscr{X}(M)$ of order λ is the set-valued mapping $J_\lambda : M \to 2^M$ defined by

$$J_\lambda(x) = \{z \in M |\ x \in \exp_z \lambda Az\}, \tag{14.15}$$

for any $x \in M$.

Remark 14.35. For any $\lambda > 0$, by definition of resolvent of a vector field the following assertions hold.

(a) The range of the resolvent J_λ is contained in the domain of A and

$$\mathrm{Fix}\ (J_\lambda) = A^{-1}(0). \tag{14.16}$$

(b) The domains of the resolvent J_λ is the range of the vector field defined by $x \mapsto \exp_x \lambda Ax$. We will denote this range as $\mathscr{R}(\exp_{.} \lambda A(\cdot))$. Then we have that

$$\mathscr{D}(J_\lambda) = \mathscr{R}(\exp_{.} \lambda A(\cdot)).$$

Out of linear spaces, the resolvent had been implicitly defined in the setting of differential manifolds, in particular, in Finsler manifolds by Hoyos [28] and in Hilbert manifolds by Iwamiya and Okochi [29]. As a matter of fact, these two definitions can be proved to coincide with the corresponding concept defined on Hadamard manifolds. However, it turns out that in the former settings, where the resolvent is defined is still unknown, whereas under certain monotonicity conditions it was proved in [41] that the resolvent has full domain in a Hadamard manifold.

Using a parallel approach to convex problems, nonmonotone problems can be transformed into monotone ones by endowing the space with a suitable Riemannian metric. See [19] for examples.

Example 14.36. Consider the vector field $A : \mathbb{R}^2 \to \mathbb{R}^2$ defined by

$$A(x_1,x_2) = (-x_1^2 + x_1 + x_2, -2x_1^3 + 2x_1^2 + 2x_1x_2 - x_1).$$

It turns out that A is not monotone in $\mathbb{R}^2$ but it is monotone in the Riemannian manifold M_G defined in *Example 14.24*.

14.3 Nonexpansive Type Mappings

This section is devoted to the properties of nonexpansive type mappings defined on a Hadamard manifold. We start by presenting the definitions of firmly nonexpansive and pseudo-contractive mappings on Hadamard manifolds [41].

Definition 14.37. Given a subset $C \subseteq M$, the mapping $T : C \to M$ is said to be

- *Nonexpansive* if for any $x, y \in C$,

$$d(T(x), T(y)) \leq d(x, y); \tag{14.17}$$

- *Firmly nonexpansive* if for any $x, y \in C$, the function $\theta : [0,1] \to [0,\infty]$ defined by

$$\theta(t) = d(\gamma_1(t), \gamma_2(t)), \tag{14.18}$$

  is nonincreasing, where γ_1 and γ_2 denote the geodesics joining x to $T(x)$ and y to $T(y)$, respectively;
- *Pseudo-contractive* if its complementary vector field is accretive; that is, given $x, y \in C$ and $r \geq 0$ we have that

$$d(x,y) \leq d(\exp_x(-r\exp_x^{-1} T(x)), \exp_y(-r\exp_y^{-1} T(y))); \tag{14.19}$$

- *α-strongly pseudo-contractive* if its complementary vector field is α-strongly accretive.

Remark 14.38. It is clear from definition that any firmly nonexpansive mapping is nonexpansive and any strongly pseudo-contractive mapping is pseudo-contractive.

Let us denote the fixed point set of T by

$$\mathrm{Fix}\,(T) := \{x \in C \,|\, x = T(x)\}. \tag{14.20}$$

From either Brouwer's theorem or the fixed point property for CAT(0) spaces (cf. [35]), the existence of fixed points of a nonexpansive mapping T is ensured provided that C is bounded. Kirk, in [36], proved the following result in the more general setting of complete CAT(0) spaces. A simpler proof can be found in [41].

Proposition 14.39. *Let $T : C \to M$ be a nonexpansive mapping defined on a closed convex set $C \subseteq M$. Then the fixed point set Fix(T) is closed and convex.*

The notion of firm nonexpansivity was previously defined on a Banach space [9, 10] and the Hilbert ball with the hyperbolic metric [26], so-called firmly nonexpansive mapping of the first kind in the latter case. In fact, the following result (cf. [41]) shows that in the framework of Hadamard manifolds this class of mappings satisfies similar properties to those ones defined on Hilbert spaces.

Proposition 14.40. *Let $T : C \subseteq M \to M$. Then the following assertions are equivalent.*

(i) T is firmly nonexpansive.
(ii) For any $x, y \in C$ and $t \in [0,1]$

$$d(T(x), T(y)) \leq d(\exp_x t\exp_x^{-1} T(x), \exp_y t\exp_y^{-1} T(y)). \tag{14.21}$$

(iii) For any $x, y \in C$

$$\left\langle \exp_{T(x)}^{-1} T(y), \exp_{T(x)}^{-1} x \right\rangle + \left\langle \exp_{T(y)}^{-1} T(x), \exp_{T(y)}^{-1} y \right\rangle \leq 0. \tag{14.22}$$

This result together with Proposition 14.19 implies that an example of firmly nonexpansive mapping is the metric projection onto a closed convex set.

The resolvent operator from Definition 14.34 establishes a strong relationship between monotone vector fields and nonexpansive mappings, in particular, firmly nonexpansive mappings, as it was stated in the following theorem; see [41].

Theorem 14.41. *Let* $A \in \mathscr{X}(M)$*. Then, for any* $\lambda > 0$,

(i) A is monotone if and only if J_λ *is single-valued and firmly nonexpansive;*
(ii) If $\mathscr{D}(A) = M$*, A is maximal monotone if and only if* J_λ *is single-valued, firmly nonexpansive and* $\mathscr{D}(J_\lambda) = M$.

Remark 14.42. Note that the previous theorem shows indeed that, for each $\lambda > 0$, any firmly nonexpansive T with full domain $\mathscr{D}(T) = M$ is the resolvent $T = J_\lambda$ of a maximal monotone vector field A.

From Theorem 14.41 and Remark 14.35, the following result which constitutes a counterpart to Minty's theorem [47] in the setting of Hadamard manifolds is deduced. Note that in this case the monotone operator is required to have full domain while in the original theorem this requirement is not needed.

Corollary 14.43. *Let* $A \in \mathscr{X}(M)$ *be monotone such that* $\mathscr{D}(A) = M$*, and let* $\lambda > 0$*. Then A is maximal monotone if and only if* $\mathscr{R}(\exp_{\cdot} \lambda A(\cdot)) = M$.

As a byproduct of Theorem 14.41 and Proposition 14.39, it follows the following result about the structure of the set of singularities of a maximal monotone vector field. A similar result was proved in [23] under the assumption that A is smooth.

Corollary 14.44. *Let* $A \in \mathscr{X}(M)$ *be monotone with closed convex domain* $\mathscr{D}(A)$ *such that* $\mathscr{D}(A) \subseteq \mathscr{D}(J_\lambda)$*. Then* $A^{-1}(0)$ *is closed and convex.*

The concept of pseudo-contractive mappings in the setting of Hadamard manifolds is defined by using the notion of complementary vector field in Definition 14.30. This definition coincides with the one introduced by Reich and Shafrir in the more general setting of hyperbolic spaces [63]. In view of Theorem 14.29, the definition of pseudo-contractive mappings can be given in terms of monotonicity.

Corollary 14.45. *Let* $T : C \subseteq M \to M$ *and* $\alpha > 0$*. Then the following assertions hold.*

(i) T is pseudo-contractive if and only if its complementary vector field is monotone.
(ii) If T is α*-strongly pseudo-contractive, then its complementary vector field is* α*-strongly monotone.*

(iii) Conversely, if the complementary vector field of T is α-strongly monotone, then T is α'-strongly pseudo-contractive, where $0 < \alpha' < \alpha$.

Remark 14.46. If T is a nonexpansive mapping, by Theorem 14.31, the complementary vector field of T is monotone. Hence, by Corollary 14.45, we deduce that any nonexpansive mapping is pseudo-contractive.

The following result about the existence of fixed points of continuous pseudo-contractive mappings on Hadamard manifolds (cf. [41]) is the counterpart of Theorem 1 in [37] proved by Kirk and Schöneberg in the setting of Hilbert spaces.

Corollary 14.47. *Let $T : M \to M$ be a continuous pseudo-contractive mapping. Let $x_0 \in M$ and $\varepsilon > 0$ such that*

$$d(x_0, T(x_0)) < d(x, T(x)), \tag{14.23}$$

for any $x \in \partial B(x_0, \varepsilon)$. Then there exists a fixed point of T in $B(x_0, \varepsilon)$.

14.4 Iterative Algorithms for Nonexpansive Type Mappings

The study of the asymptotic behavior of nonexpansive type mappings is one of the most active research areas of nonlinear analysis. Most of the investigations in this direction have focused on the case when T is a self-mapping defined on a closed convex subset C of a normed linear space. Besides Picard iteration $\{T^n(x)\}$, which converges for any initial point x when T is either a contraction or firmly nonexpansive, basically two types of algorithms has been considered: Mann and Halpern algorithms. Because of the convex structure of both iterative methods, few results have been obtained out of the setting of linear spaces. Our objective in this section is to present the convergence results of different iterative methods for nonexpansive type mappings defined on Hadamard manifolds.

14.4.1 Picard Iteration for Firmly Nonexpansive Mappings

As it happens in Banach spaces and the Hilbert ball [26, 62], the class of firmly nonexpansive mappings is characterized by the good asymptotic behavior of the sequence of Picard iterates $\{T^n x\}$ stated in the following theorem; see [16].

Theorem 14.48. *Let $C \subseteq M$ be a closed convex set and $T : C \to C$ be a firmly nonexpansive mapping such that its fixed point set $Fix(T) \neq \emptyset$. Then for each $x \in C$, the sequence of iterates $\{T^n(x)\}$ converges to a fixed point of T.*

In the case when the mapping T is just nonexpansive, we know that in general Picard iteration $\{T^n(x)\}$ does not converge. However, as it happens in Hilbert spaces, there exists an associated family of mappings $\{G_t : 0 \leq t < 1\}$, whose fixed point set coincides with the fixed point set of T.

Indeed, given $T : C \to C$ nonexpansive and $x \in C$, for any $t \in [0,1)$, let T_t be the mapping defined by

$$T_t(y) = \exp_x t \exp_x^{-1} T(y), \tag{14.24}$$

for any $y \in C$. Thus T_t is a contraction for any $t \in [0,1)$ and the Banach contraction principle implies that there exists a unique fixed point of T_t, which is being denoted by x_t. By means of the approximating curve $\{x_t\}$, for any $t \in [0,1)$, we define the mapping $G_t : C \to C$ by

$$G_t(x) =: x_t = \exp_x t \exp_x^{-1} T(G_t(x)), \tag{14.25}$$

for all $x \in C$. Then the following result holds (cf. [41]).

Proposition 14.49. *For any $t \in [0,1)$, the following statements hold.*

(i) The mapping G_t is firmly nonexpansive.
(ii) $Fix(G_t) = Fix(T)$.

Therefore, we can use the family $\{G_t\}$ for approximating a fixed point of T, considering the sequence defined by Picard iteration $x_{n+1} = G_t(x_n)$, for any $t \in [0,1)$.

The convergence of Picard iteration also lets us approximate a singularity of a maximal monotone vector field A with full domain. Indeed, since Theorem 14.41 says that the resolvent J_λ of A is firmly nonexpansive and has full domain, it follows from Theorem 14.48 that, for any $x \in M$ the sequence $\{(J_\lambda)^n(x)\}$ converges to a fixed point of J_λ, that is a singularity of A. This algorithm is actually a particular case of the proximal point algorithm defined as follows. Given $x_0 \in \mathscr{D}(A)$ and $\{\lambda_n\} \subset \mathbb{R}^+$ it generates a sequence $\{x_n\}$ by means of the recursive formula

$$0 \in A(x_{n+1}) - \lambda_n \exp_{x_{n+1}}^{-1} x_n. \tag{14.26}$$

Actually, this algorithm, which constitutes an extension of the one studied by Rockafellar in the setting of Hilbert spaces [64], was first introduced in this framework in [19] for single-valued differentiable monotone vector fields. The following result shows the convergence for set-valued maximal monotone vector fields (cf. [39]).

Theorem 14.50. *Let $A \in \mathscr{X}(M)$ be maximal monotone such that $A^{-1}(0) \neq \emptyset$ and $\mathscr{D}(A) = M$. Let $\{\lambda_n\} \subset \mathbb{R}^+$ satisfy $\sup\{\lambda_n : n \geq 0\} < \infty$. Then, for any $x_0 \in M$, the sequence $\{x_n\}$ generated by algorithm (14.26) is well-defined and converges to a singularity of A.*

It is worth mentioning that the proximal point algorithm for convex functions in Hadamard manifolds was previously studied; see, for instance, [22, 54, 55].

14.4.2 Mann Algorithm for Nonexpansive Mappings

Given a nonexpansive mapping T defined on a Banach space X, Mann iteration is the averaged algorithm defined by the recursive scheme

$$x_{n+1} = \alpha_n x_n + (1-\alpha_n)T(x_n),\ n \geq 0, \tag{14.27}$$

where x_0 is an arbitrary point in the domain of T and $\{\alpha_n\}$ is a sequence in $[0,1]$ (cf. [44]). One of the classical results, due to Reich [60], states that if the underlying space is uniformly convex and has a Fréchet differentiable norm, T has fixed points and $\sum_n \alpha_n(1-\alpha_n) = \infty$, then the sequence $\{x_n\}$ defined by Mann algorithm converges weakly to a fixed point of T. Moreover, a counterexample provided by Genel and Lindenstrauss ([24]) shows that in infinite-dimensional spaces Mann iteration does not have strong convergence in general.

Mann iteration (14.27) and some of the convergence results known in Banach spaces have been studied in the more general framework of metric spaces by Goebel–Kirk [25,34] and Reich–Shafrir [63]. They provided an iterative method for finding fixed points of nonexpansive mappings on spaces of *hyperbolic type* which includes Hadamard manifolds as a particular case. The algorithm is defined by

$$x_{n+1} \in [x_n, T(x_n)] \quad \text{such that} \quad d(x_n, T(x_n)) = (1-\alpha_n)d(x_n, x_{n+1}), \tag{14.28}$$

where $[x_n, T(x_n)]$ denotes the metric segment joining x_n to $T(x_n)$. More precisely, under the assumption that $\{\alpha_n\}$ is bounded away from 0 and 1, Reich and Shafrir proved the convergence of this iteration to a fixed point of T defined on the Hilbert ball with the hyperbolic metric.

Motivated by these results, in [40], Mann iteration (14.28) was introduced in Hadamard manifolds by means of the recursive formula

$$x_{n+1} = \exp_{x_n}(1-\alpha_n)\exp_{x_n}^{-1} T(x_n),\ \forall n \geq 0; \tag{14.29}$$

or equivalently,

$$x_{n+1} = \gamma_n(1-\alpha_n),\ \forall n \geq 0,$$

where γ_n is the geodesic joining x_n to $T(x_n)$. Then the sequence $\{x_n\}$ generated by Mann algorithm (14.29) was proved to converge to a fixed point of T when $\{\alpha_n\}$ satisfies the condition:

$$\sum_{n=0}^{\infty} \alpha_n(1-\alpha_n) = \infty. \tag{14.30}$$

Theorem 14.51. *Let $C \subseteq M$ be a closed convex set and $T: C \to C$ be a nonexpansive mapping with $Fix(T) \neq \emptyset$. Suppose that $\{\alpha_n\} \subset (0,1)$ satisfy condition (14.30). Then, for any $x_0 \in C$, the sequence $\{x_n\}$ generated by algorithm (14.29) converges to a fixed point of T.*

14.4.3 Halpern Algorithm for Nonexpansive Mappings

Halpern iteration [27] is generated in the setting of Banach spaces by the recursive formula

$$x_{n+1} = \alpha_n x + (1-\alpha_n)Tx_n,\ n \geq 0, \tag{14.31}$$

where x_0 and x are arbitrary points in the domain of a nonexpansive mapping T, and $\{\alpha_n\}$ is a sequence in $[0,1]$. Unlike Mann iteration, Halpern algorithm can be proved to have strong convergence provided that the underlying space is smooth enough and the sequence $\{\alpha_n\}$ satisfies good conditions; see [13,42] and references therein.

To solve the problem of finding a fixed point of T out of the setting of linear spaces, Kirk, in [35], provided an implicit algorithm for approximating fixed points of nonexpansive mappings. More precisely, he studied such an algorithm in a complete CAT(0) space though the following convergence result is formulated for the special case of a Hadamard manifold.

Theorem 14.52. *Suppose that $C \subseteq M$ is bounded besides closed and convex. Let $T : C \to C$ be nonexpansive, $x \in C$, and for each $t \in [0,1)$, let x_t be the unique point such that*

$$x_t = \exp_x (1-t) \exp_x^{-1} T(x_t)$$

(which exists by Banach's contraction Theorem). Then $\lim_{t\to 0} x_t = \bar{x}$, the unique nearest point to x in $Fix(T)$.

Remark 14.53. Note that this implicit algorithm is actually the approximation curves $\{x_t\} = \{G_t(x)\}$ defined in (14.25).

In an Euclidean space $\mathbb{R}^n$, this iteration scheme turns into the implicit Browder iteration (cf. [5,6])

$$x_t = tx + (1-t)T(x_t); \tag{14.32}$$

that is, x_t is the unique fixed point of the contraction $tx + (1-t)T$, for any $t \in [0,1)$. The discretization of this implicit algorithm leads to the explicit Halpern iteration (14.31). An analogue of algorithm (14.31) to approximate fixed points for nonexpansive mappings on Hadamard manifolds was studied in [40]. Let x_0, $x \in M$ and let $\{\alpha_n\} \subset [0,1]$. Consider the iteration scheme

$$x_{n+1} = \exp_x (1-\alpha_n) \exp_x^{-1} T(x_n),\ \forall n \geq 0; \tag{14.33}$$

or equivalently,

$$x_{n+1} = \gamma_n(1-\alpha_n),\ \forall n \geq 0,$$

where γ_n is the geodesic joining x to $T(x_n)$. This algorithm indeed coincides with Halpern algorithm in the particular case of an Euclidean space, and its convergence can be proved under the same conditions on the sequence $\{\alpha_n\}$:

(H1) $\lim_{n\to\infty} \alpha_n = 0$;
(H2) $\sum_{n\geq 0} \alpha_n = \infty$;

(H3) $\sum_{n\geq 0}|\alpha_{n+1}-\alpha_n|<\infty$;
(H4) $\lim_{n\to\infty}(\alpha_n-\alpha_{n-1})/\alpha_n=0$.

Theorem 14.54. *Let $C\subseteq M$ be a closed convex set, $T:C\to C$ be nonexpansive with $Fix(T)\neq\emptyset$ and $x,x_0\in C$. Suppose that $\{\alpha_n\}\in[0,1]$ satisfies (H1), (H2) and, (H3) or (H4). Then the sequence $\{x_n\}$ generated by algorithm (14.33) converges to $P_{Fix(T)}(x)$.*

The convergence of Halpern iteration in CAT(0) spaces was studied in [65] and in more general CAT(K) spaces in [58].

A numerical implementation for analyzing the behavior of Halpern as well as Mann iteration is presented in Sect. 14.5.

14.4.4 Viscosity Approximation Method for Nonexpansive Mappings

Let X be a Banach space and $C\subseteq X$ be a closed convex set. Given a nonexpansive mapping $T:C\subseteq X\to C$, a real number $t\in(0,1]$ and a contraction ψ on C, define the contraction $T_t:C\to C$ by

$$T_tx=t\psi(x)+(1-t)T(x),\ x\in C.$$

Hence, T_t has a unique fixed point which is denoted by x_t; that is, x_t is the unique solution to the fixed point equation

$$x_t=t\psi(x_t)+(1-t)T(x_t),\quad t\in(0,1]. \tag{14.34}$$

The explicit iterative discretization of (14.34) is

$$x_{n+1}=\alpha_n\psi(x_n)+(1-\alpha_n)T(x_n),\quad n\geq 0, \tag{14.35}$$

where $\{\alpha_n\}\subset[0,1]$. Note that these two iterative processes (14.34) and (14.35) have Browder and Halpern iterations as special cases by taking $\psi(y)=x\in C$ for any $y\in C$.

The viscosity approximation method of selecting a particular fixed point of a given nonexpansive mapping was proposed by Moudafi [48] in the framework of a Hilbert space. The interest in the convergence of the implicit (14.34) and explicit (14.35) algorithms is based on the fact that under suitable conditions these iterations converge strongly to the unique solution $q\in\text{Fix}(T)$ of a variational inequality which, in the case of a Hilbert space, is the following:

$$\langle (I-\psi)q,x-q\rangle\geq 0,\ \forall x\in\text{Fix}(T). \tag{14.36}$$

This fact allows us to apply this method to convex optimization, linear programming and monotone inclusions. See [43,71] and references therein for convergence results regarding viscosity approximation methods in Banach spaces.

The convergence of a viscosity method for nonexpansive mappings in the setting of a Hadamard manifold was established in [46]. For any $x_0 \in M$ and $\{\alpha_n\} \subset [0,1]$, consider the iteration scheme

$$x_{n+1} = \exp_{\psi(x_n)}\left((1-\alpha_n)\exp^{-1}_{\psi(x_n)} T(x_n)\right),\ \forall n \geq 0; \tag{14.37}$$

or equivalently,

$$x_{n+1} = \gamma_n(1-\alpha_n),\ \forall n \geq 0,$$

where γ_n is the geodesic joining $\psi(x_n)$ to $T(x_n)$. Consider hypothesis (H1)–(H4) on $\{\alpha_n\}$ used in Theorem 14.54.

Theorem 14.55. *Let $C \subseteq M$ be a closed convex set, $T : C \to C$ be nonexpansive with $Fix(T) \neq \emptyset$ and $\psi : C \to C$ a contraction. Suppose that $\{\alpha_n\} \subset [0,1]$ satisfies (H1), (H2) and, (H3) or (H4). Then the sequence $\{x_n\}$ generated by algorithm (14.37) converges to $\bar{x} \in C$, the unique fixed point of the contraction $P_{Fix(T)}\psi$.*

Moreover, the convergence point $\bar{x}$ is the unique solution of the variational inequality

$$\langle \exp_{\bar{x}}^{-1} \psi(\bar{x}), \exp_{\bar{x}}^{-1} x\rangle \leq 0,\ \forall x \in Fix(T). \tag{14.38}$$

14.4.5 Iterative Algorithm for Pseudo-Contractive Mappings

In the setting of Banach spaces, iterative methods to approximate fixed points of strongly pseudo-contractive mappings or, equivalently, singularities of strongly monotone vector fields, have been studied by many authors; see, for instance, [14, 31]. In [70], a section is devoted to define and study the convergence of an iterative scheme for strongly monotone vector fields, which is an extension to Riemannian manifolds of the one studied by Chidume (cf. [14]) in Banach spaces.

For the following theorem, it is necessary to extend the notion of L-Lipschitz continuity to single-valued vector fields in the setting of a Hadamard manifold; see [70].

Definition 14.56. Given $L > 0$, a single-valued vector field $A \in \mathscr{X}(M)$ is said to be *L-Lipschitz continuous* if

$$\|P_{y,x}A(x) - A(y)\| \leq Ld(x,y),$$

for any $x, y \in \mathscr{D}(A)$.

Theorem 14.57. *Let $A \in \mathscr{X}(M)$ be single-valued, L-Lipschitz continuous and α-strongly monotone with $\mathscr{D}(A) = M$. Given $x_0 \in M$, let $\{x_n\}$ be the sequence defined by the algorithm*

$$x_{n+1} = \exp_{x_n}(-rA(x_n)),\ \forall n \geq 0, \tag{14.39}$$

where $0 < r < \frac{2\alpha}{L^2}$. Then $\{x_n\}$ converges to the unique singularity of A.

Note that, under the hypotheses in the previous theorem, since $A \in \mathscr{X}(M)$ is single-valued and L-Lipschitz continuous, A is USC and therefore by Theorem 14.27 it is maximal monotone. Then, thanks to the strong monotonicity, Theorem 14.33 guarantees the existence and uniqueness of singularity.

Recall that a (strongly) pseudo-contractive mapping T is defined as a mapping whose complementary vector field A is (strongly) monotone. On the other hand, the fixed point set of T coincides with the set of singularities of A. Then, as a consequence of Theorem 14.57, we get the following theorem on the existence and approximation of a fixed point of a pseudo-contractive mapping.

Theorem 14.58. *Let $T : M \to M$ be a single-valued α-strongly pseudo-contractive mapping such that its complementary vector field is L-Lipschitz continuous. Then there exists a unique fixed point of T.*

Moreover, the sequence $\{x_n\}$ defined by the algorithm

$$x_{n+1} = \exp_{x_n}(r(\exp_{x_n}^{-1} T(x_n))), \ \forall n \geq 0, \tag{14.40}$$

where $0 < r \leq \frac{2\alpha}{L^2}$, converges to the fixed point of T.

Remark 14.59. Given $T : M \to M$ single-valued and α-strongly pseudo-contractive with L-Lipschitz continuous complementary vector field, in the case when $2\alpha > L^2$, by considering the constant $r = 1$, Theorem 14.58 implies the convergence of Picard iteration $\{T^n(x)\}$, for any initial point $x \in M$.

There exist other iterative methods for approximating singularities of monotone vector fields which can be applied to approximate fixed points of pseudo-contractive mappings. For instance, the proximal point algorithm (14.26), which converges for a maximal monotone vector field A with full domain, constitutes an approach for single-valued pseudo-contractive mappings with L-Lipschitz continuous complementary vector fields.

14.5 Numerical Example

To illustrate the application of these methods, in particular, Mann and Halpern iterations, the following numerical example was provided in [40].

Let $\mathbb{E}^{m,1}$ be the vector space $\mathbb{R}^{m+1}$ endowed with the symmetric bilinear form defined by

$$\langle x, y \rangle = \sum_{i=1}^{m} x_i y_i - x_{m+1} y_{m+1}, \ \forall x = (x_i), \ y = (y_i) \in \mathbb{R}^{m+1}.$$

This bilinear form is called the Lorentz metric. The *hyperbolic m-space* $\mathbb{H}^m$ is defined by

$$\{x = (x_1, \ldots, x_{m+1}) \in \mathbb{E}^{m,1} : \langle x, x \rangle = -1, x_{m+1} > 0\};$$

this is the upper sheet of the hyperboloid $\{x \in \mathbb{E}^{m,1} : \langle x, x \rangle = -1\}$. Note that $x_{m+1} \geq 1$ for any $x \in \mathbb{H}^m$, with equality if and only if $x_i = 0$ for all $i = 1, ..., m$. The metric of $\mathbb{H}^m$ is induced from the Lorentz metric $\langle \cdot, \cdot \rangle$ and it will be denoted by the same symbol. Then $\mathbb{H}^m$ is a Hadamard manifold with sectional curvature -1 (cf. [4, 23]). By using the corresponding expressions of the exponential map and its inverse, Mann and Halpern algorithms can be formulated in a simple way in the hyperbolic space $\mathbb{H}^m$. Given $y, z \in M$, set

$$r(y,z) = \operatorname{arccosh}(-\langle y, T(z) \rangle) \text{ and } V(y,z) = \frac{T(z) + \langle y, T(z) \rangle y}{\sqrt{\langle y, T(z) \rangle^2 - 1}}.$$

Then Mann algorithm (14.29) has the form

$$x_{n+1} = \cosh((1-\alpha_n) r(x_n, x_n)) x_n + \sinh((1-\alpha_n) r(x_n, x_n)) V(x_n, x_n), \ \forall n \geq 0;$$

while Halpern algorithm (14.33) has the form

$$x_{n+1} = \cosh((1-\alpha_n) r(x, x_n)) x + \sinh((1-\alpha_n) r(x, x_n)) V(x, x_n), \ \forall n \geq 0.$$

We present an example in $\mathbb{H}^3$, where these methods are implemented for some specific data.

Example 14.60. Let $M = \mathbb{H}^3$ and $T_1, T_2 : M \to M$ be the nonexpansive mappings defined by

$$T_1(x) = (-x_1, -x_2, -x_3, x_4) \text{ and } T_2(x) = (-x_1, x_2, x_3, x_4),$$

for any $x = (x_1, x_2, x_3, x_4) \in \mathbb{H}^3$. Then $\operatorname{Fix}(T_1) = \{(0,0,0,1)\}$ and

$$\operatorname{Fix}(T_2) = \{(x_1, x_2, x_3, x_4) \in \mathbb{H}^3 : x_1 = 0, x_2^2 + x_3^2 = x_4^2 - 1\}.$$

For both algorithms, we are going to consider the sequence of parameters

$$\alpha_n = \frac{1}{n+3}, \ \forall n \geq 0,$$

and the point

$$u = (0.6037924791938, 0.2721879249700, 0.1988142677611, 1.2158037413562)$$

for Halpern iteration. As initial point let us take

$$x_0^1 = (0.6944544097848, 1.0138260928014, 0.9936087133075, 1.8701252762515).$$

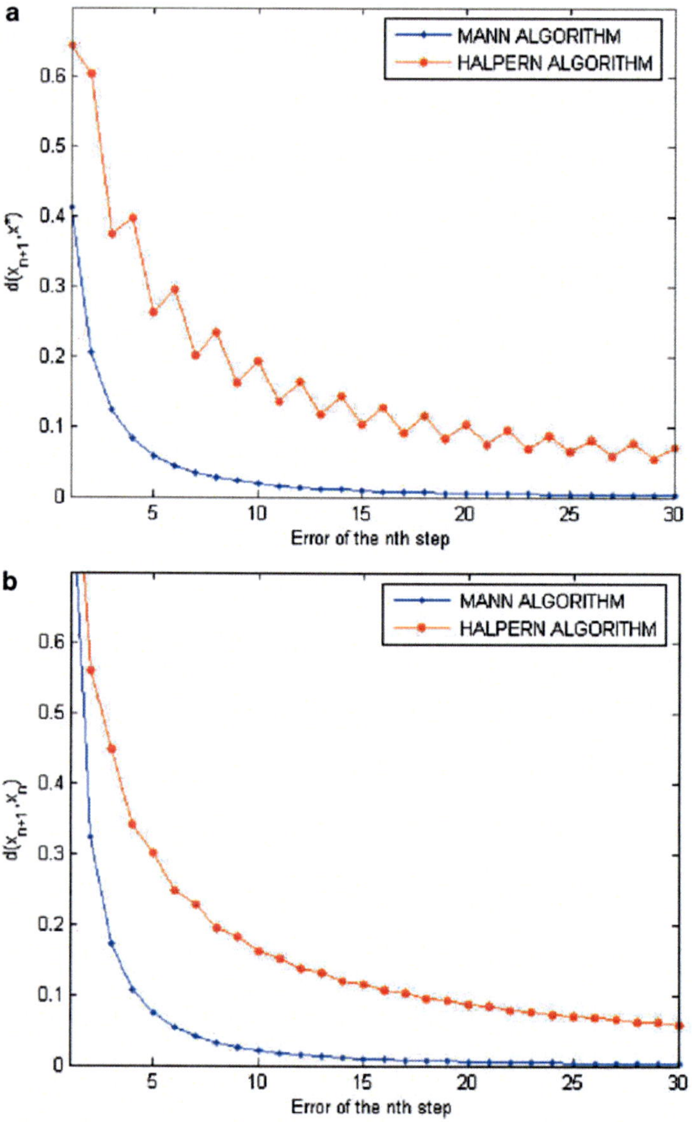

Fig. 14.1 (**a**) The error in the nth step of the Mann and Halpern algorithms, measured by means of the distance $d(x_{n+1}, x^*)$, where $x^* = (0,0,0,1)$ is the unique fixed point of the mapping T_1. (**b**) The distance between two consecutive iterates x_{n+1} and x_n, $d(x_{n+1}, x_n)$, for both the Mann and Halpern algorithms for the mapping T_2

The numerical results are illustrated in the graphics above.

In Fig. 14.1a, the error in the nth step of both algorithms is measured by means of the distance $d(x_{n+1}, x^*)$, where $x^* = (0,0,0,1)$ is the unique fixed point of the mapping T_1. On the other hand, in Fig. 14.1b, the distance between two consecutive iterates x_{n+1} and x_n, $d(x_{n+1}, x_n)$, measures the error in each step of both algorithms for the mapping T_2.

From the numerical results, as one can observe in both graphics, Mann iteration seems to converge much quicker than Halpern iteration. Moreover, as it is predicted from the theoretical results, the measure of the errors in Fig. 14.1a shows that the sequence $\{x_n\}$ generated by Mann algorithm is indeed Fejr monotone with respect to Fix (T_1). That is,

$$d(x_{n+1}, y) \leq d(x_n, y),$$

for all $y \in$ Fix (T_1); see [40]. On the other hand, Halpern algorithm generates a sequence which does not satisfy this property.

Acknowledgements This work was supported by DGES, Grant MTM2009-13997-C02-01 and Junta de Andaluca, Grant FQM-127. It was partially prepared while the second author was visiting the Department of Mathematics of UBC Okanagan in Kelowna, Canada. She is very grateful to Professor Bauschke for his wonderful hospitality.

References

1. Barbu, V.: Nonlinear differential equations of monotone types in Banach spaces. Springer Monographs in Mathematics. Springer, New York (2010)
2. Brézis, H.: Opérateurs maximaux monotones et semi-groupes de contractions dans les espaces de Hilbert. North-Holland, Amsterdam-London (1973)
3. Brézis, H., Crandall, G., Pazy, P.: Perturbations of nonlinear maximal monotone sets in Banach spaces. Comm. Pure Appl Math. **23**, 123–144 (1970)
4. Bridson, M., Haefliger, A.: Metric spaces of non-positive curvature. Springer, Berlin (1999)
5. Browder, F.E.: Existence and approximation of solutions of nonlinear variational inequalities. Proc. Nat. Acad. Sci. U.S.A. **56**, 1080–1086 (1966)
6. Browder, F.E.: Convergence of approximants to fixed points of nonexpansive nonlinear mappings in Banach spaces. Arch. Rational Mech. Anal. **24**, 82–90 (1967)
7. Browder, F.E.: Nonlinear maximal monotone operators in Banach spaces. Math. Ann. **175**, 89–113 (1968)
8. Browder, F.E., Petryshyn, W.V.: Construction of fixed points of nonlinear mappings in Hilbert space. J. Math. Anal. Appl. **20**, 197–228 (1967)
9. Bruck, R.E.: Convergence theorems for sequence of nonlinear operators in Banach spaces. Math. Z. **100**, 201–225 (1967)
10. Bruck, R.E.: Nonexpansive projections on subsets of Banach spaces. Pac. J. Math **47**, 341–355 (1973)
11. Bruck, R.E.: Asymptotic behavior of nonexpansive mappings. Contemp. Math. **18**, 1–47 (1983)
12. Bruck, R.E., Reich, S.: Nonexpansive projections and resolvents of accretive operators in Banach spaces. Houston J. Math. **3**, 459–470 (1977)
13. Chidume, C.: Geometric properties of Banach spaces and nonlinear iterations. Lecture Notes in Mathematics, 1965. Springer, London (2009)
14. Chidume, C.E.: Iterative approximation of fixed points of Lipschitzian strictly pseudo-contractive mappings. Proc. Amer. Math. Soc. **99**, 283–288 (1987)
15. Cioranescu, I.: Geometry of Banach spaces, duality mappings and nonlinear problems. Kluwer Academic Publishers, Dordrecht (1990)
16. Colao, V., López., G., Marino, G., Martín-Márquez, V.: Equilibrium problems in Hadamard manifolds. J. Math. Anal. Appl. (submitted)

17. Da Cruz Neto, J.X., Ferreira, O.P., Lucambio Pérez, L.R.: Monotone point-to-set vector fields. Balkan J. Geom. Appl. **5**, 69–79 (2000)
18. Da Cruz Neto, J.X., Ferreira, O.P., Lucambio Pérez, L.R.: Contributions to the study of monotone vector fields. Acta Math. Hungarica **94**, 307–320 (2002)
19. Da Cruz Neto, J.X., Ferreira, O.P., Lucambio Pérez, L.R., Nmeth, S.Z.: Convex- and monotone-transformable mathematical programming problems and a proximal-like point method. J. Global Optim. **35**, 53–69 (2006)
20. DoCarmo, M.P.: Riemannian Geometry. Boston, Birkhauser (1992)
21. Ferreira, O.P., Oliveira, P.R.: Subgradient algorithm on Riemannian manifolds. J. Optim. Theory Appl. **97**, 93–104 (1998)
22. Ferreira, O.P., Oliveira, P.R.: Proximal point algorithm on Riemannian manifolds. Optimization **51**, 257–270 (2002)
23. Ferreira, O.P., Lucambio Pérez, L.R., Németh, S.Z.: Singularities of monotene vector fields and an extragradient-type algorithm. J. Global Optim. **31**, 133–151 (2005)
24. Genel, A., Lindenstrauss, J.: An example concerning fixed points. Israel Journal of Math. **22**, 81–86 (1975)
25. Goebel, K., Kirk, W.A.: Iteration processes for nonexpansive mappings. Contemp. Math. **21**, 115–123 (1983)
26. Goebel, K., Reich, S.: Uniform Convexity, Hyperbolic Geometry, and Nonexpansive Mappings. Marcel Dekker, New York (1984)
27. Halpern, B.: Fixed points of nonexpanding maps. Bull. Amer. Math. Soc. **73**, 591–597 (1967)
28. Hoyos Guerrero, J.J.: Differential Equations of Evolution and Accretive Operators on Finsler Manifolds. *Ph. D. Thesis*, University of Chicago (1978)
29. Iwamiya, T., Okochi, H.: Monotonicity, resolvents and Yosida approximations of operators on Hilbert manifolds. Nonlinear Anal. **54**, 205–214 (2003)
30. Jost, J.: Nonpositive curvature: geometric and analytic aspects. Lectures in Mathematics ETH Zrich. Birkhuser, Basel (1997)
31. Kamimura, S., Takahashi, W.: Approximating solutions of maximal monotone operators in Hilbert spaces. J. Approx. Theory. **13**, 226–240 (2000)
32. Kato, T.: Nonlinear semigroups and evolution equations. J. Math. Soc. Japan **19**, 508–520 (1967)
33. Kido, K.: Strong convergence of resolvent of monotone operators in Banach spaces. Proc. Amer. Math. Soc. **103**, 755–758 (1988)
34. Kirk, W.A.: Krasnoselskii's Iteration process in hyperbolic space. Numer. Funct. Anal. Optim. **4**, 371–381 (1981/1982)
35. Kirk, W.A.: Geodesic Geometry and Fixed Point Theory. Seminar of Mathematical Analysis (Malaga/Seville, 2002/2003), 195–225, Univ. Sevilla Secr. Publ., Seville (2003)
36. Kirk, W.A.: Geodesic geometry and fixed point theory. In: II International Conference on Fixed Point Theory and Applications, 113–142, Yokohama Publ., Yokohama (2004)
37. Kirk, W.A., Schöneberg, R.: Some results on pseudo-contractive mappings. Pacific J. Math. **71**, 89–99 (1977)
38. Li, S.L., Li, C., Liu, Y.C., Yao, J.C.: Existence of solutions for variational inequalities on Riemannian manifolds. Nonlinear Anal. **71**, 5695–5705 (2009)
39. Li, C., López, G., Martín-Márquez, V.: Monotone vector fields and the proximal point algorithm on Hadamard manifolds. J. Lond. Math. Soc. **79**, 663–683 (2009)
40. Li, C., López, G., Martín-Márquez, V.: Iterative algorithms for nonexpansive mappings in Hadamard manifolds. Taiwanese J. Math. **14**, 541–559 (2010)
41. Li, C., López, G., Martín-Márquez, V., Wang, J.H.: Resolvents of set-valued monotone vector fields on Hadamard manifolds. Set-Valued Var. Anal., DOI: 10.1007/s11228-010-0169-1
42. López, G., Martín-Márquez, V., Xu, H.K.: Halpern's iteration for nonexpansive mappings. In: Nonlinear Analysis and Optimization I: Nonlinear Analysis. Contemp. Math., AMS, **513**, 187–207 (2010)
43. Maingé, P.E.: A hybrid extragradient-viscosity method for monotone operators and fixed point problems. SIAM J. Control Optim. **47**, 1499–1515 (2008)

44. Mann, W.R.: Mean value methods in iteration. Proc. Amer. Math. Soc. **4**, 506–510 (1953)
45. Martín-Márquez, V.: Nonexpansive mappings and monotone vector fields in Hadamard manifolds. Commun. Appl. Anal. **13**, 633–646 (2009)
46. Martín-Márquez, V.: Fixed point approximation methods for nonexpansive mappings: optimization problems. *Ph. D. Thesis*, University of Seville (2010)
47. Minty, G.J.: On the monotonicity of the gradient of a convex function. Pacific J. Math. **14**, 243–247 (1964)
48. Moudafi, A.: Viscosity approximation methods for fixed-points problems. J. Math. Anal. Appl. **241**, 46–55 (2000)
49. Németh, S.Z.: Five kinds of monotone vector fields. Pure Math. Appl. **9**, 417–428(1999)
50. Németh, S.Z.: Geodesic monotone vector fields. Lobachevskii J. Math. **5**, 13–28 (1999)
51. Németh, S.Z.: Monotone vector fields. Publ. Math. Debrecen **54**, 437–449 (1999)
52. Németh, S.Z.: Monotonicity of the complementary vector field of a nonexpansive map. Acta Math. Hungarica **84**, 189–197 (1999)
53. Németh, S.Z.: Variational inqualities on Hadamard manifolds. Nonlinear Anal. **52**, 1491–1498 (2003)
54. Papa Quiroz, E.A., Oliveira, P.R.: Proximal point methods for quasiconvex and convex functions with Bregman distances on Hadamard manifolds. J. Convex Anal. **16**, 49–69 (2009)
55. Papa Quiroz, E.A., Quispe, E. M., Oliveira, P.R.: Steepest descent method with a generalized Armijo search for quasiconvex functions on Riemannian manifolds. J. Math. Anal. Appl. **341**, 467–477 (2008)
56. Pascali, D., Sburlan, S.: Nonlinear Mappings of Monotone Type. Sythoff & Noordhoff, Alphen aan den Rijn, The Netherlands (1978)
57. Phelps, R.R.: Convex sets and nearest points. Proc. Amer. Math. Soc. **8**, 790–797 (1957)
58. Piątek, B.: Halpern iteration in CAT(κ) spaces. Acta Math. Sinica (English Series) **27**, 635–646 (2011)
59. Rapcsk, T.: Sectional curvature in nonlinear optimization. J. Global Optim. **40**, 375–388 (2008)
60. Reich, S.: Weak convergence theorems for nonexpansive mappings in Banach spaces. J. Math Anal. Appl. **67**, 274–276 (1979)
61. Reich, S.: Strong convergence theorems for resolvents of accretive operators in Banach spaces. J Math. Anal. Appl. **75**, 287–292 (1989)
62. Reich, S., Shafrir, I.: The asymptotic behavior of firmly nonexpansive mappings. Proc. Amer. Math. Soc. **101**, 246–250 (1987)
63. Reich, S., Shafrir, I.: Nonexpansive iterations in hyperbolic spaces. Nonlinear Anal. **15**, 537–558 (1990)
64. Rockafellar, R.T.: Monotone operators and the proximal point algorithm. SIAM J. Control Optim. **14**, 877–898 (1976)
65. Saejung, S.: Halpern's iteration in CAT(0) spaces. Fixed Point Theory Appl., Art. ID 471781, 13 pp. (2010)
66. Sakai, T.: Riemannian Geometry. Translations of Mathematical Monographs 149. American Mathematical Society, Providence, RI (1996)
67. Singer, I.: The Theory of Best Approximation and Functional Analysis. CBMS-NSF Regional Conf. Ser. in Appl. Math., 13, SIAM, Philadelphia, PA (1974)
68. Udriste, C.: Convex Functions and Optimization Methods on Riemannian Manifolds. Mathematics and Its Applications, 297. Kluwer Academic Publisher, Dordrecht (1994)
69. Walter, R.: On the metric projection onto convex sets in Riemannian spaces. Arch. Math. **25**, 91–98 (1974)
70. Wang, J.H., López, G., Martín-Márquez, V., Li, C.: Monotone and accretive vector fields on Riemannian manifolds. J. Optim. Theory Appl. **146**, 691–708 (2010) DOI: 10.1007/s10957-010-9688-z
71. Xu, H.K.: Viscosity approximation methods for nonexpansive mappings. J. Math. Anal. Appl. **298**, 279–291 (2004)
72. Zeidler, E.: Nonlinear Functional Analysis and Applications, II/B. Nonlinear Monotone Operators. Springer, New York (1990)

Chapter 15
Existence and Approximation of Fixed Points of Bregman Firmly Nonexpansive Mappings in Reflexive Banach Spaces

Simeon Reich and Shoham Sabach

Abstract We study the existence and approximation of fixed points of Bregman firmly nonexpansive mappings in reflexive Banach spaces.

Keywords Banach space · Bregman projection · Firmly nonexpansive mapping · Legendre function · Monotone operator · Resolvent · Totally convex function

AMS 2010 Subject Classification: 46T99, 47H04, 47H05, 47H09, 47H10, 47J05, 47J25, 49J40

15.1 Introduction

In this paper, X denotes a real reflexive Banach space with norm $\|\cdot\|$ and X^* stands for the (topological) dual of X endowed with the induced norm $\|\cdot\|_*$. We denote the value of the functional $\xi \in X^*$ at $x \in X$ by $\langle \xi, x \rangle$. An operator $A : X \to 2^{X^*}$ is said to be *monotone* if for any $x, y \in \operatorname{dom} A$, we have

$$\xi \in Ax \text{ and } \eta \in Ay \Longrightarrow \langle \xi - \eta, x - y \rangle \geq 0. \tag{15.1}$$

(Recall that the set $\operatorname{dom} A = \{x \in X \ : \ Ax \neq \varnothing\}$ is called the *effective domain* of such an operator A.) A monotone operator A is said to be *maximal* if graph A, the graph of A, is not a proper subset of the graph of any other monotone operator. In this paper, $f : X \to (-\infty, +\infty]$ is always a proper, lower semicontinuous and convex function, and $f^* : X^* \to (-\infty, +\infty]$ is the Fenchel conjugate of f. A *sublevel set* of f is a set of the form $\operatorname{lev}^f_{\leq}(r) = \{x \in X \ : \ f(x) \leq r\}$ for some $r \in \mathbb{R}$. We say that f is *positively homogeneous of degree* $\alpha \in \mathbb{R}$ if $f(tx) = t^{\alpha} f(x)$ for all $x \in X$ and $t > 0$. The set of nonnegative integers will be denoted by $\mathbb{N}$.

S. Reich (✉)
Department of Mathematics, The Technion – Israel Institute of Technology, 32000 Haifa, Israel
e-mail: sreich@tx.technion.ac.il

H.H. Bauschke et al. (eds.), *Fixed-Point Algorithms for Inverse Problems in Science and Engineering*, Springer Optimization and Its Applications 49,
DOI 10.1007/978-1-4419-9569-8_15,

Let C be a nonempty, closed and convex subset of a Hilbert space H. Then a mapping $T : C \to C$ is said to be *nonexpansive* if $\|Tx - Ty\| \le \|x - y\|$ for all $x, y \in C$. It turns out that nonexpansive fixed point theory can be applied to the problem of finding a point $z \in H$ satisfying

$$0 \in Az, \tag{15.2}$$

where $A : H \to 2^H$ is a maximal monotone operator. A key tool for solving this problem is the classical *resolvent* of A, which is defined by $R_A = (I + A)^{-1}$. This resolvent is not only nonexpansive but also a *firmly nonexpansive* mapping, that is

$$\|R_A x - R_A y\|^2 \le \langle R_A x - R_A y, x - y\rangle \tag{15.3}$$

for all $x, y \in H$ (the resolvent R_A has full domain H when A is maximal monotone). See [11,17,22] for more details. We also have $F(R_A) = A^{-1}(0)$, where $F(R_A)$ stands for the set of fixed points of R_A. Thus the problem of finding zeroes of maximal monotone operators in Hilbert space is reduced to that of finding fixed points of firmly nonexpansive mappings. In particular, if A is the subdifferential ∂f of f, then R_A is given by

$$R_A x = \operatorname{argmin}_{y \in H} \left\{ f(y) + \frac{1}{2}\|y - x\|^2 \right\} \tag{15.4}$$

for all $x \in H$ [23]. In this case, $F(R_A) = \left\{ z \in H \mid f(z) = \inf_{y \in H} f(y) \right\}$.

The notion of a firmly nonexpansive mapping was extended to Banach spaces in [10] and [11]; see also [17]. However, in contrast with the case of Hilbert space, the resolvent of a maximal monotone operator is not, in general, even a nonexpansive mapping in the case of Banach spaces. Many other types of resolvents have been studied. For example, Alber [1], and Kohsaka and Takahashi [19–21] initiated the study of a generalized resolvent based on the duality mapping J.

Recently, Kohsaka and Takahashi [20,21] have introduced the class of mappings of *firmly nonexpansive type*. Such a mapping T satisfies

$$\langle JTx - JTy, Tx - Ty\rangle \le \langle Jx - Jy, Tx - Ty\rangle \tag{15.5}$$

for all $x, y \in C$, where J is the duality mapping of the Banach space X, and C is a nonempty, closed and convex subset of X. It is obvious that if we return to Hilbert space, then $J = I$ and the definitions of a firmly nonexpansive mapping and a mapping of firmly nonexpansive type coincide. Kohsaka and Takahashi prove that the generalized resolvent is a mapping of firmly nonexpansive type when X is a smooth, strictly convex and reflexive Banach space.

Even earlier, Bauschke et al. [4] generalized the class of firmly nonexpansive mappings on smooth, strictly convex and reflexive Banach spaces to the case of general reflexive Banach spaces. Their mappings do not depend on the duality mapping J, but on the gradient ∇f of a well chosen function f. They call those mappings D_f-firmly nonexpansive mappings. In this paper we call them Bregman firmly nonexpansive mappings (BFNE for short) with respect to the function f. Bauschke, Borwein and Combettes prove that the resolvent based on the gradient ∇f of a well chosen function f is a BFNE mapping.

Our aim in this paper is to study the existence and approximation of fixed points of BFNE mappings in reflexive Banach spaces. In Sect. 15.2, we present several preliminary definitions and results. The third section is devoted to two properties of BFNE mappings. In the fourth section we prove two existence theorems (Theorems 15.7 and 15.8) regarding fixed points of a single BFNE mapping, as well as a common fixed point theorem (Theorem 15.12). Our approximation result is proved in Sect. 15.5 (Theorem 15.13). In the sixth and last section, we present two consequences of Theorem 15.13.

15.2 Preliminaries

15.2.1 Some Facts About Legendre Functions

Legendre functions mapping a general Banach space X into $(-\infty,+\infty]$ are defined in [3]. According to [3, Theorems 5.4 and 5.6], since X is reflexive, the function f is Legendre if and only if it satisfies the following two conditions:

(L1) The interior of the domain of f, $\operatorname{int}\operatorname{dom} f$, is nonempty, f is Gâteaux differentiable (see below) on $\operatorname{int}\operatorname{dom} f$, and

$$\operatorname{dom}\nabla f = \operatorname{int}\operatorname{dom} f; \tag{15.6}$$

(L2) The interior of the domain of f^*, $\operatorname{int}\operatorname{dom} f^*$, is nonempty, f^* is Gâteaux differentiable on $\operatorname{int}\operatorname{dom} f^*$, and

$$\operatorname{dom}\nabla f^* = \operatorname{int}\operatorname{dom} f^*. \tag{15.7}$$

Since X is reflexive, we always have $(\partial f)^{-1} = \partial f^*$ (see [7, p. 83]). This fact, when combined with conditions (L1) and (L2), implies the following equalities:

$$\nabla f = (\nabla f^*)^{-1}, \tag{15.8}$$

$$\operatorname{ran}\nabla f = \operatorname{dom}\nabla f^* = \operatorname{int}\operatorname{dom} f^* \tag{15.9}$$

and

$$\operatorname{ran}\nabla f^* = \operatorname{dom}\nabla f = \operatorname{int}\operatorname{dom} f. \tag{15.10}$$

Also, conditions (L1) and (L2), in conjunction with [3, Theorem 5.4], imply that the functions f and f^* are strictly convex on the interior of their respective domains.

Several interesting examples of Legendre functions are presented in [2] and [3]. Among them are the functions $\frac{1}{s}\|\cdot\|^s$ with $s \in (1,\infty)$, where the Banach space X is smooth and strictly convex and, in particular, a Hilbert space.

15.2.2 Two Properties of Gradients

For any convex $f : X \to (-\infty, +\infty]$ we denote by $\operatorname{dom} f$ the set $\{x \in X : f(x) < +\infty\}$. For any $x \in \operatorname{int}\operatorname{dom} f$ and $y \in X$, we denote by $f^{\circ}(x,y)$ the *right-hand derivative of f at x in the direction y*, that is,

$$f^{\circ}(x,y) := \lim_{t \searrow 0} \frac{f(x+ty) - f(x)}{t}. \tag{15.11}$$

The function f is said to be *Gâteaux differentiable at* x if $\lim_{t\to 0}(f(x+ty) - f(x))/t$ exists for any y. The function f is said to be *Fréchet differentiable at x* if this limit is attained uniformly in $\|y\| = 1$. Finally, f is said to be *uniformly Fréchet differentiable on a subset E of X* if the limit is attained uniformly for $x \in E$ and $\|y\| = 1$. We will need the following result.

Proposition 15.1 (Proposition 2.1 of [27]). *If $f : X \to \mathbb{R}$ is uniformly Fréchet differentiable and bounded on bounded subsets of X, then ∇f is uniformly continuous on bounded subsets of X from the strong topology of X to the strong topology of X^*.*

Proposition 15.2. *If $f : X \to \mathbb{R}$ is a positively homogeneous function of degree $\alpha \in \mathbb{R}$, then ∇f is a positively homogeneous function of degree $\alpha - 1$.*

Proof. By the definition of the gradient, we have

$$\begin{aligned}\nabla f(tx) &= \lim_{h\to 0} \frac{f(tx+hy) - f(tx)}{h} = \lim_{h\to 0} \frac{f(tx+thy) - f(tx)}{th} \\ &= \frac{t^{\alpha}}{t} \lim_{h\to 0} \frac{f(x+hy) - f(x)}{h} = t^{\alpha-1}\nabla f(x)\end{aligned} \tag{15.12}$$

for any $x \in X$ and all $t > 0$. ■

15.2.3 Some Facts About Totally Convex Functions

Let $f : X \to (-\infty, +\infty]$ be a convex and Gâteaux differentiable function. The function $D_f : \operatorname{dom} f \times \operatorname{int}\operatorname{dom} f \to [0, +\infty]$, defined by

$$D_f(y,x) := f(y) - f(x) - \langle \nabla f(x), y - x \rangle, \tag{15.13}$$

is called the *Bregman distance with respect to f* (*cf.* [16]). With the function f we associate the function $W^f : X^* \times X \to [0, +\infty]$ defined by

$$W^f(\xi, x) = f(x) - \langle \xi, x \rangle + f^*(\xi). \tag{15.14}$$

It is clear that $W^f(\nabla f(x), y) = D_f(y,x)$ for any $x \in \operatorname{int}\operatorname{dom} f$ and $y \in \operatorname{dom} f$.

The Bregman distance has the following important properties, called the *three-point identity*: for any $x \in \operatorname{dom} f$ and $y, z \in \operatorname{int}\operatorname{dom} f$,

$$D_f(x,y) + D_f(y,z) - D_f(x,z) = \langle \nabla f(z) - \nabla f(y), x - y \rangle, \tag{15.15}$$

and the *four-point identity*: for any $x, z \in \operatorname{int}\operatorname{dom} f$ and $y, w \in \operatorname{dom} f$,

$$D_f(y,x) - D_f(y,z) - D_f(w,x) + D_f(w,z) = \langle \nabla f(z) - \nabla f(x), y - w \rangle. \tag{15.16}$$

Recall that, according to [12, Sect. 1.2, p. 17] (see also [14]), the function f is called *totally convex at a point* $x \in \operatorname{int}\operatorname{dom} f$ if its *modulus of total convexity at* x, that is, the function $\upsilon_f : \operatorname{int}\operatorname{dom} f \times [0, +\infty) \to [0, +\infty]$, defined by

$$\upsilon_f(x,t) := \inf\left\{D_f(y,x) \ : \ y \in \operatorname{dom} f, \ \|y - x\| = t\right\}, \tag{15.17}$$

is positive whenever $t > 0$. The function f is called *totally convex* when it is totally convex at every point $x \in \operatorname{int}\operatorname{dom} f$. Examples of totally convex functions can be found, for instance, in [12, 13]. The next proposition turns out to be very useful in the proof of Theorem 15.13.

Proposition 15.3 (Proposition 2.2 of [28]). *If $x \in \operatorname{int}\operatorname{dom} f$, then the following statements are equivalent:*

(i) The function f is totally convex at x;
(ii) For any sequence $\{y_n\}_{n \in \mathbb{N}} \subset \operatorname{dom} f$,

$$\lim_{n \to +\infty} D_f(y_n, x) = 0 \Rightarrow \lim_{n \to +\infty} \|y_n - x\| = 0. \tag{15.18}$$

15.2.4 Some Facts About Bregman Firmly Nonexpansive Mappings

Let C be a nonempty, closed and convex subset of $\operatorname{int}\operatorname{dom} f$. We say that a mapping $T : C \to C$ is a *Bregman firmly nonexpansive mapping with respect to f (BFNE with respect to f* for short*)* if

$$\langle \nabla f(Tx) - \nabla f(Ty), Tx - Ty \rangle \leq \langle \nabla f(x) - \nabla f(y), Tx - Ty \rangle \tag{15.19}$$

for all $x, y \in C$. It is clear from the definition of the Bregman distance (15.13) that (15.19) is equivalent to

$$D_f(Tx,Ty) + D_f(Ty,Tx) + D_f(Tx,x) + D_f(Ty,y) \leq D_f(Tx,y) + D_f(Ty,x). \tag{15.20}$$

Bauschke, Borwein and Combettes [4, Proposition 3.8, p. 604] prove that the resolvent $\mathrm{Res}_A^f = (\nabla f + A)^{-1} \circ \nabla f$ is a BFNE mapping with respect to f whenever A is a monotone mapping.

We remark in passing that an analogous result for very general resolvents can be found in a recent paper by Bauschke et al. [6].

15.2.5 The Resolvent of A Relative to f

Let $A : X \to 2^{X^*}$ be an operator and assume that f is Gâteaux differentiable. The operator

$$\mathrm{Prt}_A^f := (\nabla f + A)^{-1} : X^* \to 2^X \tag{15.21}$$

is called the *protoresolvent* of A, or, more precisely, the *protoresolvent of A relative to f*. This allows us to define the *resolvent* of A, or, more precisely, the *resolvent of A relative to f*, introduced and studied in [4], as the operator $\mathrm{Res}_A^f : X \to 2^X$ given by $\mathrm{Res}_A^f := \mathrm{Prt}_A^f \circ \nabla f$. This operator is single-valued when A is monotone and f is strictly convex on $\operatorname{int}\operatorname{dom} f$. If $A = \partial\varphi$, where φ is a proper, lower semicontinuous and convex function, then we denote

$$\mathrm{Prox}_\varphi^f := \mathrm{Prt}_{\partial\varphi}^f \quad \text{and} \quad \mathrm{prox}_\varphi^f := \mathrm{Res}_{\partial\varphi}^f. \tag{15.22}$$

If C is a nonempty, closed and convex subset of X, then the indicator function ι_C of C, that is, the function

$$\iota_C(x) := \begin{cases} 0 & \text{if } x \in C \\ +\infty & \text{if } x \notin C, \end{cases} \tag{15.23}$$

is proper, convex and lower semicontinuous, and therefore $\partial\iota_C$ exists and is a maximal monotone operator with domain C. The operator $\mathrm{prox}_{\iota_C}^f$ is called the *Bregman projection* onto C with respect to f (*cf.* [8]) and we denote it by proj_C^f. Note that if X is a Hilbert space and $f(x) = \frac{1}{2}\|x\|^2$, then the Bregman projection of x onto C, i.e., $\operatorname{argmin}\{\|y - x\| : y \in C\}$, is the metric projection P_C.

Recall that the Bregman projection of x onto the nonempty, closed and convex set $K \subset \operatorname{dom} f$ is the necessarily unique vector $\mathrm{proj}_K^f(x) \in K$ satisfying

$$D_f\left(\mathrm{proj}_K^f(x), x\right) = \inf\{D_f(y,x) : y \in K\}. \tag{15.24}$$

Similarly to the metric projection in Hilbert spaces, Bregman projections with respect to totally convex and differentiable functions have variational characterizations.

Proposition 15.4 (Corollary 4.4 of [13]). *Suppose that f is totally convex on* int dom$f$. *Let $x \in \operatorname{int}\operatorname{dom} f$ and let $K \subset \operatorname{int}\operatorname{dom} f$ be a nonempty, closed and convex set. If $\hat{x} \in K$, then the following conditions are equivalent:*

(i) The vector $\hat{x}$ is the Bregman projection of x onto K with respect to f;
(ii) The vector $\hat{x}$ is the unique solution of the variational inequality

$$\langle \nabla f(x) - \nabla f(z), z - y \rangle \geq 0, \qquad \forall y \in K; \tag{15.25}$$

(iii) The vector $\hat{x}$ is the unique solution of the inequality

$$D_f(y,z) + D_f(z,x) \leq D_f(y,x), \qquad \forall y \in K. \tag{15.26}$$

15.3 Two Properties of Bregman Firmly Nonexpansive Mappings

In this section, we present two properties of the fixed point set $F(T)$ of a BFNE mapping. We first show that $F(T)$ is closed and convex for any BFNE mapping with respect to f when f is also Gâteaux differentiable.

Lemma 15.5. *Let $f : X \to (-\infty, +\infty]$ be a Legendre function. Let C be a nonempty, closed and convex subset of* $\operatorname{int}\operatorname{dom} f$*, and let $T : C \to C$ be a BFNE mapping with respect to f. Then $F(T)$ is closed and convex.*

Proof. It is sufficient to consider the case where $F(T)$ is nonempty. From (15.20) it follows that

$$D_f(x, Ty) + D_f(Ty, y) \leq D_f(x, y) \tag{15.27}$$

for any $x \in F(T)$ and $y \in C$. *A fortiori*,

$$D_f(x, Ty) \leq D_f(x, y) \tag{15.28}$$

for any $x \in F(T)$ and $y \in C$.

We first show that $F(T)$ is closed. To this end, let $\{x_n\}_{n \in \mathbb{N}}$ be a sequence in $F(T)$ such that $x_n \to \bar{x}$. From (15.28) it follows that

$$D_f(x_n, T\bar{x}) \leq D_f(x_n, \bar{x}) \tag{15.29}$$

for any $n \in \mathbb{N}$. Since f is continuous at $\bar{x} \in C \subset \operatorname{int}\operatorname{dom} f$ and $x_n \to \bar{x}$, it follows that

$$\begin{aligned}\lim_{n\to+\infty} D_f(x_n, T\bar{x}) &= \lim_{n\to+\infty} [f(x_n) - f(T\bar{x}) - \langle \nabla f(T\bar{x}), x_n - T\bar{x} \rangle] \\ &= [f(\bar{x}) - f(T\bar{x}) - \langle \nabla f(T\bar{x}), \bar{x} - T\bar{x} \rangle] = D_f(\bar{x}, T\bar{x})\end{aligned}$$

and

$$\lim_{n\to+\infty} D_f(x_n, \bar{x}) = D_f(\bar{x}, \bar{x}) = 0. \tag{15.30}$$

Thus, (15.29) implies that $D_f(\bar{x}, T\bar{x}) = 0$ and therefore it follows from [3, Lemma 7.3(vi), p. 642] that $\bar{x} = T\bar{x}$. Hence, $\bar{x} \in F(T)$ and this means that $F(T)$ is closed, as claimed.

Next we show that $F(T)$ is convex. For any $x, y \in F(T)$ and $t \in (0,1)$, put $z = tx + (1-t)y$. We have to show that $Tz = z$. Indeed, from the definition of the Bregman distance and (15.28) it follows that

$$\begin{aligned}
D_f(z, Tz) &= f(z) - f(Tz) - \langle \nabla f(Tz), z - Tz \rangle \\
&= f(z) - f(Tz) - \langle \nabla f(Tz), tx + (1-t)y - Tz \rangle \\
&= f(z) + tD_f(x, Tz) + (1-t)D_f(y, Tz) - tf(x) - (1-t)f(y) \\
&\le f(z) + tD_f(x, z) + (1-t)D_f(y, z) - tf(x) - (1-t)f(y) \\
&= \langle \nabla f(z), z - tx - (1-t)y \rangle = 0.
\end{aligned}$$

Again from [3, Lemma 7.3(vi), p. 642] it follows that $Tz = z$. Therefore, $F(T)$ is also convex, as asserted. ■

Next we show that if f is a Legendre function which is uniformly Fréchet differentiable on bounded subsets of X, and T is a BFNE mapping with respect to f, then the set of fixed points of T coincides with the set of its asymptotic fixed points. Recall that a point $u \in C$ is said to be an *asymptotic fixed point* [26] of T if there exists a sequence $\{x_n\}_{n \in \mathbb{N}}$ in C such that $x_n \rightharpoonup u$ and $x_n - Tx_n \to 0$. We denote the set of asymptotic fixed points of T by $\hat{F}(T)$.

Lemma 15.6. *Let $f : X \to \mathbb{R}$ be a Legendre function which is uniformly Fréchet differentiable and bounded on bounded subsets of X. Let C be a nonempty, closed and convex subset of X and let $T : C \to C$ be a BFNE mapping with respect to f. Then $F(T) = \hat{F}(T)$.*

Proof. The inclusion $F(T) \subset \hat{F}(T)$ is obvious. To show that $F(T) \supset \hat{F}(T)$, let $u \in \hat{F}(T)$ be given. Then we have a sequence $\{x_n\}_{n \in \mathbb{N}}$ in C such that $x_n \rightharpoonup u$ and $x_n - Tx_n \to 0$. Since f is uniformly Fréchet differentiable on bounded subsets of X, ∇f is uniformly continuous on bounded subsets of X (see Proposition 15.1). Hence $(\nabla f(Tx_n) - \nabla f(x_n)) \to 0$ as $n \to +\infty$ and therefore

$$\lim_{n \to +\infty} \langle \nabla f(Tx_n) - \nabla f(x_n), y \rangle = 0 \tag{15.31}$$

for any $y \in X$, and

$$\lim_{n \to +\infty} \langle \nabla f(Tx_n) - \nabla f(x_n), x_n \rangle = 0, \tag{15.32}$$

because $\{x_n\}_{n \in \mathbb{N}}$ is bounded. On the other hand, since T is a BFNE mapping with respect to f, we have

$$0 \le D_f(Tx_n, u) - D_f(Tx_n, Tu) + D_f(Tu, x_n) - D_f(Tu, Tx_n). \tag{15.33}$$

From the three-point identity (15.15) and (15.33), we now obtain

$$\begin{aligned}
D_f(u,Tu) &= D_f(Tx_n,Tu) - D_f(Tx_n,u) - \langle \nabla f(u) - \nabla f(Tu), Tx_n - u\rangle \\
&\le D_f(Tu,x_n) - D_f(Tu,Tx_n) - \langle \nabla f(u) - \nabla f(Tu), Tx_n - u\rangle \\
&= [f(Tu) - f(x_n) - \langle \nabla f(x_n), Tu - x_n\rangle] \\
&\quad - [f(Tu) - f(Tx_n) - \langle \nabla f(Tx_n), Tu - Tx_n\rangle] \\
&\quad - \langle \nabla f(u) - \nabla f(Tu), Tx_n - u\rangle \\
&= f(Tx_n) - f(x_n) - \langle \nabla f(x_n), Tu - x_n\rangle + \langle \nabla f(Tx_n), Tu - Tx_n\rangle \\
&\quad - \langle \nabla f(u) - \nabla f(Tu), Tx_n - u\rangle \\
&= -[f(x_n) - f(Tx_n) - \langle \nabla f(Tx_n), x_n - Tx_n\rangle] - \langle \nabla f(Tx_n), x_n - Tx_n\rangle \\
&\quad - \langle \nabla f(x_n), Tu - x_n\rangle + \langle \nabla f(Tx_n), Tu - Tx_n\rangle \\
&\quad - \langle \nabla f(u) - \nabla f(Tu), Tx_n - u\rangle \\
&= -D_f(x_n,Tx_n) - \langle \nabla f(Tx_n), x_n - Tx_n\rangle - \langle \nabla f(x_n), Tu - x_n\rangle \\
&\quad + \langle \nabla f(Tx_n), Tu - Tx_n\rangle - \langle \nabla f(u) - \nabla f(Tu), Tx_n - u\rangle \\
&\le -\langle \nabla f(Tx_n), x_n - Tx_n\rangle - \langle \nabla f(x_n), Tu - x_n\rangle \\
&\quad + \langle \nabla f(Tx_n), Tu - Tx_n\rangle - \langle \nabla f(u) - \nabla f(Tu), Tx_n - u\rangle \\
&= \langle \nabla f(x_n) - \nabla f(Tx_n), x_n - Tu\rangle - \langle \nabla f(u) - \nabla f(Tu), Tx_n - x_n\rangle \\
&\quad - \langle \nabla f(u) - \nabla f(Tu), x_n - u\rangle .
\end{aligned}$$

From (15.31), (15.32), and the hypotheses $x_n \rightharpoonup u$ and $x_n - Tx_n \to 0$ we get that $D_f(u,Tu) \le 0$. Consequently, $D_f(u,Tu) = 0$ and from [3, Lemma 7.3(vi), p. 642] it follows that $Tu = u$. That is, $u \in \mathrm{F}(T)$, as required. ∎

15.4 Existence of Fixed Points

In this section, we obtain necessary and sufficient conditions for BFNE mappings to have a (common) fixed point in general reflexive Banach spaces. We begin with a theorem for a single BFNE mapping. This result can be proved by combining Theorem 3.3 and Lemma 7.3(viii) of [3] with Proposition 4.1(v)(a) of [4]. However, we include a more detailed version of the proof for the reader's convenience.

Theorem 15.7. *Let $f : X \to (-\infty, +\infty]$ be a Legendre function such that ∇f^* is bounded on bounded subsets of* $\operatorname{int}\operatorname{dom} f^*$. *Let C be a nonempty, closed and convex subset of* $\operatorname{int}\operatorname{dom} f$ *and let $T : C \to C$ be a BFNE mapping with respect to f. If $F(T)$ is nonempty, then $\{T^n y\}_{n\in\mathbb{N}}$ is bounded for each $y \in C$.*

Proof. We know by (15.28) that

$$D_f(x,Ty) \le D_f(x,y) \tag{15.34}$$

for any $x \in F(T)$ and $y \in C$. Therefore

$$D_f(x,T^n y) \le D_f(x,y) \tag{15.35}$$

for any $x \in F(T)$ and $y \in C$. This inequality shows that the nonnegative sequence $\left\{D_f(x, T^n y)\right\}_{n\in\mathbb{N}}$ is bounded. Let M be an upper bound of $\left\{D_f(x, T^n y)\right\}_{n\in\mathbb{N}}$. Then

$$f(x) - \langle \nabla f(T^n y), x\rangle + f^*(\nabla f(T^n y)) = W^f(\nabla f(T^n y), x) = D_f(x, T^n y) \leq M. \tag{15.36}$$

This implies that the sequence $\{\nabla f(T^n y)\}_{n\in\mathbb{N}}$ is contained in the sublevel set $\operatorname{lev}_{\leq}^{\psi}(M - f(x))$ of the function $\psi = f^* - \langle \cdot, x\rangle$. Since the function f^* is proper and lower semicontinuous, an application of the Moreau–Rockafellar Theorem [29, Theorem 7A] shows that $\psi = f^* - \langle \cdot, x\rangle$ is coercive. Consequently, all sublevel sets of ψ are bounded. Hence, the sequence $\{\nabla f(T^n y)\}_{n\in\mathbb{N}}$ is bounded. Since the function f^* is bounded on bounded subsets of X by hypothesis, the gradient ∇f^* is also bounded on bounded subsets of X [12, Proposition 1.1.11, p. 16]. Thus the sequence $T^n y = \nabla f^*(\nabla f(T^n y))$, $n \in \mathbb{N}$, is bounded too, as claimed. ■

For a mapping $T : C \to C$, let $S_n(z) := 1/n \sum_{k=1}^{n} T^k z$ for all $z \in C$.

Theorem 15.8. *Let $f : X \to (-\infty, +\infty]$ be a Legendre function. Let C be a nonempty, closed and convex subset of* $\operatorname{int}\operatorname{dom} f$ *and let $T : C \to C$ be a BFNE mapping with respect to f. If there exists $y \in C$ such that $\|S_n(y)\| \nrightarrow \infty$ as $n \to \infty$, then $F(T)$ is nonempty.*

Proof. Suppose that there exists $y \in C$ such that $\|S_n(y)\| \nrightarrow \infty$ as $n \to \infty$. Let $x \in C$, $k \in \mathbb{N}$ and $n \in \mathbb{N}$ be given. Since T is BFNE with respect to f, we have

$$D_f\left(T^{k+1}y, Tx\right) + D_f\left(Tx, T^{k+1}y\right) \leq D_f\left(Tx, T^k y\right) + D_f\left(T^{k+1}y, x\right). \tag{15.37}$$

From the three-point identity (15.15) we get

$$\begin{aligned} D_f\left(T^{k+1}y, Tx\right) + D_f\left(Tx, T^{k+1}y\right) \leq D_f\left(Tx, T^k y\right) + D_f\left(T^{k+1}y, Tx\right) + D_f(Tx, x) \\ + \left\langle \nabla f(Tx) - \nabla f(x), T^{k+1}y - Tx\right\rangle. \end{aligned}$$

This implies that

$$\begin{aligned} 0 \leq D_f(Tx, x) + D_f\left(Tx, T^k y\right) - D_f\left(Tx, T^{k+1}y\right) \\ + \left\langle \nabla f(Tx) - \nabla f(x), T^{k+1}y - Tx\right\rangle. \end{aligned}$$

Summing these inequalities with respect to $k = 0, 1, \ldots, n-1$, we now obtain

$$\begin{aligned} 0 \leq nD_f(Tx, x) + D_f(Tx, y) - D_f(Tx, T^n y) \\ + \left\langle \nabla f(Tx) - \nabla f(x), \sum_{k=0}^{n-1} T^{k+1}y - nTx\right\rangle. \end{aligned}$$

Dividing this inequality by n, we have

$$0 \le D_f(Tx,x) + \frac{1}{n}\left[D_f(Tx,y) - D_f(Tx,T^n y)\right] + \langle \nabla f(Tx) - \nabla f(x), S_n(y) - Tx\rangle \tag{15.38}$$

and

$$0 \le D_f(Tx,x) + \frac{1}{n} D_f(Tx,y) + \langle \nabla f(Tx) - \nabla f(x), S_n(y) - Tx\rangle . \tag{15.39}$$

Since $\|S_n(y)\| \nrightarrow \infty$ as $n \to \infty$ by assumption, there exists a subsequence $\{S_{n_k}(y)\}_{k\in\mathbb{N}}$ of $\{S_n(y)\}_{n\in\mathbb{N}}$ such that $S_{n_k}(y) \rightharpoonup u \in C$. Letting $n_k \to +\infty$ in (15.39), we obtain

$$0 \le D_f(Tx,x) + \langle \nabla f(Tx) - \nabla f(x), u - Tx\rangle . \tag{15.40}$$

Setting $x = u$ in (15.40), we get from the four-point identity (15.16) that

$$\begin{aligned} 0 &\le D_f(Tu,u) + \langle \nabla f(Tu) - \nabla f(u), u - Tu\rangle \\ &= D_f(Tu,u) + D_f(u,u) - D_f(u,Tu) - D_f(Tu,u) + D_f(Tu,Tu) \\ &= -D_f(u,Tu) . \end{aligned}$$

Hence $D_f(u,Tu) \le 0$ and so $D_f(u,Tu) = 0$. It now follows from [3, Lemma 7.3(vi), p. 642] that $Tu = u$. That is, $u \in F(T)$. This completes the proof of Theorem 15.8. ∎

Remark 15.9. As can be seen from the proof, Theorem 15.8 remains true for those mappings which only satisfy (15.37). In the special case where $f = 1/2\|\cdot\|^2$, such mappings are called *non-spreading*. For more information see [21].

Remark 15.10. We remark in passing that we still do not know if the analog of Theorem 15.8 for nonexpansive mappings holds outside Hilbert space (*cf.* [24, Remark 2, p. 275]).

Corollary 15.11. *Let $f : X \to (-\infty, +\infty]$ be a Legendre function. Every nonempty, bounded, closed and convex subset of* $\operatorname{int}\operatorname{dom} f$ *has the fixed point property for BFNE self-mappings with respect to f.*

As in [21], Corollary 15.11, when combined with Lemma 15.5, yields the following result.

Theorem 15.12. *Let $f : X \to (-\infty, +\infty]$ be a Legendre function. Let C be a nonempty, bounded, closed and convex subset of* $\operatorname{int}\operatorname{dom} f$. *Let $\{T_\alpha\}_{\alpha\in A}$ be a commutative family of BFNE mappings with respect to f from C into itself. Then the family $\{T_\alpha\}_{\alpha\in A}$ has a common fixed point.*

15.5 Approximation of Fixed Points

In this section, we prove a strong convergence theorem of Browder's type for BFNE mappings with respect to a well chosen function f.

Theorem 15.13. *Let $f : X \to \mathbb{R}$ be a Legendre, totally convex function which is positively homogeneous of degree $\alpha > 1$, uniformly Fréchet differentiable and bounded on bounded subsets of X. Let C be a nonempty, bounded, closed and convex subset of X with $0 \in C$, and let T be a BFNE self-mapping with respect to f. Then the following two assertions hold:*

(i) For each $t \in (0,1)$, there exists a unique $u_t \in C$ satisfying $u_t = tTu_t$;
(ii) The net $\{u_t\}_{t\in(0,1)}$ converges strongly to $\operatorname{proj}^f_{F(T)}(\nabla f^*(0))$ as $t \to 1^-$.

Proof. (i) Fix $t \in (0,1)$ and let S_t be the mapping defined by $S_t = tT$. Since $0 \in C$ and C is convex, S_t is a mapping from C into itself. We next show that S_t is a BFNE mapping with respect to f. Indeed, if $x, y \in C$, then, since T is BFNE with respect to f, it follows from Proposition 15.2 that

$$\begin{aligned}\langle \nabla f(S_t x) - \nabla f(S_t y), S_t x - S_t y\rangle &= t^{\alpha}\langle \nabla f(Tx) - \nabla f(Ty), Tx - Ty\rangle \\ &\le t^{\alpha}\langle \nabla f(x) - \nabla f(y), Tx - Ty\rangle \\ &= t^{\alpha-1}\langle \nabla f(x) - \nabla f(y), S_t x - S_t y\rangle \\ &\le \langle \nabla f(x) - \nabla f(y), S_t x - S_t y\rangle . \quad (15.41)\end{aligned}$$

Thus, S_t is also BFNE with respect to f. Since C is bounded, it follows from Corollary 15.11 that S_t has a fixed point. We next show that $F(S_t)$ consists of exactly one point. If $u, u' \in F(S_t)$, then it follows from (15.41) that

$$\begin{aligned}\left\langle \nabla f(u) - \nabla f(u'), u - u'\right\rangle &= \left\langle \nabla f(S_t u) - \nabla f(S_t u'), S_t u - S_t u'\right\rangle \\ &\le t^{\alpha-1}\left\langle \nabla f(u) - \nabla f(u'), S_t u - S_t u'\right\rangle \\ &= t^{\alpha-1}\left\langle \nabla f(u) - \nabla f(u'), u - u'\right\rangle . \quad (15.42)\end{aligned}$$

By (15.42) and the monotonicity of ∇f, we have

$$\left\langle \nabla f(u) - \nabla f(u'), u - u'\right\rangle = 0. \quad (15.43)$$

Since f is Legendre, ∇f is strictly monotone and therefore $u = u'$. Thus, there exists a unique $u_t \in C$ such that $u_t = S_t u_t$.

(ii) Let $\{t_n\}_{n\in\mathbb{N}}$ be a sequence in $(0,1)$ such that $t_n \to 1^-$ as $n \to +\infty$. Put $x_n = u_{t_n}$ for all $n \in \mathbb{N}$. By Lemma 15.5 and Theorem 15.8, $F(T)$ is nonempty, closed and convex. Thus the Bregman projection $\operatorname{proj}^f_{F(T)}$ is well defined. To show that $u_t \to \operatorname{proj}^f_{F(T)}(\nabla f^*(0))$, it is sufficient to show that $x_n \to \operatorname{proj}^f_{F(T)}(\nabla f^*(0))$. Since C is bounded, there is a subsequence $\{x_{n_k}\}_{k\in\mathbb{N}}$ of $\{x_n\}_{n\in\mathbb{N}}$ such that $x_{n_k} \rightharpoonup v$. By the definition of x_n, we have $\|x_n - Tx_n\| = (1 - t_n)\|Tx_n\|$ for all

$n \in \mathbb{N}$. So, we have $x_n - Tx_n \to 0$ and hence $v \in \hat{F}(T)$. Lemma 15.6 now implies that $v \in F(T)$. We next show that $x_{n_k} \to v$. Let $y \in F(T)$ be given and fix $n \in \mathbb{N}$. Then, since T is BFNE with respect to f, we have

$$\langle \nabla f(Tx_n) - \nabla f(Ty), Tx_n - Ty \rangle \leq \langle \nabla f(x_n) - \nabla f(y), Tx_n - Ty \rangle . \quad (15.44)$$

That is,

$$0 \leq \langle \nabla f(x_n) - \nabla f(Tx_n), Tx_n - y \rangle . \quad (15.45)$$

Since

$$\begin{aligned} \nabla f(x_n) - \nabla f(Tx_n) &= \nabla f(t_n Tx_n) - \nabla f(Tx_n) \\ &= t_n^{\alpha-1} \nabla f(Tx_n) - \nabla f(Tx_n) = (t_n^{\alpha-1} - 1) \nabla f(Tx_n), \end{aligned}$$

we have

$$0 \leq \left\langle (t_n^{\alpha-1} - 1) \nabla f(Tx_n), Tx_n - y \right\rangle . \quad (15.46)$$

This yields

$$0 \leq \langle -\nabla f(Tx_n), Tx_n - y \rangle \quad (15.47)$$

and

$$\langle \nabla f(y) - \nabla f(Tx_n), y - Tx_n \rangle \leq \langle \nabla f(y), y - Tx_n \rangle . \quad (15.48)$$

Since $x_{n_k} \rightharpoonup v$ and $x_{n_k} - Tx_{n_k} \to 0$, it follows that $Tx_{n_k} \rightharpoonup v$. Hence from (15.48) we obtain

$$\begin{aligned} \limsup_{k \to +\infty} \left\langle \nabla f(y) - \nabla f\left(Tx_{n_k}\right), y - Tx_{n_k} \right\rangle &\leq \limsup_{k \to +\infty} \left\langle \nabla f(y), y - Tx_{n_k} \right\rangle \\ &= \langle \nabla f(y), y - v \rangle \end{aligned} \quad (15.49)$$

Substituting $y = v$ in (15.49), we get

$$0 \leq \limsup_{k \to +\infty} \left\langle \nabla f(v) - \nabla f\left(Tx_{n_k}\right), v - Tx_{n_k} \right\rangle \leq 0.$$

Thus,

$$\lim_{k \to +\infty} \left\langle \nabla f(v) - \nabla f\left(Tx_{n_k}\right), v - Tx_{n_k} \right\rangle = 0.$$

Since

$$D_f\left(v, Tx_{n_k}\right) + D_f\left(Tx_{n_k}, v\right) = \left\langle \nabla f(v) - \nabla f\left(Tx_{n_k}\right), v - Tx_{n_k} \right\rangle , \quad (15.50)$$

it follows that

$$\lim_{k \to +\infty} D_f\left(v, Tx_{n_k}\right) = \lim_{k \to +\infty} D_f\left(Tx_{n_k}, v\right) = 0. \quad (15.51)$$

Proposition 15.3 now implies that $Tx_{n_k} \to v$.

Finally, we claim that $v = \mathrm{proj}_{F(T)}^{f}(\nabla f^{*}(0))$. Since ∇f is norm-to-weak* continuous on bounded subsets, it follows that $\nabla f(Tx_{n_k}) \rightharpoonup \nabla f(v)$. Setting $n := n_k$ and letting $k \to +\infty$ in (15.47), we obtain

$$0 \leq \langle -\nabla f(v), v - y \rangle \tag{15.52}$$

for any $y \in F(T)$. Hence

$$0 \leq \langle \nabla f(\nabla f^{*}(0)) - \nabla f(v), v - y \rangle \tag{15.53}$$

for any $y \in F(T)$. Thus Proposition 15.4 implies that $v = \mathrm{proj}_{F(T)}^{f}(\nabla f^{*}(0))$. Consequently, the whole net $\{u_t\}_{t \in (0,1)}$ converges strongly to $\mathrm{proj}_{F(T)}^{f}(\nabla f^{*}(0))$ as $t \to 1^{-}$. This completes the proof of Theorem 15.13. ■

Remark 15.14. Early analogs of Theorem 15.13 for nonexpansive mappings in Hilbert and Banach spaces may be found in [9, 18, 25].

15.6 Consequences of the Approximation Result

We first specialize Theorem 15.13 to the case, where $f(x) = \frac{1}{2}\|x\|^2$ and X is a uniformly smooth and uniformly convex Banach space, and then apply it to the problem of finding zeroes of a maximal monotone operator $A : X \to 2^{X^*}$. In this case, the function $f(x) = \frac{1}{2}\|x\|^2$ is Legendre (*cf.* [3, Lemma 6.2, p.24]) and uniformly Fréchet differentiable on bounded subsets of X. According to [15, Corollary 1(ii), p. 325], since X is uniformly convex, f is totally convex. Thus we obtain the following corollary.

Corollary 15.15. *Let X be a uniformly smooth and uniformly convex Banach space. Let C be a nonempty, bounded, closed and convex subset of X with $0 \in C$, and let $T : C \to C$ be of firmly nonexpansive type. Then the following two assertions hold:*

(i) For each $t \in (0,1)$, there exists a unique $u_t \in C$ satisfying $u_t = tTu_t$;
(ii) The net $\{u_t\}_{t \in (0,1)}$ converges strongly to $\mathrm{proj}_{F(T)}^{f}(0)$ as $t \to 1^{-}$.

As a matter of fact, this corollary is known to hold even when X is only a smooth and uniformly convex Banach space [21].

As a direct consequence of Theorem 15.13 we get the following new result.

Corollary 15.16. *Let $f : X \to \mathbb{R}$ be a Legendre, totally convex function which is positively homogeneous of degree $\alpha > 1$, uniformly Fréchet differentiable and bounded on bounded subsets of X. Let C be a nonempty, bounded, closed and convex subset* *of X with* $0 \in C$. *Let* λ *be positive real number and let A be a monotone operator such that* $\operatorname{dom} A \subset C \subset (\nabla f)^{-1}(\operatorname{ran}(\nabla f + \lambda A))$. *Then the following two assertions hold:*

(i) For each $t \in (0,1)$, *there exists a unique* $u_t \in C$ *satisfying* $u_t = t\operatorname{Res}^f_{\lambda A} u_t$;
(ii) The net $\{u_t\}_{t \in (0,1)}$ *converges strongly to* $\operatorname{proj}^f_{A^{-1}(0^*)}(\nabla f^*(0))$ *as* $t \to 1^-$.

Remark 15.17. Algorithm 5.5 in [5] provides another way of constructing Bregman projections onto the zero sets of maximal monotone operators.

Acknowledgements The first author was partially supported by the Israel Science Foundation (Grant 647/07), by the Fund for the Promotion of Research at the Technion and by the Technion President's Research Fund. Both authors are grateful to the referees for many detailed and helpful comments.

References

1. Alber, Y.I.: Metric and generalized projection operators in Banach spaces: properties and applications. In: A.G. Kartsatos (ed.) Theory and Applications of Nonlinear Operators of Accretive and Monotone Type, pp. 15–50. Marcel Dekker, New York (1996)
2. Bauschke, H.H., Borwein, J.M.: Legendre functions and the method of random Bregman projections. J. Convex Anal. **4**, 27–67 (1997)
3. Bauschke, H.H., Borwein, J.M., Combettes, P.L.: Essential smoothness, essential strict convexity, and Legendre functions in Banach spaces. Comm. Contemp. Math. **3**, 615–647 (2001)
4. Bauschke, H.H., Borwein, J.M., Combettes, P.L.: Bregman monotone optimization algorithms. SIAM J. Control Optim. **42**, 596–636 (2003)
5. Bauschke, H.H., Combettes, P.L.: Construction of best Bregman approximations in reflexive Banach spaces. Proc. Amer. Math. Soc. **131**, 3757–3766 (2003)
6. Bauschke, H.H., Wang, X., Yao, L.: General resolvents for monotone operators: characterization and extension. Biomedical Mathematics: Promising Directions in Imaging, Therapy Planning and Inverse Problems, Medical Physics Publishing, Madison, WI, USA, 57–74 (2010)
7. Bonnans, J.F., Shapiro, A.: Perturbation Analysis of Optimization Problems. Springer, New York (2000)
8. Bregman, L.M.: The relaxation method of finding a common point of convex sets and its application to the solution of problems in convex programming. USSR Comput. Math. and Math. Phys. **7**, 200–217 (1967)
9. Browder, F.E.: Convergence of approximants to fixed points of nonexpansive non-linear mappings in Banach spaces. Arch. Rational. Mech. Anal. **24**, 82–90 (1967)
10. Bruck, R.E.: Nonexpansive projections on subsets of Banach spaces. Pacific J. Math. **47**, 341–355 (1973)
11. Bruck, R.E., Reich, S.: Nonexpansive projections and resolvents of accretive operators in Banach spaces. Houston J. Math. **3**, 459–470 (1977)
12. Butnariu, D., Iusem, A.N.: Totally Convex Functions for Fixed Points Computation and Infinite Dimensional Optimization. Kluwer Academic Publishers, Dordrecht (2000)
13. Butnariu, D., Resmerita, E.: Bregman distances, totally convex functions and a method for solving operator equations in Banach spaces. Abstr. Appl. Anal. **2006**, 1–39, Art. ID 84919 (2006)
14. Butnariu, D., Censor, Y., Reich, S.: Iterative averaging of entropic projections for solving stochastic convex feasibility problems. Comput. Optim. Appl. **8**, 21–39 (1997)

15. Butnariu, D., Iusem, A.N., Resmerita, E.: Total convexity for powers of the norm in uniformly convex Banach spaces. J. Convex Anal. **7**, 319–334 (2000)
16. Censor, Y., Lent, A.: An iterative row-action method for interval convex programmings. J. Optim. Theory Appl. **34**, 321–353 (1981)
17. Goebel, K., Reich, S.: Uniform Convexity, Hyperbolic Geometry, and Nonexpansive Mappings. Marcel Dekker, New York (1984)
18. Halpern, B.: Fixed points of nonexpanding maps. Bull. Amer. Math. Soc. **73**, 957–961 (1967)
19. Kohsaka, F., Takahashi, W.: Strong convergence of an iterative sequence for maximal monotone operators in a Banach space. Abstr. Appl. Anal. **3**, 239–249 (2004)
20. Kohsaka, F., Takahashi, W.: Existence and approximation of fixed points of firmly nonexpansive-type mappings in Banach spaces. SIAM J. Control Optim. **19**, 824–835 (2008)
21. Kohsaka, F., Takahashi, W.: Fixed point theorems for a class of nonlinear mappings related to maximal monotone operators in Banach spaces. Arch. Math. (Basel) **21**, 166–177 (2008)
22. Minty, G.J.: Monotone (nonlinear) operators in Hilbert space. Duke Math. J. **29**, 341–346 (1962)
23. Moreau, J.J.: Proximité et dualité dans un espace hilbertien. Bull. Soc. Math. France **93**, 273–299 (1965)
24. Reich, S.: Weak convergence theorems for nonexpansive mappings in Banach spaces. J. Math. Anal. Appl. **67**, 274–276 (1979)
25. Reich, S.: Strong convergence theorems for resolvents of accretive operators in Banach spaces. J. Math. Anal. Appl. **75**, 287–292 (1980)
26. Reich, S.: A weak convergence theorem for the alternating method with Bregman distances. In: A.G. Kartsatos (ed.) Theory and Applications of Nonlinear Operators of Accretive and Monotone Type, pp. 313–318. Marcel Dekker, New York (1996)
27. Reich, S., Sabach, S.: A strong convergence theorem for a proximal-type algorithm in reflexive Banach spaces. J. Nonlinear Convex Anal. **10**, 471–485 (2009)
28. Resmerita, E.: On total convexity, Bregman projections and stability in Banach spaces. J. Convex Anal. **11**, 1–16 (2004)
29. Rockafellar, R.T.: Level sets and continuity of conjugate convex functions. Trans. Amer. Math. Soc **123**, 46–63 (1966)

Chapter 16
Regularization Procedures for Monotone Operators: Recent Advances

J.P. Revalski

Abstract In this essentially survey article, we present some recent advances concerning two regularization procedures for monotone operators: extended and variational sums of maximal monotone operators and, the related to them, extended and variational compositions of monotone operators with linear continuous mappings.

Keywords Maximal monotone operators · Sums · Compositions · Graph-convergence · Variational sum · Extended sum · Variational composition · Yosida regularization · Subdifferential

AMS 2010 Subject Classification: 47H05, 46B10, 54C60, 26B25

16.1 Introduction

In the last almost 40 years, the monotone operators have turned out to be an important tool in the study of various problems arising in the domain of optimization, nonlinear analysis, differential equations and other related fields. Among those operators, it seems that the class of maximal monotone ones contains the mappings that possess the most desirable properties, such as, for example, local boundedness, perturbation surjectivity in reflexive spaces, generic single-valuedness and continuity in appropriate classes of Banach spaces, and others. Therefore, when dealing with natural operations on maximal monotone operators, such as, for example, pointwise sums and precompositions with linear operators, it is natural to ask whether the obtained operator is also maximal monotone. In general, when summing up two maximal monotone operators (or precomposing a maximal monotone operator with a linear continuous mapping) the resulting operator is monotone, but not necessarily maximal. In such cases, one needs additional qualification conditions to obtain

J.P. Revalski (✉)
Institute of Mathematics and Informatics, Bulgarian Academy of Sciences,
Acad. G. Bonchev Street, Block 8, 1113 Sofia, Bulgaria
e-mail: revalski@math.bas.bg

H.H. Bauschke et al. (eds.), *Fixed-Point Algorithms for Inverse Problems in Science and Engineering*, Springer Optimization and Its Applications 49,
DOI 10.1007/978-1-4419-9569-8_16,

maximality (see below more details about such conditions and the corresponding references). This lack of maximality was the reason for some researchers to try to find a kind of "generalized" notion of sum (or precomposition) which has more chances to be maximal than the usual point-wise operation. Such attempts led to notions like the parallel sum [27], a sum notion based on the Trotter–Lie formula (see [28]), the variational sum [3, 4] (see also [39, 40]), the extended sum [39, 41], the variational composition [32] and the extended composition [20, 24]. It turned out that some of these notions, as for instance, the variational and the extended ones, give maximality of the corresponding operation in situations when the usual point-wise operation cannot assure this property. This is the case, for example, of subdifferentials of convex functions or certain differential operators [3, 32, 39–41].

Our aim in this, in essence survey, article is to present in more detail the notions of extended and variational sum of maximal monotone mappings and particularly some recent important advances related to them. We also treat the related question of precompositions with linear continuous operators by showing how the results concerning this concept could be derived from corresponding results for sums via natural identifications. Some of the results presented here are formulated without being proved, but those that are key for the presentation are accompanied with detailed proofs.

The paper is organized as follows. In Sect. 16.2, we give the needed preliminary facts and results. Although the extended sum chronologically comes after the variational one, we start in Sect. 16.3 with the former, because its setting is a general Banach space. We present its basic properties, as well as the most important results related to this concept, including the case of extended sum of subdifferentials. Section 16.4 introduces and studies the variational sum of maximal monotone mappings. Its natural setting is when the underlying space is a reflexive Banach space because this concept relies on the Yosida regularization of the operators involved. In particular, we present in this part a recent result of García [21] that the variational sum contains the extended one (and thus, the usual one) – a question that has stayed open since the introduction of the variational sum in [3]. The last Sect. 16.5 treats similar questions to those from Sects. 16.3 and 16.4, related to precompositions of monotone operators with linear continuous mappings. We show that there exists a sort of equivalence between the above notions for sums and precompositions and thus most part of such results for precompositions can be derived from the corresponding facts about sums (and vice versa in many cases). In particular, by using the above mentioned result that the variational sum contains the usual one, we see how one can obtain that the variational composition contains the point-wise one–this also has remained open since the introduction of the variational composition in [32].

16.2 Preliminary Results

Throughout this article $(X, \|\cdot\|)$ will denote a real Banach space. Its topological dual will be designated as usual by X^* and we use the same symbol $\|\cdot\|$ for the dual norm in X^*. The notation $\langle \cdot, \cdot \rangle$ stands for the canonical pairing between X^* and X, and w and w^* for the weak and the weak star topology in X and in X^*, respectively.

When a set-valued operator $T : X \rightrightarrows X^*$ is given we designate by $\mathrm{Dom}\,(T)$ the *domain of* T, that is the set

$$\mathrm{Dom}\,(T) := \{x \in X : Tx \neq \emptyset\},$$

and by $\mathrm{R}(T)$ its *range*, that is

$$\mathrm{R}(T) := \{x^* \in X^* : x^* \in Tx \text{ for some } x \in X\}.$$

The *graph* of T is the following set in $X \times X^*$

$$\mathrm{Gr}(T) := \{(x,x^*) \in X \times X^* : x^* \in Tx\},$$

and obviously the projections of $\mathrm{Gr}(T)$ on X and X^* coincide with the domain and the range of T, respectively. Given an operator T, by $\overline{T}$ we will denote the operator whose images are the norm-closures of the images of T, i.e., $\overline{T}x = \overline{Tx}$, $x \in X$, and by $\overline{T}^G$ the operator whose graph is the norm-closure (in $X \times X^*$) of the graph of T, that is $\mathrm{Gr}(\overline{T}^G) = \overline{\mathrm{Gr}(T)}$. As usual, for a given T, the symbol T^{-1} is reserved for the inverse operator of T, that is $T^{-1}x^* = \{x \in X : x^* \in Tx\}$, $x^* \in X^*$. We obviously have $\mathrm{Dom}\,(T^{-1}) = \mathrm{R}(T)$, $\mathrm{R}(T^{-1}) = \mathrm{Dom}\,(T)$ and the graph of T^{-1} can be identified with $\mathrm{Gr}(T)$. Finally, when we have two operators S and T, we will write often $S \subset T$ (resp. $S = T$) as an equivalent notation for $\mathrm{Gr}(S) \subset \mathrm{Gr}(T)$ (resp. $\mathrm{Gr}(S) = \mathrm{Gr}(T)$).

A (set-valued) mapping $T : X \rightrightarrows X^*$ is called *monotone* if every two couples (x,x^*), $(y,y^*) \in \mathrm{Gr}(T)$ satisfy the inequality

$$\langle x^* - y^*, x - y\rangle \geq 0.$$

If this inequality is strict for any two couples (x,x^*), (y,y^*) from the graph of the operator such that $x \neq y$, then T is called *strictly monotone*. The operator T is *maximal monotone* if it is monotone and its graph is a maximal element with respect to the set inclusion partial order in the class of all monotone operators between X and X^*. Another way to express maximal monotonicity is the following: T is maximal monotone if and only if any couple (z,z^*) which is *monotonically related to* T, that is, satisfies $\langle z^* - y^*, z - y\rangle \geq 0$ for each $(y,y^*) \in \mathrm{Gr}(T)$, necessarily belongs to $\mathrm{Gr}(T)$, i.e., $(z,z^*) \in \mathrm{Gr}(T)$. The following facts are well known: any maximal monotone operator has convex and closed (also for the weak star topology) images; the graph of every maximal monotone mapping is closed in the product (norm) topology in $X \times X^*$; as a consequence of Zorn lemma, each monotone mapping between X and X^* can be extended to a maximal monotone mapping; finally, if $T : X \rightrightarrows X^*$ is maximal monotone then T^{-1} is maximal monotone as a mapping from X^* into X.

Classical examples of maximal monotone mappings are the subdifferentials of certain convex functions. Let is recall that given $\varepsilon \geq 0$, the *ε-subdifferential* $\partial_\varepsilon f$ of a proper convex function $f : X \to \mathbb{R} \cup \{+\infty\}$ is the operator from X to X^* defined as:

$$\partial_\varepsilon f(x) := \{x^* \in X^* : \langle x^*, y - x\rangle \leq f(y) - f(x) + \varepsilon,\ \forall y \in X\},\ \text{if } x \in \mathrm{dom}\, f,$$

and $\partial_\varepsilon f(x) = \emptyset$ if $x \notin \operatorname{dom} f$. Here, as usual, $\operatorname{dom} f$ denotes the set $\{x \in X : f(x) < +\infty\}$, which is the *effective domain* of f, and f is called *proper* if $\operatorname{dom} f \neq \emptyset$. When $\varepsilon = 0$, we denote $\partial_0 f$ simply by ∂f, and this is the well known *subdifferential* of f. It is known that if f is also lower semicontinuous and $\varepsilon > 0$ then $\operatorname{Dom}(\partial_\varepsilon f) = \operatorname{dom} f$, while ∂f, in general, may be empty at some points of $\operatorname{dom} f$ (but $\operatorname{Dom}(\partial f)$ is dense in $\operatorname{dom} f$ according to the Brøndsted–Rockafellar theorem). Observe also, in connection with our considerations in the next section, that $\partial_\varepsilon f$ is an enlargement of ∂f for any $\varepsilon > 0$, that is $\partial f(x) \subset \partial_\varepsilon f(x)$ for each $x \in X$ and $\varepsilon > 0$. Finally, a classical result of Rockafellar [44] states that the subdifferential of a proper lower semicontinuous convex function is a maximal monotone operator. Other classical examples of monotone operators come from some differential operators–see, e.g., the monograph of Zeidler [57]. Comprehensive sources about monotone operators are also the monographs of Phelps [37] and Simons [46] and the paper [38].

16.3 Extended Sum of Monotone Operators

Let $S, T : X \rightrightarrows X^*$ be two maximal (set-valued) monotone mappings. Their usual point-wise (or Minkowski) sum is defined in an obvious algebraic way: $(S+T)(x) = Sx + Tx$, $x \in X$. This operator has domain $\operatorname{Dom}(S+T) = \operatorname{Dom}(S) \cap \operatorname{Dom}(T)$ and is readily seen to be monotone. However, it is not obliged to be maximal monotone: there are simple counterexamples in the plane involving subdifferentials of convex functions – see, e.g., [37]. In the case of reflexive Banach spaces, a classical sufficient qualification condition for the maximality of the sum is that the interior of the domain of one of the operators intersects the domain of the other and was given by Rockafellar [43]. Improvements (in the sense of weakening) of this qualification condition in the reflexive case have been proposed in, e.g., [5, 11, 12, 18, 31, 34, 36, 47]. A substantial progress with the Rockafellar qualification condition in the nonreflexive case was done recently by Voisei [55] (see also the subsequent papers [9, 56]). But it is still an open question whether this condition guarantees the maximality of the sum of the operators in any Banach space. More details concerning this question can be found in Simons [46].

As we mentioned in the introduction, the lack of maximality of the sum of two maximal monotone operators pushed researchers to look for other concepts of sums with the simple idea to have more opportunities to get maximality. In this section, we will present one of them, the *extended sum* which was proposed in [39, 41].

The extended sum relies on the concept of *ε-enlargement* of a given monotone mapping, which is naturally motivated by the notion of ε-subdifferential. Let a monotone operator $T : X \rightrightarrows X^*$ and $\varepsilon \geq 0$ be given. The *ε-enlargement* of T is the operator $T^\varepsilon : X \rightrightarrows X^*$ defined as follows:

$$T^\varepsilon x := \{x^* \in X^* : \langle y^* - x^*, y - x\rangle \geq -\varepsilon \quad \text{for every } (y, y^*) \in \operatorname{Gr}(T)\}, \qquad x \in X.$$

This notion was independently mentioned in [16, 30], and detailed study has been performed in a series of papers by Burachik et al. [14–17], Svaiter [48] and other authors. Some of the basic properties of this notion readily follow from the definition, namely that T^ε has convex w^*-closed values and that it is really an enlargement of T, that is $Tx \subset T^\varepsilon x$ for any $x \in X$, because of the monotonicity of T. Moreover, $T^{\varepsilon_1} \subset T^{\varepsilon_2}$, provided $0 \le \varepsilon_1 \le \varepsilon_2$. It can be seen that T is maximal monotone exactly when $T = T^0 = \bigcap_{\varepsilon>0} T^\varepsilon$. In the particular case of the subdifferential of a proper convex lower semicontinuous function f, if we denote by $\partial^\varepsilon f$ the above enlargement, then one has $\partial_\varepsilon f \subset \partial^\varepsilon f$ and the inclusion can be strict (for instance, when $f(x) = x^2/2$, $x \in \mathbb{R}$ [16, 30]).

This enlargement satisfies, as the subdifferentials do, the Brøndsted-Rockafellar property in a reflexive Banach space – the result is due to Torralba [51] (cf. also [17]). For the same and similar properties outside the reflexive case, the reader is referred to [29,41,45]. Another useful property which is related to the ε-enlargement (and which is a consequence of the so-called transportation formula) is given by the next.

Proposition 16.1. *[15, 48] If $T : X \rightrightarrows X^*$ is a maximal monotone mapping and $\varepsilon \ge 0$, then the enlargement T^ε satisfies:*

$$\langle x^* - y^*, x - y\rangle \ge -4\varepsilon \qquad \forall (x,x^*),(y,y^*) \in \mathrm{Gr}(T^\varepsilon). \tag{16.1}$$

Operators $T : X \rightrightarrows X^*$ that satisfy inequality (16.1) (with $\varepsilon \ge 0$ instead of 4ε) are known as ε-monotone operators and were introduced and studied by Veselý [54], who showed that this class of operators preserves some of the good properties of the monotone operators, for example they are locally bounded in the interior of their domain.

The enlargement given above is related also to the *Fitzpatrick function* [19, Definition 3.1] $\varphi_T : X \times X^* \to \mathbb{R} \cup \{+\infty\}$, associated with any monotone (nontrivial) operator $T : X \rightrightarrows X^*$:

$$\begin{aligned} \varphi_T(x,x^*) &:= \sup_{(y,y^*)\in \mathrm{Gr}(T)} \left(\langle y^*, x\rangle - \langle y^*, y\rangle + \langle x^*, y\rangle\right) \\ &= \sup_{(y,y^*)\in \mathrm{Gr}(T)} \left(\langle y^* - x^*, x - y\rangle\right) + \langle x^*, x\rangle. \end{aligned}$$

This function has turned out to be quite useful in monotone operator theory (see, e.g., [6–8, 46–48, 56] just to mention a few). The function φ_T is proper, convex and lower semicontinuous. Moreover, φ_T verifies the following: for any $\varepsilon \ge 0$,

$$x^* \in T^\varepsilon(x) \Longleftrightarrow \varphi_T(x,x^*) \le \langle x^*, x\rangle + \varepsilon,$$

and, if T is maximal monotone (see [19]), then φ_T is the minimal convex function with the property

$$\langle x^*, x\rangle \le \varphi_T(x,x^*) \;\; \forall (x,x^*) \in X \times X^*, \;\; \text{and} \;\; \langle x^*, x\rangle = \varphi_T(x,x^*) \quad \forall (x,x^*) \in \mathrm{Gr}(T).$$

With these properties in hand, one can verify that:

Proposition 16.2. *[24] Let $T : X \rightrightarrows X^*$ be a monotone operator. If pr_X is the usual projection of $X \times X^*$ on X, then*

$$\mathrm{pr}_X \operatorname{dom} \varphi_T = \bigcup_{\varepsilon>0} \operatorname{Dom} T^{\varepsilon}.$$

Let us pass now to the definition of the extended sum. In what follows, the symbol $\overline{A}^{w^*}$ for a set A in the dual space X^*, will mean the closure of A with respect to the weak star topology w^* in X^*. In [25] Hiriart–Urruty and Phelps established the following formula for the subdifferential of the sum of two convex functions:

Theorem 16.3. *[25] Let $f, g : X \to \mathbb{R} \cup \{+\infty\}$ be two proper convex lower semi-continuous functions. Then for every $x \in \operatorname{dom} f \cap \operatorname{dom} g$ one has:*

$$\partial(f+g)(x) = \bigcap_{\varepsilon>0} \overline{\partial_{\varepsilon} f(x) + \partial_{\varepsilon} g(x)}^{w^*}.$$

Having this result and disposing with the notion of enlargement of a given monotone operator, the concept of extended sum comes in a natural way. Namely,

Definition 16.4. [39, 41] Let $S, T : X \rightrightarrows X^*$ be monotone operators. The extended sum of S and T, denoted by $S \underset{\text{ext}}{+} T$, is defined by

$$S \underset{\text{ext}}{+} T(x) := \bigcap_{\varepsilon>0} \overline{S^{\varepsilon}x + T^{\varepsilon}x}^{w^*}, \qquad x \in X.$$

Evidently, this sum is commutative and it contains the usual point-wise sum of S and T. Moreover, the extended sum has w^*-closed and convex images. But, as we will see later, the graph of the extended sum is not obliged to be closed. Although we can sum up in this extended way arbitrary monotone operators, it is not of substantial interest because there are examples (see, e.g., [24], Example 3.3) showing that, in general, this sum is not monotone. But if both operators are maximal monotone, then the extended sum is monotone:

Proposition 16.5 ([24] Proposition 3.4). *Let $S, T : X \rightrightarrows X^*$ be maximal monotone. Then the extended sum $S \underset{\text{ext}}{+} T$ is a monotone operator.*

Proof. Take two couples $(x, x^*), (y, y^*) \in \mathrm{Gr}(S \underset{\text{ext}}{+} T)$ and let us fix some $\varepsilon > 0$. By the definition of the extended sum there are nets $\{u^*_{x,\alpha}\}_\alpha \subset S^{\varepsilon}x$ and $\{v^*_{x,\alpha}\}_\alpha \subset T^{\varepsilon}x$ such that

$$u^*_{x,\alpha} + v^*_{x,\alpha} \xrightarrow{w^*} x^* \tag{16.2}$$

and similarly, nets $\{u^*_{y,\beta}\}_\beta \subset S^{\varepsilon}y$ and $\{v^*_{y,\beta}\}_\beta \subset T^{\varepsilon}y$ with the property

$$u^*_{y,\beta} + v^*_{y,\beta} \xrightarrow{w^*} y^*. \tag{16.3}$$

By Proposition 16.1 for any α and β, we have

$$\langle u^*_{y,\beta} - u^*_{x,\alpha}, y - x\rangle \geq -4\varepsilon \qquad \text{and} \qquad \langle v^*_{y,\beta} - v^*_{x,\alpha}, y - x\rangle \geq -4\varepsilon.$$

Thus, by summing up, passing to the limit on α and β and using (16.2) and (16.3) we obtain that

$$\langle y^* - x^*, y - x\rangle \geq -8\varepsilon.$$

Since $\varepsilon > 0$ was arbitrary, we conclude that

$$\langle y^* - x^*, y - x\rangle \geq 0.$$

Thus, $S \underset{\text{ext}}{+} T$ is monotone and this completes the proof. ∎

This proposition helps obtain easily several results from [39,41], which originally were proved directly. The first one is the following immediate corollary.

Corollary 16.6. *[39, 41] Let $S, T : X \rightrightarrows X^*$ be maximal monotone operators. If $\overline{S+T}$ is maximal monotone, then $\overline{S+T} = S \underset{\text{ext}}{+} T$. In particular, if $S+T$ is maximal monotone, then $S+T = S \underset{\text{ext}}{+} T$.*

Let us mention that in some cases the two types of sums–the usual and the extended one coincide without being a maximal monotone operator (cf. Example 16.20 below).

The next result is more interesting since it gives an important case where the extended sum is always maximal monotone, while the usual one is not, in general. Namely, without any qualification condition the subdifferential of the sum of two proper convex lower semicontinuous functions is equal to the extended sum of their subdifferentials.

Corollary 16.7. *[39,41] Let $f, g : X \to \mathbb{R} \cup \{+\infty\}$ be proper convex lower semicontinuous functions such that* $\operatorname{dom} f \cap \operatorname{dom} g \neq \emptyset$. *Then, for any $x \in X$*

$$\partial(f+g)(x) = (\partial f \underset{\text{ext}}{+} \partial g)(x)$$

Proof. Theorem 16.3 shows that $\partial f \underset{\text{ext}}{+} \partial g$ contains $\partial(f+g)$ (remember that $\partial_\varepsilon f \subset \partial^\varepsilon f$). Since ∂f and ∂g are maximal monotone, $\partial f \underset{\text{ext}}{+} \partial g$ is monotone by Proposition 16.5, hence it must coincide with $\partial(f+g)$ because the latter operator is maximal monotone. ∎

Other results in which the subdifferential of the sum of two convex functions is represented by approximations of the subdifferentials of the functions involved can be found also in [33,49,50].

The previous result hints that to have an equality between the usual and the extended sum we will need a qualification condition. Indeed, the next result shows that the usual and extended sum are equal under a qualification condition of Robinson–Rockafellar type. Let us recall that a set $A \subset X$ is said to be *absorbing in* X (we express this by writing $0 \in \operatorname{core}_X A$) if for any $h \in X$ there is $t > 0$ with $th \in A$.

Theorem 16.8 ([24], Theorem 3.7). *Let $S, T : X \rightrightarrows X^*$ be maximal monotone operators. Suppose that*

$$0 \in \operatorname{core}_X (\operatorname{pr}_X \operatorname{dom} \varphi_S - \operatorname{pr}_X \operatorname{dom} \varphi_T).$$

Then:

(1) For any $\varepsilon \geq 0$ and $x \in X$, $S^\varepsilon x + T^\varepsilon x$ is w^*-closed;
(2) $S + T = S \underset{\text{ext}}{+} T$.

The proof is based on the Krein–Shmulian Theorem and the properties of the enlargements. An analogous result for subdifferentials can be found in [49] and the case $\varepsilon = 0$ in (1) is proved in [53] with a different (formally, stronger) qualification condition. A very recent study of qualification conditions of the same nature as above is contained in the papers [10, 56]. In particular, as it was pointed out to the author by C. Zălinescu, the condition above is equivalent to the usual interior point one (the latter concerns also the condition in Theorem 16.26 below). The paper [10] contains also a weakening to get (1) in the reflexive case.

Since for any proper, lower semicontinuous convex function $h : X \to \mathbb{R} \cup \{+\infty\}$ and $\varepsilon > 0$ we have $\operatorname{dom} h = \operatorname{Dom}(\partial_\varepsilon h) \subset \operatorname{Dom}(\partial^\varepsilon h) \subset \operatorname{pr}_X \operatorname{dom} \varphi_{\partial h}$ the previous theorem easily implies the following well known result about exact sum rule under Robinson–Rockafellar condition.

Corollary 16.9. *Let $f, g : X \to \mathbb{R} \cup \{+\infty\}$ be proper convex lower semicontinuous functions. If $0 \in \operatorname{core}_X(\operatorname{dom} f - \operatorname{dom} g)$, then $\partial f + \partial g = \partial(f + g)$.*

16.4 Variational Sum of Maximal Monotone Operators

The concept of variational sum was introduced and studied for the first time by Attouch et al. in [3]. Originally, it was considered in the setting of Hilbert spaces but as it was shown later in [39, 40] the notion can be introduced in a natural way also in the setting of reflexive spaces by preserving its properties. Therefore, in this section (and in part of the next one) X will be assumed always a reflexive Banach space. According to a well known renorming theorem of Asplund [1] we may (and will) suppose that both norms, the norm in X and its dual norm in X^*, are Gâteaux differentiable away from the origin. Thus, in such a case these norms are strictly convex, that is, the corresponding unit spheres do not contain line segments. When needed (see Troyanski [52]) the renorming can be stronger, in order these norms to be also locally uniformly rotund in which case they satisfy also *the Kadec–Klee property*: if $\{x_n\}_{n\geq 1} \subset X$ converges weakly to $x \in X$ and if $\|x_n\| \to \|x\|$, then x_n converges to x strongly (and similarly for the dual norm).

With such norms the duality mapping $J_X : X \to X^*$ which is defined by

$$J_X x := \{x^* \in X^* : \langle x^*, x\rangle = \|x\|^2 = \|x^*\|^2\}, \qquad x \in X,$$

is an everywhere defined single-valued mapping which is bijective and norm-to-weak continuous. When there is no ambiguity (as it is in this section) we will write simply J instead of J_X. It is well known that this mapping is, in fact, the subdifferential of the convex function $(1/2)\|\cdot\|^2$ and thus it is a maximal monotone operator. With the chosen norms we obviously have that the duality mapping J^* of the dual is J^{-1}. If, in addition, the norms satisfy the Kadec-Klee property then J and J^* are norm-to-norm continuous. Let us recall the well known Minty–Rockafellar theorem (see, e.g., [43]): *A monotone operator $T : X \rightrightarrows X^*$ is maximal monotone if and only if for any $\lambda > 0$ the (maximal monotone) operator $T + \lambda J$ is surjective. In this case the inverse operator $(T + \lambda J)^{-1}$ is an everywhere defined single-valued strictly monotone operator which is norm-to-weak continuous.*

As the concept of extended sum relies on the notion of enlargement of the operators involved, the notion of variational sum is based on another regularization of a given monotone operator: the *Yosida regularization*. First, let us remind that given a maximal monotone operator $T : X \rightrightarrows X^*$ the *resolvent* J_λ^T of T of order $\lambda > 0$ is the operator which to each $x \in X$ assigns the (unique, according to the above cited result of Rockafellar) solution $x_\lambda = J_\lambda^T x$ to the inclusion

$$0 \in J(x_\lambda - x) + \lambda T x_\lambda. \tag{16.4}$$

The resolvent J_λ^T is an everywhere defined operator which maps X into $\mathrm{Dom}(T)$. The *Yosida regularization* T_λ of T of order $\lambda > 0$ is the operator

$$T_\lambda x := \frac{1}{\lambda} J(x - x_\lambda), \qquad x \in X, \tag{16.5}$$

which obviously is everywhere defined.

It can be easily seen that for any $\lambda > 0$ we have the following properties

$$J_\lambda^T x = x - \lambda J^{-1} T_\lambda x \qquad \forall x \in X, \tag{16.6}$$

and

$$T_\lambda x \in T(J_\lambda^T x) \qquad \forall x \in X. \tag{16.7}$$

One can verify that an equivalent purely analytical definition (which avoids the use of resolvents) of the Yosida regularization is the following one:

$$T_\lambda = (T^{-1} + \lambda J^{-1})^{-1}. \tag{16.8}$$

A sum of the type $(S^{-1}+T^{-1})^{-1}$ for given monotone operators $S,T : X \rightrightarrows X^*$ is known as the *parallel sum* of S and T. Therefore, the Yosida regularization of T of order $\lambda > 0$ is the parallel sum of T and $(1/\lambda)J$. From this equivalent definition one easily sees (by using the Minty–Rockafellar result cited above) that the Yosida regularization T_λ, $\lambda > 0$, of a given maximal monotone operator $T : X \rightrightarrows X^*$ is an everywhere defined single-valued maximal monotone operator which is norm-to-weak continuous.

When X is a Hilbert space and we identify X^* with X, then J is the identity mapping I and the above definitions take their most known forms: $J_\lambda^T = (I+\lambda T)^{-1}$ and $T_\lambda = (I - J_\lambda^T)/\lambda$.

Let us now introduce the concept of variational sum. Put $\mathscr{I} := \{(\lambda,\mu) \in \mathbb{R}^2 : \lambda,\mu \geq 0, \lambda+\mu \neq 0\}$. The general idea is the following one: given two maximal monotone operators $S,T : X \rightrightarrows X^*$ and $(\lambda,\mu) \in \mathscr{I}$, to consider the operator $S_\lambda + T_\mu$ (with the convention $S_0 = S$ and $T_0 = T$) and then to pass to an appropriate limit on λ,μ. Let us stress the fact that, since $(\lambda,\mu) \in \mathscr{I}$, at least one of the parameters is different from 0, thus at least one of the operators S_λ or T_μ is everywhere defined maximal monotone operator and therefore, according to the Rockafellar qualification condition the sum $S_\lambda + T_\mu$ will be always a maximal monotone operator.

A convergence that has turned out to be useful when monotone operators are involved is the Painlevé–Kuratowski convergence – see, e.g., [2]: one identifies the maximal monotone operators with their graphs in $X \times X^*$, endowed with some usual product norm, and then considers Painlevé–Kuratowski convergence of these graphs as closed sets. More formally, let $\{C_{\lambda,\mu} : (\lambda,\mu) \in \mathscr{I}\}$ be a family of maximal monotone operators between X and X^*. Denote by $\mathscr{F}$ the filter of all neighborhoods of the zero in $\mathscr{I}$.

- $(x,x^*) \in X \times X^*$ belongs to the *lower limit* in the sense of Painlevé–Kuratowski of the family $\{C_{\lambda,\mu}\}$ along the filter $\mathscr{F}$, which is denoted by $\liminf_{\mathscr{F}} C_{\lambda,\mu}$, if for any neighborhood U of (x,x^*) in the product topology in $X \times X^*$ there is $F \in \mathscr{F}$ so that $\mathrm{Gr}(C_{\lambda,\mu}) \cap U \neq \emptyset$ for each $(\lambda,\mu) \in F$;
- $(x,x^*) \in X \times X^*$ belongs to the *upper limit* in the sense of Painlevé–Kuratowski of the family $\{C_{\lambda,\mu}\}$ along the filter $\mathscr{F}$, which is denoted by $\limsup_{\mathscr{F}} C_{\lambda,\mu}$, if for any neighborhood U of (x,x^*) in the product topology in $X \times X^*$ and for every $F \in \mathscr{F}$ there is $(\lambda,\mu) \in F$ with $\mathrm{Gr}(C_{\lambda,\mu}) \cap U \neq \emptyset$;
- The family $\{C_{\lambda,\mu} : (\lambda,\mu) \in \mathscr{I}\}$ *(graph)-converges* to $C \subset X \times X^*$ in the sense of Painlevé–Kuratowski if $C = \liminf_{\mathscr{F}} C_{\lambda,\mu} = \limsup_{\mathscr{F}} C_{\lambda,\mu}$. We write in this case $C = \lim_{\mathscr{F}} C_{\lambda,\mu}$.

  One can check that equivalent sequential definitions of the above limits sound as follows:

- $(x,x^*) \in \liminf_{\mathscr{F}} C_{\lambda,\mu}$ iff for any sequence $\{(\lambda_n,\mu_n)\}_{n\geq 1} \in \mathscr{I}$ such that $(\lambda_n,\mu_n) \to (0,0)$ there is a sequence $\{(x_n,x_n^*)\}_{n\geq 1}$ so that $(x_n,x_n^*) \in \mathrm{Gr}(C_{\lambda_n,\mu_n})$ for any $n \geq 1$ and $(x_n,x_n^*) \to (x,x^*)$: this comes from the fact that a couple (x,x^*) is in the lower limit of $C_{\lambda,\mu}$ exactly when the distances from (x,x^*) to $C_{\lambda,\mu}$ go to zero along the filter $\mathscr{F}$;

- $(x,x^*) \in \limsup_{\mathscr{F}} C_{\lambda,\mu}$ iff there is a sequence $\{(\lambda_n,\mu_n)\}_{n\geq 1} \in \mathscr{I}$ such that $(\lambda_n,\mu_n) \to (0,0)$ and a sequence $\{(x_n,x_n^*)\}_{n\geq 1}$ so that $(x_n,x_n^*) \in \mathrm{Gr}(C_{\lambda_n,\mu_n})$ for any $n \geq 1$ and $(x_n,x_n^*) \to (x,x^*)$.

The interest to this convergence for maximal monotone operators is motivated by the following well known (and easily verifiable) facts: the lower limit of the family of maximal monotone operators $\{C_{\lambda,\mu} : (\lambda,\mu) \in \mathscr{I}\}$ (when it exists) is always a monotone operator (with closed graph by definition); and, if C is a maximal monotone operator then $C = \lim_{\mathscr{F}} C_{\lambda,\mu}$ if and only if $C \subset \liminf_{\mathscr{F}} C_{\lambda,\mu}$. The latter follows from the fact that any point of $\limsup_{\mathscr{F}} C_{\lambda,\mu}$ is monotonically related to $\liminf_{\mathscr{F}} C_{\lambda,\mu}$.

Now we are ready to give the definition of the variational sum and some of its basic properties.

Definition 16.10. [3, 39, 40] Let S and T be maximal monotone operators in the reflexive Banach space X. The variational sum of S and T, denoted by $S \underset{v}{+} T$, is the operator between X and X^* having the following graph:

$$S \underset{v}{+} T := \liminf_{\mathscr{F}}(S_\lambda + T_\mu).$$

A first list of basic properties which can be derived directly from the definition is the following one:

Proposition 16.11 (see, e.g., [40, Proposition 4.6]). *Let X be a reflexive Banach space and $S,T : X \rightrightarrows X^*$ be maximal monotone operators. Then*

(1) $\mathrm{Dom}\,(S) \cap \mathrm{Dom}\,(T) \subset \mathrm{Dom}\,(S \underset{v}{+} T) \subset \overline{\mathrm{Dom}\,(S)} \cap \overline{\mathrm{Dom}\,(T)}$;

(2) $S \underset{v}{+} T$ *is a monotone operator with closed graph;*

(3) If $S \underset{v}{+} T$ *is maximal then* $S \underset{v}{+} T = \lim_{\mathscr{F}}(S_\lambda + T_\mu)$;

(4) $S \underset{v}{+} T = T \underset{v}{+} S$.

Indeed, (2)–(4) and the second inclusion in (1) are direct consequences from the definitions and the remarks above. Precisely speaking, the first inclusion in (1) was proved in [40], provided the norm in X^* satisfies also the Kadec–Klee property, but as we will see in Theorem 16.16 it is true without supposing this property.

A couple of remarks are in order here: first of all observe that the definition of the variational sum at a certain point, takes into account the behavior of the operators involved also in nearby points, while in the definition of the extended sum this is not the case. The second remark concerns the comparison with the usual sum: while in the case of extended sum it is immediately seen that the usual point-wise sum is included in the extended one, here in the case of variational sum, it is not clear from the definition whether we obtain, in general, a bigger sum. The question was resolved in [21] – see below Theorem 16.16.

When dealing with the variational sum, sometimes we need more workable equivalent definitions involving solutions to resolvent type inclusions. The following one (condition (b) in the next proposition) was mentioned for the first time in [3] in the Hilbert space setting. The equivalence below relies on a technique, which originates from the work of Brezis, Crandall, and Pazy [13] and then it was developed for the case of the variational sum (when we have two perturbed operators) by Attouch, Baillon and Théra in [3, 4] for the Hilbert space setting. The extension of this technique to the case of reflexive spaces is in [40] and a further development is contained in [21].

Proposition 16.12. *Let S, T be maximal monotone operators in the reflexive Banach space X. Then the following are equivalent:*

(a) $(x, x^*) \in \mathrm{Gr}(S \underset{v}{+} T)$;

(b) For any $(\lambda, \mu) \in \mathscr{I}$ *the (unique) solution* $x_{\lambda,\mu}$ *of the inclusion*

$$x^* \in J(x_{\lambda,\mu} - x) + S_\lambda x_{\lambda,\mu} + T_\mu x_{\lambda,\mu} \tag{16.9}$$

converges to x *as* $\lambda, \mu \to 0$;

Proof. Since $\|J(x_{\lambda,\mu} - x)\| = \|x_{\lambda,\mu} - x\|$, the implication (b)$\Longrightarrow$(a) is immediate. The argument that (a) implies (b) was not given explicitly in [3] (only the boundedness of the filtered family was obtained). The proof which we give here, uses an argument from [21, Lemma 3.1], where, in particular, condition (16.10) is also established. Let us mention that the inclusion (16.9) has solutions for any couple $(x, x^*) \in X \times X^*$.

Lemma 16.13. *Let* $(x, x^*) \in X \times X^*$ *be any couple and suppose that* $\mathrm{Dom}\,(S \underset{v}{+} T) \neq \emptyset$. *Then the solutions of the inclusion* (16.9) *for* (x, x^*) *remain bounded as* $\lambda, \mu \to 0$. *Moreover, for any* $(y, y^*) \in \mathrm{Gr}(S \underset{v}{+} T)$ *and any sequence* $\{x_{\lambda_n,\mu_n}\}_{n \geq 1}$, *with* $(\lambda_n, \mu_n) \to 0$, *such that* x_{λ_n,μ_n} *converges weakly to some* $\bar{x}$, *we have*

$$\frac{1}{2}\|y - x\|^2 + \langle x^* - y^*, \bar{x} - y \rangle \geq \frac{1}{2} \limsup_n \|x_{\lambda_n,\mu_n} - x\|^2. \tag{16.10}$$

The boundedness of the filtered family $\{x_{\lambda,\mu} : (\lambda, \mu) \in \mathscr{I}\}$ when $(\lambda, \mu) \to (0,0)$ was proved in the Hilbert space setting by Attouch et al. in [3] (see the proof of Theorem 6.1 there) provided $\mathrm{Dom}\,(S) \cap \mathrm{Dom}\,(T) \neq \emptyset$. The same fact in reflexive spaces for the family $\{x_{\lambda,\mu}\}$ of solutions of the variant of (16.9) in which, instead of the term $J(x_{\lambda,\mu} - x)$, we consider $Jx_{\lambda,\mu} - Jx$, is in [40] (the proof of Theorem 4.12 and Remark 4.8). It can be easily seen that the boundedness of the family of solutions to (16.9) and the boundedness of the solutions of the latter variant of (16.9) are equivalent. However, in the absence of the Kadec–Klee property of the norms, it is better to work with the solutions of (16.9).

In the proof below we use the reasoning from [21, Lemma 3.1] not because it allows showing the boundedness of the family above in the (formally) more general case when $\mathrm{Dom}\,(S \underset{v}{+} T) \neq \emptyset$ (we do not have examples of operators for which $\mathrm{Dom}\,(S) \cap \mathrm{Dom}\,(T) = \emptyset$ and $\mathrm{Dom}\,(S \underset{v}{+} T) \neq \emptyset$) but more important, it gives in addition the inequality (16.10) which helps obtaining properties related to the variational sum, without supposing that the norm in X satisfies the Kadec–Klee property.

Proof of Lemma 16.13. Let $(y, y^*) \in \mathrm{Gr}(S \underset{v}{+} T)$ and let U be the open ball in $X \times X^*$ around (y, y^*) with radius 1. By the definition of the variational sum there is $F \in \mathscr{F}$ such that for any $(\lambda, \mu) \in F$ we have $U \cap \mathrm{Gr}(S_\lambda + T_\mu) \neq \emptyset$. For any $(\lambda, \mu) \in F$ take an arbitrary $(y_{\lambda,\mu}, y^*_{\lambda,\mu}) \in U \cap \mathrm{Gr}(S_\lambda + T_\mu)$. This means $y^*_{\lambda,\mu} \in S_\lambda y_{\lambda,\mu} + T_\mu y_{\lambda,\mu}$ and thus the monotonicity of $S_\lambda + T_\mu$ and the fact that $x_{\lambda,\mu}$ is a solution to (16.9) entail that for any $(\lambda, \mu) \in F$ we have

$$\langle x^* - J(x_{\lambda,\mu} - x) - y^*_{\lambda,\mu}, x_{\lambda,\mu} - y_{\lambda,\mu} \rangle \geq 0.$$

Therefore, for each $(\lambda, \mu) \in F$ the following is true

$$\langle x^* - y^*_{\lambda,\mu}, x_{\lambda,\mu} - y_{\lambda,\mu} \rangle \geq \langle J(x_{\lambda,\mu} - x), x_{\lambda,\mu} - y_{\lambda,\mu} \rangle.$$

On the other hand, the fact that J is the subdifferential of the function $(1/2)\|\cdot\|^2$ allows us to obtain that

$$\begin{aligned} \langle J(x_{\lambda,\mu} - x), y_{\lambda,\mu} - x_{\lambda,\mu} \rangle &= \langle J(x_{\lambda,\mu} - x), y_{\lambda,\mu} - x - (x_{\lambda,\mu} - x) \rangle \\ &\leq \frac{1}{2}\|y_{\lambda,\mu} - x\|^2 - \frac{1}{2}\|x_{\lambda,\mu} - x\|^2 \end{aligned}$$

for every $(\lambda, \mu) \in F$. The latter two inequalities easily yield the next one

$$\frac{1}{2}\|y_{\lambda,\mu} - x\|^2 + \langle x^* - y^*_{\lambda,\mu}, x_{\lambda,\mu} - y_{\lambda,\mu} \rangle \geq \frac{1}{2}\|x_{\lambda,\mu} - x\|^2 \qquad \forall (\lambda, \mu) \in F. \quad (16.11)$$

Since for $(\lambda, \mu) \in F$ the families $\{y_{\lambda,\mu}\}$ and $\{y^*_{\lambda,\mu}\}$ remain bounded then there are some $\alpha, \beta \geq 0$ satisfying $\alpha + \beta\|x_{\lambda,\mu} - x\| \geq (1/2)\|x_{\lambda,\mu} - x\|^2$ for every $(\lambda, \mu) \in F$ and this shows that the family $\{x_{\lambda,\mu} : (\lambda, \mu) \in \mathscr{I}\}$ is bounded when $(\lambda, \mu) \to (0, 0)$.

As to (16.10), take a sequence $\{(\lambda_n, \mu_n)\}_{n \geq 1}$ from $\mathscr{I}$ which converges to $(0, 0)$ and such that x_{λ_n,μ_n} converges weakly to some $\bar{x}$. Let $(y, y^*) \in \mathrm{Gr}(S \underset{v}{+} T)$. By the definition of the lower limit, there is a sequence $\{(y_n, y^*_n)\}_{n \geq 1}$ such that $(y_n, y^*_n) \in \mathrm{Gr}(S_{\lambda_n} + T_{\mu_n})$ for every $n \geq 1$ and $(y_n, y^*_n) \to (y, y^*)$. Since obviously for large enough n we have $(\lambda_n, \mu_n) \in F$, then plugging these y_n, y^*_n and the corresponding x_{λ_n,μ_n} in the inequality (16.11) above and passing to the limit on n we obtain (16.10). The proof of the lemma is completed. ■

Now let us come back to the proof of Proposition 16.12. As we mentioned, we only have to prove that (a)⟹(b). Let $(x,x^*) \in \mathrm{Gr}(S \underset{v}{+} T)$ and let $\{x_{\lambda,\mu} : (\lambda,\mu) \in \mathscr{I}\}$ be the family of solutions to (16.9). To prove that $x_{\lambda,\mu}$ converges strongly to x as $(\lambda,\mu) \to (0,0)$, it is enough to show that for every sequence $\{(\lambda_n,\mu_n)\}_{n\geq 1} \subset \mathscr{I}$ which converges to $(0,0)$, the sequence $\{x_{\lambda_n,\mu_n}\}_{n\geq 1}$ has a subsequence which converges strongly to x. Indeed, let $\{(\lambda_n,\mu_n)\}_{n\geq 1}$ be a sequence in $\mathscr{I}$ which converges to $(0,0)$. Since according to the above lemma $\{x_{\lambda_n,\mu_n}\}_{n\geq 1}$ is bounded then it has a subsequence (we do not relabel and denote this subsequence again by $\{x_{\lambda_n,\mu_n}\}_{n\geq 1}$) which converges weakly to some $\bar{x}$. Apply (16.10) with $(y,y^*) = (x,x^*)$. This gives $\limsup_n \|x_{\lambda_n,\mu_n} - x\|^2 \leq 0$ which shows that $\{x_{\lambda_n,\mu_n}\}_{n\geq 1}$ converges strongly to x. This completes the proof of Proposition 16.12. ■

Another application of (16.10) and the equivalent definitions above is the following

Corollary 16.14 ([21] Proposition 3.2). *Let $S,T : X \rightrightarrows X^*$ be maximal monotone operators in the reflexive Banach space X. Then the values of $S \underset{v}{+} T$ are convex.*

We come now to the question of comparison of the variational sum with the usual one (or the extended one). So far only partial results related to this question have been known: for example if the operator $\overline{S+T}^G$ is maximal then it was known that it coincides with the variational sum [3, 39, 40]. Or, if either the extended sum or the variational one was supposed to be maximal, then it contained the other– [39, 40]. The problem is solved by García in [21] where the author showed that in the setting of reflexive spaces the variational sum contains the extended one (and hence the usual one). To prove this result we need the following simple, but useful, lemma (see [32, Lemma 3.1] and [21, Lemma 3.5]):

Lemma 16.15. *Let $T : X \rightrightarrows X^*$ be a maximal monotone operator. Then, for every $\lambda,\varepsilon \geq 0$, $w^* \in T^\varepsilon w$ and $u^* \in T_\lambda u$ we have*

$$\langle u^* - w^*, u - w\rangle + \frac{\lambda}{4}\|w^*\|^2 \geq -\varepsilon.$$

Proof. By (16.6) and (16.7) we have $u^* \in T(u - \lambda J^{-1}u^*)$ (for $\lambda = 0$ this is obviously true) and thus by the definition of the ε-enlargement we obtain

$$\langle u^* - w^*, u - \lambda J^{-1}u^* - w\rangle \geq -\varepsilon.$$

This yields

$$\begin{aligned}\langle u^* - w^*, u - w\rangle &\geq \lambda\langle u^* - w^*, J^{-1}u^*\rangle - \varepsilon\\ &\geq \lambda(\|u^*\|^2 - \|w^*\|\,\|u^*\|) - \varepsilon\\ &\geq -\frac{\lambda}{4}\|w^*\|^2 - \varepsilon.\end{aligned}$$

■

Before giving the next result, let us mention that in the setting of reflexive spaces where the weak and the weak star topology in X^* are the same, the extended sum of two monotone operators $S,T : X \rightrightarrows X^*$ at a point $x \in X$ is, in fact, the intersection of the norm closures of the sets of the type $S^\varepsilon x + T^\varepsilon x$, $\varepsilon > 0$, because of the convexity of the ε-enlargements. The proof of the theorem below differs a bit from the original one from [21, Theorem 3.6] to be more direct.

Theorem 16.16 ([21] Theorem 3.6). *Let $S,T : X \rightrightarrows X^*$ be maximal monotone operators in the reflexive Banach space X. Then*

$$S \underset{\text{ext}}{+} T(x) \subset S \underset{v}{+} T(x) \qquad \forall x \in X.$$

Proof. Take an arbitrary couple (x,x^*) from $\mathrm{Gr}(S \underset{\text{ext}}{+} T)$ and let $\{(\lambda_n,\mu_n)\}_{n\geq 1}$ be a sequence from $\mathscr{I}$ which converges to $(0,0)$. For any $n \geq 1$, let x_n be the (unique) solution of the inclusion (16.9) for the couple (x,x^*). That is, for any $n \geq 1$, there are $u_n^* \in S_{\lambda_n} x_n$ and $v_n^* \in T_{\mu_n} x_n$ such that

$$x^* = J(x_n - x) + u_n^* + v_n^* \qquad \forall n \geq 1. \tag{16.12}$$

According to the definitions the proof will be completed if we show that the sequence $\{x_n\}_{n\geq 1}$ converges to x (because in this case obviously $u_n^* + v_n^* \to x^*$). To this end, let us take an arbitrary $\varepsilon > 0$ and fix it. By the definition of the extended sum and the remark before the theorem, there are $s_\varepsilon^* \in S^\varepsilon x$ and $t_\varepsilon^* \in T^\varepsilon x$ so that $\|x^* - (s_\varepsilon^* + t_\varepsilon^*)\| < \varepsilon$. Let us now apply Lemma 16.15 first for the operator S, the couples (x,s_ε^*), (x_n,u_n^*), λ_n and ε and then for the operator T, the couples (x,t_ε^*), (x_n,v_n^*), μ_n and ε. This gives:

$$\langle u_n^* - s_\varepsilon^*, x_n - x\rangle + \frac{\lambda_n}{4}\|s_\varepsilon^*\|^2 \geq -\varepsilon \qquad \forall n \geq 1$$

and

$$\langle v_n^* - t_\varepsilon^*, x_n - x\rangle + \frac{\mu_n}{4}\|t_\varepsilon^*\|^2 \geq -\varepsilon \qquad \forall n \geq 1,$$

which after summing up yields

$$\langle (u_n^* + v_n^*) - (s_\varepsilon^* + t_\varepsilon^*), x_n - x\rangle + \frac{\lambda_n}{4}\|s_\varepsilon^*\|^2 + \frac{\mu_n}{4}\|t_\varepsilon^*\|^2 \geq -2\varepsilon \qquad \forall n \geq 1.$$

According to (16.12), this means that

$$\langle x^* - J(x_n - x) - (s_\varepsilon^* + t_\varepsilon^*), x_n - x\rangle + \frac{\lambda_n}{4}\|s_\varepsilon^*\|^2 + \frac{\mu_n}{4}\|t_\varepsilon^*\|^2 \geq -2\varepsilon \qquad \forall n \geq 1.$$

and since $\langle J(x_n - x), x_n - x\rangle = \|x_n - x\|^2$ the latter entails

$$\langle x^* - (s_\varepsilon^* + t_\varepsilon^*), x_n - x\rangle + \frac{\lambda_n}{4}\|s_\varepsilon^*\|^2 + \frac{\mu_n}{4}\|t_\varepsilon^*\|^2 \geq \|x_n - x\|^2 - 2\varepsilon \qquad \forall n \geq 1.$$

Remember now that $\varepsilon > 0$ was fixed, and thus also s_ε^* and t_ε^*, which do not depend on n. Therefore, the first conclusion from the last inequality is that the sequence $\{x_n\}_{n\geq 1}$ is bounded. Let $M > 0$ be an upper bound of $\{\|x_n - x\|\}_{n\geq 1}$. Then the last inequality shows also (after passing to the limit on n) that

$$M\varepsilon \geq \limsup_n \|x_n - x\|^2 - 2\varepsilon.$$

And since $\varepsilon > 0$ was arbitrary this entails that the sequence $\{x_n\}_{n\geq 1}$ converges to x. The proof is completed. ■

This result seems in a certain sense natural having in mind the definitions of the two notions which show that the variational sum takes into account the behavior of the operators at nearby points of the point of reference, while the extended sum does not do so.

Some corollaries are in order here. The first one is that the variational sum is larger than the usual one.

Corollary 16.17. *[21] Let $S,T : X \rightrightarrows X^*$ be maximal monotone operators in the reflexive Banach space X. Then $S+T \subset S \underset{v}{+} T$.*

Therefore, because the variational sum is always with closed graph, we have

Corollary 16.18. *[3, 21, 39, 40] Let $S,T : X \rightrightarrows X^*$ be maximal monotone operators in the reflexive Banach space X. If $\overline{S+T}^G$ is maximal monotone, then*

$$\overline{S+T}^G = S \underset{v}{+} T.$$

In particular, if $S+T$ is maximal, then $S+T = S \underset{\text{ext}}{+} T = S \underset{v}{+} T$.

The fact that $\overline{S+T}^G = S \underset{v}{+} T$ provided $\overline{S+T}^G$ is maximal was proved for Hilbert spaces in [3, Theorem 6.1] and for reflexive spaces in [40, Theorem 4.2] (we should mention that our Theorem 4.2 from [40] was formulated for the operator $\overline{S+T}$, instead of $\overline{S+T}^G$, but the proof in Theorem 4.2 from [40] shows, in fact more, that $\overline{S+T}^G = S \underset{v}{+} T$). However, in our paper [40] we were supposing, in addition, that the norms satisfy the Kadec–Klee property.

Finally, having in mind also Corollary 16.7 we have the following corollary from Theorem 16.16 (this was first proved in [3] in the Hilbert space setting; see also [26]).

Corollary 16.19 ([40], Theorem 5.1, Corollary 5.2). *Let $f,g : X \to \mathbb{R} \cup \{+\infty\}$ be two proper lower semicontinuous convex functions defined in the reflexive Banach space X such that* $\operatorname{dom} f \cap \operatorname{dom} g \neq \emptyset$. *Then*

$$\partial(f+g) = \partial f \underset{\text{ext}}{+} \partial g = \partial f \underset{v}{+} \partial g.$$

The fact that the variational sum of subdifferentials of convex functions is the same as the subdifferential of the sum of the functions was used in [3, 4] to study certain Schrödinger equations related to problems from quantum mechanics.

We finish this section by giving an example (built by using an example from Phelps [37, p. 29]) which shows that the variational sum, not only is, in general, strictly larger than the usual point-wise sum (as suggests the previous corollary) but also that, in general, it is strictly larger than the extended sum as well. This is an example where the usual point-wise sum and the extended one coincide without being a maximal monotone operator.

Example 16.20. (See [24, Example 3.11] and [21, Example 3.13]) Let X be the Hilbert space $l_2 \times l_2$ with the usual scalar product and norm generated by it and identify X^* with X. Let $\operatorname{Dom}(T) := D \times D$ with

$$D := \{\{x_n\}_{n\in\mathbb{N}} \in l_2 : \{2^n x_n\}_{n\in\mathbb{N}} \in l_2\},$$

so that $\operatorname{Dom}(T)$ is a dense linear subspace of X, and let $T : \operatorname{Dom}(T) \to X$ be defined by

$$T(\{x_n\},\{y_n\}) := (\{2^n y_n\}, -\{2^n x_n\}).$$

Then $T_1 := T$ and $T_2 := -T$ are linear anti-symmetric operators with common (dense in X) domain $\operatorname{Dom}(T)$ and in addition they are maximal monotone. Moreover, as it is shown in [24], $T_1 \underset{\text{ext}}{+} T_2 = T_1 + T_2$. On the other hand, having in mind that the variational sum has closed graph, it follows that $\operatorname{Dom}(T_1 \underset{v}{+} T_2) = l_2 \times l_2$ and $T_1 \underset{v}{+} T_2(x) = 0$ for any $x \in X$. That is $T_1 + T_2 = T_1 \underset{\text{ext}}{+} T_2 \neq T_1 \underset{v}{+} T_2$.

It remains an open question whether the variational sum of two maximal monotone operators S and T is a maximal monotone operator, provided $\operatorname{Dom}(S) \cap \operatorname{Dom}(T) \neq \emptyset$, or there are counterexamples of this. Apart from the case of subdifferentials, where we know that the variational sum is maximal (Corollary 16.19), and the case when the usual sum (or its graph closure) is maximal monotone (Corollary 16.18), the only result related to the latter question which is known is that, in finite dimensions, if the variational sum has a unique maximal monotone extension, then they both coincide, i.e., in this case the variational sum is also a maximal monotone operator [21, Corollary 3.4].

16.5 Precompositions of Maximal Monotone Operators with Linear Continuous Mappings

In this section, we will discuss another operation on monotone operators that has been studied from similar points of view as the sum of operators. This operation is the precomposition of a given maximal monotone operator with a linear and continuous mapping. We will see that there exists a sort of equivalence between this operation and the operation of sum, which extends also to the generalized notions of sums and compositions.

More precisely, let X be a Banach space, $T : X \rightrightarrows X^*$ be a maximal monotone operator and suppose that we have also a linear and continuous operator $A : Y \to X$ defined in another Banach space Y and with values in X. Let A^* denote as usual the adjoint of A, i.e. $A^* : X^* \to Y^*$ is given by the relation $\langle A^*x^*, y\rangle = \langle x^*, Ay\rangle$ for $x^* \in X^*$ and $y \in Y$. Then, the point-wise composition $A^*TA : Y \rightrightarrows Y^*$ is easily seen to be a monotone operator with domain $\mathrm{Dom}\,(A^*TA) = A^{-1}(\mathrm{Dom}\,(T))$. Such compositions can be observed in some partial differential equations in divergence form or in certain problems arising from mathematical economics (cf. e.g., [31,32,42]). The composition A^*TA is monotone, but the maximal monotonicity of T is not enough to assure maximal monotonicity of the composition. We need, as in the case of sums, qualification conditions (of the same nature as for sums) to have maximality (cf. e.g. [11,31,36,42,46]).

It is well known that compositions as above are closely related to sums of operators in the sense that the precompositions could be used to express sums of operators and vice versa, sums can be used to obtain compositions. To illustrate this, let $T_1, T_2 : X \rightrightarrows X^*$ be two monotone operators and define $A : X \to X \times X$ by $Ax = (x,x)$ and $T : X \times X \rightrightarrows X^* \times X^*$ by $T(x,y) = T_1x \times T_2y$, $x,y \in X$. Then the operator T is monotone (and maximal if T_1, T_2 are maximal) and, moreover, $T_1 + T_2 = A^*TA$. This shows how to express sums as precompositions.

Conversely, if we have a monotone operator $T : X \rightrightarrows X^*$ and a continuous linear operator $A : Y \to X$, where X and Y are Banach spaces, then let $Y \times X$ be endowed with some usual product norm and consider the operators $\widetilde{S}_A, \widetilde{T} : Y \times X \rightrightarrows Y^* \times X^*$ defined as follows

$$\begin{aligned} \widetilde{S}_A &:= \partial i_{\mathrm{Gr}(A)} \\ \widetilde{T}(y,x) &:= \{0\} \times Tx, \text{ for } (y,x) \in Y \times X. \end{aligned} \tag{16.13}$$

Here, as usual, $i_{\mathrm{Gr}(A)}$ means the indicator function of (the linear closed space) $\mathrm{Gr}(A)$ in $Y \times X$. Since $i_{\mathrm{Gr}(A)}$ is proper convex and lower semicontinuous, the operator $\widetilde{S}_A$ is maximal monotone with domain $\mathrm{Gr}(A)$, and $\widetilde{T}$ is (maximal) monotone, provided T is so, with domain $Y \times \mathrm{Dom}\,(T)$. The following relations are true:

$$\widetilde{S}_A(y,Ay) = \{(A^*x^*, -x^*) : x^* \in X^*\}, \qquad \forall y \in Y, \tag{16.14}$$

and that

$$y^* \in A^*TAy \Longleftrightarrow (y^*,0) \in (\widetilde{S}_A + \widetilde{T})(y,Ay), \tag{16.15}$$

or equivalently

$$y^* + A^*x^* \in A^*TAy \Longleftrightarrow (y^*,x^*) \in (\widetilde{S}_A + \widetilde{T})(y,Ay). \tag{16.16}$$

These relations show how to express precompositions via sums.

Summarizing, we see that there exists a sort of equivalence between these two operations on monotone operators and therefore, it is not surprising that we can mutually deduce results related to one of the notions from the corresponding results for the other. We will see in this section that this kind of equivalence holds also when the extended notions of sums are involved.

Having in mind which regularization procedures have been used to define the concepts of generalized sums, the notions of extended composition and variational composition have natural analogous definitions. We start below again with the extended composition since it can be given in any Banach space.

16.5.1 Extended Composition of a Linear Mapping with a Monotone Operator

In this subsection, X and Y are real Banach spaces and the majority of the results are taken from [24].

The next concept was studied in case of subdifferentials in [20]. A similar, but different, notion was investigated in [35].

Definition 16.21. [24] The extended composition of a linear continuous mapping $A : Y \to X$ and a monotone operator $T : X \rightrightarrows X^*$ is the operator $(A^*TA)_{\text{ext}} : Y \rightrightarrows Y^*$ defined by

$$(A^*TA)_{\text{ext}}(y) := \cap_{\varepsilon>0}\overline{A^*T^{\varepsilon}Ay}^{w^*}, \qquad y \in Y.$$

The extended composition is obviously with w^*-closed and convex images. It extends the usual point-wise composition and as it is seen below, in the case when T is maximal monotone, it is also a monotone operator.

Our next result shows that the equivalence from (16.16) (and (16.15)) remains valid also for the extended notions of sum and compositition. Let us mention that the operator $\widetilde{S}_A$ is not properly enlargeable, that is $\widetilde{S}_A^{\varepsilon} = \widetilde{S}_A$ for any $\varepsilon > 0$ (see, e.g., [24, Proposition 2.3]). Non enlargeable operators are studied also in [14].

Theorem 16.22 ([24] Theorem 4.2). *Let $T : X \rightrightarrows X^*$ be monotone and let $A : Y \to X$ be linear and continuous. Then*

(a) For any $\varepsilon \geq 0$ we have: $y^* + A^*x^* \in A^*T^{\varepsilon}Ay \Longleftrightarrow (y^*, x^*) \in (\widetilde{S}_A + \widetilde{T}^{\varepsilon})(y, Ay)$;
(b) $y^* + A^*x^* \in (A^*TA)_{\text{ext}}(y) \Longleftrightarrow (y^*, x^*) \in (\widetilde{S}_A \underset{\text{ext}}{+} \widetilde{T})(y, Ay)$.

The proof of this statement consists of an appropriate use of the definitions and (16.14)–(16.16). The above theorem allows us to derive several properties of the extended composition from the corresponding ones for extended sums. For example, a straightforward use of Theorem 16.22 and Proposition 16.5 gives:

Proposition 16.23 ([24] Proposition 4.3). *Let $T : X \rightrightarrows X^*$ be a maximal monotone operator and $A : Y \to X$ be linear and continuous. Then the extended composition $(A^*TA)_{\text{ext}}$ is a monotone operator.*

And the next corollary is immediate.

Corollary 16.24 ([24] Corollary 4.4). *Let $T : X \rightrightarrows X^*$ be a maximal monotone operator and $A : Y \to X$ be linear and continuous. If $\overline{A^*TA}$ is maximal monotone then we have $\overline{A^*TA} = (A^*TA)_{\text{ext}}$. In particular, if A^*TA is maximal monotone, then we have $A^*TA = (A^*TA)_{\text{ext}}$.*

The corresponding version of Corollary 16.7 reads as follows:

Corollary 16.25. *[20] Let $f : X \to \mathbb{R} \cup \{+\infty\}$ be a proper convex lower semicontinuous function and $A : Y \to X$ be a continuous linear operator with $\mathrm{R}(A) \cap \mathrm{dom} f \neq \emptyset$. Then*

$$\partial(f \circ A) = (A^* \partial f A)_{\text{ext}}.$$

Proof. Put $T := \partial f$. Then, for $(y,x) \in Y \times X$, $\widetilde{T}(y,x) = \{0\} \times \partial f(x) = \partial F(y,x)$ where $F : Y \times X \to \mathbb{R} \cup \{+\infty\}$ is defined by $F(y,x) := f(x)$, $(y,x) \in Y \times X$. It is easily seen that

$$y^* \in \partial(f \circ A)(y) \Longleftrightarrow (y^*, 0) \in \partial(i_{\mathrm{Gr}(A)} + F)(y, Ay).$$

By Corollary 16.7, $\partial(i_{\mathrm{Gr}(A)} + F) = \widetilde{S}_A \underset{\text{ext}}{+} \widetilde{T}$. Using this in the equivalence relation above and also Theorem 16.22, we obtain

$$y^* \in \partial(f \circ A)(y) \Longleftrightarrow (y^*, 0) \in (\widetilde{S}_A \underset{\text{ext}}{+} \widetilde{T})(y, Ay) \Longleftrightarrow y^* \in (A^* \partial f A)_{\text{ext}}(y),$$

and this completes the proof. ■

In the latter result, as it was the case when we considered extended sums, we have a situation when the usual point-wise composition of a maximal monotone operator with a continuous linear operator is not necessarily maximal monotone, while their extended composition is always maximal, without any qualification condition.

If we are interested in qualification conditions under which the usual and the extended composition coincide, as one can expect, we have a natural analogue of Theorem 16.8 (see also [10, 56] and the remarks after Theorem 16.8). Namely, the following result holds:

Theorem 16.26 ([24] Theorem 4.6). *Let $T : X \rightrightarrows X^*$ be a maximal monotone operator and let $A : Y \to X$ be linear and continuous. Assume that*

$$0 \in \mathrm{core}_X(\mathrm{R}(A) - \mathrm{pr}_X \mathrm{dom}\, \varphi_T).$$

Then:

(1) For each $\varepsilon \geq 0$ and $y \in Y$, $A^*T^\varepsilon Ay$ is w^*-closed;
(2) $A^*TA = (A^*TA)_{\text{ext}}$.

The proof uses the corresponding result for sums (Theorem 16.8) and Theorem 16.22. Here as well, the following well-known result comes as an immediate corollary from the previous theorem and Corollary 16.25.

Corollary 16.27. *Let $f : X \to \mathbb{R} \cup \{+\infty\}$ be a proper convex lower semicontinuous function and let $A : Y \to X$ be linear and continuous. If $0 \in \operatorname{core}_X(\mathrm{R}(A) - \operatorname{dom} f)$, then $A^*\partial f A = \partial(f \circ A)$.*

To complete the picture in this subsection, we give two results (whose proofs, although not immediate, we omit) which show that, first, symmetrically, extended sums can be viewed as extended compositions, as it is the case for the usual operations; and second, that the maximality is preserved when passing to the corresponding representing operations. Namely, we have first that

Theorem 16.28 ([24] Theorem 4.8). *Let $T_1, T_2 : X \rightrightarrows X^*$ be maximal monotone. Define $A : X \to X \times X$ as $Ax := (x,x)$ and $T : X \times X \rightrightarrows X^* \times X^*$ as $T(x,y) := T_1 x \times T_2 y$, $x, y \in X$. Then:*

(1) For any $\varepsilon \geq 0$ and $x \in X$, $A^*T^{\varepsilon}Ax \subset T_1^{\varepsilon}x + T_2^{\varepsilon}x \subset A^*T^{2\varepsilon}Ax$;
(2) For any $x \in X$, $(T_1 \underset{\text{ext}}{+} T_2)(x) = (A^*TA)_{\text{ext}}(x)$.

It is seen from this theorem that the maximality of the extended sum is equivalent to the maximality of its representation as extended composition. As expected, the situation here again is symmetric: the maximality of the extended composition is equivalent to the maximality of its representation as extended sum:

Proposition 16.29 ([24] Proposition 4.9). *Let $T : X \rightrightarrows X^*$ be maximal monotone and let $A : Y \to X$ be linear and continuous. Then, $(A^*TA)_{\text{ext}}$ is maximal monotone (as an operator from Y to Y^*) if and only if $\widetilde{S_A} \underset{\text{ext}}{+} \widetilde{T}$ is maximal monotone (as an operator from $Y \times X$ to $Y^* \times X^*$).*

16.5.2 Variational Composition of a Linear Mapping with a Monotone Operator

In this subsection, we will discuss another way of having a generalized composition – the variational one. As in the case of variational sum our setting will be a real reflexive Banach space X for which the norm as well as its dual norm are at least Gâteaux differentiable (away from the origin). In this case, the duality mapping J_X of X will possess the nice properties from Sect. 16.4. Moreover, given a maximal monotone operator $T : X \rightrightarrows X^*$, we will dispose with the corresponding resolvent J_λ^T and Yosida regularization T_λ of order $\lambda > 0$ which satisfy the properties listed in Sect. 16.4.

Now, let us consider another real reflexive Banach space Y and a linear and continuous operator $A : Y \to X$. As for X, the norms in Y and its dual will be supposed

Gâteaux differentiable away from the origin (and thus strictly convex). In such a case the duality mapping J_Y of Y will share the same properties as J_X. Let $T : X \rightrightarrows X^*$ be a maximal monotone operator. Then, for any $\lambda > 0$ the operator $A^*T_\lambda A : Y \to Y^*$ is not only (single-valued and) monotone but also maximal due to the fact that T_λ is an everywhere defined single-valued maximal monotone operator (see, e.g., [42, 46]). Thus, one can use the same idea as for sums and pass to a limit in order to define a variational notion of the composition of A with T. Namely, endowing $Y \times Y^*$ with some usual product norm, we can give the following definition:

Definition 16.30 ([32] Definition 2.1). Let Y,X be reflexive Banach spaces, $A : Y \to X$ be continuous and linear, and let $T : X \rightrightarrows X^*$ be maximal monotone. The *variational composition* $(A^*TA)_{\text{var}} : Y \rightrightarrows Y^*$ of A and T is the mapping

$$(A^*TA)_{\text{var}} = \liminf_{\lambda \downarrow 0} A^*T_\lambda A,$$

where the limit is in the sense of graphs.

Precisely speaking, $(y,y^*) \in \liminf_{\lambda\downarrow 0} A^*T_\lambda A$ if and only if for every neighborhood U of (y,y^*) (in the product topology in $Y \times Y^*$) there is $\lambda_0 > 0$ so that for each $\lambda \in (0,\lambda_0)$ we have $\mathrm{Gr}(A^*T_\lambda A) \cap U \neq \emptyset$. Which in its turn is equivalent to: for any sequence $\{\lambda_n\}_{n\geq 1}$ such that $\lambda_n \downarrow 0$ there is a sequence $\{(y_n,y_n^*)\}_{n\geq 1}$ such that $(y_n,y_n^*) \in \mathrm{Gr}(A^*T_{\lambda_n}A)$ for each $n \geq 1$ and $\{(y_n,y_n^*)\}_{n\geq 1}$ converges to (y,y^*).

The lim sup notion is defined in an obvious way: $\limsup_{\lambda\downarrow 0} A^*T_\lambda A$ is the operator whose graph consists of couples $(y,y^*) \in Y \times Y^*$ such that for any neighborhood U of (y,y^*) (in the product topology in $Y \times Y^*$) and for any $\lambda > 0$ there is $\mu \in (0,\lambda)$ with $\mathrm{Gr}(A^*T_\mu A) \cap U \neq \emptyset$. And this is equivalent to the following sequential definition: there is a sequence $\{\lambda_n\}_{n\geq 1}$ such that $\lambda_n \downarrow 0$ and a sequence $\{(y_n,y_n^*)\}_{n\geq 1}$ such that $(y_n,y_n^*) \in \mathrm{Gr}(A^*T_{\lambda_n}A)$ for each $n \geq 1$ and $\{(y_n,y_n^*)\}_{n\geq 1}$ converges to (y,y^*). We write, $(A^*TA)_{\text{var}} = \lim_{\lambda\downarrow 0} A^*T_\lambda A$, when $\limsup_{\lambda\downarrow 0} A^*T_\lambda A = \liminf_{\lambda\downarrow 0} A^*T_\lambda A$.

Most of the following properties are immediate:

Proposition 16.31. *[32] Let Y,X be reflexive Banach spaces, $A : Y \to X$ be continuous and linear, and let $T : X \rightrightarrows X^*$ be maximal monotone. Then:*

(1) The variational composition $(A^*TA)_{\text{var}}$ is a monotone operator with closed graph;
(2) $\mathrm{Dom}\,(A^*TA) \subset \mathrm{Dom}\,(A^*TA)_{\text{var}}$;
(3) If $(A^*TA)_{\text{var}}$ is a maximal monotone operator, then $(A^*TA)_{\text{var}} = \lim_{\lambda\downarrow 0} A^*T_\lambda A$.

Condition (2) above was shown to be true in [32] provided the norms in X and X^* satisfy also Kadec-Klee property, but as it is seen from Corollary 16.35 below, we do not need this additional property in order to have (2).

The following result is proved as Proposition 16.12, using also the corresponding variant of Lemma 16.13.

Proposition 16.32. *Let X and Y be reflexive Banach spaces, $T : X \rightrightarrows X^*$ be a maximal monotone mapping and $A : Y \to X$ be a linear continuous operator. Then the following are equivalent:*

(a) $(y,y^*) \in (A^*TA)_{\text{var}}$;
(b) For any $\lambda > 0$ *the (unique) solution* y_λ *of the inclusion (in fact, equality)*

$$y^* \in J_Y(y_\lambda - y) + A^*T_\lambda A y_\lambda \tag{16.17}$$

converges to y *as* $\lambda \to 0$;

The analogy between the variational sum of two operators and the variational composition is not so complete, since the concept of variational sum which we presented in the previous section involves two independent parameters when regularizing the operators, while the variational composition uses only one parameter. But still, there is a kind of analogy and we can use it to derive properties of the variational composition from known properties of the variational sum.

Namely, given two maximal monotone operators $S,T : X \rightrightarrows X^*$, between a reflexive space X and its dual, we can define a (formally different from the notion from Sect. 16.4) concept of variational sum by considering non symmetric sums of the type $S+T_\lambda$ (or $S_\lambda + T$) for $\lambda > 0$ and then passing to the limit. Such asymmetric sums have been already considered in the papers [3] and [13]. Another possibility is to consider symmetric sums with the same parameter $S_\lambda + T_\lambda$, $\lambda > 0$, and again to pass to an appropriate limit (as it was done in [32]). It is easily seen that the limits $\liminf_{\lambda\downarrow 0}(S_\lambda + T_\lambda)$ and $\liminf_{\lambda\downarrow 0}(S + T_\lambda)$, where liminf is understood as above in the definition of the variational composition, give monotone operators with closed graphs and according to the definitions both are, in general, larger than the variational composition $S \underset{v}{+} T$.

For some operators all these notions give the same result as the variational sum. For example if $S = \partial f$ and $T = \partial g$ are subdifferentials of proper lower semicontinuous convex functions in X such that $\operatorname{dom} f \cap \operatorname{dom} g \neq \emptyset$ then, according to Corollary 16.19, the variational sum $S \underset{v}{+} T$, and thus also $\liminf_{\lambda\downarrow 0}(S + T_\lambda)$ and $\liminf_{\lambda\downarrow 0}(S_\lambda + T_\lambda)$ which are monotone and larger than $S \underset{v}{+} T$, will be equal to the subdifferential $\partial(f+g)$. Another particular case when all these sums coincide is of course when $S+T$ (or, more general, when $\overline{S+T}^G$) is a maximal monotone operator (see Corollary 16.18 above; cf. also [3]).

In some extremal cases we may have different operators obtained by the above notions. For instance, this is the case in the following example, which was communicated to the author by García [22]: let $X = \mathbb{R}$, $S = \partial i_{\{-1\}}$ and $T = \partial i_{\{1\}}$. We have $\operatorname{Dom}(S) \cap \operatorname{Dom}(T) = \emptyset$ and one can easily check that for $\lambda > 0$ we have $S_\lambda(x) = (x+1)/\lambda$ and $T_\lambda(x) = (x-1)/\lambda$, $x \in \mathbb{R}$. It is readily seen that in this case the usual and the variational sum are the trivial empty operator, while $(0,0)$ is in the graph of $\liminf_{\lambda\downarrow 0}(S_\lambda + T_\lambda)$ (in fact, the latter is maximal monotone with graph $\{0\} \times \mathbb{R}$). Just to see how different can be the sums in such a degenerate case, let us mention that in this example the operator $\liminf_{\lambda\downarrow 0}(S + T_\lambda)$ is with graph $\{-1\} \times \mathbb{R}$ and $\liminf_{\lambda\downarrow 0}(S_\lambda + T)$ with graph $\{1\} \times \mathbb{R}$. We do not dispose with non-degenerate examples showing that some of the new sums above are strictly bigger than the variational one.

The following proposition (given also in [23]) can be proved exactly as Proposition 16.12 using the corresponding variant of Lemma 16.13. Convergence of solutions like in (b) were studied first in [13]. Let us mention that the analogous proposition, concerning $\liminf_{\lambda\downarrow 0}(S_\lambda + T_\lambda)$, also holds.

Proposition 16.33. *Let S,T be maximal monotone operators in the reflexive Banach space X. Then the following are equivalent:*

(a) $(x,x^*) \in \liminf_{\lambda\downarrow 0}(S+T_\lambda)$;
(b) For any $\lambda > 0$ the (unique) solution x_λ of the inclusion

$$x^* \in J_X(x_\lambda - x) + Sx_\lambda + T_\lambda x_\lambda \tag{16.18}$$

converges to x as $\lambda \to 0$;

Disposing with the latter concepts, one can see that if we are given two operators $T_1, T_2 : X \rightrightarrows X^*$, and if we consider as above the operator $T := T_1 \times T_2 : X \times X \rightrightarrows X^* \times X^*$ and the linear continuous operator $A : X \to X \times X$ determined by $Ax = (x,x), x \in X$, then $(A^*TA)_{\mathrm{var}} = \liminf_{\lambda\downarrow 0}(T_{1,\lambda} + T_{2,\lambda})$ (see [32]). Here, we endow the product $X \times X$ with the usual square norm in order to have Gâteaux differentiability of the norm in $X \times X$ and of the dual norm in $X^* \times X^*$. In this case, the duality mapping for $X \times X$ is $J_X \times J_X$.

Reciprocally, let $T : X \rightrightarrows X^*$ be a maximal monotone operator and $A : Y \to X$ be a continuous linear mapping. Consider the operators $\widetilde{S}_A, \widetilde{T} : Y \times X \rightrightarrows Y^* \times X^*$ defined in (16.13). If we equip $Y \times X$ with the usual square product norm, it will be a reflexive Banach space with both Gâteaux differentiable norm and dual norm (the latter is also the square norm generated by the norms in Y^* and X^*). Moreover, the duality mapping for $Y \times X$ will be $J_Y \times J_X$. In such a case the Yosida regularisation of $\widetilde{T}$ is simply given by $\widetilde{T}_\lambda(y,x) = (0, T_\lambda x)$, $(y,x) \in Y \times X$, $\lambda > 0$. This together with (16.14) readily entail the next proposition (given also in [23]) which shows that the property from (16.15) concerning point-wise compositions and sums extends also to the case of the variational composition and a kind of variational sum.

Proposition 16.34. *Let X and Y be reflexive Banach spaces, $T : X \rightrightarrows X^*$ be maximal monotone and $A : Y \to X$ be linear and continuous. Then,*

(1) For any $\lambda > 0$

$$y^* = (A^*T_\lambda A)(y) \Longleftrightarrow (y^*,0) \in (\widetilde{S}_A + \widetilde{T}_\lambda)(y, Ay)$$

(2) And therefore,

$$y^* \in (A^*TA)_{\mathrm{var}}(y) \Longleftrightarrow (y^*,0) \in \liminf_{\lambda\downarrow 0}(\widetilde{S}_A + \widetilde{T}_\lambda)(y, Ay).$$

A consequence of Proposition 16.34 is another fact that has not been known so far: whether the variational composition contains the usual one. Knowing already that this holds for the case of sums and having in mind that $\liminf_{\lambda\downarrow 0}(\widetilde{S} + \widetilde{T}_\lambda)$ is, in general, larger than $(\widetilde{S} \underset{v}{+} \widetilde{T})$, we have the following corollary as a direct consequence of Theorem 16.22(b), Theorem 16.16 and Proposition 16.34 (the result is contained also in the manuscript [23]).

Corollary 16.35. *Let Y, X be reflexive Banach spaces, $A : Y \to X$ be a continuous and linear operator and $T : X \rightrightarrows X^*$ be a maximal monotone operator. Then, $(A^*TA)_{\text{ext}} \subset (A^*TA)_{\text{var}}$. In particular, $A^*TA \subset (A^*TA)_{\text{var}}$.*

And therefore, since $(A^*TA)_{\text{var}}$ has always closed graph, we obtain

Corollary 16.36. *[32] Let Y, X be reflexive Banach spaces, $A : Y \to X$ be a continuous and linear operator and $T : X \rightrightarrows X^*$ be a maximal monotone operator. If $\overline{A^*TA}^G$ is maximal monotone, then $\overline{A^*TA}^G = (A^*TA)_{\text{var}}$. In particular, if A^*TA is maximal monotone, then $A^*TA = (A^*TA)_{\text{var}}$.*

The next corollary is a consequence of Corollaries 16.25 and 16.35 and gives a nontrivial case when the variational composition is a maximal monotone operator, while the usual one is not, in general.

Corollary 16.37. *[32] Let Y, X be reflexive Banach spaces, $A : Y \to X$ be a continuous and linear operator and $f : X \to \mathbb{R} \cup \{+\infty\}$ be a proper convex lower semicontinuous function such that* $\mathrm{R}(A) \cap \mathrm{dom} f \neq \emptyset$. *Then, $\partial(f \circ A) = (A^*\partial f A)_{\text{var}}$.*

Let us finally mention that the variational composition was used in [32] to study the solutions of certain partial differential equations in divergence form.

Acknowledgements The author would like to thank Radu Boţ, Yboon García, Marc Lassonde and Constantin Zălinescu, whose valuable remarks after a careful reading of an earlier version of the manuscript, helped to present a more complete picture concerning the notions and the results in this article. The author is also grateful to two anonymous referees for their detailed remarks.

This article was prepared while the author was *professeur associé* in the group LAMIA in the Department of Mathematics and Informatics of the Université des Antilles et de la Guyane, Guadeloupe, France.

The author has been partially supported by the Bulgarian National Fund for Scientific Research, under grant DO02-360/2008.

References

1. Asplund, E.: Averaged norms. Israel J. Math. **5**, 227–233 (1967)
2. Attouch, H.: Variational Convergence for Functions and Operators. Math. Series, Pitman, London (1984)
3. Attouch, H., Baillon, J.-B., Théra, M.: Variational sum of monotone operators. J. Convex Anal. **1**, 1–29 (1994)
4. Attouch, H., Baillon, J.-B., Théra, M.: Weak solutions of evolution equations and variational sum of maximal monotone operators. SEA Bull. Math. **19**, 117–126 (1995)
5. Attouch, H., Riahi, H., Théra, M.: Somme ponctuelle d'opérateurs maximaux monotones. Serdica Math. J. **22**, 267–292 (1996)

6. Bauschke, H.H.: Fenchel duality, Fitzpatrick functions and the extension of firmly nonexpansive mappings. Proc. Amer. Math. Soc. **135**, 135–139 (2007)
7. Bauschke, H.H., Borwein, J.M., Wang, X.: Fitzpatrick functions and continuous linear monotone operators. SIAM J. Optim. **18**, 789–809 (2007)
8. Borwein, J.M.: Maximal monotonicity via convex analysis. J. Convex Anal. **13**, 561–586 (2006)
9. Borwein, J.M.: Maximality of sums of two maximal monotone operators in general Banach space. Proc. Amer. Math. Soc. **135**, 3917–3924 (2007)
10. Boţ, R.I., Csetnek, E.R.: On two properties of enlargements of maximal monotone operators. J. Convex Anal. **16**, 713–725 (2009)
11. Boţ, R.I., Grad, S.-M., Wanka, G.: Maximal monotonicity for the precomposition with a linear operator. SIAM J. Optim. **17**, 1239–1252 (2006)
12. Boţ, R.I., Csetnek, E.R., Wanka, G.: A new condition for maximal monotonicity via representative functions. Nonlinear Anal., TMA **67**, 2390–2402 (2007)
13. Brezis, H., Crandall, M.G., Pazy, A.: Perturbations of nonlinear maximal monotone sets in Banach space. Comm. Pure Appl. Math. **XXIII**, 123–144 (1970)
14. Burachik, R.S., Iusem, A.: On non-enlargable and fully enlargable monotone operators. J. Convex Anal. **13**, 603–622 (2006)
15. Burachik, R.S., Svaiter, B.F.: ε-Enlargements in Banach spaces. Set-Valued Anal. **7**, 117–132 (1999)
16. Burachik, R.S., Iusem, A.N., Svaiter, B.F.: Enlargements of maximal monotone operators with applications to variational inequalities. Set-Valued Anal. **5**, 159–180 (1997)
17. Burachik, R.S., Sagastizábal, C.A., Svaiter, B.F.: ε-Enlargements of maximal monotone operators: Theory and Applications. In: M. Fukushima and L. Qi (eds) *Nonsmooth, Piecewise Smooth, Semismooth and Smoothing Methods*, Kluwer Academic Publishers, 25–43 (1998)
18. Chu, L.-J.: On the sum of monotone operators. Michigan Math. J. **43**, 273–289 (1996)
19. Fitzpatrick, S.P.: Representing monotone operators by convex functions. Workshop/Miniconference on Functional Analysis and Optimization. Austral. Nat. Univ., Canberra, 59–65 (1988)
20. Fitzpatrick, S.P., Simons, S.: The conjugates, compositions and marginals of convex functions. J. Convex Anal. **8**, 423–446 (2001)
21. García, Y.: New properties of the variational sum of monotone operators. J. Convex Anal. **16**, 767–778 (2009)
22. García, Y.: Personal communication
23. García, Y., Lassonde, M.: Representable monotone operators and limits of sequences of maximal monotone operators. To appear in Set-Valued Analysis and its Applications
24. García, Y., Lassonde, M., Revalski, J.P.: Extended sums and extended compositions of monotone operators. J. Convex Anal. **13**, 721–738 (2006)
25. Hiriart-Urruty, J.-B., Phelps, R.R.: Subdifferential calculus using ε-subdifferentials. J. Funct. Anal. **118**, 154–166 (1993)
26. Jourani, A.: Variational sum of subdifferentials of convex functions. In: C. Garcia, C. Olivé and M. Sanroma (eds.) Proc. of the Fourth Catalan Days on Applied Mathematics. Tarragona Press University, Tarragona, 71–80 (1998)
27. Kubo, F.: Conditional expectations and operations derived from network connections. J. Math. Anal. Appl. **80**, 477–489 (1981)
28. Lapidus, M.: Formules de Trotter et calcul opérationnnel de Feynman. Thèse d'Etat, Université Paris VI (1986)
29. Marques Alves, M., Svaiter, B.: Brøndsted-Rockafellar property and maximality of monotone operators representable by convex functions in nonreflexive Banach spaces. J. Convex Anal. **15**, 693–706 (2008)
30. Martinez-Legaz, J.E., Théra, M.: ε-Subdifferentials in terms of subdifferentials. Set-Valued Anal. **4**, 327–332 (1996)

31. Pennanen, T.: Dualization of generalized equations of maximal monotone type. SIAM J. Optim. **10**, 809–835 (2000)
32. Pennanen, T., Revalski, J.P., Théra, M.: Variational composition of a monotone mapping with a linear mapping with applications to PDE with singular coefficients. J. Funct. Anal. **198**, 84–105 (2003)
33. Penot, J.-P.: Subdifferential calculus without qualification conditions. J. Convex Anal. **3**, 1–13 (1996)
34. Penot, J.-P.: The relevance of convex analysis for the study of monotonicity. Nonlinear Anal., TMA **58**, 855–871 (2004)
35. Penot, J.-P.: Natural closure, natural compositions and natural sums of monotone operators. J. Math. Pures et Appl. **89**, 523–537 (2008)
36. Penot, J.-P., Zălinescu, C.: Some problems about the representations of monotone operators by convex functions. ANZIAM J. **47**, 1–20 (2005)
37. Phelps, R.R.: Convex Functions, Monotone Operators and Differentiability. Lecture Notes in Mathematics **1364**, Springer, Berlin (1993)
38. Phelps, R.R.: Lectures on Maximal Monotone Operators. Extracta Mathematicae **12**, 193–230 (1997)
39. Revalski, J.P., Théra, M.: Generalized sums of monotone operators. Comptes Rendus de l'Académie des Sciences, Paris **t. 329**, Série I, 979–984 (1999)
40. Revalski, J.P., Théra, M.: Variational and extended sums of monotone operators. In: M. Théra and R. Tichatschke (eds.) *Ill-posed Variational Problems and Regularization Techniques*. Lecture Notes in Economics and Mathematical Systems, Springer, Vol. **477**, 229–246 (1999)
41. Revalski, J.P., Théra, M.: Enlargements and sums of monotone operators. Nonlinear Anal., TMA **48**, 505–519 (2002)
42. Robinson, S.M.: Composition duality and maximal monotonicity. Math. Program. **85A**, 1–13 (1999)
43. Rockafellar, R.T.: On the maximality of sums of nonlinear monotone operators. Trans. Amer. Math. Soc. **149**, 75–88 (1970)
44. Rockafellar, R.T.: On the maximal monotonicity of subdifferential mappings. Pacific J. Math. **33**, 209–216 (1970)
45. Simons, S.: Maximal monotone multifunctions of Brøndsted-Rockafellar type. Set-Valued Anal. **7**, 255–294 (1999)
46. Simons, S.: From Hahn-Banach to Monotonicity. Lecture Notes in Mathematics, Vol. **1693** (2nd edn.). Springer, Berlin (2008)
47. Simons, S., Zălinescu, C.: Fenchel duality, Fitzpatrick functions and maximal monotonicity. J. Nonlinear Convex Anal. **6** 1–22 (2005)
48. Svaiter, B.F.: Fixed points in the family of convex representations of a maximal monotone operator. Proc. Amer. Math. Soc. **131**, 3851–3859 (2003)
49. Thibault, L.: A general sequential formula for subdifferentials of sums of convex functions defined on Banach spaces. In: R. Durier and C. Michelot (eds.), *Recent Developments in Optimization*, Lecture Notes in Economics and Mathematical Systems, Springer, Berlin, Vol. **429**, 340–345 (1995)
50. Thibault, L.: Limiting subdifferential calculus with applications to integration and maximal monotonicity of subdifferential. In: M. Théra (ed.) *Constructive, experimental, and nonlinear analysis*, CMS Conference Proceedings, Vol. **27**, 279–289 (2000)
51. Torralba, D.: Convergence épigraphique et changements d'échelle en analyse variationnelle et optimisation. Thèse de l'Université de Montpellier II (1996)
52. Troyanski, S.: On locally uniformly convex and differentiable normes in certain nonseparable Banach spaces. Stuida Math. **37**, 173–180 (1971)
53. Verona, A., Verona, M.: Regular maximal monotone operators and the sum theorem. J. Convex Anal. **7**, 115–128 (2000)
54. Veselý, L.: Local uniform boundedness principle for families of ε-monotone operators. Nonlinear Anal., TMA **24**, 1299–1304 (1994)

55. Voisei, M.D.: A maximality theorem for sum of maximal monotone operators in non-reflexive Banach spaces. Math. Sci. Res. **10**, 36–41 (2006)
56. Voisei, M.D., Zălinescu, C.: Maximal monotonicity criteria for the composition and the sum under weak interiority conditions. Math. Program. **123**, 265–283 (2010)
57. Zeidler, E.: Nonlinear Functional Analysis and its Applications. Vol. II/B Nonlinear Monotone Operators, Springer, Berlin (1990)

Chapter 17
Minimizing the Moreau Envelope of Nonsmooth Convex Functions over the Fixed Point Set of Certain Quasi-Nonexpansive Mappings

Isao Yamada, Masahiro Yukawa, and Masao Yamagishi

Abstract The first aim of this paper is to present a useful toolbox of quasi-nonexpansive mappings for convex optimization from the viewpoint of using their fixed point sets as constraints. Many convex optimization problems have been solved through elegant translations into fixed point problems. The underlying principle is to operate a certain quasi-nonexpansive mapping T iteratively and generate a convergent sequence to its fixed point. However, such a mapping often has infinitely many fixed points, meaning that a selection from the fixed point set $\mathrm{Fix}(T)$ should be of great importance. Nevertheless, most fixed point methods can only return an "unspecified" point from the fixed point set, which requires many iterations. Therefore, based on common sense, it seems unrealistic to wish for an "optimal" one from the fixed point set. Fortunately, considering the collection of quasi-nonexpansive mappings as a toolbox, we can accomplish this challenging mission simply by the *hybrid steepest descent method*, provided that the cost function is smooth and its derivative is Lipschitz continuous. A question arises: *how can we deal with "nonsmooth" cost functions*?

The second aim is to propose a nontrivial integration of the ideas of the *hybrid steepest descent method* and the *Moreau-Yosida regularization*, yielding a useful approach to the challenging problem of nonsmooth convex optimization over $\mathrm{Fix}(T)$. The key is the use of smoothing of the original nonsmooth cost function by its *Moreau-Yosida regularization* whose derivative is always Lipschitz continuous. The field of application of hybrid steepest descent method can be extended to the minimization of the ideal smooth approximation over $\mathrm{Fix}(T)$. We present the mathematical ideas of the proposed approach together with its application to a combinatorial optimization problem: the minimal antenna-subset selection problem under a highly nonlinear capacity-constraint for efficient multiple input multiple output (MIMO) communication systems.

I. Yamada (✉)
Department of Communications and Integrated Systems, Tokyo Institute of Technology, S3-60, Tokyo, 152-8550 Japan
e-mail: isao@sp.ss.titech.ac.jp

H.H. Bauschke et al. (eds.), *Fixed-Point Algorithms for Inverse Problems in Science and Engineering*, Springer Optimization and Its Applications 49,
DOI 10.1007/978-1-4419-9569-8_17,

Keywords Nonsmooth convex optimization · Moreau envelope · Hybrid steepest descent method

AMS 2010 Subject Classification: 47H10, 47H09, 49M20, 65K10

17.1 Introduction

How can we exploit various types of information efficiently in convex optimization? This has been one of the fundamental questions of paramount importance from both practical and theoretical viewpoints. We present a new insight into this question with (1) *fixed point characterizations* of constraint sets and (2) the *Moreau–Yosida regularization* of a nonsmooth convex function. To contrast our contribution with existing approaches, let us briefly introduce a stream of research developments, including classical and state-of-the-art techniques, for treating (multiple) constraints.

17.1.1 Treatments of Constraints in Convex Optimization

A general convex optimization problem is formulated as follows: minimize a convex function $f \in \Gamma_0(\mathscr{H})$ over a closed convex subset C of a real Hilbert space $\mathscr{H}$. Here, $\Gamma_0(\mathscr{H})$ stands for the class of all lower semicontinuous convex functions from $\mathscr{H}$ to $(-\infty,\infty]$ which are not identically equal to $+\infty$. Suppose, for instance, that f is differentiable with its derivative Lipschitz continuous and P_C, the metric projection onto C (see Fact 17.2(c)), can be computed *efficiently*. In this special case, we may use Goldstein's *projected gradient method* [72]. However, this classical approach cannot satisfy the increasing demand for nonsmooth convex optimization under more general constraints.

A couple of unified approaches covering many existing schemes involve the following formulation [42, 46, 57, 66, 90, 104, 126]: minimize $f_1 + f_2$ for $f_i \in \Gamma_0(\mathscr{H})$, $i = 1,2$. For example, under a certain *qualification condition* on f_1 and f_2, the *Douglas–Rachford splitting*-type algorithm (see Examples 17.6(c) and 17.12(f)) [42, 57, 90] approximates a minimizer of $f_1 + f_2$ with successive use of

$$\mathrm{prox}_{\gamma f_i} : \mathscr{H} \to \mathscr{H} : x \mapsto \arg\min_{y \in \mathscr{H}} \left\{ f_i(y) + \frac{1}{2\gamma}\|x - y\|^2 \right\}, \tag{17.1}$$

which is well-defined as a single valued mapping called the *proximity operator* or *proximal mapping* [46, 98, 99, 109] of index $\gamma \in (0,\infty)$ of f_i $(i = 1,2)$ (see Sect. 17.2.1). This approach can handle the problem considered in the previous paragraph by letting $f_1 := f$ and $f_2 := \iota_C$ which denotes the indicator function

$$(\forall x \in \mathscr{H}) \quad \iota_C(x) := \begin{cases} 0, & \text{if } x \in C; \\ \infty, & \text{otherwise.} \end{cases}$$

In fact, the proximity operator of ι_C for any $\gamma \in (0,\infty)$ coincides with P_C. We emphasize, however, that the approach in [42, 46, 57, 126] practically requires an efficient scheme to compute the proximity operators, and obtaining such a scheme itself is often a challenging issue to address for each application individually.

This certainly motivates the recent active studies on computational schemes for proximity operators of various types of functions in $\Gamma_0(\mathscr{H})$ [42, 46, 62], which include the pre-composition of $g \in \Gamma_0(\mathscr{H})$ with a frame synthesis affine operator [31, 42, 62, 118]. Another development is the extension of the Douglas–Rachford splitting-type scheme to the case of multiple convex functions [43, 44, 67, 68]; i.e., minimize $\Sigma_{i=1}^m f_i$ for $m > 2$ and $f_i \in \Gamma_0(\mathscr{H})$ $(i = 1,2,\ldots,m)$, through the Pierra-type product-space reformulation [105, 106]. This extension enables us to deal with the case where a constraint set C can be expressed as the intersection of a finite number of closed convex sets C_i ($i \in \mathscr{I}$, *assuming that* P_{C_i} *can be computed efficiently*). Indeed, we can minimize a nonsmooth convex function $f := \Sigma_{j\in\mathscr{J}} f_j$ over C by applying the extended scheme to $\Sigma_{i\in\mathscr{I}} \iota_{C_i} + \Sigma_{j\in\mathscr{J}} f_j$. The use of the expression $C = \bigcap_{i\in\mathscr{I}} C_i$ shares similarity with the commonly used strategy in the simpler contexts of the convex feasibility problems (see, e.g., [7, 18, 29, 34, 48]). However, again, this approach has an obvious limitation, as there are many applications, including the one addressed in this work, in which the constraint set $C \subset \mathscr{H}$ can hardly be expressed as the intersection of (a finite number of) *simple* closed convex sets. The *fixed point characterization* throws us a rope to escape from the dilemma, as explained in the following.

17.1.2 Fixed Point Characterizations of Closed Convex Sets

A mapping $T : \mathscr{H} \to \mathscr{H}$ is called *quasi-nonexpansive* if this mapping has its nonempty fixed point set $\mathrm{Fix}(T) := \{x \in \mathscr{H} \mid T(x) = x\} \neq \emptyset$ and $\|T(x) - z\| \leq \|x - z\|$ $(\forall x \in \mathscr{H}, \forall z \in \mathrm{Fix}(T))$. In this case, the fixed point set $\mathrm{Fix}(T)$ is guaranteed to be closed convex in $\mathscr{H}$ (see Proposition 17.3 in Sect. 17.2.2). In the context of recent studies on the convex feasibility problems as well as the unified treatment of certain nonsmooth optimization schemes, many powerful ideas have been found to deal with a closed convex set C as *the fixed point set of an efficiently computable quasi-nonexpansive mapping* [7, 8, 35, 132, 136, 139].

For example, if the set $S := \operatorname*{arg\,min}_{x\in\mathscr{H}} \{f_1(x) + f_2(x)\}$ is nonempty in the above context of minimizing $f_1 + f_2$ for $f_i \in \Gamma_0(\mathscr{H})$, $i = 1,2$, the set S is a closed convex set which is usually hard to be expressed as the intersection of (a finite number of) simple closed convex sets. On the other hand, in a variety of scenarios, the set S can be expressed as the fixed point set of a nonexpansive mapping [46] or as the image of a proximity operator of the fixed point set of another nonexpansive mapping [42], where these nonexpansive mappings can be computed efficiently (see Sect. 17.2.2 for basic ideas to design a mapping that has a desirable fixed point set).

Another quite useful example is found in the characterization of the nonempty level set $\mathrm{lev}_{\leq 0}(g) := \{x \in \mathscr{H} \mid g(x) \leq 0\}$ of $g \in \Gamma_0(\mathscr{H})$ as the fixed point set of the subgradient projection $T_{\mathrm{sp}(g)}$ relative to g. The subgradient projection operator is *firmly quasi-nonexpansive* ([127, Lemma 2.8], [8, 136]) and has been playing important roles as a low complexity approximation of the metric projection onto $\mathrm{lev}_{\leq 0}(g)$ in many scenarios; e.g., in signal and image processing applications [37, 41, 140], the metric projection is often hard to compute (see Proposition 17.7 and Example 17.9 for designing better approximations than the subgradient projection). In [114, 115, 125, 135, 140], the subgradient projection was used to elude from the load for solving large scale systems of equations in an adaptive signal processing or adaptive online classification problems. In [41], the subgradient projection was used to suppress the *total variation* of the restored image.

The idea of dealing with a closed convex set as the fixed point set of a nonexpansive mapping has been applied successfully in creations of many powerful optimization schemes with the strong support of the innovative discovery of the *Mann iterative process* [54, 75, 94], which is an extremely simple algorithm to generate a (weakly) convergent sequence to a fixed point of a general nonexpansive mapping. Moreover recent notable extensions, e.g., [39], of the algorithm have a guarantee of convergence under much weaker conditions than those found in [54, 75, 94] and applied in the unifications, e.g., in [42, 46]. In short, these previous studies aim to find an *arbitrary* point in the fixed point set of a nonexpansive mapping. The next stage which we should clear is the following: find an *optimal* point in some sense in the fixed point set. The following subsection introduces some existing methods for this problem with a touch of motivation of the current study.

17.1.3 Existing Methods on the Advanced Stage

We now consider the problem of minimizing a convex function over the fixed point set of a certain quasi-nonexpansive mapping. There seems to be only few types of algorithms that can deal with this problem in a computationally manageable way. Among others, the *hybrid steepest descent method* (HSDM) (see, e.g., [33, 49, 84, 92, 101, 102, 122, 130, 136–139, 151]) has been developed as an algorithm to achieve such a goal originally by extending a fixed point iteration [6, 36, 78, 89, 129]; the so-called *Halpern-type iteration* or *anchor method*, which is able to find from a given point the nearest fixed point of a nonexpansive mapping. The HSDM has two distinguished features. First, it has a mathematical guarantee of convergence to the solution to the convex optimization over the fixed point set. Second, it only requires at each iteration simple computation of a gradient descent operator and a quasi-nonexpansive mapping, of which the fixed point set defines the constraint set of the optimization problem. Indeed, the method has been applied successfully to signal and image processing problems (see, e.g., [79, 113, 117, 118, 122, 152]).

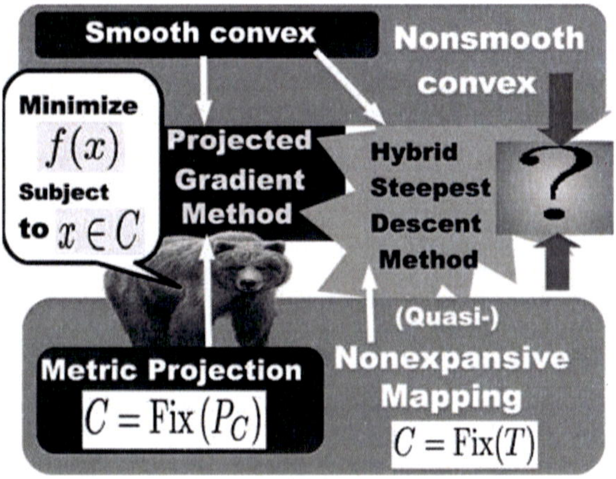

Fig. 17.1 Treatment of constraint sets as fixed point sets of nonlinear mappings

By extending the ideas in [80], another algorithm, which we refer to as the *generalized Haugazeau's algorithm*, was developed for minimizing a *strictly convex* function in $\Gamma_0(\mathscr{H})$ over the fixed point set of a certain quasi-nonexpansive mapping [38]. In particular, this algorithm was specialized in a clear way for finding the nearest fixed point of a certain quasi-nonexpansive mapping [8] and applied successfully to an image recovery problem [41]. If we focus on the case of a non-strictly convex function, the generalized Haugazeau's algorithm is not applicable, while some convergence theorems of the HSDM suggest its sound applicability *provided that the derivative of the function is Lipschitzian*. Due to the Lipschitz-continuity assumption, however, it still remains an open problem to minimize a *nonsmooth* convex function over the fixed point set of a quasi-nonexpansive mapping (see Fig. 17.1).

17.1.4 Contributions of This Paper

So far we do not have in general any promising (computationally manageable) algorithm for the solution to the minimization problem of a *nonsmooth* convex function over the fixed point set of a quasi-nonexpansive mapping. We therefore present a nontrivial application of the HSDM to approach the problem. Our attention is to the notable fact that any function $f \in \Gamma_0(\mathscr{H})$ can be approximated with any accuracy by

$$\begin{aligned} {}^{\gamma}f : \mathscr{H} \to \mathbb{R} : \quad & x \mapsto \min_{y \in \mathscr{H}} \left\{ f(y) + \frac{1}{2\gamma}\|x-y\|^2 \right\} \\ &= f\left(\operatorname{prox}_{\gamma f}(x)\right) + \frac{1}{2\gamma}\left\|x - \operatorname{prox}_{\gamma f}(x)\right\|^2, \end{aligned} \tag{17.2}$$

which is called the *Moreau envelope*[1] (or the *Moreau–Yosida regularization*[2]) of index $\gamma \in (0,\infty)$ of f. The Moreau envelope ${}^{\gamma}f$ is a smooth approximation of f with surprisingly beautiful properties. In particular, the most attractive property for us is that the Moreau envelope ${}^{\gamma}f$ has a Lipschitz continuous gradient over $\mathscr{H}$ (see Sect. 17.3.1). Moreover, if $\arg\min_{x\in\mathscr{H}} f(x) \neq \emptyset$, the set of all global minimizers of f is equal to that of the Moreau envelope (see Fact 17.2). These distinctive features suggest that the Moreau–Yosida regularization and the proximity operator are the keys bridging the gap between the analyses of smooth and nonsmooth convex functions. For example, these features have been utilized to develop efficient algorithms specialized for unconstrained nonsmooth convex optimization problems (see, e.g., [65, 108]). In addition to this direct use, the practical value of the Moreau envelope has been examined implicitly or explicitly as a smooth relaxation of the absolute value function in many applications (see Sect. 17.3.1).

In this study, we propose to approach the nonsmooth optimization problem

$$\text{minimize } f(x) \text{ subject to } x \in \mathrm{Fix}(T) \tag{17.3}$$

by solving its smooth relaxation

$$\text{minimize } {}^{\gamma}f(x) \text{ subject to } x \in \mathrm{Fix}(T) \tag{17.4}$$

with the HSDM. Here, $f \in \Gamma_0(\mathscr{H})$ (which in particular we consider to be nonsmooth) and $T : \mathscr{H} \to \mathscr{H}$ is a quasi-nonexpansive mapping (Note: The solution sets for (17.3) and (17.4) are not the same in general although they coincide specially in the simplest unconstrained case, i.e., $\mathrm{Fix}(T) = \mathscr{H}$). Thanks to (1) the beautiful properties of the Moreau envelope and (2) the flexibility in expressing a constraint set as the fixed point set of a quasi-nonexpansive mapping, the proposed approach enjoys wide applicability.

The rest of this paper is organized as follows. For readers' convenience, Sect. 17.2 presents a short tour in computational convex analysis which contains (1) elements of convex analysis, (2) the fixed point theory of quasi-nonexpansive mapping including a basic algorithm to approximate a fixed point of the mapping, and (3) elements of the variational inequality problems (VIPs). It also introduces briefly one role of quasi-nonexpansive mapping in signal processing. In Sect. 17.3, we will introduce the essence of the Moreau–Yosida regularization and the HSDM. Then we will show how to join the two concepts to approach the minimization problem of a nonsmooth convex function over the fixed point set of certain quasi-nonexpansive mappings. In Sect. 17.4, we demonstrate the effectiveness of the proposed approach in its application to the minimal antenna-subset selection problem under a highly nonlinear capacity-constraint for efficient multiple input multiple

[1] Nice introductions to the Moreau envelope are found, e.g., in [46, 109].

[2] As will be seen in (17.23), the derivative $\nabla {}^{\gamma}f$ is given as the *Yosida approximation* [142] of the subdifferential ∂f of f.

output (MIMO) communication systems; the convex relaxation of the problem is the ℓ_1 norm minimization under the constraint. Finally, in Sect. 17.5, we conclude this paper with some remarks on other possible advanced applications of the HSDM.

17.2 A Short Tour in Computational Convex Analysis

17.2.1 Selected Elements of Convex Analysis

In the following, we list minimum notions in convex analysis, which are necessary for our discussion (see, e.g., [7, 10, 46, 48, 59, 82, 109, 121, 134, 148] for detailed account on these notions). Let $\mathcal{H}$ be a real Hilbert space equipped with an inner product $\langle \cdot, \cdot \rangle$ and its induced norm $\| \cdot \|$.

Definition 17.1 (Basics in Convex Analysis).

(a) (Convex set) A set $C \subset \mathcal{H}$ is called convex if $\lambda x + (1-\lambda)y \in C$ for every $x, y \in C$ and every $\lambda \in [0,1]$. If a set $C \subset \mathcal{H}$ is closed as well as convex, it is called closed convex.

(b) (Convex function, Proper function) A function $f : \mathcal{H} \to (-\infty, \infty] := \mathbb{R} \cup \{\infty\}$ is called convex if

$$(\forall x, y \in \mathcal{H}, \forall \lambda \in (0,1)) \quad f(\lambda x + (1-\lambda)y) \leq \lambda f(x) + (1-\lambda) f(y). \quad (17.5)$$

In particular, a convex function $f : \mathcal{H} \to (-\infty, \infty]$ is called proper if

$$\mathrm{dom}(f) := \{x \in \mathcal{H} \mid f(x) < \infty\} \neq \emptyset.$$

A function $f \in \Gamma_0(\mathcal{H})$ is called strictly convex if

$$(x \neq y, \lambda \in (0,1)) \Rightarrow f(\lambda x + (1-\lambda)y) < \lambda f(x) + (1-\lambda) f(y).$$

(c) (Lower semicontinuous function) A function $f : \mathcal{H} \to (-\infty, \infty]$ is called lower semicontinuous if the set $\mathrm{lev}_{\leq \alpha}(f) := \{x \in \mathcal{H} \mid f(x) \leq \alpha\}$ is closed for every $\alpha \in \mathbb{R}$ (Note: If f is continuous over $\mathcal{H}$, f is lower semicontinuous). The set of all proper lower semicontinuous convex functions is denoted by $\Gamma_0(\mathcal{H})$.

(d) (Coercivity) A function $f \in \Gamma_0(\mathcal{H})$ is called *coercive* if

$$\|x\| \to \infty \Rightarrow f(x) \to \infty.$$

In this case, the existence of a minimizer of f, i.e., $\{x^\star \in \mathcal{H} \mid f(x^\star) \leq f(x) \quad (\forall x \in \mathcal{H})\} \neq \emptyset$, is guaranteed.

Fact 17.2 (Fundamental Tools for Convex Optimization).

(a) (Subgradient, Subdifferential, Legendre–Fenchel conjugate) Given $f \in \Gamma_0(\mathscr{H})$, the *subdifferential* of f at x is defined as the set of all *subgradients* of f at x:

$$\partial f(x) := \{u \in \mathscr{H} \mid \langle y-x, u\rangle + f(x) \leq f(y), \forall y \in \mathscr{H}\}.$$

Therefore, $0 \in \partial f(x) \Leftrightarrow f(x) = \min_{y\in\mathscr{H}} f(y)$. If f is continuous at $x \in \mathscr{H}$, $\partial f(x)$ is a nonempty closed convex set. Moreover, if f is Gâteaux differentiable[3] at x, the subdifferential at x is a singleton as $\partial f(x) = \{\nabla f(x)\}$ [10, 82, 134]. The subdifferential is regarded as a set-valued mapping $\partial f : \mathscr{H} \to 2^{\mathscr{H}}$, which is called bounded if it maps bounded sets to bounded sets [14] (Note: $2^{\mathscr{H}}$ stands for the collection of all subsets of $\mathscr{H}$).
Remark that the subdifferential of f at $x \in \mathscr{H}$ can be defined alternatively as $\partial f(x) := \{u \in \mathscr{H} \mid f(x) + f^*(u) = \langle x, u\rangle\}$, where $f^* \in \Gamma_0(\mathscr{H})$ is defined by

$$(\forall u \in \mathscr{H}) \quad f^*(u) := \sup_{x\in\mathscr{H}} \{\langle x, u\rangle - f(x)\}$$

and it is called the *conjugate* (also named *Legendre–Fenchel conjugate*, or *Legendre–Fenchel transform*) of f.

(b) (Proximity operator) The proximity operator of index $\gamma \in (0, \infty)$ of $f \in \Gamma_0(\mathscr{H})$ is defined (as in (17.1)) by

$$\mathrm{prox}_{\gamma f} : \mathscr{H} \to \mathscr{H} : x \mapsto \arg\min_{y\in\mathscr{H}} \left\{ f(y) + \frac{1}{2\gamma}\|x-y\|^2 \right\}, \tag{17.6}$$

where the existence and the uniqueness of the minimizer are guaranteed respectively by the coercivity and the strict convexity of $f(\cdot) + \frac{1}{2\gamma}\|x - \cdot\|^2$. Equivalently, for every $x \in \mathscr{H}$, $\mathrm{prox}_{\gamma f}(x)$ is characterized as a unique point satisfying

$$\{\mathrm{prox}_{\gamma f}(x)\} = \{z \in \mathscr{H} \mid z + \gamma \partial f(z) \ni x\}, \tag{17.7}$$

---

[3] (Gâteaux and Fréchet derivatives of function) Let U be an open subset of $\mathscr{H}$. Then a function $f : U \to \mathbb{R}$ is called Gâteaux differentiable at $x \in U$ if there exists $a(x) \in \mathscr{H}$ such that $\lim_{\delta\to 0} \frac{f(x+\delta h) - f(x)}{\delta} = \langle a(x), h\rangle$ $(\forall h \in \mathscr{H})$. In this case, $\nabla f(x) := a(x)$ is called Gâteaux derivative (or gradient) of f at x.

On the other hand, a function $f : U \to \mathbb{R}$ is called *Fréchet* differentiable over U if for each $u \in U$ there exists $a(u) \in \mathscr{H}$ such that

$$f(u+h) = f(u) + \langle a(u), h\rangle + o(\|h\|) \text{ for all } h \in \mathscr{H},$$

where $r(h) = o(\|h\|)$ means $\lim_{h\to 0} r(h)/\|h\| = 0$. In this case, $\nabla f : U \to \mathscr{H}$ defined by $\nabla f(u) = a(u)$ is called Fréchet derivative of f over U. If f is Fréchet differentiable over U, f is also Gâteaux differentiable over U and both derivatives coincide. Moreover, if f is Gâteaux differentiable with continuous derivative ∇f over U, then f is also Fréchet differentiable over U.

i.e.,

$$\operatorname{prox}_{\gamma f}(x) = (I + \gamma \partial f)^{-1}(x), \tag{17.8}$$

which is again equivalent to

$$(\forall y \in \mathscr{H}) \quad \left\langle y - \operatorname{prox}_{\gamma f}(x), \frac{x - \operatorname{prox}_{\gamma f}(x)}{\gamma} \right\rangle + f(\operatorname{prox}_{\gamma f}(x)) \leq f(y).$$

The proximity operator is firmly nonexpansive, i.e., $\operatorname{rprox}_{\gamma f} := 2\operatorname{prox}_{\gamma f} - I\colon \mathscr{H} \to \mathscr{H}$ is nonexpansive (see Sect. 17.2.2 for the definition of nonexpansivity of a mapping):

$$(\forall x, y \in \mathscr{H}) \quad \|(2\operatorname{prox}_{\gamma f} - I)x - (2\operatorname{prox}_{\gamma f} - I)y\| \leq \|x - y\|.$$

Moreover, if $\arg\min_{x \in \mathscr{H}} f(x) \neq \emptyset$, the set of all minimizers of f is equal to that of the Moreau envelope and also expressed as the fixed point set of $\operatorname{prox}_{\gamma f} : \mathscr{H} \to \mathscr{H}$; i.e.,

$$\arg\min_{x \in \mathscr{H}} f(x) = \arg\min_{x \in \mathscr{H}} {}^{\gamma}f(x) = \operatorname{Fix}\left(\operatorname{prox}_{\gamma f}\right).$$

(c) (Metric projection onto closed convex sets) Given a nonempty closed convex set $C \subset \mathscr{H}$ and any point $x \in \mathscr{H}$, there exists a unique point $P_C(x) \in C$ satisfying

$$d_C(x) := \min_{z \in C} \|x - z\| = \|x - P_C(x)\|.$$

The mapping $\mathscr{H} \ni x \mapsto P_C(x) \in C$ is called the metric projection (or convex projection) onto C and obviously $P_C(x) = \operatorname{prox}_{\gamma \iota_C}(x)$ $(\forall \gamma \in (0, \infty), \forall x \in \mathscr{H})$, hence P_C is firmly nonexpansive with $\operatorname{Fix}(P_C) = C \neq \emptyset$ (see Example 17.6(a) and Fig. 17.2). Moreover, $P_C : \mathscr{H} \to C$ is characterized by

$$x^{\star} \in C \text{ satisfies } \langle x - x^{\star}, z - x^{\star} \rangle \leq 0 \; (\forall z \in C) \;\Leftrightarrow\; x^{\star} \in C \text{ satisfies } x^{\star} = P_C(x). \tag{17.9}$$

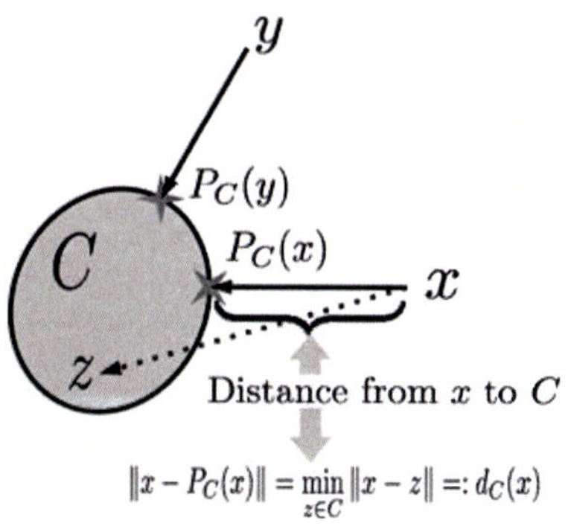

Fig. 17.2 Convex Projection: Metric projection onto a closed convex set C

(d) (Expression of a closed convex set I) Given (possibly infinitely many) closed convex sets $C_i \subset \mathscr{H}$ ($i \in \mathscr{I}$: an index set), their intersection $\bigcap_{i \in \mathscr{I}} C_i$ is again a closed convex set (Note: This property is a natural nonlinear generalization of the elementary fact that the intersection of multiple subspaces is again a subspace in a vector space).

(e) (Expression of a closed convex set II) Given a function $f \in \Gamma_0(\mathscr{H})$, the set $\operatorname{lev}_{\leq 0}(f)$, which is called the (zero-)level set of f, is closed convex. Conversely, given a closed convex set $C \subset \mathscr{H}$, there exists a continuous convex function $f : \mathscr{H} \to \mathbb{R}$ satisfying $C = \operatorname{lev}_{\leq 0}(f)$. The function $d_C : \mathscr{H} \to [0, \infty)$ in (c) is obviously such an example.

17.2.2 *Quasi-Nonexpansive Mappings and Their Fixed Point Sets*

Suppose that a mapping $T : \mathscr{H} \to \mathscr{H}$ has at least one fixed point. Then the mapping $T : \mathscr{H} \to \mathscr{H}$ is called quasi-nonexpansive (or Fejér) [7, 8, 54, 127] if T satisfies for every $x \in \mathscr{H}$ and every $z \in \operatorname{Fix}(T)$

$$\|T(x) - z\| \leq \|x - z\|. \tag{17.10}$$

The identity operator $I : \mathscr{H} \to \mathscr{H}$ is also a quasi-nonexpansive mapping which satisfies of course $\operatorname{Fix}(I) = \mathscr{H}$.

We introduce special subclasses of quasi-nonexpansive mappings below (see also Fig. 17.3). A quasi-nonexpansive mapping T is said to be attracting if T satisfies for every $x \notin \operatorname{Fix}(T)$ and every $z \in \operatorname{Fix}(T)$

$$\|T(x) - z\| < \|x - z\|.$$

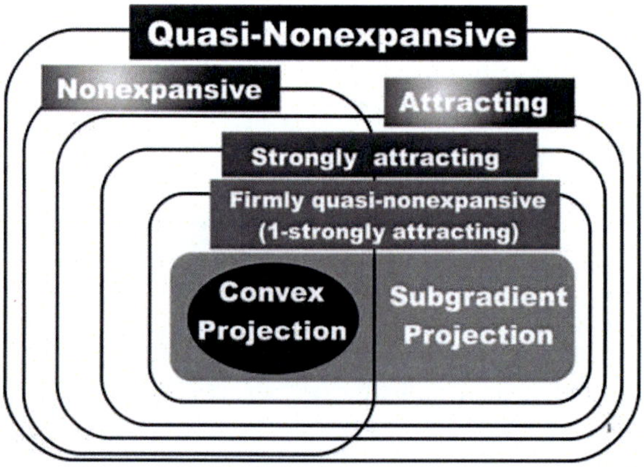

Fig. 17.3 Quasi-nonexpansive mapping and its subclasses (A nonexpansive mapping is also quasi-nonexpansive if this mapping has at least one fixed point)

In particular, an attracting mapping T is called α-strongly attracting if there exists some $\alpha > 0$ satisfying for every $x \in \mathscr{H}$ and every $z \in \mathrm{Fix}(T)$

$$\alpha\|x - T(x)\|^2 \leq \|x - z\|^2 - \|T(x) - z\|^2.$$

The above inequality offers a lower bound for improvement by T of approximation accuracy of a point x to all fixed points z of T.

A quasi-nonexpansive mapping $T : \mathscr{H} \to \mathscr{H}$ is said to be α-averaged [4, 7] if there exist some $\alpha \in (0,1)$ and some quasi-nonexpansive mapping N such that $T = (1-\alpha)I + \alpha N$. In this case, T satisfies an obvious relation $\mathrm{Fix}(T) = \mathrm{Fix}(N)$. Moreover, T is strongly attracting (see Proposition 17.3(b) below). In particular, if T is $\frac{1}{2}$-averaged, T is called a firmly quasi-nonexpansive mapping [136] (the class of firmly quasi-nonexpansive mappings is specially denoted by $\mathfrak{T}$ [8]).

On the other hand, a mapping $T : \mathscr{H} \to \mathscr{H}$ is called Lipschitz continuous with a Lipschitz constant κ or shortly κ-Lipschitzian if there exists some $\kappa > 0$ satisfying for every $x, y \in \mathscr{H}$

$$\|T(x) - T(y)\| \leq \kappa\|x - y\|.$$

In particular, if there exists some $\kappa < 1$, T is called a contraction (or a strictly contractive) mapping. In this case, the Banach–Picard's contraction mapping theorem guarantees the unique existence of the fixed point of T, and it is not hard to see that T is α-averaged for any $\alpha \in \left[\frac{\kappa+1}{2}, 1\right)$. If the mapping T is 1-Lipschitzian, T is called a nonexpansive mapping [7, 70, 71, 121] and in this case, T is also quasi-nonexpansive if $\mathrm{Fix}(T) \neq \emptyset$. In contrast to the case of the existence of $\kappa < 1$, the existence of $\kappa = 1$ is insufficient to guarantee the existence of a fixed point in view of the following example: $T : \mathbb{R} \ni x \mapsto x + 1 \in \mathbb{R}$.

The following Proposition 17.3(a) guarantees that the closedness and convexity of the fixed point set of any quasi-nonexpansive mapping. This property is very fortunate to express a constraint set, in convex optimization, as the fixed point set of a quasi-nonexpansive mapping. For example, Proposition 17.3(a) together with Fact 17.2(d),(e) suggests that a closed convex set can be expressed as the intersection of possibly infinitely many simpler closed convex sets, each of which can be expressed as the fixed point set of an efficiently computable quasi-nonexpansive mapping. Moreover, by Proposition 17.3(b), given a quasi-nonexpansive mapping $N : \mathscr{H} \to \mathscr{H}$, we can construct a strongly attracting quasi-nonexpansive mapping $T := (1-\alpha)I + \alpha N$ ($\alpha \in (0,1)$) with $\mathrm{Fix}(T) = \mathrm{Fix}(N)$. Therefore, the quasi-nonexpansive mapping (or even more specifically the attracting mapping) has a great deal of potential not only as an computational tool for monotone approximation to the closed convex set but also as an alternative mathematical expression of the closed convex set as its fixed point set.

Proposition 17.3 (Fundamental Properties of Quasi-Nonexpansive Mapping).

(a) Let $T : \mathscr{H} \to \mathscr{H}$ be a quasi-nonexpansive mapping. Then $\mathrm{Fix}(T)$ *can be expressed as (see for example* [8, 136]*):*

$$\mathrm{Fix}(T) = \bigcap_{y \in \mathscr{H}} \left\{ x \in \mathscr{H} \mid \langle y - T(y), x \rangle \leq \frac{\|y\|^2 - \|T(y)\|^2}{2} \right\}.$$

This tells us that $\mathrm{Fix}(T)$ *can be expressed as the intersection of infinitely many closed half spaces, hence the closedness and convexity of* $\mathrm{Fix}(T)$ *are guaranteed by Fact 17.2(d).*

(b) A quasi-nonexpansive mapping $T : \mathscr{H} \to \mathscr{H}$ is α-averaged for some $\alpha \in (0,1)$ if and only if T is $\left(\frac{1-\alpha}{\alpha}\right)$-strongly attracting [136]*. Therefore, a quasi-nonexpansive mapping T is $\frac{1}{2}$-averaged if and only if it is 1-strongly attracting.*

In Proposition 17.4 below, (a) and (b) are slight refinement of similar results in [7, Propositions 2.10 and 2.12]. By applying the properties in Proposition 17.4, we can construct a new quasi-nonexpansive mapping whose fixed point set is the intersection of the fixed point sets of given multiple quasi-nonexpansive mappings in Examples 17.6 and 17.9 in Sect. 17.2.3. Note that Proposition 17.4(c) holds even when $\mathrm{Fix}(T_1) \cap \mathrm{Fix}(T_2) = \emptyset$.

Proposition 17.4 (Algebraic Properties of Quasi-Nonexpansive Mapping).

(a) (Convex combination [136]*) Suppose that $T_i : \mathscr{H} \to \mathscr{H}$ $(i = 1,2)$ are quasi-nonexpansive mappings satisfying $\mathrm{Fix}(T_1) \cap \mathrm{Fix}(T_2) \neq \emptyset$. Then for any $w \in (0,1)$, the mapping $T := wT_1 + (1-w)T_2$ is quasi-nonexpansive and satisfies $\mathrm{Fix}(T) = \mathrm{Fix}(T_1) \cap \mathrm{Fix}(T_2)$. In particular, if each T_i $(i = 1,2)$ is $\alpha_i (> 0)$-strongly attracting, then T is $\left(\frac{(\alpha_1+1)(\alpha_2+1)}{(1-w)\alpha_1 + w\alpha_2 + 1} - 1\right)$-strongly attracting.*

(b) (Composition [136]*) Let $T_1 : \mathscr{H} \to \mathscr{H}$ be a quasi-nonexpansive mapping and $T_2 : \mathscr{H} \to \mathscr{H}$ an attracting quasi-nonexpansive mapping satisfying $\mathrm{Fix}(T_1) \cap \mathrm{Fix}(T_2) \neq \emptyset$. Then $T := T_2 T_1$ is quasi-nonexpansive and $\mathrm{Fix}(T) = \mathrm{Fix}(T_1) \cap \mathrm{Fix}(T_2)$. In particular, if each T_i $(i = 1,2)$ is $\alpha_i (> 0)$-strongly attracting, then T is $\left(\frac{\alpha_1 \alpha_2}{\alpha_1 + \alpha_2}\right)$-strongly attracting.*

(c) (Operations for averaged nonexpansive mappings [101, 136]*) Suppose that each $T_i : \mathscr{H} \to \mathscr{H}$ $(i = 1,2)$ is α_i-averaged nonexpansive for some $\alpha_i \in [0,1)$. Then for every $w \in [0,1]$, the mapping $(1-w)T_1 + wT_2$ is $\{(1-w)\alpha_1 + w\alpha_2\}$-averaged nonexpansive. Moreover, $T_1 T_2$ is α-averaged nonexpansive for $\alpha := \frac{\alpha_1 + \alpha_2 - 2\alpha_1\alpha_2}{1 - \alpha_1\alpha_2} \in [0,1)$.*

Finally, for intuitive understanding, we explain briefly how the attracting mapping is connected in essence with signal processing.

Remark 17.5 (A Role of Attracting Mapping in Signal Processing). Monotone approximation to an unknown desirable information to be estimated, say *estimandum*, is one of the most favorable properties for signal processing algorithms. In particular, in adaptive filtering or adaptive system identification problems (e.g., adaptive

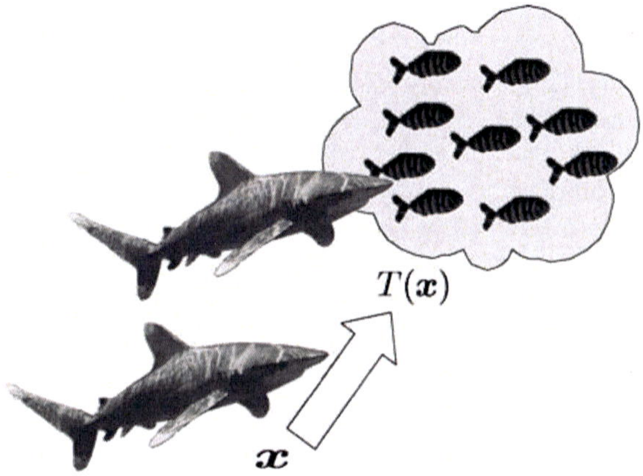

Fig. 17.4 What is the best possible strategy for a starving shark? Maximal satisfaction is expected by approaching monotonically every fish. This is realized by an attracting mapping

channel equalization, adaptive echo cancellation, etc.), the algorithms are required, at each time, to offer a tentative approximation of an estimandum. By utilizing a priori knowledge as well as the latest statistical knowledge obtained from observed data, the algorithm is desired to update the previous estimate to a better one which is closer to the *estimandum*. A practical scenario to realize such a monotone approximation is divided into the following two steps: (Step 1) define a set, say a *target set*, which is sufficiently small but contains candidates consistent with all available knowledge on the *estimandum*, and (Step 2) realize a mapping T which shifts any point not in the target set strictly closer to every point in the target set and does not move any point in the target set (see Fig. 17.4). If the *estimandum* surely belongs to the target set, the above scenario automatically realizes a monotone approximation to the *estimandum*. The mapping satisfying the condition in Step 2 is called *attracting mapping*. Obviously, a point does not move by the mapping if and only if it is already in the target set. Therefore, the target set must be the fixed point set of the attracting mapping. This observation suggests that *a key to realize a successful signal processing algorithm is how to design an attracting mapping of which the fixed point set is the target set.* On the other hand, as seen in Proposition 17.3(a), *the fixed point set of any attracting mapping is a closed convex set.* This simple but valuable observation tells us that for realizing monotone approximation, the attracting mapping is certainly ideal, and in this case, the target set is ensured to be a closed convex set. Moreover, if multiple attracting mappings with a common fixed point are given, we can define in constructive ways a new attracting mapping whose fixed point set is the intersection of the fixed point sets of the given mappings (see Proposition 17.4), which is extremely fortunate for the refinement of the target set in Step 1. Therefore, the attracting mapping has a great deal of potential to be not only a computational tool for monotone approximation to the closed convex set but also an alternative mathematical expression of the closed convex set as its fixed point set.

In [140], it has been clarified that the adaptive filtering algorithms based on orthogonal projections [81, 112, 128] exploit the above feature of the attracting mapping implicitly. This discovery leads to a unified scheme called the *adaptive projected subgradient method* (APSM) [114, 133, 135]; this scheme is a time-varying extension of the Polyak's subgradient algorithm, which was developed for a nonsmooth convex optimization problem with a fixed target value, to the case where the convex objective itself keeps changing in the whole process. Under this simple umbrella of the APSM, a unified convergence analysis has been established for a wide range of adaptive algorithms. Moreover, the APSM has been serving as a guiding principle to create various powerful adaptive algorithms for acoustic systems [144, 147], wireless communication systems [25, 26, 146], distributed learning for diffusion network [27], online learning in Reproducing Kernel Hilbert Spaces [115, 116, 125], etc. Moreover, a steady-state mean-square performance analysis of a simplest example of the APSM has been established in [123]; the analysis is based on the *energy conservation argument* [112] developed specially for performance analyses of adaptive filtering algorithms.

17.2.3 Toolbox of Quasi-Nonexpansive Mapping

We list particularly useful quasi-nonexpansive mappings called in this paper *design tool mappings*. With the aid of Proposition 17.4, the design tool mappings can be used as tools to design a new quasi-nonexpansive mapping whose fixed point set is the intersection of their fixed point sets.

Example 17.6 (Design Tool Mappings).

(a) (Metric projection/Convex projection) Given a nonempty closed convex set C in $\mathcal{H}$, the metric projection $P_C : \mathcal{H} \to C$ is a firmly nonexpansive mapping with $\mathrm{Fix}(P_C) = C$ (see Fact 17.2(c)). The firm nonexpansivity of P_C implies that P_C is also a 1-strongly attracting nonexpansive mapping (see Proposition 17.3(b)). Furthermore, the function $\varphi_1 : x \mapsto d_C^2(x) := \|x - P_C(x)\|^2$ is convex and Gâteaux differentiable over $\mathcal{H}$ with its derivative $\nabla\varphi_1(x) = 2(x - P_C(x))$ $(\forall x \in \mathcal{H})$.

(b) (Proximal forward–backward splitting operator [46, 66, 104, 126]) Suppose that

$$S := \arg\min_{x\in\mathcal{H}} \{f_1(x) + f_2(x)\}$$

is nonempty for $f_1, f_2 \in \Gamma_0(\mathcal{H})$, where f_2 is Gâteaux differentiable on $\mathcal{H}$ with its gradient $\nabla f_2 : \mathcal{H} \to \mathcal{H}$. Then $x^\star \in \mathcal{H}$ satisfies $x^\star \in S$ if and only if $x^\star \in \mathcal{H}$ is a fixed point of the *proximal forward–backward splitting operator*: $\mathrm{prox}_{\mu f_1}(I - \mu\nabla f_2)$ for any $\mu > 0$, i.e., $x^\star = \mathrm{prox}_{\mu f_1}(I - \mu\nabla f_2)(x^\star)$. If in addition ∇f_2 is κ-Lipschitzian for some $\kappa > 0$, the proximal forward–backward splitting operator $\mathrm{prox}_{\mu f_1}(I - \mu\nabla f_2)$ with $\mu \in \left(0, \frac{2}{\kappa}\right]$ is nonexpansive. Moreover, this operator is $\frac{1}{2-\gamma}$-averaged nonexpansive if $\mu \in (0, \frac{2\gamma}{\kappa}] \subset \left(0, \frac{2}{\kappa}\right)$

(Note: (1) The nonexpansivity of the proximal forward–backward splitting operator with $\mu \in \left(0, \frac{2}{\kappa}\right]$ is confirmed by the nonexpansivity of $\mathrm{prox}_{\mu f_1}$ and the nonexpansivity of $I - \mu\nabla f_2 = (1 - \frac{\mu}{\frac{2}{\kappa}})I + \frac{\mu}{\frac{2}{\kappa}}\left(I - \frac{2}{\kappa}\nabla f_2\right)$ [see Fact 17.15 in Sect. 17.2.5]. (2) The averaged nonexpansivity of the operator with $\mu \in (0, \frac{2\gamma}{\kappa}] \subset (0, \frac{2}{\kappa})$ is confirmed by applying Proposition 17.4(c) to the firm nonexpansivity of $\mathrm{prox}_{\mu f_1}$ and the γ-averaged nonexpansivity of $I - \mu\nabla f_2 = (1-\gamma)I + \gamma(I - \frac{\mu}{\gamma}\nabla f_2))$. In particular, setting $f_1 := i_C$ for a closed convex set $C \subset \mathcal{H}$ reproduces the characterization of the minimizers of f_2 over C by the fixed point set of the $\frac{1}{2-\gamma}$-averaged nonexpansive mapping $P_C(I - \mu\nabla f_2)$ for $\mu \in (0, \frac{2\gamma}{\kappa}] \subset (0, \frac{2}{\kappa})$ [20, 40, 139]. This is essentially same as the fixed point characterization of the VIP as found in Fact 17.14 in Sect. 17.2.5.

(c) (Douglas–Rachford splitting operator [42, 57, 90]) Let $f_1, f_2 \in \Gamma_0(\mathcal{H})$ satisfy

$$S := \arg\min_{x \in \mathcal{H}} \{f_1(x) + f_2(x)\} \neq \emptyset.$$

Under the following *qualification condition*:

$$\left.\begin{array}{l} \mathrm{cone}\,(\mathrm{dom}(f_1) - \mathrm{dom}(f_2)) := \displaystyle\bigcup_{\lambda>0} \{\lambda x \mid x \in \mathrm{dom}(f_1) - \mathrm{dom}(f_2)\} \\ \text{is a closed subspace of } \mathcal{H}, \text{ where} \\ \mathrm{dom}(f_1) - \mathrm{dom}(f_2) := \{x_1 - x_2 \in \mathcal{H} \mid x_i \in \mathrm{dom}(f_i)\ (i = 1, 2)\}, \end{array}\right\} \quad (17.11)$$

the Douglas–Rachford splitting-type algorithm uses in principle the following characterization: for any $\gamma \in (0, \infty)$

$$x^\star \in \mathcal{H} \text{ minimizes } f_1 + f_2 \quad \Leftrightarrow \begin{cases} x^\star = \mathrm{prox}_{\gamma f_2}(y), \\ y \in \mathrm{Fix}\left(\mathrm{rprox}_{\gamma f_1}\mathrm{rprox}_{\gamma f_2}\right), \end{cases} \quad (17.12)$$

which means that S can be expressed as the image of $\mathrm{prox}_{\gamma f_2}$ of the fixed point set of the nonexpansive mapping $\mathrm{rprox}_{\gamma f_1}\mathrm{rprox}_{\gamma f_2}$ (Note: The firm nonexpansivity of $\mathrm{prox}_{\gamma f_i}$, $i = 1, 2$, guarantees the nonexpansivity of $\mathrm{rprox}_{\gamma f_i}$ (see Fact 17.2(b))).

(d) (Subgradient projection) Suppose that a continuous convex function $f : \mathcal{H} \to \mathbb{R}$ satisfies $\mathrm{lev}_{\leq 0}(f) \neq \emptyset$. Let $f'(x) \in \partial f(x)$ ($\forall x \in \mathcal{H}$) be a selection from the subdifferential $\partial f(x)$ (Note: In this paper, we use the notation $\nabla f(x)$ for a Gâteaux differentiable function f to distinguish from $f'(x)$ for a nonsmooth one). Then a mapping $T_{\mathrm{sp}(f)} : \mathcal{H} \to \mathcal{H}$ defined by

$$T_{\mathrm{sp}(f)} : x \mapsto \begin{cases} x - \frac{f(x)}{\|f'(x)\|^2} f'(x), & \text{if } f(x) > 0; \\ x, & \text{otherwise,} \end{cases}$$

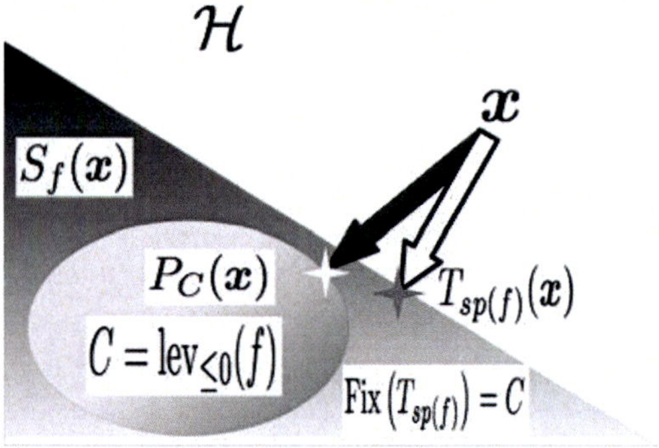

Fig. 17.5 Subgradient projection as an approximation of metric projection (see (17.13) for the definition of $S_f(x)$)

is called a *subgradient projection relative to* f. For $f(x) > 0$, $T_{\mathrm{sp}(f)}(x)$ is given by the metric projection of x onto the closed half-space $\{y \in \mathscr{H} \mid \langle y - x, f'(x) \rangle + f(x) \leq 0\} \supset \mathrm{lev}_{\leq 0}(f)$. Therefore, $T_{\mathrm{sp}(f)}$ is a 1-strongly attracting quasi-nonexpansive mapping with $\mathrm{Fix}(T_{\mathrm{sp}(f)}) = \mathrm{lev}_{\leq 0}(f)$ (see, e.g., [127, Lemma 2.8], [8] and Fig. 17.5), hence Proposition 17.3(b) implies that $2T_{\mathrm{sp}(f)} - I$ is quasi-nonexpansive. The metric projection onto a closed convex set C can also be interpreted as a subgradient projection relative to a continuous convex function $d_C : x \mapsto \|x - P_C(x)\|$, i.e., $T_{\mathrm{sp}(d_C)} = P_C$. This fact is confirmed by

$$\partial d_C(x) = \begin{cases} \{z \in \mathscr{H} \mid \|z\| \leq 1, \langle z, y - x \rangle \leq 0, \forall y \in C\} \ni 0, & \text{if } x \in C; \\ \frac{x - P_C(x)}{d(x,C)}, & \text{otherwise.} \end{cases}$$

If we can use more information on the function $f \in \Gamma_0(\mathscr{H})$, we may define other strongly attracting mappings that realize better approximation to the set $\mathrm{lev}_{\leq 0}(f)$, as shown below.

Proposition 17.7 (A Generalization of Subgradient Projection [103]). *Let* $f : \mathscr{H} \to \mathbb{R}$ *be a continuous convex function with* $\mathrm{lev}_{\leq 0}(f) \neq \emptyset$ *and* $f' : \mathscr{H} \to \mathscr{H}$ *a selection of the subdifferential* $\partial f : \mathscr{H} \to 2^{\mathscr{H}}$*, i.e.,* $f'(x) \in \partial f(x)$, $\forall x \in \mathscr{H}$*. Let* $\xi : \mathscr{H} \to \mathbb{R}$ *be a function satisfying* $\xi(x) \geq f(x), \forall x \in \mathscr{H}$*. Suppose that*

$$S_\xi(x) := \begin{cases} \{y \in \mathscr{H} \mid \langle y - x, f'(x) \rangle + \xi(x) \leq 0\}, & \text{if } f(x) \geq 0; \\ \mathscr{H}, & \text{otherwise,} \end{cases} \tag{17.13}$$

satisfies (O-i) $S_\xi(x) \supset \mathrm{lev}_{\leq 0}(f)$*, and (O-ii)* $x \notin \mathrm{lev}_{\leq 0}(f) \Rightarrow x \notin S_\xi(x)$*. Then the projection onto* $S_\xi(x)$*, i.e.,*

$$T_{\mathrm{dsp},\xi} : x \mapsto \begin{cases} x - \frac{\xi(x)}{\|f'(x)\|^2} f'(x), & \text{if } f(x) > 0; \\ x, & \textit{otherwise}, \end{cases} \tag{17.14}$$

is firmly quasi-nonexpansive with $\mathrm{Fix}(T_{\mathrm{dsp},\xi}) = \mathrm{lev}_{\leq 0}(f)$.

Remark 17.8 ($T_{\mathrm{dsp},\xi}$ as a Deeper Outer Approximation). By the definition of subgradient, $S_f(x)$ satisfies the conditions (O-i) and (O-ii) in Proposition 17.7. In this special case, we have $T_{\mathrm{dsp},f} = T_{\mathrm{sp}(f)}$, hence $T_{\mathrm{dsp},\xi} : \mathscr{H} \to \mathscr{H}$ is a generalization of the *subgradient projection relative to* f. If $\xi(x) > f(x) > 0$, we have $S_\xi(x) \subsetneq S_f(x)$, i.e., $S_\xi(x)$ is a *deeper* outer approximation (of $\mathrm{lev}_{\leq 0}(f)$ w.r.t. x) than $S_f(x)$. Several constructions of such $\xi(x)(> f(x))$ have been discussed for example in [85, Example 3.4], [103, 141].

Example 17.9 (Deepest Outer Approximation with Available Information).

(a) (Best quadratic lower bound with Lipschitz constant of gradient operator [141]) Suppose that (1) $f \in \Gamma_0(\mathscr{H})$ is Gâteaux differentiable on $\mathscr{H}$ with its gradient $\nabla f : \mathscr{H} \to \mathscr{H}$ which is κ-Lipschitzian over $\mathscr{H}$, and (2) $\mathrm{lev}_{\leq 0}(f) \neq \emptyset$ and $f(x) \geq -\rho$ ($\exists \rho \geq 0$, $\forall x \in \mathscr{H}$). Fix $z \in \mathscr{H} \setminus \mathrm{lev}_{\leq 0}(f)$ arbitrarily, and let $g_{0,z}(x) := \langle x - z, \nabla f(z)\rangle + f(z)$ ($\forall x \in \mathscr{H}$). Then the function $g_{1,z} : \mathscr{H} \to \mathbb{R}$:

$$g_{1,z}(x) := \begin{cases} g_{0,z}(x), & \text{if } a \leq g_{0,z}(x); \\ \frac{1}{2}\frac{(g_{0,z}(x)-b)^2}{a-b}, & \text{if } b \leq g_{0,z}(x) \leq a; \\ -\rho, & \text{if } g_{0,z}(x) \leq b, \end{cases} \tag{17.15}$$

where $a := -\rho + \frac{\|\nabla f(z)\|^2}{2\kappa}$ and $b := -\rho - \frac{\|\nabla f(z)\|^2}{2\kappa}$, satisfies $g_{0,z}(x) \leq g_{1,z}(x) \leq f(x)$ ($\forall x \in \mathscr{H}$). This implies $\xi(y) := g_{1,z}(y) - \langle y - z, \nabla f(z)\rangle \geq g_{0,z}(y) - \langle y - z, \nabla f(z)\rangle = f(z)$ ($\forall y \in \mathscr{H}$), hence

$$\begin{aligned} \mathrm{lev}_{\leq 0}(f) \subset \mathrm{lev}_{\leq 0}(g_{1,z}) &= \{y \in \mathscr{H} \mid \langle y - z, \nabla f(z)\rangle + \xi(z) \leq 0\} = S_\xi(z) \\ &\subset \{y \in \mathscr{H} \mid \langle y - z, \nabla f(z)\rangle + f(z) \leq 0\} = \mathrm{lev}_{\leq 0}(g_{0,z}) \\ &= S_f(z). \end{aligned} \tag{17.16}$$

Moreover, $g_{1,z}$ satisfies

(i) $g_{1,z}(x)|_{x=z} = f(z)$ and $\nabla g_{1,z}(x)|_{x=z} = \nabla f(z)$,
(ii) $f(x) \geq g_{1,z}(x) \geq -\rho$ ($\forall x \in \mathscr{H}$) and $\|\nabla g_{1,z}(x) - \nabla g_{1,z}(y)\| \leq \kappa \|x - y\|$ ($\forall x, y \in \mathscr{H}$).

(b) (Deepest outer approximating half-space of level set of a quadratic function [103]) Suppose that a quadratic function $f(x) := \|Ax - b\|^2 - \rho$ ($\forall x \in \mathscr{H}$) satisfies $\mathrm{lev}_{\leq 0}(f) \neq \emptyset$, where $A : \mathscr{H} \to \mathscr{H}'$ is a bounded linear operator ($\mathscr{H}'$ is a real Hilbert space whose inner product and its induced norm are also denoted by $\langle \cdot, \cdot \rangle$ and $\|\cdot\|$ respectively), $b \in \mathscr{H}'$ and $\rho \in \mathbb{R}$. Fix $z \in \mathscr{H} \setminus \mathrm{lev}_{\leq 0}(f)$ arbitrarily and let $\xi_\tau(z) := 2\left(f(z) - \tau - \sqrt{(f(z) - \tau)(-\tau)}\right)$ for any $\tau \in [-\rho, \inf_{y \in \mathscr{H}} f(y)]$. Then $S_{\xi_\tau}(z) \subset \mathscr{H}$ satisfies

(i) $\mathrm{lev}_{\leq 0}(f) \subset S_{\xi_\tau}(z) \subsetneq S_f(z)$ for any $\tau \in [-\rho, \inf_{y \in \mathscr{H}} f(y)]$,
(ii) $\widetilde{S}_{\xi_{\tau_{\min}}}(z) \cap \mathrm{lev}_{\leq 0}(f) \neq \emptyset$ for $\tau_{\min} := \min_{y \in \mathscr{H}} f(y)$, where $\widetilde{S}_{\xi_{\tau_{\min}}}(z)$ is the boundary hyperplane of $S_{\xi_{\tau_{\min}}}(z)$.

17.2.4 Iterative Approximation of a Fixed Point of Quasi-Nonexpansive Mapping

By introducing a real number sequence $(\alpha_n)_{n\geq 0} \subset [0,1]$, the algorithm in the Banach–Picard's contraction mapping theorem has been extended to

$$x_{n+1} := (1-\alpha_n)x_n + \alpha_n T(x_n), \tag{17.17}$$

where T is a quasi-nonexpansive mapping. To guarantee the weak convergence[4] of $(x_n)_{n\geq 0}$ to a fixed point of T, the demiclosedness of $I-T$ at $0 \in \mathscr{H}$ is required in addition to some condition on $(\alpha_n)_{n\geq 0}$, where, in general, a mapping $G: \mathscr{H} \to \mathscr{H}$ is said to be *demiclosed* at $y \in \mathscr{H}$ if weak convergence of a sequence $(x_n)_{n\geq 0} \subset \mathscr{H}$ to $x \in \mathscr{H}$ and strong convergence of $(G(x_n))_{n\geq 0}$ to $y \in \mathscr{H}$ imply $G(x) = y$. It is well known [19] that the mapping $I-T$ is demiclosed at every point $y \in \mathscr{H}$ if $T: \mathscr{H} \to \mathscr{H}$ is nonexpansive. Moreover, if a continuous convex function $f: \mathscr{H} \to \mathbb{R}$ satisfies $\mathrm{lev}_{\leq 0}(f) \neq \emptyset$ and its subdifferential $\partial f: \mathscr{H} \to 2^{\mathscr{H}}$ is bounded in the sense of Fact 17.2(a), the mapping $I - T_{\mathrm{sp}(f)}$ is demiclosed at $0 \in \mathscr{H}$ (see [127, Lemma 2.9], [8]).

The convergence theorem of the Algorithm (17.17), which is called the *Mann iterative process*, is summarized as follows.

Proposition 17.10 (Mann Iterative Process). *Let $T: \mathscr{H} \to \mathscr{H}$ be a quasi-nonexpansive mapping. Then for any initial point $x_0 \in \mathscr{H}$, the sequence $(x_n)_{n\geq 0} \subset \mathscr{H}$, generated by (17.17), converges weakly to a point in* $\mathrm{Fix}(T)$, *which depends on the choices of $x_0 \in \mathscr{H}$ and the real number sequence $(\alpha_n)_{n\geq 0} \subset [0,1]$, under either of the following conditions.*

(a) $I-T$ is demiclosed at $0 \in \mathscr{H}$ and $(\alpha_n)_{n\geq 0}$ is bounded away from 0 *and* 1, *i.e., there exist $\varepsilon_1, \varepsilon_2 > 0$ satisfying $(\alpha_n)_{n\geq 0} \subset [\varepsilon_1, 1-\varepsilon_2]$* [54].
(b) T is nonexpansive and $\sum_{n\geq 0} \alpha_n(1-\alpha_n) = \infty$ [75].

Remark 17.11 (Several Forms of Mann-type Iterates).

(a) The iterative algorithm shown in (17.17) is commonly referred to as "Mann iterative process" because this has an alternative expression of

$$x_{n+1} := \sum_{j=1}^{n} a_{n,j} u_j \quad \text{and} \quad u_{n+1} := T(x_n) \tag{17.18}$$

given in [94] if $(a_{n,j})_{0\leq j\leq n, n\geq 0} \subset [0,1]$ satisfies $a_{n+1,j} = (1-a_{n+1,n+1})a_{n,j}$ and $\alpha_n = a_{n+1,n+1}$ $(n = 0,1,2,\ldots)$.

---

[4] (Strong and weak convergences) A sequence $(x_n)_{n\geq 0}$ in a real Hilbert space $\mathscr{H}$ is said to converge strongly to a point $x \in \mathscr{H}$ if the real number sequence $(\|x_n - x\|)_{n\geq 0}$ converges to 0, and to converge weakly to $x \in \mathscr{H}$ if the real number sequence $(\langle x_n - x, y\rangle)_{n\geq 0}$ converges to 0 for every $y \in \mathscr{H}$. If $(x_n)_{n\geq 0}$ converges strongly to x, then $(x_n)_{n\geq 0}$ converges weakly to x. The converse is true if $\mathscr{H}$ is finite dimensional, hence in finite dimensional case we do not need to distinguish these convergences.

(b) Suppose in particular (1) that T is α-averaged quasi-nonexpansive for some $\alpha \in (0,1)$, i.e., $N := \frac{T-(1-\alpha)I}{\alpha}$ is quasi-nonexpansive, and (2) that $I-T$ is demi-closed at $0 \in \mathcal{H}$. Then the sequence $(x_n)_{n=0}^{\infty} \subset \mathcal{H}$ generated by any initial point $x_0 \in \mathcal{H}$ and

$$x_{n+1} := T(x_n) = (1-\alpha)x_n + \alpha N(x_n)$$

converges weakly to a point in $\mathrm{Fix}(N) = \mathrm{Fix}(T)$.

(c) If $N : \mathcal{H} \to \mathcal{H}$ is firmly nonexpansive, i.e., $T := 2N - I$ is nonexpansive with $\mathrm{Fix}(T) = \mathrm{Fix}(N)$, the iteration (17.17) can be expressed equivalently as

$$x_{n+1} := \left(1 - \frac{t_n}{2}\right) x_n + \frac{t_n}{2}(2N-I)(x_n) = (1-t_n)x_n + t_n N(x_n),$$

where the conditions for $(\alpha_n)_{n\geq 0} \subset [0,1]$ in Proposition 17.10(a),(b) are replaced, respectively, by $(t_n)_{n\geq 0} = (2\alpha_n)_{n\geq 0} \subset [2\varepsilon_1, 2-2\varepsilon_2]$ and $\sum_{n\geq 0} t_n (2-t_n) = \infty$. This is a simplest case of a weak convergence theorem shown in [39] under much weaker conditions to cope with the numerical errors possibly unavoidable in the iterative computations.

(d) Suppose that $T : \mathbb{R}^m \to \mathbb{R}^m$ is continuous as well as attracting (In this case, the mapping T is said to be paracontractive). Then for any initial point $x_0 \in \mathbb{R}^m$, the sequence $(x_n)_{n\geq 0}$ generated by $x_{n+1} := T(x_n)$ converges to a point in $\mathrm{Fix}(T)$ [60] (Note: This idea has been extended to the case of *Bregman distance* [28]).

We have found many useful algorithms whose primitive convergence properties can be examined simply by Proposition 17.10.

Example 17.12 (Mann Iterative Process Found in Applications).

(a) (POCS: Projections onto convex sets [18, 77, 119, 143]) Suppose that $C_i \subset \mathcal{H}$ $(i = 1,2,\ldots,m)$ are closed convex sets satisfying $\bigcap_{i=1}^{m} C_i \neq \emptyset$. Define $\frac{\lambda_i}{2}$-averaged nonexpansive mappings $T_i : \mathcal{H} \to \mathcal{H}$ $(i = 1,2,\ldots,m)$, with $\lambda_i \in (0,2)$, by $T_i := I + \lambda_i(P_{C_i} - I) = \left(1 - \frac{\lambda_i}{2}\right) I + \frac{\lambda_i}{2}(2P_{C_i} - I)$, which obviously satisfy $\mathrm{Fix}(T_i) = C_i$ (see Example 17.6(a)). Moreover, by Proposition 17.4 (c) and (b), $T = T_m T_{m-1} \cdots T_1$ is averaged nonexpansive with $\mathrm{Fix}(T) = \bigcap_{i=1}^{m} C_i \neq \emptyset$. Applying Remark 17.11(b) to T, we verify that the sequence $(x_n)_{n\geq 0}$ generated by any $x_0 \in \mathcal{H}$ and $x_{n+1} := T(x_n)$ $(n = 0,1,2,\ldots)$ converges weakly to a point in $\bigcap_{i=1}^{m} C_i \neq \emptyset$. This scheme is the so-called *projections onto convex sets (POCS)* and applicable to *convex feasibility problems*.

(b) (Proximal forward–backward splitting method [46, 66, 104, 126]) Suppose that

$$S := \arg\min_{x \in \mathcal{H}} \{f_1(x) + f_2(x)\}$$

is nonempty for $f_1, f_2 \in \Gamma_0(\mathscr{H})$, where f_2 is Gâteaux differentiable on $\mathscr{H}$ with its κ-Lipschitzian gradient $\nabla f_2 : \mathscr{H} \to \mathscr{H}$. Then, for any $\mu \in \left(0, \frac{2}{\kappa}\right)$, the sequence $(x_n)_{n\geq 0}$ generated by any initial point $x_0 \in \mathscr{H}$ and $x_{n+1} := \operatorname{prox}_{\mu f_1}(I - \mu\nabla f_2)(x_n)$ converges weakly to a point in S. This scheme is the so-called *proximal forward–backward splitting method* which can be interpreted as a direct application of Remark 17.11(b) to Example 17.6(b).

(c) (Projected gradient method [72, 87]) Let $C \subset \mathscr{H}$ be a closed convex set and $f : \mathscr{H} \to \mathbb{R}$ a Gâteaux differentiable convex function satisfying $\arg\min_{x\in C} f(x) \neq \emptyset$. Suppose that the derivative $\nabla f : \mathscr{H} \to \mathscr{H}$ is κ-Lipschitzian over $\mathscr{H}$ for some $\kappa > 0$. Then for any $\mu \in \left(0, \frac{2}{\kappa}\right)$, the sequence $(x_n)_{n\geq 0}$ generated by any initial point $x_0 \in \mathscr{H}$ and $x_{n+1} := P_C(x_n - \mu\nabla f(x_n))$ converges weakly to a point in $\arg\min_{x\in C} f(x)$. This scheme is the so-called *projected gradient method*, which can be interpreted as a direct application of Example 17.12(b) to $f_1 = \iota_C$ and $f_2 := f$.

(d) (PPM: Parallel projection method [34, 40]) Suppose that $K \subset \mathscr{H}$ and $C_i \subset \mathscr{H}$ $(i = 1, 2, \ldots, m)$ are nonempty closed convex sets possibly having $K \cap \left(\bigcap_{i=1}^m C_i\right) = \emptyset$. Suppose also that the *mean squared distance function*: $\Phi_{\mathrm{ms}}(x) := \frac{1}{2}\sum_{i=1}^m w_i d_{C_i}^2(x)$ has its minimizer over K, i.e., $K_{\Phi_{\mathrm{ms}}} := \arg\min_{x\in K} \Phi_{\mathrm{ms}}(x) \neq \emptyset$, where $w_i > 0$ $(i = 1, 2, \ldots, m)$ and $\sum_{i=1}^m w_i = 1$. Then the sequence $(x_n)_{n=0}^{\infty}$ generated by any $\mu \in (0, 2)$, any $x_0 \in \mathscr{H}$ and

$$x_{n+1} := P_K\left((1-\mu)x_n + \mu\sum_i w_i P_{C_i}(x_n)\right)$$

converges weakly to a point in $K_{\Phi_{\mathrm{ms}}}$. This scheme is the so-called PPM and applicable to *inconsistent convex feasibility problems*. The PPM can be interpreted as a direct application of Example 17.12(c) to $f(x) = \Phi_{\mathrm{ms}}(x)$.

(e) (Projected Landweber method [58, 76]/CQ-algorithm [20, 21]) Let $\mathscr{H}_o$ be a real Hilbert space equipped with an inner product $\langle\cdot,\cdot\rangle_o$ and its induced norm $\|\cdot\|_o$. Suppose that the operator $A : \mathscr{H} \to \mathscr{H}_o$ is linear and bounded, i.e., $\|A\| := \sup_{x\in\mathscr{H}\setminus\{0\}} \dfrac{\|A(x)\|_o}{\|x\|} < \infty$, and that a closed convex set $C \subset \mathscr{H}$ and $b \in \mathscr{H}_o$ satisfy $\mathscr{S}_1 := \arg\min_{x\in C} \|A(x) - b\|_o^2 \neq \emptyset$. Then for any $\mu \in \left(0, 2\|A\|^{-2}\right)$, the sequence $(x_n)_{n\geq 0}$ generated by any point $x_0 \in \mathscr{H}$ and

$$x_{n+1} := P_C(x_n - \mu A^*A(x_n) + \mu A^*(b))$$

converges weakly to a point in $\mathscr{S}_1$, where $A^* : \mathscr{H}_o \to \mathscr{H}$ is the adjoint operator of A [10, 48, 86, 134, 142]. This scheme is the so-called *projected Landweber method* and applicable to convexly constrained inverse problems. The projected Landweber method can be interpreted as a direct application of Example 17.12(c) to $f(x) = \frac{1}{2}\|A(x) - b\|_o^2$.

On the other hand, for given a pair of closed convex sets $C \subset \mathscr{H}$ and $Q \subset \mathscr{H}_o$, the problem for finding a point $x \in \mathscr{H}$ satisfying $x \in C$ and $A(x) \in Q$ is called the *split feasibility problem* (SFP). Since the SFP is reduced to a problem for finding a point in

$$\mathscr{S}_2 := \arg\min_{x\in C} \|P_Q A(x) - A(x)\|_o^2 \neq \emptyset,$$

a direct application of Example 17.12(c) to $f(x) = \frac{1}{2}\|P_Q A(x) - A(x)\|_o^2$ leads to the algorithm: $x_{n+1} := P_C(x_n - \mu A^*(I - P_Q)A(x_n))$, which generates a weakly convergent sequence $(x_n)_{n\geq 0}$ to a point in $\mathscr{S}_2$ for any $\mu \in (0, 2\|A\|^{-2})$ and any point $x_0 \in \mathscr{H}$. This scheme is the so-called *CQ-algorithm* and applicable to SFP (Note: The Mann iterative process has been applied to many other types of inverse problems. For example, an elliptic Cauchy problem was solved in [61] with Proposition 17.10(b) as a fixed point problem for a nonexpansive affine operator in a Hilbert space).

(f) (Douglas–Rachford splitting method [42, 57, 90]) Let $f_1, f_2 \in \Gamma_0(\mathscr{H})$ satisfy

$$S := \arg\min_{x\in\mathscr{H}} \{f_1(x) + f_2(x)\} \neq \emptyset.$$

Under the condition (17.11), the sequence $(x_n)_{n=0}^{\infty}$ generated by

$$x_{n+1} := (1-\alpha_n)x_n + \alpha_n \mathrm{rprox}_{\gamma f_1} \mathrm{rprox}_{\gamma f_2}(x_n), \tag{17.19}$$

for any $x_0 \in \mathscr{H}$, any $\gamma \in (0, \infty)$ and any $(\alpha_n)_{n\geq 0} \subset [0,1]$ satisfying $\sum_{n\geq 0} \alpha_n (1-\alpha_n) = \infty$, converges weakly to a point in $(\mathrm{prox}_{\gamma f_2})^{-1}(S)$. The scheme (17.19) can be interpreted as a direct application of Proposition 17.10(b) to Example 17.6(c). Moreover, with use of $(t_n)_{n\geq 0} := (2\alpha_n)_{n\geq 0} \subset [0,2]$ satisfying $\sum_{n\geq 0} t_n(2-t_n) = \infty$, the Scheme (17.19) can be expressed equivalently as

$$x_{n+1} := x_n + t_n \left\{ \mathrm{prox}_{\gamma f_1}\left(2\mathrm{prox}_{\gamma f_2}(x_n) - x_n\right) - \mathrm{prox}_{\gamma f_2}(x_n) \right\}, \tag{17.20}$$

which is a simplest example of the so-called *Douglas–Rachford splitting type algorithm* in [42, Theorem 20]. In particular, if $\dim(\mathscr{H}) < \infty$, the nonexpansivity of $\mathrm{prox}_{\gamma f_2}$ and the weak convergence of $(x_n)_{n=0}^{\infty}$ by (17.19) [or by (17.20)] to a point, say

$$y^\star \in (\mathrm{prox}_{\gamma f_2})^{-1}(S) (\Leftrightarrow \mathrm{prox}_{\gamma f_2}(y^\star) \in S),$$

guarantee

$$\|\mathrm{prox}_{\gamma f_2}(x_n) - \mathrm{prox}_{\gamma f_2}(y^\star)\| \leq \|x_n - y^\star\| \to 0 \quad (n \to \infty).$$

(g) (Subgradient method [107]) Let $f : \mathscr{H} \to \mathbb{R}$ be a continuous convex function satisfying $\mathrm{lev}_{\leq 0}(f) \neq \emptyset$. Define a sequence $(x_n)_{n\geq 0} \subset \mathscr{H}$ with any initial point $x_0 \in \mathscr{H}$ and

$$x_{n+1} := \begin{cases} x_n - \lambda_n \frac{f(x_n)}{\|f'(x_n)\|^2} f'(x_n), & \text{if } f(x_n) > 0; \\ x_n, & \text{otherwise,} \end{cases} \tag{17.21}$$

where $f'(x_n) \in \partial f(x_n)$ for $f(x_n) > 0$, and $(\lambda_n)_{n\geq 0} \in (0,2)$ is bounded away from 0 and 2. Then the iteration (17.21) can be expressed as

$$x_{n+1} = \left[\left(1 - \frac{\lambda_n}{2}\right) I + \frac{\lambda_n}{2} T\right](x_n),$$

where $T := 2T_{\mathrm{sp}(f)} - I$. In particular, if the subdifferential $\partial f : \mathscr{H} \to 2^{\mathscr{H}}$ is bounded in the sense of Fact 17.2(a), $I - T$ is demiclosed at $0 \in \mathscr{H}$ (see the first paragraph of Sect. 17.2.4); hence, Proposition 17.10(a) guarantees the weak convergence of $(x_n)_{n=0}^{\infty}$ to a point in $\mathrm{Fix}(T) = \mathrm{Fix}(T_{\mathrm{sp}(f)}) = \mathrm{lev}_{\leq 0}(f)$.

This method is very useful for the following convex feasibility problems. Suppose that continuous convex functions $f_i : \mathscr{H} \to \mathbb{R}$ $(i = 1,2,\dots,m)$ satisfy $\bigcap_{i=1}^{m} \mathrm{lev}_{\leq 0}(f_i) \neq \emptyset$. Then, by defining a single convex function $f : \mathscr{H} \to \mathbb{R}$ satisfying $\mathrm{lev}_{\leq 0}(f) = \bigcap_{i=1}^{m} \mathrm{lev}_{\leq 0}(f_i)$, for example by $f(x) := \max_{i=1}^{m} f_i(x)$ or by $f(x) := \sum_{i=1}^{m} w_i f_i^{+}(x)$ with $f_i^{+}(x) = \max\{f_i(x), 0\}$ and $w_i > 0$ $(i = 1,2,\dots,m)$, we can reformulate the problem of finding a point in the nonempty intersection of the closed convex sets $\mathrm{lev}_{\leq 0}(f_i)$ to the problem of finding a point in $\mathrm{lev}_{\leq 0}(f)$. Indeed, if $f_i'(x_n) \in \partial f_i(x_n)$ $(i = 1,2,\dots,m)$ are available to compute $f'(x_n) \in \partial f(x_n)$ with the well-known calculus rules [82] and $\partial f : \mathscr{H} \to 2^{\mathscr{H}}$ is bounded, we can generate a weakly convergent sequence to a point in $\mathrm{lev}_{\leq 0}(f)$ by applying (17.21) to f.

Moreover, if $P_{\mathrm{lev}_{\leq 0}(f_i)} : \mathscr{H} \to \mathrm{lev}_{\leq 0}(f_i)$ $(i = 1,2,\dots,m)$ are available, an application of (17.21) with the aid of Example 17.6(a) to

$$f(x) := \frac{1}{2}\sum_{i=1}^{m} w_i d^2_{\mathrm{lev}_{\leq 0}(f_i)}(x) = \frac{1}{2}\sum_{i=1}^{m} w_i \|x - P_{\mathrm{lev}_{\leq 0}(f_i)}(x)\|^2$$

leads immediately to a version of the parallel projection algorithm [7,29,34] for convex feasibility problems.

17.2.5 Monotonicity of Derivatives of Convex Functions, Variational Inequality Problems

A mapping $F : \mathscr{H} \to \mathscr{H}$ is called (1) *monotone* over $S \subset \mathscr{H}$ if $\langle F(u) - F(v), u - v\rangle \geq 0$ for all $u, v \in S$. In particular, a mapping F which is monotone over $S \subset \mathscr{H}$ is called (2) *paramonotone* over S if $\langle F(u) - F(v), u - v\rangle = 0 \Leftrightarrow F(u) = F(v)$ for all $u, v \in S$; (3) η*-inverse strongly monotone* (or *firmly monotone*) over S if there exists $\eta > 0$ such that $\langle F(u) - F(v), u - v\rangle \geq \eta\,\|F(u) - F(v)\|^2$ for all $u, v \in S$ [91]; (4) η*-strongly monotone* over S if there exists $\eta > 0$ such that $\langle F(u) - F(v), u - v\rangle \geq \eta\,\|u - v\|^2$ for all $u, v \in S$ [150].

Given $F : \mathscr{H} \to \mathscr{H}$ which is monotone over a nonempty closed convex set $C \subset \mathscr{H}$, the *VIP*(F,C) is defined as follows: find $u^\star \in C$ such that $\langle u - u^\star, F(u^\star)\rangle \geq 0$

for all $u \in C$. If a function $f \in \Gamma_0(\mathscr{H})$ is Gâteaux differentiable over an open set $U \supset C$, then the derivative ∇f is paramonotone over C [30]. In this case, the solution set of $VIP(\nabla f, C)$ is nothing but the set $\arg\min_{x \in C} f(x)$ provided that it is nonempty (see, e.g., [59, Proposition II.2.1] and [134, Theorem 7.7]).

The following facts are quite useful for translating a convex optimization problem into a fixed point problem.

Fact 17.13 (Properties of VIP). [30, 59] Let $F : \mathscr{H} \to \mathscr{H}$ be monotone and continuous over a nonempty closed convex set $C \subset \mathscr{H}$. Then

(a) $u^\star$ is a solution of $VIP(F,C)$ if and only if, for all $u \in C$, $\langle F(u), u - u^\star \rangle \geq 0$.
(b) Suppose that (1) F is paramonotone over C, (2) $u^\star \in C$ is a solution of $VIP(F,C)$ and (3) $u \in C$ satisfies $\langle F(u), u - u^\star \rangle = 0$. Then u is also a solution of $VIP(F,C)$.

The characterization in (17.9) of the convex projection P_C yields at once an alternative interpretation of the VIP as a fixed point problem.

Fact 17.14 (VIP as a Fixed Point Problem). Given $F : \mathscr{H} \to \mathscr{H}$ which is monotone over a nonempty closed convex set C, the following three statements are equivalent.

(a) $u^\star \in C$ is a solution of VIP(F,C); i.e.,

$$\langle v - u^\star, F(u^\star) \rangle \geq 0 \text{ for all } v \in C.$$

(b) For an arbitrarily fixed $\mu > 0$, $u^\star \in C$ satisfies

$$\langle v - u^\star, (u^\star - \mu F(u^\star)) - u^\star \rangle \leq 0 \text{ for all } v \in C.$$

(c) For an arbitrarily fixed $\mu > 0$,

$$u^\star \in \operatorname{Fix}\left(P_C\left(I - \mu F\right)\right). \tag{17.22}$$

Fact 17.15 (Baillon–Haddad Theorem [3, 9, 56, 73, 91]). Let $f \in \Gamma_0(\mathscr{H})$ be *Gâteaux* differentiable with its gradient $\nabla f : \mathscr{H} \to \mathscr{H}$. Then the following three statements are equivalent.

(a) ∇f is κ-Lipschitzian over $\mathscr{H}$.
(b) ∇f is $1/\kappa$-inverse strongly monotone over $\mathscr{H}$.
(c) $I - \frac{2}{\kappa}\nabla f : \mathscr{H} \to \mathscr{H}$ is nonexpansive over $\mathscr{H}$.

Remark 17.16 (On Fact 17.15).

(a) The equivalence of Facts 17.15(b) and (c) is confirmed by a simple algebra.
(b) Fact 17.15(c) guarantees that κ-Lipschitz continuity of ∇f implies $\frac{\mu\kappa}{2}$-averaged nonexpansivity of $I - \mu \nabla f = \left(1 - \frac{\mu\kappa}{2}\right) I + \frac{\mu\kappa}{2}\left(I - \frac{2}{\kappa}\nabla f\right)$ for any $\mu \in \left(0, \frac{2}{\kappa}\right)$.

17.3 Minimizing Moreau Envelope by Hybrid Steepest Descent Method

17.3.1 Moreau Envelope and Its Derivative

The Moreau envelope has surprisingly nice properties as follows.

Fact 17.17 (Distinctive Properties of Moreau Envelope (see, e.g., [46, 98, 99, 109])). Given a function $f \in \Gamma_0(\mathcal{H})$, the Moreau envelope ${}^{\gamma}f : \mathcal{H} \to \mathbb{R}$ of f of index $\gamma \in (0,\infty)$ in (17.2) satisfies the following.

(a) (Lower bound) $(\forall \gamma \in (0,\infty), \forall x \in \mathcal{H})\ f(x) \geq {}^{\gamma}f(x)$.

(b) (Convergence) The function ${}^{\gamma}f$ converges pointwise to f on $\operatorname{dom}(f)$ as $\gamma \to 0$, i.e.,

$$\lim_{\gamma \downarrow 0} {}^{\gamma}f(x) = f(x) \quad (\forall x \in \operatorname{dom}(f)).$$

Moreover, if f is uniformly continuous on a bounded set $S \subset \operatorname{dom}(f)$, ${}^{\gamma}f$ converges uniformly to f on S, i.e., $\lim_{\gamma \downarrow 0} \sup_{x \in S} |{}^{\gamma}f(x) - f(x)| = 0$. In particular, if f is continuous on a compact set $S \subset \operatorname{dom}(f)$, the *Heine's theorem* [1, Theorem 4.47] guarantees the uniform convergence of ${}^{\gamma}f$ to f on S.

(c) (Lipschitz continuity of Fréchet derivative) ${}^{\gamma}f : \mathcal{H} \to \mathbb{R}$ is Fréchet differentiable and its derivative is given by

$$\nabla {}^{\gamma}f(x) = \frac{x - \operatorname{prox}_{\gamma f}(x)}{\gamma} = \frac{x - (I + \gamma \partial f)^{-1}(x)}{\gamma}, \tag{17.23}$$

hence $\nabla {}^{\gamma}f(x)$ is $\frac{1}{\gamma}$-Lipschitzian (Note: The firm nonexpansivity of $I - \operatorname{prox}_{\gamma f}$ is guaranteed by the nonexpansivity of $2(I - \operatorname{prox}_{\gamma f}) - I = -\operatorname{rprox}_{\gamma f}$).

The benefits of the Moreau envelope in applied sciences have been examined for the absolute value function $|\cdot| : \mathbb{R} \to [0,\infty)$. By a simple algebra, we verify that the Moreau envelope of the absolute value function is given explicitly by

$${}^{\gamma}|t| := \begin{cases} \frac{1}{2\gamma}t^2, & \text{if } |t| \leq \gamma; \\ |t| - \frac{1}{2}\gamma, & \text{otherwise.} \end{cases} \tag{17.24}$$

As pointed out in [15, 96], this is clearly equal, up to a scaling factor γ, to the so-called *Huber's M cost function* [83]

$$\rho : \mathbb{R} \to \mathbb{R},\ t \mapsto \begin{cases} \frac{1}{2}t^2, & \text{if } |t| \leq \gamma; \\ \gamma|t| - \frac{1}{2}\gamma^2, & \text{otherwise,} \end{cases} \tag{17.25}$$

in the context of robust linear estimation theory. The Huber's M cost function has been used in an estimation problem:

$$\text{find } x^\star \in \arg\min_{x\in\mathbb{R}^n} \sum_{i=1}^{m} \rho\left((Ax-b)_i\right), \tag{17.26}$$

where $A \in \mathbb{R}^{m\times n}$ represents the underlying linear model, $b \in \mathbb{R}^m$ is the data vector, and $x \in \mathbb{R}^n$ is the parameter vector. A solution to (17.26), often referred to as an *M-estimator*, is known as a robust alternative to the least squares (LS) estimator that is unfortunately sensitive against occurrence of outliers in the ill-conditioned linear regression systems. Computational algorithms for the problem (17.26) are found, for example, in [16, 88, 93, 96]. In particular, a computational algorithm was given in [16] to a convexly constrained version of the problem (17.26) provided that the metric projection onto the constraint set is possible to compute efficiently.

The Huber's M cost function has also been used in many inverse problems [2, 24, 79, 100] as an excellent *robust convex penalty function* that grows linearly for t far from zero; hence, it achieves least sensitivity to large outliers of large residual. We can also observe that the derivative of ${}^{\gamma}|\cdot|$ is always $\frac{1}{\gamma}$-Lipschitzian over $\mathbb{R}$ as mentioned in Fact 17.17(c), while the derivative of a straightforward smooth convex approximation $|\cdot|^p : \mathbb{R} \to [0,\infty)$ for any $1 < p < 2$ can never be Lipschitz continuous over $\mathbb{R}$ due to

$$\lim_{t\downarrow 0}\left(p(p-1)t^{p-2}\right) = \infty.$$

This means that the Moreau–Yosida regularization offers a unified systematic strategy to realize a beautiful parametrized smooth convex approximation for general convex functions in $\Gamma_0(\mathcal{H})$. Nevertheless, the use of the Moreau envelope and the proximity operator has been very limited for many years in real-world applications. This is mainly due to the evident computational difficulty in the definition (17.2), i.e., we have to minimize a possibly nonsmooth convex function $f(\cdot) + \frac{1}{2\gamma}\|x-\cdot\|^2$ for each $x \in \mathcal{H}$ to obtain the $\mathrm{prox}_{\gamma f}(x) \in \mathcal{H}$. Although this computational difficulty has never been resolved in general, the effectiveness of the proximity operator has been confirmed in relatively simple finite dimensional scenarios, where $f \in \Gamma_0(\mathbb{R}^n)$ can be expressed in terms of $f_i \in \Gamma_0(\mathbb{R})$ $(i = 1, 2, \ldots, n)$ by

$$f : \mathbb{R}^n \to \mathbb{R} : (x_1, \ldots, x_n) \mapsto \sum_{i=1}^{n} f_i(x_i), \tag{17.27}$$

hence

$$\mathrm{prox}_{\gamma f}(x_1, \ldots, x_n) = \left(\mathrm{prox}_{\gamma f_1}(x_1), \ldots, \mathrm{prox}_{\gamma f_n}(x_n)\right).$$

In such a case, the computation of $\mathrm{prox}_{\gamma f}(x_1, \ldots, x_n)$ is reduced to finding the unique minimizer $\mathrm{prox}_{\gamma f_i}(x_i)$ of each univariate convex function $f_i(\cdot) + \frac{1}{2\gamma}|x_i - \cdot|^2$ $(i = 1, \ldots, n)$.

Next we list such useful examples including the soft-thresholding operator, which was developed originally for denoising [53]. Fortunately, the proximity operators of these examples have closed form expressions (Note: Many other useful formulae on the proximity operator are found, for example, in [45, 46]).

Example 17.18 (Closed Form Expressions of Some Proximity Operators [46]*).*

(a) If $f \in \Gamma_0(\mathbb{R})$ is defined by

$$f : x \mapsto \begin{cases} -\ln(x) & \text{if } x > 0; \\ \infty & \text{if } x \leq 0, \end{cases}$$

we have for any $\gamma \in (0, \infty)$

$$\operatorname{prox}_{\gamma f}(x) = \frac{1}{2}\left(x + \sqrt{x^2 + 4\gamma}\right).$$

(b) Let $\{e_k\}_{k=1}^n$ be an orthonormal basis of $\mathbb{R}^n$ where the standard inner product is defined. Define a function $f \in \Gamma_0(\mathbb{R}^n)$ by $f : \mathbb{R}^n \ni x \mapsto \sum_{k=1}^n f_k(\langle x, e_k\rangle) \in (-\infty, \infty]$, where $f_k \in \Gamma_0(\mathbb{R})$ satisfies $f_k(x_k) \geq 0$ $(\forall x_k \in \mathbb{R})$ and $f_k(0) = 0$ $(k = 1, 2, \ldots, n)$. Then we have

$$\operatorname{prox}_f(x) = \sum_{k=1}^n \left(\operatorname{prox}_{f_k}(\langle x, e_k\rangle)\right) e_k \quad (x \in \mathbb{R}^n).$$

(c) In particular, if we define, as a special example of (b),

$$f : \mathbb{R}^n \ni x \mapsto \sum_{k=1}^n \omega_k |\langle x, e_k\rangle| \in \mathbb{R},$$

with constant weights $\omega_k > 0$ $(k = 1, 2, \ldots, n)$, we have

$$\operatorname{prox}_f(x) = \sum_{k=1}^n \operatorname{sgn}(\langle x, e_k\rangle) \max\{|\langle x, e_k\rangle| - \omega_k, 0\} e_k \quad (x \in \mathbb{R}^n).$$

The proximity operator in Example 17.18(c) is called the *soft-thresholding/ shrinkage* [51, 53] and has been used widely for example in noise removal problems and in sparse matrix completion problems [22, 32, 47]. As seen from Example 17.18(c) for $n = 1$, the derivative of the Moreau envelope ${}^{\gamma}f(x)$ of the absolute value function $f(x) = |x|$ can be computed with lower complexity than the derivative of $f_\varepsilon(x) := \sqrt{x^2 + \varepsilon}$ $(\varepsilon > 0)$ of which the use as a smooth approximation of f has been found in the literature.

To compute the proximity operator of $f \in \Gamma_0(\mathbb{R}^n)$ in more complex cases where the decomposition of f as in (17.27) is hard, fundamental theorems in convex analysis have been utilized implicitly or explicitly; e.g., the *Fenchel–Rockafeller* duality theorem [109] was used in [31,46,62] to compute the proximity operator of certain functions such as the *total variation function*. The following proposition and its corollary explain directly such strategies.

Proposition 17.19 (Expression of Proximity Operator by Legendre–Fenchel Transform). *Let $\varphi \in \Gamma_0(\mathbb{R}^m)$, $L \in \mathbb{R}^{m\times n}$ and $d \in \mathrm{int}(S)$, where*

$$S := L(\mathbb{R}^n) - \mathrm{dom}(\varphi) := \{Lx - y \in \mathbb{R}^m \mid x \in \mathbb{R}^n \text{ and } y \in \mathrm{dom}(\varphi)\}$$

and $\mathrm{int}(S)$ stands for the interior of S. Define $\widetilde{\varphi} \in \Gamma_0(\mathbb{R}^n)$ by $\widetilde{\varphi} : x \mapsto \varphi(Lx - d)$. Then for arbitrarily fixed $x \in \mathbb{R}^n$ and $\gamma \in (0,\infty)$,

$$\mathrm{prox}_{\gamma\widetilde{\varphi}}(x) := \arg\min_{z\in\mathbb{R}^n}\left(\widetilde{\varphi}(z) + \frac{1}{2\gamma}\|x-z\|^2\right) = \arg\min_{z\in\mathbb{R}^n}\left(\varphi(Lz-d) + \frac{1}{2\gamma}\|x-z\|^2\right)$$

can be expressed, with $\bar{y} \in \arg\min_{y\in\mathbb{R}^m}\left(\varphi^*(y) + \langle d,y\rangle + \frac{1}{2\gamma}\|\gamma L^t y - x\|^2\right)$, by

$$\mathrm{prox}_{\gamma\widetilde{\varphi}}(x) = x - \gamma L^t \bar{y},$$

where $L^t \in \mathbb{R}^{n\times m}$ denotes the transpose of a matrix L and φ^* the (Legendre–Fenchel) conjugate of φ (see Fact 17.2(a)).

Proof. Clearly, $\phi(z) := \frac{1}{2\gamma}\|x-z\|^2$ ($\forall z \in \mathbb{R}^n$) has $\mathrm{dom}(\phi) = \mathbb{R}^n$, which implies $d \in \mathrm{int}(S) \Leftrightarrow -d \in \mathrm{int}(-L(\mathrm{dom}(\phi)) + \mathrm{dom}(\varphi))$, where $-L(\mathrm{dom}(\phi)) + \mathrm{dom}(\varphi) = \{-L(x) + y \in \mathbb{R}^m \mid x \in \mathrm{dom}(\phi) \text{ and } y \in \mathrm{dom}(\varphi)\}$. It is also obvious that the conjugate of $\phi \in \Gamma_0(\mathbb{R}^n)$ is given by $\phi^*(u) = \frac{1}{2\gamma}(\|\gamma u + x\|^2 - \|x\|^2)$ ($\forall u \in \mathbb{R}^n$) with $\mathrm{dom}(\phi^*) = \mathbb{R}^n$, which implies

$$\begin{aligned} 0 &\in \mathrm{int}\left(-L^t(\mathrm{dom}(\varphi^*)) - \mathrm{dom}(\phi^*)\right) \\ &= \{-L^t(y) - u \mid y \in \mathrm{dom}(\varphi^*) \text{ and } u \in \mathrm{dom}(\phi^*)\} \\ &= \mathbb{R}^n. \end{aligned} \tag{17.28}$$

Therefore, by applying the *Fenchel-type duality scheme* (see for example [109, Example 11.41]), we deduce

$$-L^*\bar{y} \in \partial\phi(\mathrm{prox}_{\gamma\widetilde{\varphi}}(x)) = \left\{\nabla\phi(\mathrm{prox}_{\gamma\widetilde{\varphi}}(x))\right\} = \left\{\frac{1}{\gamma}\left(\mathrm{prox}_{\gamma\widetilde{\varphi}}(x) - x\right)\right\},$$

where

$$\begin{aligned} \bar{y} &\in \arg\max_{y\in\mathbb{R}^m}\{\langle -d,y\rangle - \varphi^*(y) - \phi^*(-L^t y)\} \\ &= \arg\min_{y\in\mathbb{R}^m}\left\{\varphi^*(y) + \langle d,y\rangle + \frac{1}{2\gamma}(\|-\gamma L^t y + x\|^2 - \|x\|^2)\right\}. \end{aligned}$$

■

Corollary 17.20 (Proximity Operator of Affinely Pre-composed ℓ_1-Norm Function). *Let $\varphi \in \Gamma_0(\mathbb{R}^m)$ be defined by $\varphi : (y_1,\dots,y_m) \mapsto \sum_{i=1}^m |y_i|$. Then for any $L \in \mathbb{R}^{m\times n}$ and $d \in \mathbb{R}^m$, the proximity operator of the function $\widetilde{\varphi} : x \mapsto \varphi(Lx-d)$ of index $\gamma \in (0,\infty)$ is given by* $\mathrm{prox}_{\gamma\widetilde{\varphi}} : \mathbb{R}^n \to \mathbb{R}^n, x \mapsto x - \gamma L^t \bar{y}$, *where*

$$\bar{y} \in \arg\min_{y\in C} \left(\langle d, y\rangle + \frac{1}{2\gamma}\|\gamma L^t y - x\|^2 \right) \tag{17.29}$$

with

$$C := \{ y = (y_1,\dots,y_m) \in \mathbb{R}^m \mid |y_i| \le 1 \quad (i = 1,\dots,m)\}. \tag{17.30}$$

Proof. By $\mathrm{dom}(\varphi) = \mathbb{R}^m$, we have $S = L(\mathbb{R}^n) - \mathrm{dom}(\varphi) = \mathbb{R}^m$ and $d \in \mathrm{int}(S) = \mathbb{R}^m$. Moreover, by [17, Example 3.26], the conjugate of φ is given by $\varphi^* = \iota_C$. Therefore, $\bar{y} \in \mathbb{R}^m$ in Proposition 17.19 can be characterized by

$$\begin{aligned}\bar{y} &\in \arg\min_{y\in\mathbb{R}^m} \left(\iota_C(y) + \langle d, y\rangle + \frac{1}{2\gamma}\|\gamma L^t y - x\|^2 \right) \\ &= \arg\min_{y\in C} \left(\langle d, y\rangle + \frac{1}{2\gamma}\|\gamma L^t y - x\|^2 \right).\end{aligned}$$

■

The computation of the proximity operator $\mathrm{prox}_{\gamma\iota_C} = P_C$ is immediate for C in (17.30), i.e.,

$$P_C : \mathbb{R}^m \to C,\ (x_1,\dots,x_m) \mapsto (y_1,\dots,y_m), \quad \text{where } y_i := \begin{cases} x_i & \text{if} |x_i| \le 1 \\ \frac{x_i}{|x_i|} & \text{if} |x_i| > 1 \end{cases}, \tag{17.31}$$

which implies that the solution of the smooth minimization problem (17.29) can be approximated efficiently, for example, by the *projected gradient method* [72] or many other improved algorithms (see, e.g., [11, 12]).

17.3.2 Hybrid Steepest Descent Method

As seen in Sect. 17.3.1, minimization of the Moreau–Yosida regularization of a possibly nonsmooth convex function $\Phi \in \Gamma_0(\mathcal{H})$ can be reduced to minimization of a smooth convex function whose gradient is Lipschitz continuous. In this section, we consider the following problem for minimizing such a smooth convex function over the fixed point set of certain quasi-nonexpansive mappings.

Problem 17.21 (Convex Optimization over the Fixed Point Set of Nonlinear Mapping). Let $T : \mathcal{H} \to \mathcal{H}$ be a quasi-nonexpansive mapping whose fixed point set $\mathrm{Fix}(T) = \{x \in \mathcal{H} \mid T(x) = x\}$ is nonempty. Suppose that $\Theta \in \Gamma_0(\mathcal{H})$ is Gâteaux differentiable with the gradient $\nabla\Theta$, which is κ-Lipschitzian over

$T(\mathcal{H}) := \{T(x) \in \mathcal{H} \mid x \in \mathcal{H}\}$. Then the problem is: find a point in the solution set

$$\Omega := \left\{x^\star \in \mathrm{Fix}(T) \mid \Theta(x^\star) = \min_{x \in \mathrm{Fix}(T)} \Theta(x)\right\}$$
$$= \{x^\star \in \mathrm{Fix}(T) \mid \langle x - x^\star, \nabla\Theta(x^\star)\rangle \geq 0 \quad (\forall x \in \mathrm{Fix}(T))\} \neq \emptyset. \quad (17.32)$$

The *HSDM* (see, e.g., [33, 49, 84, 92, 101, 102, 122, 130, 136–139, 151]) :

$$u_{n+1} := T(u_n) - \lambda_{n+1}\nabla\Theta\left(T(u_n)\right), \quad (17.33)$$

is an extremely simple algorithmic solution to Problem 17.21, where $(\lambda_n)_{n\geq 1} \subset [0,\infty)$ is a slowly decreasing nonnegative sequence. Among many convergence analyses on the algorithm (17.33), we introduce the following simple ones.

Theorem 17.22 (HSDM for Quasi-Nonexpansive Mappings).

I. *(Strong convergence for nonexpansive mapping* [132, 139]*) Let* $T : \mathcal{H} \to \mathcal{H}$ *be a nonexpansive mapping with* $\mathrm{Fix}(T) \neq \emptyset$. *Suppose that the gradient* $\nabla\Theta$ *is* κ*-Lipschitzian and* η*-strongly monotone over* $T(\mathcal{H})$*, which guarantees* $|\Omega| = 1$. *Then, by using any sequence* $(\lambda_n)_{n\geq 1} \subset [0,\infty)$ *satisfying (W1)* $\lim_{n\to\infty} \lambda_n = 0$, *(W2)* $\sum_{n\geq 1} \lambda_n = \infty$, *(W3)* $\sum_{n\geq 1} |\lambda_n - \lambda_{n+1}| < \infty$ *[or* $(\lambda_n)_{n\geq 1} \subset (0,\infty)$ *satisfying (L1)* $\lim_{n\to\infty} \lambda_n = 0$, *(L2)* $\sum_{n\geq 1} \lambda_n = \infty$, *(L3)* $\lim_{n\to\infty}(\lambda_n - \lambda_{n+1})\lambda_{n+1}^{-2} = 0$*], the sequence* $(u_n)_{n\geq 0}$ *generated, for arbitrary* $u_0 \in \mathcal{H}$, *by* (17.33) *converges strongly to the uniquely existing point* $u^\star \in \Omega$ *in* (17.32).

II. *(Nonstrictly convex optimization I* [101, 102]*) Assume* $\dim(\mathcal{H}) < \infty$. *Suppose that (i)* $T : \mathcal{H} \to \mathcal{H}$ *is an attracting nonexpansive mapping with bounded* $\mathrm{Fix}(T) \neq \emptyset$, *(ii)* $\nabla\Theta$ *is* κ*-Lipschitzian over* $T(\mathcal{H})$. *If the following condition (a) or (b) is fulfilled, then* $\Omega \neq \emptyset$ *automatically holds and the sequence* $(u_n)_{n\geq 0}$ *generated by* (17.33)*, for arbitrary* $u_0 \in \mathcal{H}$, *satisfies* $\lim_{n\to\infty} d\left(u_n, \Omega\right) = 0$.

   (a) *The nonnegative sequence* $(\lambda_n)_{n\geq 1}$ *in* (17.33) *satisfies (W1), (W2) and* $(\lambda_n)_{n\geq 1} \in \ell_2$, *i.e.,* $\sum_{n\geq 1} \lambda_n^2 < \infty$.

   (b) (i) *T is asymptotically shrinking; i.e., there exists* $R > 0$ *satisfying*

$$\sup_{\|u\|\geq R} \frac{\|T(u)\|}{\|u\|} < 1$$

   (In this case, the nonemptiness and boundedness of $\mathrm{Fix}(T)$ *automatically hold (see* [101]*)), and*

   (ii) *the nonnegative sequence* $(\lambda_n)_{n\geq 1}$ *in* (17.33) *satisfies (W1) and (W2).*

III. (Nonstrictly convex optimization II [136]*) Assume* $\dim(\mathcal{H}) < \infty$*. Suppose* $f : \mathcal{H} \to \mathbb{R}$ *is a continuous convex function with* $\mathrm{lev}_{\leq 0}(f) \neq \emptyset$*. Let* $f' : \mathcal{H} \to \mathcal{H}$ *be a selection of the subdifferential* ∂f *and let* f' *be bounded on any bounded set. Assume (i)* $\xi(x) \geq f(x)$, $\forall x \in \mathcal{H}$*, and (ii)* $S_\xi(x)$ *(in Proposition* 17.7*) satisfies (O-i) and (O-ii) for all* $x \in \mathcal{H}$*. Let* $T_\alpha := (1-\alpha)I + \alpha T_{\mathrm{dsp},\xi}$, $\forall \alpha \in (0,2)$*, where* $T_{\mathrm{dsp},\xi}$ *is defined in* (17.14)*. Let* $K \subset \mathcal{H}$ *be a bounded closed convex set satisfying* $K \cap \mathrm{lev}_{\leq 0}(f) \neq \emptyset$*, which implies that* $T := P_K T_\alpha$ *satisfies* $\mathrm{Fix}(T) = K \cap \mathrm{lev}_{\leq 0}(f) \neq \emptyset$*. Suppose that* $\Theta \in \Gamma_0(\mathcal{H})$ *is Gâteaux differentiable over* K *where the gradient* $\nabla\Theta$ *is* κ*-Lipschitzian. Then* $\Omega \neq \emptyset$ *automatically holds and the sequence* $(u_n)_{n\geq 0}$ *generated by* (17.33)*, for any* $u_0 \in \mathcal{H}$ *and* $\alpha \in (0,2)$*, satisfies* $\lim_{n\to\infty} d(u_n, \Omega) = 0$ *if* $(\lambda_n)_{n\geq 1} \subset [0,\infty)$ *is chosen to satisfy (W1) and (W2).*

The algorithm (17.33) was established originally as a generalization of the following fixed point iteration [6, 78, 89, 129] so-called *Halpern-type iteration* (or *anchor method*):

$$u_{n+1} := \lambda_{n+1} a + (1-\lambda_{n+1}) T(u_n), \tag{17.34}$$

which converges strongly to $P_{\mathrm{Fix}(T)}(a)$ for a nonexpansive mapping $T : \mathcal{H} \to \mathcal{H}$ and $a \in \mathcal{H}$.

Remark 17.23 (Conditions on $(\lambda_n)_{n\geq 1} \subset [0,\infty)$ *in (*17.33*)).*

(a) (Necessary condition [78]) $\lim_{n\to\infty} \lambda_n = 0$ and $\sum_{n\geq 1} \lambda_n = \infty$ are necessary to ensure the convergence of $(u_n)_{n\geq 0}$ to a point in Ω. Indeed, in the simple case of $\mathcal{H} := \mathbb{R}$, $T(x) := 1$ ($\forall x \in \mathbb{R}$) and $\Theta(x) = \frac{1}{2}x^2$ ($\forall x \in \mathbb{R}$), the method (17.33) is reduced to

$$u_{n+1} := (1-\lambda_{n+1}) T(u_n) = 1 - \lambda_{n+1}, \quad n = 0,1,2,\ldots,$$

hence $\lim_{n\to\infty} \lambda_n = 0$ is necessary for $\lim_{n\to\infty} u_n = 1 \in \mathrm{Fix}(T) = \{1\}$. Moreover, in the case of $\mathcal{H} := \mathbb{R}$, $T(x) := -x$ ($\forall x \in \mathbb{R}$) and $\Theta(x) = \frac{1}{2}x^2$ ($\forall x \in \mathbb{R}$), the method (17.33), for $u_0 = 1$, is reduced to

$$u_{n+1} := (1-\lambda_{n+1}) T(u_n) = (-1)^n \prod_{i=0}^{n} (1-\lambda_{i+1}), \quad n = 0,1,2,\ldots,$$

from which $\prod_{i=0}^{\infty}(1-\lambda_{i+1}) = 0$ ($\Leftrightarrow \sum_{n=1}^{\infty} \lambda_n = \infty$ when $\lim_{n\to\infty} \lambda_n = 0$ and $(\lambda_n)_{n\geq 1} \subset [0,1)$) is necessary for $\lim_{n\to\infty} u_n = 0 \in \mathrm{Fix}(T) = \{0\}$.

(b) (Sufficient condition) For the formula (17.34), the set of conditions (L1)–(L3) for $(\lambda_n)_{n\geq 1} \subset (0,1]$ was introduced in [89] while (W1)–(W3) for $(\lambda_n)_{n\geq 1} \subset [0,1]$ was introduced in [129]. [Note: $\lambda_n := 1/n^\rho$ for $0 < \rho < 1$ is a simple example of the sequence $(\lambda_n)_{n\geq 1}$ satisfying (L1)–(L3). The set of conditions (W1)–(W3) allows the case $\lambda_n = \frac{1}{n}$]. The condition (L3) was relaxed to $\lim_{n\to\infty} \frac{\lambda_n}{\lambda_{n+1}} = 1$ in [130], which allows the case $\lambda_n = \frac{1}{n}$. Moreover, if T is an averaged nonexpansive mapping, it was shown in [84] that the (W1) and (W2) for $(\lambda_n)_{n\geq 1}$ are sufficient to guarantee the strong convergence of (17.33) to the unique point in Ω under the scenario of Theorem 17.22 I (The sufficiency of (W1) and (W2) to guarantee the strong convergence of (17.34) to $P_{\mathrm{Fix}(T)}(a)$ was shown in [120]).

Remark 17.24 (HSDM as an Extension of the Proximal Forward–Backward Splitting).

(a) Under the same conditions imposed in Theorem 17.22 I, the sequence $v_n := T(u_n)$ $(n = 0,1,2,\ldots)$ generated, for any $v_0 := T(u_0) \in T(\mathscr{H})$, by

$$v_{n+1} := T\left(I - \lambda_{n+1}\nabla\Theta\right)(v_n), \tag{17.35}$$

satisfies

$$0 \leq \|v_n - u^\star\| = \|T(u_n) - u^\star\| \leq \|u_n - u^\star\| \to 0 \quad (n \to \infty). \tag{17.36}$$

The formula (17.35) is regarded as a (partial) generalization of the *proximal forward-backward splitting* in Example 17.12(b). Moreover, we can deduce from (17.35) a generalization [139, Remark 2.17(a)] of an algorithm in [110] (a version of *projected Landweber method* [13, 58, 76]) developed for the convexly constrained least-squares problems.

(b) If the strict convexity of $\Theta \in \Gamma_0(\mathscr{H})$ is assumed additionally in Theorems 17.22 II and III, the solution set becomes a singleton $\Omega = \{u^\star\}$. In such a case, $\lim_{n\to\infty} d(u_n, \Omega) = 0$ in Theorems 17.22 II and III is equivalent to

$$\lim_{n\to\infty} \|u_n - u^\star\| = 0,$$

hence the relation (17.36) is again applicable to the sequence $(v_n)_{n\geq 0}$ generated by (17.35), which guarantees the convergence of $(v_n)_{n\geq 0}$ to $u^\star$. In [64, Theorem 2], a similar algorithm to (17.35) is found specially for $T = T_{\mathrm{sp}(f)}$.

Clearly, we can apply the HSDM (17.33) in Theorem 17.22 (or its alternative form (17.35)) to minimization of $\Theta : \mathscr{H} \to \mathbb{R} : x \mapsto {}^\gamma\Phi(x)$, which is the Moreau–Yosida regularization of a possibly nonsmooth convex function $\Phi \in \Gamma_0(\mathscr{H})$ of the index $\gamma > 0$, over the fixed point set of a certain quasi-nonexpansive mapping $T : \mathscr{H} \to \mathscr{H}$. In such a scenario, the $\frac{1}{\gamma}$-Lipschitz continuity of the gradient $\nabla\Theta : \mathscr{H} \to \mathscr{H}$ is guaranteed automatically by Fact 17.17(c), which is the only requirement for $\nabla\Theta$ in Theorem 17.22 II and III. By applying Propositions 17.3, 17.4, and 17.7, to various mappings in Examples 17.6 and 17.9, we can design many efficiently computable quasi-nonexpansive mappings as T whose fixed point set $\mathrm{Fix}(T)$ is desirable as the constraint set.

17.4 Application to Minimal Antenna-Subset Selection Problem for MIMO Communication Systems

We have proposed a promising approach by an integration of the ideas of the HSDM and the *Moreau–Yosida regularization* to the challenging nonsmooth convex optimization over the fixed point set of certain quasi-nonexpansive mappings. We present in this section its nontrivial application to a minimal antenna-subset selection problem for efficient MIMO systems (Note: The contents of this section have partially been presented in [145]).

17.4.1 Backgrounds and Motivations

Multiple antenna systems, broadly termed MIMO systems, have given significant impacts to a wide range of research fields including communications, signal processing, and information theory because of its potential to increase the data rate without additional bandwidth [63, 124]. The gain, however, comes at the price of hardware and signal processing complexity, power consumption, etc. [95]. One of the main causes for the complexity-increase is the cost of multiple RF (radio frequency) chains. Antenna selection has been considered as an attractive approach to reduce the hardware complexity without severely losing the advantages of MIMO systems (see [55, 69, 97, 111] and references therein). In particular, it has been shown that the antenna selection retains the diversity degree compared to the full-complexity system [74, 97]. The complexity reduction is achieved by equipping fewer RF chains than the antenna elements at the receiver/transmitter, and the same number of antennas as the RF chains are selected so that the achieved channel capacity is maximized.

Differently from the prior works, we consider power-limited systems in which it is desired to consume the minimum amount of power with the designated channel capacity achieved. At the receiver, for instance, each antenna element requires a "power-consuming" RF chain that comprises a low noise amplifier, a frequency down-converter (a mixer), and an analog-to-digital converter. Also, the signal processing complexity may seriously increase with the number of antenna elements.[5] Therefore, it would be a natural requirement to select the minimal antenna subset that achieves the designated channel capacity; the cardinality of such a subset depends highly on the channel state, signal-to-noise ratio (SNR), etc. Unfortunately, the problem of minimal antenna-subset selection is regarded as ℓ_0-norm[6]

[5] When the multiple antennas are exploited for spatial multiplexing or the space-time trellis codes are adopted, the complexity increases sometimes exponentially [95].

[6] The cardinality of the nonzero components in $x := (x_1, x_2, \ldots, x_N) \in \mathbb{R}^N$ is often denoted by $\|x\|_0 \in \mathbb{N}$ and called commonly the ℓ_0-norm of x (or the *Hamming weight* of x in Coding Theory) although $\|\cdot\|_0$ does not satisfy either the conditions for norm or quasinorm.

minimization under highly nonlinear constraint: hence, it is hard to solve the problem directly because of its combinatorial nature when the number of antennas increases.

In this section, we present an alternative algorithmic solution for reaching an approximate solution by relaxing twice the ℓ_0-norm cost function in the original problem. The first relaxation is the standard ℓ_1-relaxation of ℓ_0-norm found widely in the recent approximation techniques for sparse optimization problems. Indeed, although the first relaxed problem can be handled as a convex optimization, it is still hard to solve directly due to the *nonsmoothness* of the new ℓ_1-norm cost function coupled with the *highly nonlinear* capacity-constraint. Therefore, the second relaxation is the Moreau envelope of ℓ_1-norm, which is a computationally manageable cost function under the capacity constraint.

The proposed algorithm is based on an application of Theorem 17.22 III (a version of the HSDM for the subgradient projection operator [136]) to the doubly relaxed problem: minimize the Moreau envelope of the ℓ_1-norm subject to the capacity constraint.

17.4.2 System Model and Problem Statement

For an MIMO system with N_{T} transmit antennas and N_{R} receive antennas, the received signal can be represented as

$$r_i := \sqrt{E_s} G s_i + n_i \in \mathbb{C}^{N_{\mathrm{R}}}. \tag{17.37}$$

Here, r_i represents the ith sample of the signals measured at the N_{R} receive antennas, $s_i \in \mathbb{C}^{N_{\mathrm{T}}}$ the ith symbol transmitted from the N_{T} transmit antennas, $E_s > 0$ the average energy at each receive antenna, $G \in \mathbb{C}^{N_{\mathrm{R}} \times N_{\mathrm{T}}}$ the channel matrix whose (p,q)th component represents the channel characteristics between the pth receive antenna and the qth transmit antenna, and n_i the additive white Gaussian noise with energy $N_0/2$ per complex dimension. We make the standard assumptions that the channel has frequency-flat fading and G is perfectly known at the receiver.[7] Also, we assume that G is totally unknown at the transmitter, therefore choosing s_i such that its covariance matrix is $I_{N_{\mathrm{T}}}/N_{\mathrm{T}}$ [55]; we denote by I_m the $m \times m$ identity matrix. In this case, it is known that the channel capacity (mutual information) is given as follows [63]:

$$c_{\text{full}} := \log_2 \det \left(I_{N_{\mathrm{T}}} + \frac{\rho}{N_{\mathrm{T}}} G^H G \right) \text{ bps/Hz}, \tag{17.38}$$

where $\rho := E_s/N_0$ is the average SNR; $(\cdot)^H$ stands for the Hermitian transpose.

[7] The channel could be moderately frequency-selective [97, 111].

We focus on the receive antenna selection. Let $\underline{c} \in (0, c_{\text{full}})$ denote the designated channel capacity to be ensured. The problem is to select the minimal antenna subset that achieves the capacity $\underline{c}$. Let $x := [x_1, x_2, \cdots, x_{N_\text{R}}]^t \in \{0,1\}^{N_\text{R}}$ represent an antenna subset in such a way that $x_j = 1$ ($x_j = 0$) indicates that the jth antenna is selected (not selected). Then, the channel capacity with the antenna subset represented by x is given by

$$c(x) := \log_2 \det \left(I_{N_\text{T}} + \frac{\rho}{N_\text{T}} G^H X G \right) \text{ bps/Hz}, \tag{17.39}$$

where $X := \text{diag}(x)$. The minimal antenna-subset selection problem is thus formulated as follows:

$$\min_{x \in \{0,1\}^{N_\text{R}}} \|x\|_0 \quad \text{s.t. } c(x) \geq \underline{c}, \tag{17.40}$$

where $\|\cdot\|_0$ denotes the ℓ_0-norm that counts the number of nonzero components. The problem in (17.40) is mathematically challenging, because it is nonlinearly-constrained sparse optimization. In general, finding its optimal solution involves exhaustive search. In the following, we present an efficient algorithmic solution using convex and differentiable relaxations of the ℓ_0 norm.

17.4.3 Convex and Differentiable Relaxations

To alleviate the difficulty in the combinatorial nature of the problem, we reformulate (17.40) into

$$\min_{x \in [0,1]^{N_\text{R}}} \psi(x) := \|x\|_1 \quad \text{s.t. } \varphi(x) := \underline{c} - c(x) \leq 0, \tag{17.41}$$

which is $\|\cdot\|_1$ minimization.[8] Because the function c is concave on $\mathbb{R}_+^{N_\text{R}}$ [17, 55], φ is convex on $\mathbb{R}_+^{N_\text{R}}$; $\mathbb{R}_+$ denotes the set of all nonnegative real numbers.

Unfortunately, we can still not find any computationally efficient solver for the reformulated problem in (17.41) because (1) the function ψ is (convex but) neither smooth nor strictly-convex and (2) the metric projection onto the constraint set (i.e., the zero level set of φ) is not efficiently computable (For instance, the generalized Haugazeau's scheme [38] cannot be applied directly because of the non-strict-convexity of ψ). Therefore, we reformulate (17.41), in $\mathcal{H} := \mathbb{R}^{N_\text{R}}$ where the standard inner product and its induced norm are defined, by using the Moreau-Yosida regularization. Defining $\psi_\omega : \mathbb{R}^{N_\text{R}} \to \mathbb{R}_+, x \mapsto \omega \|x\|_1$, for an arbitrary constant $\omega > 0$ ($\psi = \psi_{\omega|\omega=1}$), our optimization problem to solve is given as follows:

$$\min_{x \in [0,1]^{N_\text{R}}} {}^\gamma\psi_\omega(x) \quad \text{s.t. } \varphi(x) \leq 0 \quad (\gamma > 0). \tag{17.42}$$

---

[8] In recent years, it has been proven both theoretically and experimentally that *sparse recovery* is possible in many cases by means of the ℓ_1-norm [23, 52].

17.4.4 Proposed Antenna-Subset Selection Algorithm

The key of the previous subsection is the second relaxation which replaces the nonsmooth ψ by ${}^{\gamma}\psi_{\omega}(x)$ having a *Lipschitz continuous* derivative. Our basic strategy is the following: (1) compute the solution $x^{\star}$ to the problem in (17.42) by HSDM (Theorem 17.22 III) and (2) choose the antenna subset associated with the indices of (the minimum number of) the largest components of $x^{\star}$ such that the designated capacity $\underline{c}$ is achieved. Letting $\mathscr{I} := \{1,2,\cdots,N_{\mathrm{R}}\}$, the proposed algorithm is given as below.

Algorithm 17.25.

(i) For an initial vector $x_0 \in \mathbb{R}^{N_{\mathrm{R}}}$, generate $(x_k)_{k=1}^{Q}$ recursively by HSDM (Q: the prespecified number of iterations), and let $x_Q =: [x_Q^{(1)}, x_Q^{(2)}, \cdots, x_Q^{(N_{\mathrm{R}})}]^t$.

(ii) Compute the arithmetic mean $\bar{x}_Q$ of x_Q.

(iii) Choose the indices corresponding to the components no smaller than $\bar{x}_Q$ as a temporary antenna subset.

Let $\mathscr{J} := \emptyset$.
for $j \in \mathscr{I}$
  if $x_Q^{(j)} \geq \bar{x}_Q$
    $\mathscr{J} := \mathscr{J} \cup \{j\}$
  end
end

(iv) Choose the minimal antenna subset.

Let $x_{\mathscr{J}} \in \{0,1\}^{N_{\mathrm{R}}}$ be the vector representing the antenna subset $\mathscr{J}$ (see Sect. 17.4.2).
if $c(x_{\mathscr{J}}) < \underline{c}$
  while $c(x_{\mathscr{J}}) < \underline{c}$
    $j \in \arg\max_{\iota \in \mathscr{I} \setminus \mathscr{J}} x_Q^{(\iota)}$
    $\mathscr{J} := \mathscr{J} \cup \{j\}$
  end
else
  $\widehat{\mathscr{J}} := \mathscr{J}$
  while $c(x_{\widehat{\mathscr{J}}}) \geq \underline{c}$
    $\mathscr{J} := \widehat{\mathscr{J}}$
    $j \in \arg\min_{\iota \in \widehat{\mathscr{J}}} x_Q^{(\iota)}$
    $\widehat{\mathscr{J}} := \widehat{\mathscr{J}} \setminus \{j\}$
  end
end

(v) Output $\mathscr{J}$ as the selected antenna subset. □

The following subsection is devoted to explain precisely how to solve the problem in (17.42) by HSDM.

17.4.5 Optimization by Hybrid Steepest Descent Method

The problem in (17.42) has two constraints: the *capacity constraint*

$$x \in \mathrm{lev}_{\leq 0}(\varphi) := \left\{x \in \mathbb{R}^{N_{\mathrm{R}}} : \varphi(x) \leq 0\right\}$$

and the *box constraint*

$$x \in \mathscr{K} := [0,1]^{N_{\mathrm{R}}} = \left\{x \in \mathbb{R}^{N_{\mathrm{R}}} : 0 \leq x_j \leq 1,\ \forall j \in \mathscr{I}\right\},$$

where $\mathscr{K} \cap \mathrm{lev}_{\leq 0}(\varphi) \neq \emptyset$ is confirmed by $\varphi(1_{N_{\mathrm{R}}}) = \underline{c} - c_{\mathrm{full}} < 0$ for $1_{N_{\mathrm{R}}} := [1,1,\cdots,1] \in \mathscr{K}$.

Note that $P_{\mathscr{K}}$ can be computed easily while the computation of $P_{\mathrm{lev}_{\leq 0}(\varphi)}$ is not a simple task at all. Fortunately, an application of Theorem 17.22 III to $\Theta := {}^{\gamma}\psi_{\omega}$, $f := \varphi$ and $K := \mathscr{K}$ guarantees for any $x_0 \in \mathbb{R}^{N_{\mathrm{R}}}$ that the recursion

$$x_{k+1} := (I - \lambda_{k+1}\nabla^{\gamma}\psi_{\omega})\left(\widehat{T}_{\alpha}(x_k)\right),\ k \geq 0, \tag{17.43}$$

with

$$\widehat{T}_{\alpha} := P_{\mathscr{K}}\left[(1-\alpha)I + \alpha T_{\mathrm{sp}(\varphi)}\right],\ \alpha \in (0,2), \tag{17.44}$$

generates a sequence of points converging to a solution to (17.42).

Since φ is differentiable on $\mathbb{R}_{+}^{N_{\mathrm{R}}}$, its gradient $\nabla\varphi(x) := \left[\frac{\partial\varphi(x)}{\partial x_1}, \frac{\partial\varphi(x)}{\partial x_2}, \cdots, \frac{\partial\varphi(x)}{\partial x_{N_{\mathrm{R}}}}\right]^t$ is the unique subgradient at any $x \in \mathbb{R}_{+}^{N_{\mathrm{R}}}$; i.e., $\partial\varphi(x) = \{\nabla\varphi(x)\}$. Letting $G^H =: [g_1\ g_2\ \cdots g_{N_{\mathrm{R}}}]$, we have

$$I_{N_{\mathrm{T}}} + \frac{\rho}{N_{\mathrm{T}}}G^H X G = I_{N_{\mathrm{T}}} + \sum_{j=1}^{N_{\mathrm{R}}} x_j\left(\frac{\rho}{N_{\mathrm{T}}} g_j g_j^H\right), \tag{17.45}$$

which is positive definite. Therefore, $\forall x \in \mathbb{R}_{+}^{N_{\mathrm{R}}}$, $\forall j \in \mathscr{I}$, we have

$$\begin{aligned}\frac{\partial\varphi(x)}{\partial x_j} &= -\frac{1}{\ln 2}\mathrm{tr}\left[\left(I_{N_{\mathrm{T}}} + \frac{\rho}{N_{\mathrm{T}}}G^H X G\right)^{-1}\frac{\rho}{N_{\mathrm{T}}} g_j g_j^H\right] \\ &= -\frac{\rho}{N_{\mathrm{T}}\ln 2} g_j^H\left(I_{N_{\mathrm{T}}} + \frac{\rho}{N_{\mathrm{T}}}G^H X G\right)^{-1} g_j,\end{aligned} \tag{17.46}$$

where tr $[\cdot]$ stands for the trace of matrix. Note that, since $(I_{N_{\mathrm{T}}} + \frac{\rho}{N_{\mathrm{T}}}G^H X G)^{-1}$ is positive definite, $g_j^H(I_{N_{\mathrm{T}}} + \frac{\rho}{N_{\mathrm{T}}}G^H X G)^{-1}\ g_j > 0$, $\forall j \in \mathscr{I}$, thus $\frac{\partial\varphi(x)}{\partial x_j} < 0$, $\forall j \in \mathscr{I}$; $g_j \neq 0$ is silently assumed without loss of generality.

Finally, $\nabla^\gamma \psi_\omega (= \frac{1}{\gamma}(I - \mathrm{prox}_{\gamma\psi_\omega}))$ is computed simply by

$$\mathrm{prox}_{\gamma\psi_\omega} : \mathbb{R}^{N_\mathrm{R}} \ni x \mapsto \sum_{j=1}^{N_\mathrm{R}} \mathrm{sgn}(\langle x, e_j \rangle) \max\{|\langle x, e_j \rangle| - \gamma\omega, 0\} e_j, \tag{17.47}$$

where e_j, $j = 1, 2, \cdots, N_\mathrm{R}$, specially denotes the unit vector that has only one nonzero element at the jth position.

Remark 17.26 (On the recursion (17.43)). The operator $I - \lambda_{k+1}\nabla^\gamma \psi_\omega$ in (17.43) can be written as $I + \frac{\lambda_{k+1}}{\gamma}(\mathrm{prox}_{\gamma\psi_\omega} - I)$. From (17.47), $\mathrm{prox}_{\gamma\psi_\omega}$ attracts to zero such components of $\widehat{T}_\alpha(x_k)$ that are not greater than $\gamma\omega$. Therefore, $I - \lambda_{k+1}\nabla^\gamma \psi_\omega$ also has a similar zero-attracting function, thereby promoting the sparsity. The parameters γ and λ_{k+1} should satisfy $\lambda_{k+1}/\gamma \leq 1$ so that all the components of x_{k+1} are kept nonnegative. Also γ and ω should satisfy $\gamma\omega < 1$ for preventing the situation where all the components are attracted to zero. We mention that a constant value for all λ_ks (as shown below) may be used, because the strict convergence is not necessarily required in the proposed algorithm. The computational complexity of the proposed algorithm is given approximately by $QN_{\min}^2(2N_{\max} + \underline{\mathrm{N}})$, where $N_{\min} := \min\{N_\mathrm{R}, N_\mathrm{T}\}$, $N_{\max} := \max\{N_\mathrm{R}, N_\mathrm{T}\}$, and $\underline{\mathrm{N}} := \min\{N_{\min}, N_{\max}/2\}$. Hence, the proposed algorithm is efficient particularly when N_T is sufficiently small compared to N_R. Note that there exists no other method available for the minimal antenna-subset selection problem (17.40).

Remark 17.27 (Equivalent expression of the problem (17.41)). Noting the range of x, the problem (17.41) can equivalently be formulated as follows:

$$\min_{x \in [0,1]^{N_\mathrm{R}}} \widetilde{\psi}(x) := 1_{N_\mathrm{R}}^t x \quad \text{s.t. } \varphi(x) \leq 0. \tag{17.48}$$

Unfortunately, although the gradient $\nabla \widetilde{\psi}(x) = 1_{N_\mathrm{R}}$ is surely Lipschitz continuous, it is *not* possible, unlike the case of (17.42), to conclude immediately that (17.48) can be solved by applying Theorem 17.22 III for the following reason. Indeed, the HSDM recursion for (17.48) is given by

$$x_{k+1} := (I - \lambda_{k+1}\nabla \widetilde{\psi}(x))(\widehat{T}_\alpha(x_k)) = \widehat{T}_\alpha(x_k) - \lambda_{k+1} 1_{N_\mathrm{R}}, \ k \geq 0. \tag{17.49}$$

A simple inspection of (17.49) clarifies that λ_{k+1} should be no smaller than the minimum component of $\widehat{T}_\alpha(x_k)$ because the function φ, which is included in the operator $\widehat{T}_\alpha$, is convex only on $\mathbb{R}_+^{N_\mathrm{R}}$. Therefore, to guarantee the convergence by Theorem 17.22 III, careful design of the step size parameter λ_k is required at each iteration step.

17.4.6 Numerical Examples

Simulations are performed to show the efficacy of the proposed minimal antenna-subset selection algorithm. We consider the Rayleigh channel where the elements of G are independently drawn from a complex zero-mean Gaussian distribution of the unit variance. In all the simulations, the HSDM parameters are set to $\alpha = 1$, $\gamma = 1.2$, $\omega = 0.8$, $Q = 20$, and $\lambda_k = 1$ ($\forall k = 1,2,\cdots,Q$). In our experiments, the proposed algorithm is insensitive to the choice of the parameters within $\gamma\omega < 1$ and $\lambda_k/\gamma \leq 1$ (see Remark 17.26). All the simulated points are calculated by averaging over 2000 independent realizations of the channel matrix G.

First, Fig. 17.6 depicts the results for $N_{\mathrm{R}} = 16$, $N_{\mathrm{T}} = 4$, and $\underline{c} = 10, 20$. Figure 17.6a describes the average number $\bar{L}_{\mathrm{R}}$ of antennas selected by the proposed algorithm. As a reference, we also plot the optimal solution to the original problem in (17.40); the optimal is computed by computationally exhaustive full search. It is seen that the results of the proposed algorithm are comparable to the optimal; this suggests the reasonability of the relaxations introduced in Sect. 17.4.3. Figure 17.6b describes the ergodic capacity of the proposed algorithm. With L_{R} denoting the number of antennas selected by the proposed algorithm, we also plot $C_{\max}(L_{\mathrm{R}})$, the maximum achievable capacity with the subset of L_{R} antennas, which is computed by exhaustive search. It is seen that the performance of the proposed algorithm is approximately the same as $C_{\max}(L_{\mathrm{R}})$; this is the side effects of the proposed algorithm. In summary, the results demonstrate that the proposed

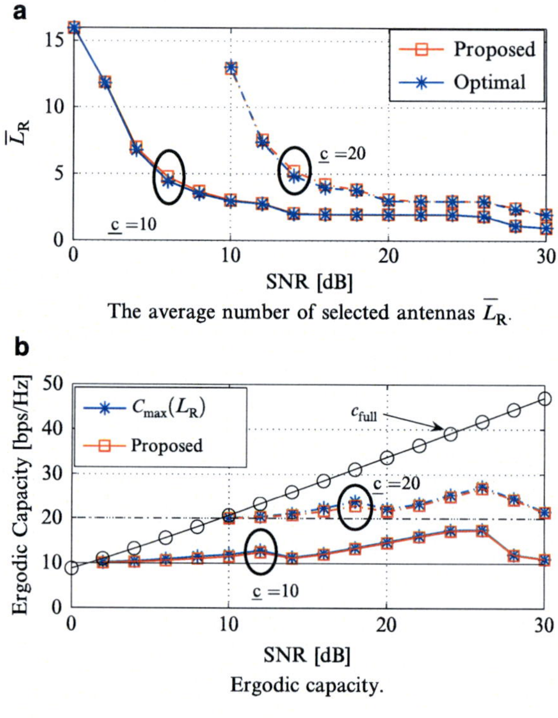

The average number of selected antennas $\bar{L}_{\mathrm{R}}$.

Ergodic capacity.

Fig. 17.6 Comparisons with the optimal selection for $N_{\mathrm{R}} = 16$, $N_{\mathrm{T}} = 4$, and $\underline{c} = 10, 20$

a

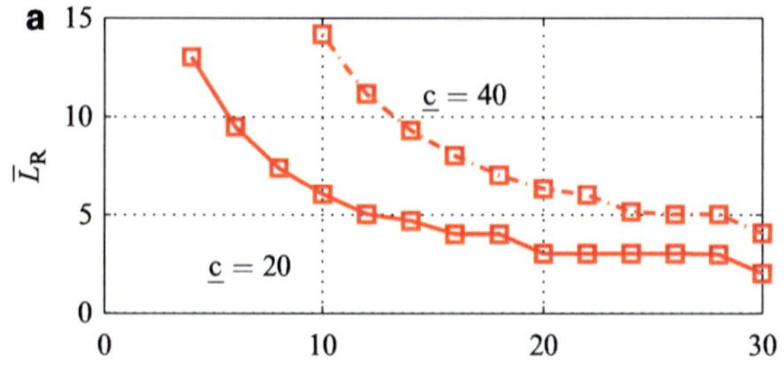

$N_R = 16$, $N_T = 16$, and $\underline{c} = 20$, 40.

b

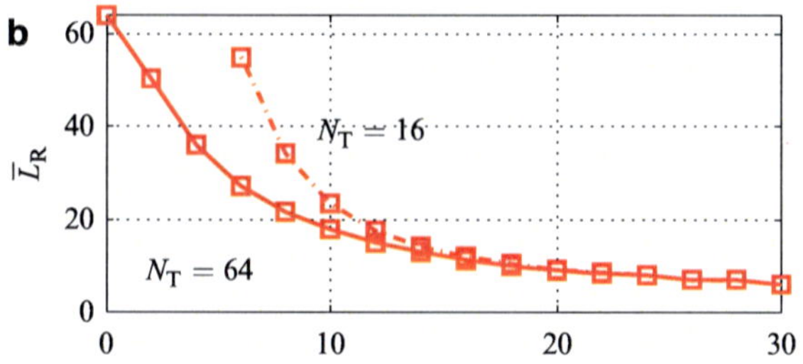

$N_R = 64$, $N_T = 16$, 64, and $\underline{c} = 60$.

Fig. 17.7 Performance for a large number of antennas

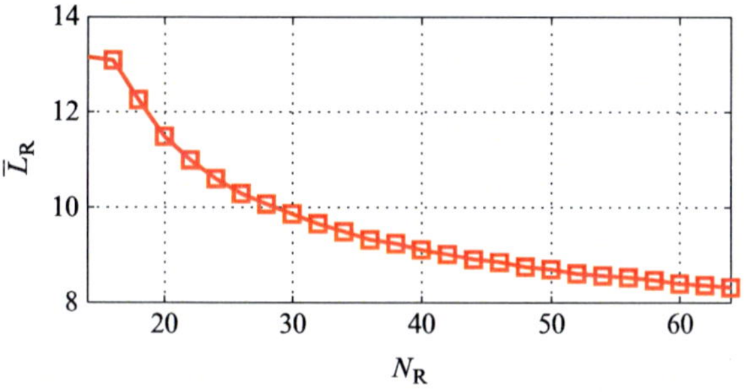

Fig. 17.8 N_R vs. $\bar{L}_R$ for SNR$=$ 10 dB, $N_T = 4$, and $\underline{c} = 20$

algorithm realizes (1) the near-minimal antenna subset and (2) the near-maximum capacity achievable with the same number of antennas as selected by the algorithm.

Second, Fig. 17.7 illustrates the results for (a) $N_R = 16$, $N_T = 16$, and $\underline{c} = 20, 40$ and (b) $N_R = 64$, $N_T = 16, 64$, and $\underline{c} = 60$. From Figs. 17.6a and 17.7, it is seen that the number of antennas to be used can significantly be reduced particularly for high SNR. Moreover, in Fig. 17.7b, we observe no distinct difference between $N_T = 16$ and $N_T = 64$ for SNR higher than 15 dB. Finally, Fig. 17.8 plots $\bar{L}_R$ against N_R for SNR$=$ 10 dB, $N_T = 4$, and $\underline{c} = 20$. The result shows that an increase of the number of antenna elements equipped could yield reduction of the number of antennas used.

17.5 Concluding Remarks

In this paper, we have introduced the essence of the great applicability of the convex optimization over the fixed point set of quasi-nonexpansive mapping. First, we have shown that the fixed point characterization gives us the powerful toolbox to address the problem of finding an "optimal" point from the fixed point set. Second, we have proposed the integration of the HSDM and the *Moreau-Yosida regularization* by highlighting its distinctive properties as a smooth approximation of a nonsmooth convex function. The novel integration with the gifted toolbox has opened a path to dealing with the challenging nonsmooth convex optimization problems under the *cumbersome* constraint of the fixed point set, which are naturally desired yet have been unexplored in mathematical sciences and engineerings. We have demonstrated the effectiveness of the proposed approach in its application to the minimal antenna-subset selection problem under a highly nonlinear capacity constraint for efficient MIMO communication systems.

This paper has focused on the nonsmooth convex optimization problems over the fixed point set. We remark, however, that the HSDM has many other possible advanced applications. For example, by letting $T := \mathrm{rprox}_{\gamma f_1}\mathrm{rprox}_{\gamma\iota_C}$ for $f_1 \in \Gamma_0(\mathcal{H})$ and a closed convex set $C \subset \mathcal{H}$, we have the characterization: $G := \arg\min_{x\in C} f_1(x) = \{P_C(z) \mid z \in \mathrm{Fix}(T)\}$ (see Example 17.6(c)). This means that we can minimize a convex function $f_2 : C \to \mathbb{R}$ over the constraint set G by applying the HSDM to $\Theta : \mathcal{H} \to \mathbb{R} : x \mapsto f_2(P_C(x))$ and T provided that the derivative of Θ is Lipschitzian.

Acknowledgements The first author thank Heinz Bauschke, Patrick Combettes and Russell Luke for their kind encouragement and invitation of the first author to the dream meeting: *The Interdisciplinary Workshop on Fixed-Point Algorithms for Inverse Problems in Science and Engineering* in November 1–6, 2009 at the Banff International Research Station.

References

1. Apostol, T.M.: Mathematical Analysis, 2nd ed. Addison-Wesley (1974)
2. Ascher, U.M., Haber, E., Huang, H.: On effective methods for implicit piecewise smooth surface recovery. SIAM J. Sci. Comput. **28**, 339–358 (2006)
3. Baillon, J.-B., Haddad, G.: Quelques propriétés des opérateurs angle-bornés et n-cycliquement monotones. Isr. J. Math. **26**, 137–150 (1977)
4. Baillon, J.-B., Bruck, R.E., Reich, S.: On the asymptotic behavior of nonexpansive mappings and semigroups in Banach spaces. Houst. J. Math. **4**, 1–9 (1978)
5. Barbu, V., Precupanu, Th.: Convexity and Optimization in Banach Spaces, 3rd Ed. D. Reidel Publishing Company (1986)
6. Bauschke, H.H.: The approximation of fixed points of compositions of nonexpansive mappings in Hilbert space. J. Math. Anal. Appl. **202**, 150–159 (1996)
7. Bauschke, H.H., Borwein, J.M.: On projection algorithms for solving convex feasibility problems. SIAM Rev. **38**, 367–426 (1996)
8. Bauschke, H.H., Combettes, P.L.: A weak-to-strong convergence principle for Fejér monotone methods in Hilbert space. Math. Oper. Res. **26**, 248–264 (2001)

9. Bauschke, H.H., Combettes, P.L.: The Baillon-Haddad theorem revisited. J. Convex Anal. **17**, 781–787 (2010)
10. Bauschke, H.H., Combettes, P.L.: Convex Analysis and Monotone Operator Theory in Hilbert Spaces. Springer (2011)
11. Beck, A., Teboulle, M.: A fast iterative shrinkage-thresholding algorithm for linear inverse problems. SIAM J. Imaging Sciences **2**, 183–202 (2009)
12. Beck, A., Teboulle, M.: Fast gradient-based algorithms for constrained total variation image denoising and deblurring problems. IEEE Trans. Image Process. **18**, 2419–2434 (2009)
13. Bertero, M., Boccacci, P.: Introduction to Inverse Problems in Imaging. IOP (1998)
14. Borwein, J.M., Fitzpatrick, S., Vanderwerff, J.: Examples of convex functions and classifications of normed spaces. J. Convex Anal. **1**, 61–73 (1994)
15. Bougeard, M.L.: Connection between some statistical estimation criteria, lower-C_2 functions and Moreau-Yosida approximates. In: Bulletin International Statistical Institute 47th session **1**, INSEE Paris Press, pp. 159–160 (1989)
16. Bougeard, M.L., Caquineau, C.D.: Parallel proximal decomposition algorithms for robust estimation. Ann. Oper. Res. **90**, 247–270 (1999)
17. Boyd, S., Vandenberghe, L.: Convex Optimization. Cambridge Univ. Press (2004)
18. Bregman, L.M.: The method of successive projection for finding a common point of convex sets. Soviet Math. Dokl. **6**, 688–692 (1965)
19. Browder, F.E.: Convergence theorems for sequences of nonlinear operators in Banach spaces. Math. Z. **100**, 201–225 (1967)
20. Byrne, C.L.: A unified treatment of some iterative algorithms in signal processing and image reconstruction. Inverse Probl. **20**, 103–120 (2004)
21. Byrne, C.L.: Applied Iterative Methods. A K Peters, Ltd., Wellesley, Massachusettes (2007)
22. Cai, J.F., Candés, E.J., Shen, Z.: A singular value thresholding algorithm for matrix completion. SIAM J. Optim. **20**, 1956–1982 (2010)
23. Candés, E.J., Wakin, M.B.: An introduction to compressive sampling. IEEE Signal Process. Mag. **25**, 21–30 (2008)
24. Capel, D., Zisserman, A.: Computer vision applied to super resolution. IEEE Signal Process. Mag. **20**, 75–86 (2003)
25. Cavalcante, R., Yamada, I.: Multiaccess interference suppression in orthogonal space-time block coded MIMO systems by adaptive projected subgradient method. IEEE Trans. Signal Process. **56**, 1028–1042 (2008)
26. Cavalcante, R., Yamada, I.: A flexible peak-to-average power ratio reduction scheme for OFDM systems by the adaptive projected subgradient method. IEEE Trans. Signal Process. **57**, 1456–1468 (2009)
27. Cavalcante, R., Yamada, I., Mulgrew, B.: An adaptive projected subgradient approach to learning in diffusion networks. IEEE Trans. Signal Process. **57**, 2762–2774 (2009)
28. Censor, Y., Reich, S.: Iterations of paracontractions and firmly nonexpansive operators with applications to feasibility and optimization. Optimization **37**, 323–339 (1996)
29. Censor, Y., Zenios, S.A.: Parallel Optimization: Theory, Algorithm, and Optimization. Oxford University Press (1997)
30. Censor, Y., Iusem, A.N., Zenios, S.A.: An interior point method with Bregman functions for the variational inequality problem with paramonotone operators. Math. Program. **81**, 373–400 (1998)
31. Chambolle, A.: An algorithm for total variation minimization and applications. J. Math. Imaging Vis. **20**, 89–97 (2004)
32. Chambolle, A., DeVore, R.A., Lee, N.Y., Lucier, B.J.: Nonlinear wavelet image processing: Variational problems, compression, and noise removal through wavelet shrinkage. IEEE Trans. Image Process. **7**, 319–335 (1998)
33. Chidume, C.: Geometric Properties of Banach Spaces and Nonlinear Iterations (Chapter 7: Hybrid steepest descent method for variational inequalities). Lecture Notes in Mathematics **1965**, Springer (2009)
34. Combettes, P.L.: Foundation of set theoretic estimation. Proc. IEEE. **81**, 182–208 (1993)

35. Combettes, P.L.: Inconsistent signal feasibility problems: least squares solutions in a product space. IEEE Trans. Signal Process. **42**, 2955–2966 (1994)
36. Combettes, P.L.: Construction d'un point fixe commun à une famille de contractions fermes. C.R. Acad. Sci.Paris Sèr. I Math. **320**, 1385–1390 (1995)
37. Combettes, P.L.: Convex set theoretic image recovery by extrapolated iterations of parallel subgradient projections. IEEE Trans. Image Process. **6**, 493–506 (1997)
38. Combettes, P.L.: Strong convergence of block-iterative outer approximation methods for convex optimization. SIAM J. Control Optim. **38**, 538–565 (2000)
39. Combettes, P.L.: Solving monotone inclusions via compositions of nonexpansive averaged operators. Optimization **53**, 475–504 (2004)
40. Combettes, P.L., Bondon, P.: Hard-constrained inconsistent signal feasibility problems. IEEE Trans. Signal Process. **47**, 2460–2468 (1999)
41. Combettes, P.L., Pesquet, J.-C.: Image restoration subject to a total variation constraint. IEEE Trans. Image Process. **13**, 1213–1222 (2004)
42. Combettes, P.L., Pesquet, J.-C.: A Douglas-Rachford splitting approach to nonsmooth convex variational signal recovery. IEEE J. Sel. Top. Signal Process. **1**, 564–574 (2007)
43. Combettes, P.L., Pesquet, J.-C.: A proximal decomposition method for solving convex variational inverse problems. Inverse Probl. **24** (2008)
44. Combettes, P.L., Pesquet, J.-C.: Split convex minimization algorithm for signal recovery. Proc. 2009 IEEE ICASSP (Taipei), 685–688 (2009)
45. Combettes, P.L., Pesquet, J.-C.: Proximal splitting methods in signal processing. In: H. H. Bauschke, R. Burachik, P. L. Combettes, V. Elser, D. R. Luke, H. Wolkowicz (eds.) Fixed-Point Algorithms for Inverse Problems in Science and Engineering, Springer (2010)
46. Combettes, P.L., Wajs, V.R.: Signal recovery by proximal forward-backward splitting. SIAM Multiscale Model. Simul. **4**, 1168–1200 (2005)
47. Daubechies, I., Defrise, M., Mol, C.D.: An iterative thresholding algorithm for linear inverse problems with a sparsity constraint. Comm. Pure Appl. Math. **57**, 1413–1457 (2004)
48. Deutsch, F., Best Approximation in Inner Product Spaces. Springer, New York (2001)
49. Deutsch, F., Yamada, I.: Minimizing certain convex functions over the intersection of the fixed point sets of nonexpansive mappings. Numer. Funct. Anal. Optim. **19**, 33–56 (1998)
50. Dolidze, Z.O.: Solutions of variational inequalities associated with a class of monotone maps. Ekonomika i Matem. Metody **18**, 925–927 (1982)
51. Donoho, D.L.: De-noising by soft-thresholding. IEEE Trans. Inf. Theory **41**, 613–627 (1995)
52. Donoho, D.L.: Compressed sensing. IEEE Trans. Inf. Theory **52**, 1289–1306 (2006)
53. Donoho, D.L., Johnstone, I.M.: Ideal spatial adaptation via wavelet shrinkage. Biometrika **81**, 425–455 (1994)
54. Dotson, Jr, W.G.: On the Mann iterative process. Trans. Amer. Math. Soc. **149**, 65–73 (1970)
55. Dua, A., Medepalli, K., Paulraj, A.J.: Receive antenna selection in MIMO systems using convex optimization. IEEE Trans. Wirel. Commun. **5**, 2353–2357 (2006)
56. Dunn, J.C.: Convexity, monotonicity, and gradient processes. J. Math. Anal. Appl. **53**, 145–158 (1976)
57. Eckstein, J., Bertsekas, D.P.: On the Douglas-Rachfold splitting method and proximal point algorithm for maximal monotone operators. Math. Program. **55**, 293–318 (1992)
58. Eicke, B.: Iteration methods for convexly constrained ill-posed problems in Hilbert space. Numer. Funct. Anal. Optim. **13**, 413–429 (1992)
59. Ekeland, I., Temam, R.: Convex Analysis and Variational Problems. Classics in Applied Mathematics **28**, SIAM (1999)
60. Elsner, L., Koltracht, L., Neumann, M.: Convergence of sequential and asynchronous nonlinear paracontractions. Numer. Math. **62**, 305–319 (1992)
61. Engle, H.W., Leitão, A.: A Mann iterative regularization method for elliptic Cauchy problems. Numer. Funct. Anal. Optim. **22**, 861–884 (2001)
62. Fadili, M.J., Starck, J.-L.: Monotone operator splitting for optimization problems in sparse recovery. Proc. 2009 IEEE ICIP, Cailo (2009)
63. Foschini, G.J., Gans, M.J.: On limits of wireless communications in a fading environment when using multiple antennas. Wirel. Pers. Commun. **6**, 311–335 (1998)

64. Fukushima, M.: A relaxed projection method for variational inequalities. Math. Program. **35**, 58–70 (1986)
65. Fukushima, M., Qi, L.: A globally and superlinearly convergent algorithm for nonsmooth convex minimization. SIAM J. Optim. **6**, 1106–1120 (1996)
66. Gabay, D.: Applications of the method of multipliers to variational inequalities. In : M. Fortin and R. Glowinski (eds.) Augmented Lagrangian Methods: Applications to the solution of boundary value problems, North-Holland, Amsterdam (1983)
67. Gandy, S., Yamada, I.: Convex optimization techniques for the efficient recovery of a sparsely corrupted low-rank matrix. Journal of Math-for-Industry **2**(2010B-5), 147–156 (2010)
68. Gandy, S., Recht, B., Yamada, I.: Tensor completion and low-n-rank tensor recovery via convex optimization. Inverse Probl. **27**(2), 025010 (2011)
69. Gharavi-Alkhansari, M., Gershman, A.B.: Fast antenna subset selection in MIMO systems. IEEE Trans. Signal Process. **52**, 339–347 (2004)
70. Goebel, K., Kirk, W.A.: Topics in Metric Fixed Point Theory. Cambridge Univ. Press. (1990)
71. Goebel, K., Reich, S.: Uniform Convexity, Hyperbolic Geometry, and Nonexpansive Mappings. New York and Basel Dekker (1984)
72. Goldstein, A.A.: Convex programming in Hilbert space. Bull. Amer. Math. Soc. **70**, 709–710 (1964)
73. Golshtein, E.G., Tretyakov, N.V.: Modified Lagrangians and Monotone Maps in Optimization. Wiley (1996)
74. Gorokhov, A., Gore, D.A., Paulraj, A.J.: Receive antenna selection for MIMO spatial multiplexing: theory and algorithms. IEEE Trans. Signal Process. **51**, 2796–2807 (2003)
75. Groetsch, C.W.: A note on segmenting Mann iterates. J. Math. Anal. Appl. **40**, 369–372 (1972)
76. Groetsch, C.W.: Inverse Problems in Mathematical Sciences. Wiesbaden-Vieweg (1993)
77. Gubin, L.G., Polyak, B.T., Raik, E.V.: The method of projections for finding the common point of convex sets. USSR Comput. Maths. Phys. **7**, 1–24 (1967)
78. Halpern, B.: Fixed points of nonexpanding maps. Bull. Amer. Math. Soc. **73**, 957–961 (1967)
79. Hasegawa, H., Ohtsuka, T., Yamada, I., Sakaniwa, K.: An edge-preserving super-precision for simultaneous enhancement of spacial and grayscale resolutions. IEICE Trans. Fundam. Electron. Commun. Comput. Sci. **E91-A**, 673–681 (2008)
80. Haugazeau, Y.: Sur les Inéquations variationnelles et la Minimisation de Fonctionnelles Convexes. Thèse, Universite de Paris (1968)
81. Haykin, S.: Adaptive Filter Theory, 4th edn. Prentice Hall (2002)
82. Hiriart-Urruty, J.-B., Lemaréchal, C.: Convex Analysis and Minimization Algorithms. Springer (1993)
83. Huber, P.J.: Robust estimation of a location parameter. Ann. Math. Statist. **35**, 73–101 (1964)
84. Iemoto, S., Takahashi, W.: Strong convergence theorems by a hybrid steepest descent method for countable nonexpansive mappings in Hilbert spaces. Sci. Math. Jpn. **69** (online: 2008-49), 227–240 (2009)
85. Kiwiel, K.C.: Block-iterative surrogate projection methods for convex feasibility problems. Linear Alg. Appl. **215**, 225–259 (1995)
86. Kreyszig, E.: Introductory Functional Analysis with Applications. Wiley Classics Library, Wiley, New York (1989)
87. Levitin, E.S., Polyak, B.T.: Constrained minimization method. USSR Comput. Maths. Phys. **6**, 1–50 (1966)
88. Li, W., Swetits, J.J.: The linear ℓ_1 estimator and the Huber M-estimator. SIAM J. Optim. **8**, 457–475 (1998)
89. Lions, P.L.: Approximation de points fixes de contractions. C. R. Acad. Sci. Paris Sèrie A-B **284**, 1357–1359 (1977)
90. Lions, P.L., Mercier, B.: Splitting algorithms for the sum of two nonlinear operators. SIAM J. Numer. Anal. **16**, 964–979 (1979)
91. Liu, F., Nashed, M.Z.: Regularization of nonlinear ill-posed variational inequalities and convergence rates. Set-Valued Anal. **6**, 313–344 (1998)

92. Mainge, P.E.: Extension of the hybrid steepest descent method to a class of variational inequalities and fixed point problems with nonself-mappings. Numer. Funct. Anal. Optim. **29**, 820–834 (2008)
93. Mangasarian, O.L., Muscicant, D.R.: Robust linear and support vector regression. IEEE Trans. Pattern Anal. Mach. Intell. **22**, 950–955 (2000)
94. Mann, W.: Mean value methods in iteration. Proc. Amer. Math. Soc. **4**, 506–510 (1953)
95. Mehta, N.B., Molisch, A.F.: MIMO System Technology for Wireless Communications, chapter 6, CRC Press (2006)
96. Michelot, C., Bougeard, M.L.: Duality results and proximal solutions of the Huber M-estimator problem. Appl. Math. Optim. **30**, 203–221 (1994)
97. Molisch, A.F., Win, M.Z.: MIMO systems with antenna selection. IEEE Microw. Mag. **5**, 46–56 (2004)
98. Moreau, J.J.: Fonctions convexes duales et points proximaux dans un espace hilbertien. C. R. Acad. Sci. Paris Ser. A Math. **255**, 2897–2899 (1962)
99. Moreau, J.J.: Proximité et dualité dans un espace hilbertien. Bull. Soc. Math. France **93**, 273–299 (1965)
100. Nikolova, M.: Minimizing of cost functions involving nonsmooth data-fidelity terms – Application to the processing of outliers. SIAM J. Numer. Anal. **40**, 965–994 (2002)
101. Ogura, N., Yamada, I.: Non-strictly convex minimization over the fixed point set of the asymptotically shrinking nonexpansive mapping. Numer. Funct. Anal. Optim. **23**, 113–137 (2002)
102. Ogura, N., Yamada, I.: Non-strictly convex minimization over the bounded fixed point set of nonexpansive mapping. Numer. Funct. Anal. Optim. **24**, 129–135 (2003)
103. Ogura, N., Yamada, I.: A deep outer approximating half space of the level set of certain Quadratic Functions. J. Nonlinear Convex Anal. **6**, 187–201 (2005)
104. Passty, G.B.: Ergodic convergence to a zero of the sum of monotone operators in Hilbert space. J. Math. Anal. Appl. **72**, 383–390 (1979)
105. Pierra, G.: Eclatement de contraintes en parallèle pour la minimisation d'une forme quadratique. Lecture Notes in Computer Science **41**, 200–218, Springer (1976)
106. Pierra, G.: Decomposition through formalization in a product space. Math. Program. **28**, 96–115 (1984)
107. Polyak, B.T.: Minimization of unsmooth functionals. USSR Comput. Maths. Phys. **9**, 14–29 (1969)
108. Rockafellar, R.T.: Monotone operators and proximal point algorithm. SIAM J. Control Optim. **14**, 877–898 (1976)
109. Rockafellar, R.T., Wets, R.J.-B.: Variational Analysis, 1st edn. Springer (1998)
110. Sabharwal, A., Potter, L.C.: Convexly constrained linear inverse problems: Iterative least-squares and regularization. IEEE Trans. Signal Process. **46**, 2345–2352 (1998)
111. Sanayei, S., Nosratinia, A.: Antenna selection in MIMO systems. IEEE Commun. Mag. **42**, 68–73 (2004)
112. Sayed, A.H.: Fundamentals of Adaptive Filtering. Wiley-IEEE Press (2003)
113. Slavakis, K., Yamada, I.: Robust wideband beamforming by the hybrid steepest descent method. IEEE Trans. Signal Process. **55**, 4511–4522 (2007)
114. Slavakis, K., Yamada, I., Ogura, N.: The adaptive projected subgradient method over the fixed point set of strongly attracting nonexpansive mappings. Numer. Funct. Anal. Optim. **27**, 905–930 (2006)
115. Slavakis, K., Theodoridis, S., Yamada, I.: Online kernel-based classification using adaptive projection algorithms. IEEE Trans. Signal Process. **56**, 2781–2796 (2008)
116. Slavakis, K., Theodoridis, S., Yamada, I.: Adaptive constrained filtering in reproducing kernel Hilbert spaces: the beamforming case. IEEE Trans. Signal Process. **57**, 4744–4764 (2009)
117. Starck, J.-L., Murtagh, F.: Astronomical Image and Data Analysis, 2nd.edn. Springer (2006)
118. Starck, J.-L., Murtagh, F., Fadili, J.M.: Sparse Image and Signal Processing – Wavelets, Curvelets, Morphological Diversity. Cambridge Univ. Press (2010)
119. Stark, H., Yang, Y.: Vector Space Projections – A Numerical Approach to Signal and Image Processing, Neural Nets, and Optics. Wiley (1998)

120. Suzuki, T.: A sufficient and necessary condition for Halpern-type strong convergence to fixed points of nonexpansive mappings. Proc. Amer. Math. Soc. **135**, 99–106 (2007)
121. Takahashi, W.: Nonlinear Functional Analysis – Fixed Point Theory and its Applications. Yokohama Publishers (2000)
122. Takahashi, N., Yamada, I.: Parallel algorithms for variational inequalities over the Cartesian product of the intersections of the fixed point sets of nonexpansive mappings. J. Approx. Theory **153**, 139–160 (2008)
123. Takahashi, N., Yamada, I.: Steady-state mean-square performance analysis of a relaxed set-membership NLMS algorithm by the energy conservation argument. IEEE Trans. Signal Process. **57**, 3361–3372 (2009)
124. Telatar, I.E.: Capacity of multi-antenna Gaussian channels. Eur. Trans. Telecomm. **10**, 585–595 (1999)
125. Theodoridis, S., Slavakis, K., Yamada, I.: Adaptive learning in a world of projections – A unifying framework for linear and nonlinear classification and regression tasks. IEEE Signal Processing Mag. **28**, 97–123 (2011)
126. Tseng, P.: Applications of a splitting algorithm to decomposition in convex programming and variational inequalities. SIAM J. Control Optim. **29**, 119–138 (1991)
127. Vasin, V.V., Ageev, A.L.: Ill-Posed Problems with A Priori Information. VSP (1995)
128. Widrow, B., Stearns, S.D.: Adaptive Signal Processing. Prentice Hall (1985)
129. Wittmann, R.: Approximation of fixed points of nonexpansive mappings. Arch. Math. **58**, 486–491 (1992)
130. Xu, H.K., Kim, T.H.: Convergence of hybrid steepest descent methods for variational inequalities. J. Optim. Theory Appl. **119**, 185–201 (2003)
131. Yamada, I.: Approximation of convexly constrained pseudoinverse by Hybrid Steepest Descent Method. Proc. 1999 IEEE ISCAS, Florida (1999)
132. Yamada, I.: The hybrid steepest descent method for the variational inequality problem over the intersection of fixed point sets of nonexpansive mappings. In: D. Butnariu, Y. Censor, S. Reich (eds.) Inherently Parallel Algorithm for Feasibility and Optimization and Their Applications, Elsevier, 473–504 (2001)
133. Yamada, I.: Adaptive projected subgradient method: A unified view for projection based adaptive algorithms. The Journal of IEICE **86**, 654–658 (2003) (in Japanese)
134. Yamada, I.: Kougaku no Tameno Kansu Kaiseki (Functional Analysis for Engineering), Suurikougaku-Sha/Saiensu-Sha (2009)
135. Yamada, I., Ogura, N.: Adaptive projected subgradient method for asymptotic minimization of sequence of nonnegative convex functions. Numer. Funct. Anal. Optim. **25**, 593–617 (2004)
136. Yamada, I., Ogura, N.: Hybrid steepest descent method for variational inequality problem over the fixed point set of certain quasi-nonexpansive mappings. Numer. Funct. Anal. Optim. **25**, 619–655 (2004)
137. Yamada, I., Ogura, N., Yamashita, Y., Sakaniwa, K.: An extension of optimal fixed point theorem for nonexpansive operator and its application to set theoretic signal estimation. Technical Report of IEICE **DSP96-106**, 63–70 (1996)
138. Yamada, I., Ogura, N., Yamashita, Y., Sakaniwa, K.: Quadratic optimization of fixed points of nonexpansive mappings in Hilbert space. Numer. Funct. Anal. Optim. **19**, 165–190 (1998)
139. Yamada, I., Ogura, N., Shirakawa, N.: A numerically robust hybrid steepest descent method for the convexly constrained generalized inverse problems. In: Z. Nashed, O. Scherzer (eds.) Inverse Problems, Image Analysis, and Medical Imaging, *Contemporary Mathematics* **313**, 269–305 (2002)
140. Yamada, I., Slavakis, K., Yamada, K.: An efficient robust adaptive filtering algorithm based on parallel subgradient projection techniques. IEEE Trans. Signal Process. **50**, 1091–1101 (2002)
141. Yamagishi, M., Yamada, I.: A deep monotone approximation operator based on the best quadratic lower bound of convex functions. IEICE Trans. Fundam. Electron. Commun. Comput. Sci. **E91-A**, 1858–1866 (2008)

142. Yosida, K.: Functional Analysis, 4th edn. Springer (1974)
143. Youla, D.C., Webb, H.: Image restoration by the method of convex projections: Part 1 – Theory. IEEE Trans. Med. Imaging **1**, 81–94 (1982)
144. Yukawa, M., Yamada, I.: Pairwise optimal weight realization – acceleration technique for set-theoretic adaptive parallel subgradient projection algorithm. IEEE Trans. Signal Process. **54**, 4557–4571 (2006)
145. Yukawa, M., Yamada, I.: Minimal antenna-subset selection under capacity constraint for power-efficient MIMO systems: a relaxed ℓ_1-minimization approach. Proc. 2010 IEEE ICASSP, Dallas (2010)
146. Yukawa, M., Cavalcante, R., Yamada, I.: Efficient blind MAI suppression in DS/CDMA by embedded constraint parallel projection techniques. IEICE Trans. Fundam. Electron. Commun. Comput. Sci. **E88-A**, 2427–2435 (2005)
147. Yukawa, M., Slavakis, K., Yamada, I.: Adaptive parallel quadratic-metric projection algorithms. IEEE Trans. Audio Speech Lang. Process. **15**, 1665–1680 (2007)
148. Zălinescu, C.: Convex Analysis in General Vector Spaces. World Scientific (2002)
149. Zeidler, E.: Nonlinear Functional Analysis and its Applications, III – Variational Methods and Optimization. Springer (1985)
150. Zeidler, E.: Nonlinear Functional Analysis and its Applications, II/B – Nonlinear Monotone Operators. Springer (1990)
151. Zeng, L.C., Schaible, S., Yao, J.C.: Hybrid steepest descent methods for zeros of nonlinear operators with applications to variational inequalities. J. Optim. Theory Appl. **141**, 75–91 (2009)
152. Zhang, B., Fadili, J.M., Starck, J.-L.: Wavelet, ridgelet, and curvelets for Poisson noise removal. IEEE Trans. Image Process. **17**, 1093–1108 (2008)

Chapter 18
The Brézis-Browder Theorem Revisited and Properties of Fitzpatrick Functions of Order n

Liangjin Yao

Abstract In this paper, we study maximal monotonicity of linear relations (set-valued operators with linear graphs) on reflexive Banach spaces. We provide a new and simpler proof of a result due to Brézis–Browder which states that a monotone linear relation with closed graph is maximal monotone if and only if its adjoint is monotone. We also study Fitzpatrick functions and give an explicit formula for Fitzpatrick functions of order n for monotone symmetric linear relations.

Keywords Adjoint · Convex function · Convex set · Fenchel conjugate · Fitzpatrick function · Linear relation · Maximal monotone operator · Multifunction · Monotone operator · Set-valued operator · Symmetric operator.

AMS 2010 Subject Classification: 47A06, 47H05

18.1 Introduction

Monotone operators play important roles in convex analysis and optimization [12, 15, 22, 24–26, 32, 33]. In 1978, Brézis–Browder gave some characterizations of a monotone operator with closed linear graph [14, Theorem 2] in reflexive Banach spaces. The Brézis–Browder Theorem states that a monotone linear relation with closed graph is maximal monotone if and only if its adjoint is monotone if and only if its adjoint is maximal monotone, which demonstrates the connection between the monotonicity of a linear relation and that of its adjoint. In this paper, we give a new and simpler proof of the hard part of the Brézis–Browder Theorem (Theorem 18.5): a monotone linear relation with closed graph is maximal monotone if its adjoint is monotone. The proof relies on a recent characterization of maximal monotonicity due to Simons and Zălinescu. Our proof does not require any renorming.

L. Yao (✉)
Department of Mathematics, Irving K. Barber School, University of British Columbia, Kelowna, B.C. V1V 1V7, Canada
e-mail: ljinyao@interchange.ubc.ca

H.H. Bauschke et al. (eds.), *Fixed-Point Algorithms for Inverse Problems in Science and Engineering*, Springer Optimization and Its Applications 49,
DOI 10.1007/978-1-4419-9569-8_18,

We suppose throughout this note that X is a real reflexive Banach space with norm $\|\cdot\|$, that X^* is its continuous dual space with norm $\|\cdot\|_*$ and dual product $\langle\cdot,\cdot\rangle$. We now introduce some notation. Let $A\colon X \rightrightarrows X^*$ be a *set-valued operator* or *multifunction* whose graph is defined by

$$\operatorname{gra} A := \{(x,x^*) \in X \times X^* \mid x^* \in Ax\}.$$

The *inverse operator* of A, $A^{-1}\colon X^* \rightrightarrows X$, is given by $\operatorname{gra} A^{-1} := \{(x^*,x) \in X^* \times X \mid x^* \in Ax\}$; the *domain* of A is $\operatorname{dom} A := \{x \in X \mid Ax \neq \varnothing\}$. The *Fitzpatrick function* of A (see [19]) is given by

$$F_A\colon (x,x^*) \mapsto \sup_{(a,a^*)\in \operatorname{gra} A} \big(\langle x,a^*\rangle + \langle a,x^*\rangle - \langle a,a^*\rangle\big). \tag{18.1}$$

For every $n \in \{2,3,\ldots\}$, the *Fitzpatrick function of A of order n* (see [1, Definition 2.2 and Proposition 2.3]) is defined by

$$F_{A,n}(x,x^*) := \sup_{\{(a_1,a_1^*),\cdots,(a_{n-1},a_{n-1}^*)\}\subseteq \operatorname{gra} A} \left(\langle x,x^*\rangle + \left(\sum_{i=1}^{n-2}\langle a_{i+1}-a_i,a_i^*\rangle\right)\right.$$
$$\left. +\langle x-a_{n-1},a_{n-1}^*\rangle + \langle a_1-x,x^*\rangle\right).$$

Clearly, $F_{A,2} = F_A$. We set $F_{A,\infty} = \sup_{n\in\{2,3,\cdots\}} F_{A,n}$.

If Z is a real reflexive Banach space with dual Z^* and a set $S \subseteq Z$, we denote $S^\perp$ by $S^\perp := \{z^* \in Z^* \mid \langle z^*,s\rangle = 0, \forall s \in S\}$. Then the *adjoint* of A, denoted by A^*, is defined by

$$\operatorname{gra} A^* := \{(x,x^*) \in X \times X^* \mid (x^*,-x) \in (\operatorname{gra} A)^\perp\}.$$

Note that A is said to be a *linear relation* if $\operatorname{gra} A$ is a linear subspace of $X \times X^*$. (See [18] for further information on linear relations.) Recall that A is said to be *monotone* if for all $(x,x^*),(y,y^*) \in \operatorname{gra} A$ we have

$$\langle x-y,x^*-y^*\rangle \geq 0,$$

and A is *maximal monotone* if A is monotone and A has no proper monotone extension (in the sense of graph inclusions). We say $(x,x^*) \in X \times X^*$ is *monotonically related* to $\operatorname{gra} A$ if (for every $(y,y^*) \in \operatorname{gra} A$) $\langle x-y,x^*-y^*\rangle \geq 0$. Recently, linear relations have been become an interesting object and comprehensively studied in Monotone Operator Theory: see [1–3, 5–10, 23, 29–31]. We can now precisely describe the Brézis–Browder Theorem. Let A be a monotone linear relation with closed graph. Then

$$A \text{ is maximal monotone} \Leftrightarrow A^* \text{ is maximal monotone}$$
$$\Leftrightarrow A^* \text{ is monotone.}$$

The original proof of Brézis–Browder Theorem is based on the application of Zorn Lemma by constructing a series of finite-dimensional subspaces, which is complicated. Our goal of this paper is to give a simpler proof of Brézis–Browder Theorem and to derive more properties of Fitzpatrick functions of order n. The paper is organized as follows. The first main result (Theorem 18.5) is proved in Sect. 18.2 providing a new and simpler proof of the Brézis–Browder Theorem. In Sect. 18.3, some explicit formulas for Fitzpatrick functions are given. Recently, *Fitzpatrick functions of order n* [1] have turned out to be a useful tool in the study of n-cyclic monotonicity (see [1, 3, 4, 13]). Theorem 18.14 gives an explicit formula for Fitzpatrick functions of order n associated with symmetric linear relations, which generalizes and simplifies [1, Example 4.4] and [3, Example 6.4].

Our notation is standard. The notation $A\colon X \to X^*$ means that A is a *single-valued* mapping (with full domain) from X to X^*. Given a subset C of X, $\overline{C}$ is the norm closure of C. The *indicator function* $\iota_C : X \to \left]-\infty,+\infty\right]$ of C is defined by

$$x \mapsto \begin{cases} 0, & \text{if } x \in C; \\ +\infty, & \text{otherwise.} \end{cases} \tag{18.2}$$

Let $x \in X$ and $C^* \subseteq X^*$. We write $\langle x, C^*\rangle := \{\langle x, c^*\rangle \mid c^* \in C^*\}$. If $\langle x, C^*\rangle = \{a\}$ for some constant $a \in \mathbb{R}$, then we write $\langle x, C^*\rangle = a$ for convenience. For a function $f\colon X \to \left]-\infty,+\infty\right]$, $\operatorname{dom} f = \{x \in X \mid f(x) < +\infty\}$ and $f^*\colon X^* \to [-\infty,+\infty] : x^* \mapsto \sup_{x\in X}(\langle x, x^*\rangle - f(x))$ is the *Fenchel conjugate* of f. Recall that f is said to be *proper* if $\operatorname{dom} f \neq \varnothing$. If f is convex, $\partial f\colon X \rightrightarrows X^* : x \mapsto \{x^* \in X^* \mid (\forall y \in X)\ \langle y - x, x^*\rangle + f(x) \leq f(y)\}$ is the *subdifferential operator* of f. Denote J by the duality map, i.e., the subdifferential of the function $\frac{1}{2}\|\cdot\|^2$, by [22, Example 2.26],

$$Jx := \{x^* \in X^* \mid \langle x^*, x\rangle = \|x^*\|_* \cdot \|x\|, \text{ with } \|x^*\|_* = \|x\|\}.$$

18.2 A New Proof of the Brézis–Browder Theorem

Fact 18.1 (Simons). (See [26, Lemma 19.7 and Sect. 22].) Let $A : X \rightrightarrows X^*$ be a monotone operator such that $\operatorname{gra} A$ is convex with $\operatorname{gra} A \neq \varnothing$. Then the function

$$g\colon X \times X^* \to \left]-\infty,+\infty\right] : (x,x^*) \mapsto \langle x, x^*\rangle + \iota_{\operatorname{gra} A}(x,x^*) \tag{18.3}$$

is proper and convex.

Fact 18.2 (Simons-Zălinescu). (See [27, Theorem 1.2] or [25, Theorem 10.6].) Let $A\colon X \rightrightarrows X^*$ be monotone. Then A is maximal monotone if and only if

$$\operatorname{gra} A + \operatorname{gra}(-J) = X \times X^*.$$

Remark 18.3. When J and J^{-1} are single valued, Fact 18.2 yields Rockafellar's characterization of maximal monotonicity of A. See [27, Theorem 1.3] and [26, Theorem 29.5 and Remark 29.7].

Now we state the Brézis–Browder Theorem.

Theorem 18.4 (Brézis–Browder). *(See [14, Theorem 2].) Let $A\colon X \rightrightarrows X^*$ be a monotone linear relation with closed graph. Then the following statements are equivalent.* (The hard part is to show (iii)⇒(i)).

(i) A is maximal monotone.
(ii) A^* is maximal monotone.
(iii) A^* is monotone.

Proof. (i)⇒(iii): Suppose to the contrary that A^* is not monotone. Then there exists $(x_0, x_0^*) \in \operatorname{gra} A^*$ such that $\langle x_0, x_0^* \rangle < 0$. Now we have

$$\begin{aligned}\langle -x_0 - y, x_0^* - y^* \rangle &= \langle -x_0, x_0^* \rangle + \langle y, y^* \rangle + \langle x_0, y^* \rangle + \langle -y, x_0^* \rangle \\ &= \langle -x_0, x_0^* \rangle + \langle y, y^* \rangle > 0, \quad \forall (y, y^*) \in \operatorname{gra} A. \qquad (18.4)\end{aligned}$$

Thus, $(-x_0, x_0^*)$ is monotonically related to $\operatorname{gra} A$. By maximal monotonicity of A, $(-x_0, x_0^*) \in \operatorname{gra} A$. Then $\langle -x_0 - (-x_0), x_0^* - x_0^* \rangle = 0$, which contradicts (18.4). Hence, A^* is monotone.

(iii)⇒(i): See Theorem 18.5 below.
(i)⇔(ii): Apply directly (iii)⇔(i) by using $A^{**} = A$ (since $\operatorname{gra} A$ is closed). ■

In Theorem 18.5, we provide a new and simpler proof to show the hard part (iii)⇒(i) in Theorem 18.4. The proof was inspired by that of [33, Theorem 32.L].

Theorem 18.5. *Let $A\colon X \rightrightarrows X^*$ be a monotone linear relation with closed graph. Suppose A^* is monotone. Then A is maximal monotone.*

Proof. By Fact 18.2, it suffices to show that $X \times X^* \subseteq \operatorname{gra} A + \operatorname{gra}(-J)$. For this, let $(x, x^*) \in X \times X^*$ and we define $g : X \times X^* \to \left]-\infty, +\infty\right]$ by

$$(y, y^*) \mapsto \frac{1}{2}\|y^*\|_*^2 + \frac{1}{2}\|y\|^2 + \langle y^*, y \rangle + \iota_{\operatorname{gra} A}(y - x, y^* - x^*).$$

Since $\operatorname{gra} A$ is closed, g is lower semicontinuous on $X \times X^*$. Note that $(y, y^*) \mapsto \langle y^*, y \rangle + \iota_{\operatorname{gra} A}(y - x, y^* - x^*) = \langle y^*, y \rangle + \iota_{\operatorname{gra} A + (x, x^*)}(y, y^*)$. By Fact 18.1, g is convex and coercive. According to [32, Theorem 2.5.1(ii)], g has minimizers. Suppose that (z, z^*) is a minimizer of g. Then $(z - x, z^* - x^*) \in \operatorname{gra} A$, that is,

$$(x, x^*) \in \operatorname{gra} A + (z, z^*). \qquad (18.5)$$

On the other hand, since (z, z^*) is a minimizer of g, $(0, 0) \in \partial g(z, z^*)$. By a result of Rockafellar (see [17, Theorem 2.9.8] and [32, Theorem 3.2.4(ii)]), there exist

$(z_0^*, z_0) \in \partial(\iota_{\text{gra}A}(\cdot - x, \cdot - x^*))(z, z^*) = \partial\iota_{\text{gra}A}(z - x, z^* - x^*) = (\text{gra}A)^\perp$, and $(v, v^*) \in X \times X^*$ with $v^* \in Jz, z^* \in Jv$ such that

$$(0,0) = (z^*, z) + (v^*, v) + (z_0^*, z_0).$$

Then

$$\big(-(z+v), z^* + v^*\big) \in \text{gra}A^*.$$

Since A^* is monotone,

$$\langle z^* + v^*, z + v\rangle = \langle z^*, z\rangle + \langle z^*, v\rangle + \langle v^*, z\rangle + \langle v^*, v\rangle \leq 0. \tag{18.6}$$

Note that since $\langle z^*, v\rangle = \|z^*\|_*^2 = \|v\|^2$, $\langle v^*, z\rangle = \|v^*\|_*^2 = \|z\|^2$, by (18.6), we have

$$\frac{1}{2}\|z\|^2 + \frac{1}{2}\|z^*\|_*^2 + \langle z^*, z\rangle + \frac{1}{2}\|v^*\|_*^2 + \frac{1}{2}\|v\|^2 + \langle v, v^*\rangle \leq 0.$$

Hence, $z^* \in -Jz$. By (18.5), $(x, x^*) \in \text{gra}A + \text{gra}(-J)$. Thus, $X \times X^* \subseteq \text{gra}A + \text{gra}(-J)$. Hence, A is maximal monotone. ∎

Remark 18.6. Haraux provides a very simple proof of Theorem 18.5 in Hilbert spaces in [20, Theorem 10], but the proof could not be adapted to reflexive Banach spaces.

18.3 Fitzpatrick Functions and Fitzpatrick Functions of Order *n*

Now we introduce some properties of monotone linear relations.

Fact 18.7. (See [6].) Assume that $A : X \rightrightarrows X^*$ is a monotone linear relation. Then the following hold.

(i) The function $\text{dom}A \to \mathbb{R} : y \mapsto \langle y, Ay\rangle$ is convex.
(ii) $\text{dom}A \subseteq (A0)^\perp$. For every $x \in (A0)^\perp$, the function $\text{dom}A \to \mathbb{R} : y \mapsto \langle x, Ay\rangle$ is linear.

Proof. (i): See [6, Proposition 2.3]. (ii): See [6, Proposition 2.2(i)(iii)]. ∎

Definition 18.8. Suppose $A : X \rightrightarrows X^*$ is a linear relation. We say A is *symmetric* if $\text{gra}A \subseteq \text{gra}A^*$.

By the definition of A^*, we have $(\forall x, y \in \text{dom}A)$ $\langle x, Ay\rangle$ is single valued and $\langle x, Ay\rangle = \langle y, Ax\rangle$.

For a monotone linear relation $A\colon X \rightrightarrows X^*$ (where A is not necessarily symmetric), it will be convenient to define (as in, e.g., [3])

$$q_A\colon x \in X \mapsto \begin{cases} \frac{1}{2}\langle x, Ax\rangle, & \text{if } x \in \operatorname{dom} A; \\ +\infty, & \text{otherwise.} \end{cases} \tag{18.7}$$

By Fact 18.7(i), q_A is well defined and is at most single-valued and convex. According to the definition of q_A, $\operatorname{dom} q_A = \operatorname{dom} A$. Moreover, by $(0,0) \in \operatorname{gra} A$ and A is monotone, we have that $q_A \geq 0$.

The following generalizes a result of Phelps–Simons (see [23, Theorem 5.1]) from symmetric monotone linear operators to symmetric monotone linear relations. We write $\overline{f}$ for the lower semicontinuous hull of f.

Proposition 18.9. *Let $A : X \rightrightarrows X^*$ be a monotone symmetric linear relation. Then*

(i) q_A is convex, and $\overline{q_A} + \iota_{\operatorname{dom} A} = q_A$.
(ii) $\operatorname{gra} A \subseteq \operatorname{gra} \partial\overline{q_A}$. If A is maximal monotone, then $A = \partial\overline{q_A}$.

Proof. Let $x \in \operatorname{dom} A$.

(i): Since A is monotone, q_A is convex. Let $y \in \operatorname{dom} A$. Since A is monotone, by Fact 18.7(ii),

$$0 \leq \frac{1}{2}\langle Ax - Ay, x - y\rangle = \frac{1}{2}\langle Ay, y\rangle + \frac{1}{2}\langle Ax, x\rangle - \langle Ax, y\rangle, \tag{18.8}$$

we have $q_A(y) \geq \langle Ax, y\rangle - q_A(x)$. Take lower semicontinuous hull on y and then deduce that $\overline{q_A}(y) \geq \langle Ax, y\rangle - q_A(x)$. For $y = x$, we have $\overline{q_A}(x) \geq q_A(x)$. On the other hand, $\overline{q_A}(x) \leq q_A(x)$. Altogether, $\overline{q_A}(x) = q_A(x)$. Thus, (i) holds.

(ii): Let $y \in \operatorname{dom} A$. By (18.8) and (i),

$$q_A(y) \geq q_A(x) + \langle Ax, y - x\rangle = \overline{q_A}(x) + \langle Ax, y - x\rangle. \tag{18.9}$$

Since $\operatorname{dom}\overline{q_A} \subseteq \overline{\operatorname{dom} q_A} = \overline{\operatorname{dom} A}$, by (18.9), $\overline{q_A}(z) \geq \overline{q_A}(x) + \langle Ax, z - x\rangle$, $\forall z \in \operatorname{dom}\overline{q_A}$. Hence $Ax \subseteq \partial\overline{q_A}(x)$. If A is maximal monotone, $A = \partial\overline{q_A}$. Thus (ii) holds. ∎

Definition 18.10 (Fitzpatrick family). Let $A\colon X \rightrightarrows X^*$ be a maximal monotone operator. The associated *Fitzpatrick family* $\mathscr{F}_A$ consists of all functions $F\colon X \times X^* \to]-\infty, +\infty]$ that are lower semicontinuous and convex, and that satisfy $F \geq \langle\cdot,\cdot\rangle$, and $F = \langle\cdot,\cdot\rangle$ on $\operatorname{gra} A$.

Following [21], it will be convenient to set $F^{\intercal}\colon X^* \times X \to]-\infty, +\infty]\colon (x^*, x) \mapsto F(x, x^*)$, when $F\colon X \times X^* \to]-\infty, +\infty]$, and similarly for a function defined on $X^* \times X$.

Fact 18.11 (Fitzpatrick). (See [19, Theorem 3.10] or [16, Corollary 4.1].) Let $A\colon X \rightrightarrows X^*$ be a maximal monotone operator. Then for every $(x, x^*) \in X \times X^*$,

$$F_A(x, x^*) = \min\{F(x, x^*) \mid F \in \mathscr{F}_A\} \quad \text{and} \quad F_A^{*\intercal}(x, x^*) = \max\{F(x, x^*) \mid F \in \mathscr{F}_A\}. \tag{18.10}$$

Proposition 18.12. *Let $A\colon X \rightrightarrows X^*$ be a maximal monotone and symmetric linear relation. Then*

$$F_A(x,x^*) = \frac{1}{2}\overline{q_A}(x) + \frac{1}{2}\langle x,x^*\rangle + \frac{1}{2}q_A^*(x^*), \quad \forall (x,x^*) \in X\times X^*.$$

Proof. Define function $k : X\times X^* \to \,]-\infty,+\infty]$ by

$$(z,z^*) \mapsto \frac{1}{2}\overline{q_A}(z) + \frac{1}{2}\langle z,z^*\rangle + \frac{1}{2}q_A^*(z^*).$$

Claim 1. $F_A = k$ on $\operatorname{dom}A \times X^*$.

Let $(x,x^*) \in X\times X^*$, and suppose that $x \in \operatorname{dom}A$. Then

$$\begin{aligned}
F_A(x,x^*) &= \sup_{(y,y^*)\in\operatorname{gra}A}\Big(\langle x,y^*\rangle + \langle y,x^*\rangle - \langle y,y^*\rangle\Big)\\
&= \sup_{y\in\operatorname{dom}A}\Big(\langle x,Ay\rangle + \langle y,x^*\rangle - 2q_A(y)\Big)\\
&= \frac{1}{2}\,q_A(x) + \sup_{y\in\operatorname{dom}A}\Big(\langle Ax,y\rangle + \langle y,x^*\rangle - \frac{1}{2}\,q_A(x) - 2q_A(y)\Big)\\
&= \frac{1}{2}\,q_A(x) + \frac{1}{2}\sup_{y\in\operatorname{dom}A}\Big(\langle Ax,2y\rangle + \langle 2y,x^*\rangle - q_A(x) - 4q_A(y)\Big)\\
&= \frac{1}{2}\,q_A(x) + \frac{1}{2}\sup_{z\in\operatorname{dom}A}\Big(\langle Ax,z\rangle + \langle z,x^*\rangle - q_A(x) - q_A(z)\Big)\\
&= \frac{1}{2}\,q_A(x) + \frac{1}{2}\sup_{z\in\operatorname{dom}A}\Big(\langle z,x^*\rangle - q_A(z-x)\Big)\\
&= \frac{1}{2}\,q_A(x) + \frac{1}{2}\langle x,x^*\rangle + \frac{1}{2}\sup_{z\in\operatorname{dom}A}\Big(\langle z-x,x^*\rangle - q_A(z-x)\Big)\\
&= \frac{1}{2}q_A(x) + \frac{1}{2}\langle x,x^*\rangle + \frac{1}{2}q_A^*(x^*)\\
&= k(x,x^*) \quad \text{(by Proposition 18.9(i))}.
\end{aligned}$$

Claim 2. k is convex and proper lower semicontinuous on $X\times X^*$.

Since F_A is convex, $\frac{1}{2}q_A + \frac{1}{2}\langle\cdot,\cdot\rangle + \frac{1}{2}q_A^*$ is convex on $\operatorname{dom}A\times X^*$. Now we show that k is convex. Let $\{(a,a^*),(b,b^*)\} \subseteq \operatorname{dom}k$, and $t \in\,]0,1[$. Then we have $\{a,b\} \subseteq \operatorname{dom}\overline{q_A} \subseteq \overline{\operatorname{dom}A}$. Thus, there exist $(a_n),(b_n)$ in $\operatorname{dom}A$ such that $a_n \to a, b_n \to b$ with $q_A(a_n) \to \overline{q_A}(a), q_A(b_n) \to \overline{q_A}(b)$. Since $\frac{1}{2}q_A + \frac{1}{2}\langle\cdot,\cdot\rangle + \frac{1}{2}q_A^*$ is convex on $\operatorname{dom}A\times X^*$, we have

$$\begin{aligned}
&\left(\frac{1}{2}q_A + \frac{1}{2}\langle\cdot,\cdot\rangle + \frac{1}{2}q_A^*\right)(ta_n + (1-t)b_n, ta^* + (1-t)b^*)\\
&\quad \le t\left(\frac{1}{2}q_A + \frac{1}{2}\langle\cdot,\cdot\rangle + \frac{1}{2}q_A^*\right)(a_n,a^*) + (1-t)\left(\frac{1}{2}q_A + \frac{1}{2}\langle\cdot,\cdot\rangle + \frac{1}{2}q_A^*\right)(b_n,b^*).
\end{aligned} \tag{18.11}$$

Take lim inf on both sides of (18.11) to see that

$$k\big(ta+(1-t)b, ta^*+(1-t)b^*\big) \leq tk(a,a^*)+(1-t)k(b,b^*).$$

Hence, k is convex on $X \times X^*$. Thus, k is convex and proper lower semicontinuous.

Claim 3. $F_A = k$ on $X \times X^*$. To this end, we first observe that

$$\operatorname{dom}\partial k^* = \operatorname{gra}A^{-1}. \tag{18.12}$$

We have

$$\begin{aligned}
(w^*,w) \in \operatorname{dom}\partial k^* &\Leftrightarrow (w^*,w) \in \operatorname{dom}\partial(2k)^* \\
&\Leftrightarrow (a,a^*) \in \partial(2k)^*(w^*,w), \quad \exists (a,a^*) \in X \times X^* \\
&\Leftrightarrow (w^*,w) \in \partial(2k)(a,a^*), \quad \exists (a,a^*) \in X \times X^* \\
&\Leftrightarrow (w^*-a^*, w-a) \in \partial(\overline{q_A} \oplus q_A^*)(a,a^*), \quad \exists (a,a^*) \in X \times X^* \qquad (18.13) \\
&\Leftrightarrow w^*-a^* \in \partial\overline{q_A}(a),\ w-a \in \partial q_A^*(a^*), \quad \exists (a,a^*) \in X \times X^* \\
&\Leftrightarrow w^*-a^* \in \partial\overline{q_A}(a),\ a^* \in \partial\overline{q_A}(w-a), \quad \exists (a,a^*) \in X \times X^* \\
&\Leftrightarrow w^*-a^* \in Aa,\ a^* \in A(w-a), \quad \exists (a,a^*) \in X \times X^* \qquad (18.14) \\
&\Leftrightarrow (w,w^*) \in \operatorname{gra}A \Leftrightarrow (w^*,w) \in \operatorname{gra}A^{-1},
\end{aligned}$$

where (18.13) follows from [32, Theorem 3.2.4(vi)(ii)] and (18.14) from Proposition 18.9(ii).

Next, we observe that

$$k^{*\intercal}(z,z^*) = \langle z,z^*\rangle, \quad \forall (z,z^*) \in \operatorname{gra}A. \tag{18.15}$$

Since $k(z,z^*) \geq \langle z,z^*\rangle$ and

$$k(z,z^*) = \langle z,z^*\rangle \Leftrightarrow \overline{q_A}(z)+q_A^*(z^*) = \langle z,z^*\rangle \Leftrightarrow z^* \in \partial\overline{q_A}(z) = Az$$

by Proposition 18.9(ii), Fact 18.11 implies that $F_A \leq k \leq F_A^{*\intercal}$. Hence $F_A \leq k^{*\intercal} \leq F_A^{*\intercal}$. Then by Fact 18.11, (18.15) holds.

Now using (18.15), (18.12) and a result by Borwein (see [11, Theorem 1] or [32, Theorem 3.1.4(i)]), we have $k = k^{**} = (k^* + \iota_{\operatorname{dom}\partial k^*})^* = (\langle\cdot,\cdot\rangle + \iota_{\operatorname{gra}A^{-1}})^* = F_A$. ■

Fact 18.13 (Recursion). (See [4, Proposition 2.13].) Let $A : X \rightrightarrows X^*$ be monotone, and let $n \in \{2,3,\ldots\}$. Then

$$F_{A,n+1}(x,x^*) = \sup_{(a,a^*)\in\operatorname{gra}A} \big(F_{A,n}(a,x^*) + \langle x-a,a^*\rangle\big), \quad \forall (x,x^*) \in X \times X^*.$$

Theorem 18.14. *Let $A : X \rightrightarrows X^*$ be a maximal monotone and symmetric linear relation, let $n \in \{2,3,\ldots\}$, and let $(x,x^*) \in X \times X^*$. Then*

$$F_{A,n}(x,x^*) = \frac{n-1}{n}\overline{q_A}(x) + \frac{n-1}{n}q_A^*(x^*) + \frac{1}{n}\langle x,x^*\rangle, \tag{18.16}$$

consequently, $F_{A,n}(x,x^*) = \frac{2(n-1)}{n} F_A(x,x^*) + \frac{2-n}{n}\langle x,x^*\rangle$. *Moreover,*

$$F_{A,\infty} = \overline{q_A} \oplus q_A^* = 2F_A - \langle\cdot,\cdot\rangle. \tag{18.17}$$

Proof. Let $(x,x^*) \in X \times X^*$. The proof is by induction on n. If $n = 2$, then the result follows for Proposition 18.12.

Now assume that (18.16) holds for $n \geq 2$. Using Fact 18.13, we see that

$$\begin{aligned}
F_{A,n+1}(x,x^*) &= \sup_{(a,a^*)\in\operatorname{gra}A} \left(F_{A,n}(a,x^*) + \langle x-a,a^*\rangle\right) \\
&= \sup_{(a,a^*)\in\operatorname{gra}A} \left(\frac{n-1}{n} q_A^*(x^*) + \frac{n-1}{n}\overline{q_A}(a) + \frac{1}{n}\langle a,x^*\rangle + \langle x-a,a^*\rangle\right) \\
&= \frac{n-1}{n} q_A^*(x^*) + \sup_{(a,a^*)\in\operatorname{gra}A} \left(\frac{n-1}{2n}\langle a,a^*\rangle + \left\langle a,\frac{1}{n}x^*\right\rangle + \langle x,a^*\rangle - \langle a,a^*\rangle\right), \qquad (18.18) \\
&= \frac{n-1}{n} q_A^*(x^*) + \sup_{(a,a^*)\in\operatorname{gra}A} \left(\left\langle a,\frac{1}{n}x^*\right\rangle + \langle x,a^*\rangle - \frac{n+1}{2n}\langle a,a^*\rangle\right) \\
&= \frac{n-1}{n} q_A^*(x^*) + \frac{2n}{n+1} \sup_{(a,a^*)\in\operatorname{gra}A} \left(\left\langle \frac{n+1}{2n}a,\frac{1}{n}x^*\right\rangle + \left\langle x,\frac{n+1}{2n}a^*\right\rangle \right. \\
&\quad \left. - \left\langle \frac{n+1}{2n}a,\frac{n+1}{2n}a^*\right\rangle\right) \\
&= \frac{n-1}{n} q_A^*(x^*) + \frac{2n}{n+1} \sup_{(b,b^*)\in\operatorname{gra}A} \left(\left\langle b,\frac{1}{n}x^*\right\rangle + \langle x,b^*\rangle - \langle b,b^*\rangle\right) \\
&= \frac{n-1}{n} q_A^*(x^*) + \frac{2n}{n+1} F_A\left(x,\frac{1}{n}x^*\right) \\
&= \frac{n-1}{n} q_A^*(x^*) + \frac{n}{n+1} q_A^*\left(\frac{1}{n}x^*\right) + \frac{n}{n+1}\overline{q_A}(x) + \frac{1}{n+1}\langle x^*,x\rangle \qquad (18.19) \\
&= \frac{n-1}{n} q_A^*(x^*) + \frac{1}{(n+1)n} q_A^*(x^*) + \frac{n}{n+1}\overline{q_A}(x) + \frac{1}{n+1}\langle x^*,x\rangle \\
&= \frac{n}{n+1} q_A^*(x^*) + \frac{n}{n+1}\overline{q_A}(x) + \frac{1}{n+1}\langle x,x^*\rangle, \qquad (18.20)
\end{aligned}$$

which is the result for $n+1$, where (18.18) follows from Proposition 18.9(i) and (18.19) from Proposition 18.12. Thus, by Proposition 18.12,

$$F_{A,n}(x,x^*) = \frac{2(n-1)}{n} F_A(x,x^*) + \frac{2-n}{n}\langle x,x^*\rangle.$$

By (18.16), $\operatorname{dom} F_{A,n} = \operatorname{dom}(\overline{q_A} \oplus q_A^*)$. Now suppose that $(x,x^*) \in \operatorname{dom} F_{A,n}$. By $\overline{q_A}(x) + q_A^*(x^*) - F_{A,n}(x,x^*) = \frac{1}{n}\left(\overline{q_A}(x) + q_A^*(x^*) - \langle x,x^*\rangle\right) \geq 0$ and

$$F_{A,n}(x,x^*) \to (\overline{q_A} \oplus q_A^*)(x,x^*),\ n \to \infty.$$

Thus, (18.17) holds. ■

Remark 18.15. Theorem 18.14 generalizes and simplifies [1, Example 4.4] and [3, Example 6.4]. See Corollary 18.17.

Remark 18.16. Formula Identity (18.16) does not hold for nonsymmetric linear relations. See [3, Example 2.8] for an example when A is skew linear operator and (18.16) fails.

Corollary 18.17. *Let $A : X \to X^*$ be a maximal monotone and symmetric linear operator, let $n \in \{2,3,\ldots\}$, and let $(x,x^*) \in X \times X^*$. Then*

$$F_{A,n}(x,x^*) = \frac{n-1}{n} q_A(x) + \frac{n-1}{n} q_A^*(x^*) + \frac{1}{n}\langle x,x^*\rangle, \tag{18.21}$$

and,

$$F_{A,\infty} = q_A \oplus q_A^*. \tag{18.22}$$

If X is a Hilbert space, then

$$F_{\mathrm{Id},n}(x,x^*) = \frac{n-1}{2n}\|x\|^2 + \frac{n-1}{2n}\|x^*\|^2 + \frac{1}{n}\langle x,x^*\rangle, \tag{18.23}$$

and,

$$F_{\mathrm{Id},\infty} = \frac{1}{2}\|\cdot\|^2 \oplus \frac{1}{2}\|\cdot\|^2. \tag{18.24}$$

Definition 18.18. Let $F_1, F_2 \colon X \times X^* \to \,]-\infty,+\infty]$. Then the *partial inf-convolution* $F_1 \Box_2 F_2$ is the function defined on $X \times X^*$ by

$$F_1 \Box_2 F_2 \colon (x,x^*) \mapsto \inf_{y^* \in X^*} \big(F_1(x,x^*-y^*) + F_2(x,y^*)\big).$$

Theorem 18.19 (nth order Fitzpatrick function of the sum). *Let $A,B \colon X \rightrightarrows X^*$ be maximal monotone and symmetric linear relations, and let $n \in \{2,3,\cdots\}$. Suppose that* $\operatorname{dom} A - \operatorname{dom} B$ *is closed. Then $F_{A+B,n} = F_{A,n} \Box_2 F_{B,n}$. Moreover, $F_{A+B,\infty} = F_{A,\infty} \Box_2 F_{B,\infty}$.*

Proof. By [28, Theorem 5.5] or [30], $A+B$ is maximal monotone. Hence, $A+B$ is a maximal monotone and symmetric linear relation. Let $(x,x^*) \in X \times X^*$. Then by Theorem 18.14,

$$\begin{aligned}
&F_{A,n} \Box_2 F_{B,n}(x,x^*) \\
&= \inf_{y^* \in X^*} \bigg(\frac{2(n-1)}{n} F_A(x,y^*) + \frac{2-n}{n}\langle x,y^*\rangle + \frac{2(n-1)}{n} F_B(x,x^*-y^*) \\
&\qquad + \frac{2-n}{n}\langle x,x^*-y^*\rangle \bigg) \\
&= \frac{2-n}{n}\langle x,x^*\rangle + \inf_{y^* \in X^*} \frac{2(n-1)}{n}\big(F_A(x,y^*) + F_B(x,x^*-y^*)\big)
\end{aligned}$$

$$= \frac{2-n}{n}\langle x,x^*\rangle + \frac{2(n-1)}{n} F_A \Box_2 F_B(x,x^*)$$
$$= \frac{2-n}{n}\langle x,x^*\rangle + \frac{2(n-1)}{n} F_{A+B}(x,x^*), \quad \text{(by [6, Theorem 5.10])}$$
$$= F_{A+B,n}(x,x^*) \quad \text{(by Theorem 18.14)}.$$

Similarly, using (18.17), we have $F_{A+B,\infty} = F_{A,\infty} \Box_2 F_{B,\infty}$. ■

Remark 18.20. Theorem 18.19 generalizes [3, Theorem 5.4].

Acknowledgements The author thanks Dr. Heinz Bauschke and Dr. Xianfu Wang for valuable discussions. The author also thanks the two anonymous referees for their careful reading and their pertinent comments.

References

1. Bartz, S., Bauschke, H.H., Borwein, J.M., Reich, S., Wang, X.: Fitzpatrick functions, cyclic monotonicity and Rockafellar's antiderivative. Nonlinear Anal. **66**, 1198–1223 (2007)
2. Bauschke, H.H., Borwein, J.M.: Maximal monotonicity of dense type, local maximal monotonicity, and monotonicity of the conjugate are all the same for continuous linear operators. Pacific J. Math. **189**, 1–20 (1999)
3. Bauschke, H.H., Borwein, J.M., Wang, X.: Fitzpatrick functions and continuous linear monotone operators. SIAM J. Optim. **18**, 789–809 (2007)
4. Bauschke, H.H., Lucet, Y., Wang, X.: Primal-dual symmetric antiderivatives for cyclically monotone operators. SIAM J. Control Optim. **46**, 2031–2051 (2007)
5. Bauschke, H.H., Wang, X., Yao, L.: An answer to S. Simons' question on the maximal monotonicity of the sum of a maximal monotone linear operator and a normal cone operator. Set-Valued Var. Anal. **17**, 195–201 (2009)
6. Bauschke, H.H., Wang, X., Yao, L.: Monotone linear relations: maximality and Fitzpatrick functions. J. Convex Anal. **16**, 673–686 (2009)
7. Bauschke, H.H., Wang, X., Yao, L.: Autoconjugate representers for linear monotone operators. Math. Program. Ser. B **123**, 5–24 (2010)
8. Bauschke, H.H., Wang, X., Yao, L.: Examples of discontinuous maximal monotone linear operators and the solution to a recent problem posed by B.F. Svaiter. J. Math. Anal. Appl. **370**, 224–241 (2010)
9. Bauschke, H.H., Wang, X., Yao, L.: On Borwein-Wiersma Decompositions of monotone linear relations. SIAM J. Optim. **20**, 2636–2652 (2010)
10. Bauschke, H.H., Wang, X., Yao, L.: On the maximal monotonicity of the sum of a maximal monotone linear relation and the subdifferential operator of a sublinear function. To appear in *Proceedings of the Haifa Workshop on Optimization Theory and Related Topics.* Contemp. Math., Amer. Math. Soc., Providence, RI (2010). http://arxiv.org/abs/1001.0257v1
11. Borwein, J.M.: A note on ε-subgradients and maximal monotonicity. Pacific J. Math. **103**, 307–314 (1982)
12. Borwein, J.M., Vanderwerff, J.D.: Convex Functions. Cambridge University Press (2010)
13. Boţ, R.I., Csetnek, E.R.: On extension results for n-cyclically monotone operators in reflexive Banach spaces. J. Math. Anal. Appl. **367**, 693–698 (2010)
14. Brézis, H., Browder, F.E.: Linear maximal monotone operators and singular nonlinear integral equations of Hammerstein type. In: Nonlinear analysis (collection of papers in honor of Erich H. Rothe), Academic, 31–42 (1978)

15. Burachik, R.S., Iusem, A.N.: Set-Valued Mappings and Enlargements of Monotone Operators. Springer (2008)
16. Burachik, R.S., Svaiter, B.F.: Maximal monotone operators, convex functions and a special family of enlargements. Set-Valued Anal. **10**, 297–316 (2002)
17. Clarke, F.H.: Optimization and Nonsmooth Analysis. SIAM, Philadelphia (1990)
18. Cross, R.: Multivalued Linear Operators. Marcel Dekker (1998)
19. Fitzpatrick, S.: Representing monotone operators by convex functions. In: Workshop/Miniconference on Functional Analysis and Optimization (Canberra 1988), Proceedings of the Centre for Mathematical Analysis **20**, 59–65. Australian National University, Canberra, Australia (1988)
20. Haraux, A.: Nonlinear Evolution Equations – Global Behavior of Solutions. Springer, Berlin (1981)
21. Penot, J.-P.: The relevance of convex analysis for the study of monotonicity. Nonlinear Anal. **58**, 855–871 (2004)
22. Phelps, R.R.: Convex functions, Monotone Operators and Differentiability, 2nd edn. Springer (1993)
23. Phelps, R.R., Simons, S.: Unbounded linear monotone operators on nonreflexive Banach spaces. J. Convex Anal. **5**, 303–328 (1998)
24. Rockafellar, R.T., Wets, R.J-B.: Variational Analysis. Springer (2004)
25. Simons, S.: Minimax and Monotonicity. Springer (1998)
26. Simons, S.: From Hahn-Banach to Monotonicity. Springer (2008)
27. Simons, S., Zălinescu, C.: A new proof for Rockafellar's characterization of maximal monotone operators. Proc. Amer. Math. Soc. **132**, 2969–2972 (2004)
28. Simons, S., Zălinescu, C.: Fenchel duality, Fitzpatrick functions and maximal monotonicity. J. Nonlinear Convex Anal. **6**, 1–22 (2005)
29. Svaiter, B.F.: Non-enlargeable operators and self-cancelling operators. J. Convex Anal. **17**, 309–320 (2010)
30. Voisei, M.D.: The sum theorem for linear maximal monotone operators. Math. Sci. Res. J. **10**, 83–85 (2006)
31. Voisei, M.D., Zălinescu, C.: Linear monotone subspaces of locally convex spaces. Set-Valued Var. Anal. **18**, 29–55 (2010)
32. Zălinescu, C.: Convex Analysis in General Vector Spaces. World Scientific Publishing (2002)
33. Zeidler, E.: Nonlinear Functional Analysis and its Application, Vol II/B Nonlinear Monotone Operators. Springer, Berlin (1990)